AF249080

CONTINUUM THERMODYNAMICS

Part II: Applications and Examples

Series on Advances in Mathematics for Applied Sciences – Vol. 85

CONTINUUM THERMODYNAMICS

Part II: Applications and Examples

Bettina Albers
Technical University of Berlin, Germany

Krzysztof Wilmanski
Technical University of Berlin, Germany &
Rose School Pavia, Italy

NEW JERSEY · LONDON · SINGAPORE · BEIJING · SHANGHAI · HONG KONG · TAIPEI · CHENNAI

Published by

World Scientific Publishing Co. Pte. Ltd.

5 Toh Tuck Link, Singapore 596224

USA office: 27 Warren Street, Suite 401-402, Hackensack, NJ 07601

UK office: 57 Shelton Street, Covent Garden, London WC2H 9HE

British Library Cataloguing-in-Publication Data
A catalogue record for this book is available from the British Library.

Series on Advances in Mathematics for Applied Sciences — Vol. 85
CONTINUUM THERMODYNAMICS
Part II: Applications and Examples

ISBN 978-981-4412-37-7

Printed in Singapore

Preface

In 2008 Krzysztof Wilmanski published Part I of this book, which deals with the *Foundations* of *Continuum Thermodynamics*. In this book Part II on *Applications and Exercises* and Part III on *Numerical Methods* were announced. Departing from the original plan, Krzysztof Wilmanski and I started to write the present book, Part II, together. Some progress was made before Krzysztof was diagnosed with cancer. He learned with initial disbelief of the devastating diagnosis only three and a half months before his death. I can still hear him expressing his wish that the spots revealed on the X-rays were harmless and medically treatable. In spite of the stomach pain that prompted his initial visit to the doctor, he still had many plans and ideas for the future. One of them was to finish the present book. After intensive efforts to arrest the progress of his disease, it was clear that it could not be cured. With a race against time, finishing this book became very important to both of us. Unfortunately, time ran out too soon for Krzysztof.

Our book project came to a halt after Krzysztof's quite unexpected death on 26 August 2012. I promised Krzysztof before his death and also his family and friends at his funeral that I would finish our joint work. However, at first I was paralyzed – Krzysztof's passing touched me deeply and it still does. I have fond memories of the first time I met Krzysztof in Essen in 1992, where I studied Civil Engineering. He was my teacher in courses on tensor analysis, elasticity theory, continuum mechanics and some other topics. I was excited by his enthusiastic style of teaching, and especially impressed by his spontaneity. He could derive even complicated examples on the blackboard without notes. Near the end of my studies, Krzysztof already had a nearly thirty year career making original contributions on: the axiomatic and kinetic foundations of continuum thermodynamics, mixture theory, phase transformations in solids, non-Newtonian fluids, crystal plasticity and the evolution of textures. At that time, he was just beginning his research on porous media and he awoke my interest in this field. This was my first contact with science and it impressed me so much that I started my PhD studies – advised by Krzysztof – in his research group in Berlin. In 1996, Krzysztof had become the head of the *Continuum Mechanics* research group at the Weierstrass Institute for Applied Analysis and Stochastics. A pleasant time of intense cooperation followed and scientifically resulted in twelve joint publications. But also privately Krzysztof was an extremely nice contemporary. He had a warm-hearted nature and was an interested listener who offered help in all matters – scientific and also private. I am very grateful for having had the opportunity to benefit from the immense knowledge of this distinguished scientist and for the time we had together. Of course, this book is the cherry on the top of our joint work. It is a pity that we could not finish the book together, and I sincerely hope that I have completed it in accordance with Krzysztof's intentions.

I have to admit that the completion of this book posed a serious challenge for me. From the beginning it was clear that Part II should have the same structure as Part I, with chapters with the same numbers in the two parts being directly related. Of course,

Krzysztof and I decided on the contents of the book together. We had planned, for instance, that Krzysztof would write Chapter 8 on *Extended Thermodynamics*. In Part I he had introduced the version of Müller/Ruggeri/Liu. In Part II he intended to discuss the counterpart by Jou/Casas-Vázquez/Lebon. However, he could not realize this plan and it was my job to do so. I am greatly indebted to David Jou for proofreading my version of the chapter (which is an excerpt of his own book) and for providing me with valuable constructive comments.

The contexts in this book are interdisciplinary in nature. Both solids and fluids, and also their combination – porous media – are tackled. Also, the range of applications addressed is broad. Many chapters describe our own research, but we also discuss achievements of other colleagues. For example, we include: the stability of flows analyzed by S. Chandrasekhar, D. D. Joseph and Z. Wesołowski; some thermodynamical problems introduced by H. S. Carslaw and J. C. Jaeger; the model and a numerical example for composite beams with embedded shape memory alloys by M. S. Kuczma; the model with double porosity describing seepage in fissured rocks by G. I. Barenblatt; and an example on media with multi-porosity (i.e. namely swelling) by J. Huyghe. A short introduction and an example in the highly topical field of biomechanics is based on work by G. Holzapfel. The description of different coordinate systems uses the work of H. Margenau and G. M. Murphy. Throughout the text reference is made to works of other colleagues. I am deeply indebted to all of my colleagues whose ideas have expanded the range of applications and examples of continuum thermodynamics mentioned in this book.

A second stroke of fate prevented me from working on this book during the end of last year. My dear husband, Dieter, succumbed to cancer on 16 November 2013, less than a month after having been diagnosed. Again, I am deeply wounded and often think more about our life together than about science. However, my desire to hold this completed book in my hands has been strong. I dedicate it to Dieter who always encouraged me to work in science, even at the expense of limitations on our personal lives.

I appreciate very much not only the financial support of the Einstein Foundation Berlin but also the freedom of research which I have as an Einstein Junior Fellow. This fellowship has allowed me to complete the book in time for it to be published on 1 March 2015, Krzysztof's 75th birthday.

Berlin, in July 2014   Bettina Albers

Contents

Appendices   373

Chapter 1

Introduction

The intention by writing Part II of the book on continuum thermodynamics was the deepening of some issues covered in Part I as well as a development of certain skills in dealing with practical problems of macroscopic processes. However, the main motivation for this part is the presentation of main facets of thermodynamics which appear when interdisciplinary problems are considered. There are many monographs on the subjects of solid mechanics and thermomechanics, on fluid mechanics and on coupled fields but most of them cover only special problems in great details which are characteristic for the chosen field. It is rather seldom that relations between these fields are discussed. This concerns, for instance, large deformations of the skeleton of porous materials with diffusion (e.g. lungs), couplings of deformable particles with the fluid motion in suspensions, couplings of adsorption processes and chemical reactions in immiscible mixtures with diffusion, various multi-component aspects of the motion, e.g. of avalanches, such as segregation processes, etc.

As mentioned in the preface, Part II has the same structure as Part I. This means that the basic notions, the foundations and some further considerations related to a certain topic which is treated in the present part can be found in the corresponding chapter in Part I. However, we are endeavored to present a summary of the basics also at the beginning of the chapters in Part II. The theoretical expositions in Part I often go beyond the necessary theory used in practical examples.

In some applications we rely on the work of other scientists. In order not to adulterate such theories and examples, we have predominantly kept the original notation. Therefore it may happen that the notation throughout the book is not uniform in all places. However, where possible we tried to conform the notation of Part II to that of Part I.

Contents of the book

In the following three chapters some foundations of continuum mechanics are repeated.

Chapter 2 deals with the geometry of deformations of solids. We introduce reference and current configurations, the function of motion and the corresponding deformation gradient. Different measures of deformation are compared and a generalized measure is presented and illustrated. The polar decomposition of the deformation gradient is addressed. This important property is taken up in Appendix A.1 where mathematical basics are shown on an example of polar decomposition. Furthermore, geometrical aspects of universal solutions are discussed. Such solutions are the foundation of experimental verification of constitutive relations for many materials under static conditions. In the literature classified families are introduced and corresponding deformation tensors are shown. The description comes about in rectangular Cartesian, cylindrical polar or spherical polar coordinate systems. These and some other coordinate systems are the topic of Appendix A.2.

In Chapter 3 the time dependence of motion is also accounted for. The kinematics of continua in both material or Lagrangian and in spatial or Eulerian description is presented. Also the transformation properties of vectors and tensors are shortly addressed. For porous media the Lagrangian description is often used with the skeleton as reference. We show an example clarifying the Lagrangian description of relative motion.

Chapter 4 is concerned with balance equations. Global and local balance equations for regular and singular points are summarized and the relation between Cauchy and Piola-Kirchhoff stresses is pointed out.

In Chapter 5 we focus for the first time on material behavior. The discussion of ideal fluids is preceded by the introduction of the d'Alembert paradox. We discuss its origin within the frame of the general momentum conservation law. Due to its role in the theory of boundary layers and in the linear modeling of porous materials, in regard to ideal fluids a simple example is presented which is connected with the d'Alembert paradox and the added mass effect. Afterward the Navier-Stokes equation for the description of viscous fluids, its thermodynamical properties and the uniqueness of solutions are addressed. We mention two types of viscous flow, namely lamellar and creeping flows. One section is devoted to the boundary layer theory basing on ideas of Ludwig Prandtl. These considerations are followed by the investigation of Maxwell and N-th grade fluids. Several examples for viscometric flows, namely plate-and-plate, cone-and-plate, Couette flow and Poiseuille flow are considered. In the section on nonlinear elastic solids rubber-like materials are studied. As examples of homogeneous deformations, for isochoric extension and simple shear, stresses for compressible and incompressible materials are investigated. We introduce Ericksen's Theorem and close the section with an example on heterogeneous deformations, namely pure torsion of a circular cylinder. In the following section viscoelastic solids are considered. They belong to the broad class of simple materials in which the set of constitutive variables consists of the deformation gradient, the temperature and the gradient of temperature. Formally, their response in all processes is determined by the response to all homogeneous thermokinetic processes. Examples of simple rheological models indicate that the constitutive relations have the form of evolution equations. We introduce the Kramers-Kronig relation which states that viscoelastic materials are inherently dispersive, i.e. the propagation speed of the mechanical disturbance is frequency dependent. Furthermore, we introduce the correspondence principle. It states that the complex viscoelastic moduli can be replaced by those of the elasticity theory. This reveals the possibility of converting numerous static solutions of elasticity into quasi-static solutions of viscoelasticity. We demonstrate the application of the correspondence principle on a simple example, the axial symmetric problem of a cylinder under given radial loading on both lateral surfaces.

Chapter 6 is devoted to the stability analysis of continua. We show a few characteristic examples, as for instance, the stability properties of some flows of fluids (the torsional Couette flow and the Rayleigh-Bénard problem), large static deformations of solids (stability of a nonlinear elastic strip) and thermodynamic equilibrium states of some continuous systems (second-grade fluids). The concern of the stability analysis of thermomechanical systems is not only with applications and particular engineering problems. It yields as well important information on fundamental properties of thermodynamical models.

In particular, it prescribes the ranges of material parameters in which such models are physically acceptable and relevant.

In Chapter 7 some thermodynamical problems are presented and solved. The first section concerns heat conduction problems described in Cartesian, cylindrical and spherical coordinates. While for some problems the solution involves only one space variable and the time, in others two or more space variables are involved. There are different methods to obtain solutions. The simplest case is present if the solutions can be expressed as a product of solutions of one-variable problems. Moreover, multiple Fourier series or their generalizations or Green's function can be used. Another possibility is the direct application of the Laplace transformation method. All the mentioned methods are treated in the Appendix. Fourier transforms are the topic of Appendix B.1, Laplace transforms of Appendix B.2. Appendix C is devoted to Green functions – both the static and dynamic cases of isotropic elastic materials are considered. Finally, in Appendix D Bessel functions and the Bessel equation are introduced because they are needed in Chapter 7 to describe the solutions of some problems. Of course, the solutions for steady temperature are less complex than those for variable temperature. Examples for both situations are presented. Since they have considerable importance in practice, in the next section anisotropic media are investigated. As an example, the conduction in a thin crystal plate is studied – the general theory of flow, without any assumptions on symmetry is developed. The third section of this chapter is devoted to thermal boundary layers. They appear in many practical applications such as phase transformations (melting, evaporation, solidification, condensation, etc.) but also problems of heat transfer between civil engineering constructions and environment, air conditioning systems, etc. contain field equations whose boundary conditions concern transition regions in which thermal boundary layers appear. Composite beams with embedded shape memory alloy form the last thermodynamical problem demonstrated in this chapter. Shape memory alloys constitute a class of functional, smart materials which have found many technological applications and offer innovative solutions in the design of adaptive structures. They may undergo a temperature- or stress-induced martensitic phase transformation resulting in the shape memory effect and pseudoelastic behavior.

Chapter 8 reveals an introduction to Extended Thermodynamics in the version of Jou-Casas-Vázquez-Lebon. They were primarily motivated by the non-equilibrium statistical mechanics and, in particular, by the so-called Fluctuation-Dissipation Theorem. The ideas of this microscopic theory are presented, and, in conclusion we comment on common points and on main differences between this version of Extended Thermodynamics and that of Müller-Liu-Ruggeri which was the subject of Chapter 8 of Part I. As was done in Part I the example of ideal gases is studied.

It is well known that dislocations are the source of plastic deformation. In Chapter 9 we present some properties of discrete dislocations as well as a continuum model of these defects in crystalline materials. Dislocations are line-defects characterized by the Burgers vector which, in turn, is defined in a crystal by the Burgers contour. However, dislocations do not only play a role in crystal materials but also in the modeling of rupture of tectonic plates yielding earthquakes. This application is briefly presented at the end of Chapter 9. Various other defects may exist which are not that easily to describe. In spite of their practical importance, such defects are not described in this book. However, in some way we are coming back to such a problem in Chapter 12 where in the description of freezing and thawing processes also the creation of microcracks is of importance.

Chapter 10 is on acoustic waves. We start with a general discussion on parabolic and hyperbolic models. Afterward the propagation of acoustic waves in nonlinear materials with memory is accounted for. An approximate solution in the vicinity of the front is constructed. Several dynamical problems of continua are the topic of Section 4. Not only waves in fluids and fluid layers with different boundaries are addressed but also waves in linear solids are considered. Both bulk and surface waves appear in different situations under consideration. Also cylindrical surfaces are mentioned because they appear quite often in geotechnics – for instance, they occur in the analysis of wave in boreholes. A particular class of waves, leaky waves, is mentioned in Section 5. These are such waves whose energy is transferred on some other modes. The last section of this chapter is devoted to bulk and surface waves in viscoelastic solids. The motivation for these investigations is the modeling of soils and rocks by means of viscoelastic materials. Such models are not multi-component but the diffusion process yields naturally a viscous character of the material modeled by a single component continuum.

In Chapter 11 interactions of ponderable bodies with electromagnetic fields are examined. Modern technologies yield discoveries of effects and devices whose description requires more sophisticated models than these considering thermomechanical properties alone. Many questions of the construction of macroscopic models can be answered only with the help of modern continuum thermodynamics. Some of these questions have already been mentioned in Part I. In this chapter we extend the subject and discuss also some issues of plasmas. To this aim at first the Maxwell theory of electromagnetism is summarized. Afterward the coupling of thermomechanical and electromagnetic fields is addressed. Finally, several magnetohydrodynamical models of plasmas are introduced and the stability of plasmas is discussed.

Chapters 12 and 13 deal with porous materials. While in Chapter 12 a few problems of special behavior of multi-component porous materials are shown, Chapter 13 focuses on chosen examples for which the model with the balance equation of porosity introduced by K. Wilmanski is used.

Thus, Chapter 12 begins with a summary of the theory of immiscible mixtures which is the basis of many models used for porous materials, and with an outline of the description of multi-component porous materials in Lagrangian and in Eulerian way. In the next section two-component models with constitutive relations for the porosity are introduced. Both models for incompressible and compressible components are considered and their thermodynamic admissibility is inspected. A further section of this chapter is devoted to double- and multi-porosity models. First, the original field of application of double porosity is given attention to: fissured rocks. Such materials consist of pores and permeable blocks, the blocks separated from each other by a system of fissures. Thus, the coefficient of fissuring of the rock builds one porosity and the porosity of the individual blocks is the second one. We repeat here Barenblatt's model for such media which, obviously, not only contains two different porosities but also two permeabilities, two pressures etc. A second example of multi-porosity models is presented, namely swelling media which appear mainly in bioengineering. We are concerned with ionized porous structures imbibed with electrolyte solutions in which interfacial phenomena often determine the macroscopic behavior. A further biomechanical example is presented: topics of the next section are soft tissues. They behave anisotropically because their fibers have preferred directions. They undergo large deformations and some of them show viscoelastic behavior. We point out the structure of these materials, introduce a model and show an example.

Chapter 13 starts with a summary of the balance equation of porosity and associated models. In the following sections these models are applied. The first examples are freezing and thawing processes. An iterative procedure for the calculation of the mechanical properties is shown. It incorporates two different stages of the process according to the actual temperature. First, isothermal diffusion in the poroelastic range without freezing is considered while the second range contains the process of freezing. The model for the latter range is based on the Gurson-Tvergaard-Needleman theory for plastic deformations. The measure of damage is described by the extent of the porosity changes caused by freezing. Section 3 is concerned with the linear stability of a 1D flow under transversal disturbance with adsorption. The disturbances satisfy equations of the model for multi-component systems with adsorption. It is considered that a fluid/adsorbate mixture flows through the channels of a skeleton. In this case a kinematic nonlinearity acts against the permeability of the medium. Adsorption processes contribute in a nonlinear way to the field equations and essentially influence the stability properties. In Section 4 we study the wave propagation in porous media with anisotropic permeability. We investigate a model in which the stress-strain relations are isotropic but the tortuosity is not. The anisotropy of tortuosity yields essential changes of the attenuation of the waves depending on the direction of propagation in relation to the principal directions of tortuosity and on the mode of the wave. Also in the last section the wave propagation is discussed. A linear model for three-component media is shown. In such media the speeds and attenuations of the waves depend not only on the frequency but also on the degree of saturation. The capillary pressure between the pore fluids is one of the most important quantities entering the hyperbolic model.

In the end, in Appendix E, the basic physical units are listed which are used throughout the book in the presented examples and applications.

Chapter 2

Geometry of deformations of solids

2.1 Summary: Geometry

For a better understanding, some basics of continuum mechanics concerning the geometry are summarized in this section. For further reading on continuum mechanics we refer to the books of Wilmanski (e.g. Part I or [437]), Liu [231] or Marsden/Hughes [242]. A comprehensive overview of the development of continuum mechanics throughout the 20th century is given in the recent book [245] by Maugin.

2.1.1 Configurations

In continuum modeling a physical object can be geometrically described by a compact measurable subset of the Euclidean space. This so-called *continuum* or *body* fills at any time t a part of the Euclidean space, the so-called *space of motion*. Even in an infinitesimal neighborhood of one point of the continuum there are countless others. These points do neither possess mass nor other physical properties, thus they cannot be called particles. Even though they received the – somewhat puzzling – notation of *material points*. This is done in order to distinguish them from *positions in the space of motion*. The configuration of the points at time $t = t_0$ is called the *reference configuration* $\mathcal{B}_0$. The motion of the material points $\mathbf{X}$ belonging to $\mathcal{B}_0$ into a current configuration $\mathcal{B}_t$ is described by

$$\forall \mathbf{X} \in \mathcal{B}_0, t: \quad \mathbf{x} = \mathbf{f}(\mathbf{X}, t) \in \mathcal{B}_t. \tag{2.1}$$

The positions of material points $\mathbf{X}$ in the *current configuration* $\mathcal{B}_t$ are denoted by $\mathbf{x}$, and $\mathbf{f}(\mathbf{X}, \cdot)$ is the *motion* of the material point $\mathbf{X} \in \mathcal{B}_0$.

The material points $\mathbf{X} \in \mathcal{B}_0$ are indicated by three coordinates $\left\{X^K\right\}_{K=1,2,3}$, the so-called *Lagrange-coordinates*. The corresponding unit basis vectors are denoted by $\mathbf{e}_K$, $K = 1, 2, 3$. For the description of the positions $\mathbf{x}$ of the material points instead the so-called *Euler-coordinates* $\left\{x^k\right\}_{k=1,2,3}$ are used which characterize the space of motion without reference to the body and its motion. The corresponding unit basis vectors are $\mathbf{e}_k$, $k = 1, 2, 3$.

2.1.2 Deformation gradient

The first derivative of $\mathbf{f}$ with respect to $\mathbf{X}$ is the *deformation gradient* $\mathbf{F}$

$$\mathbf{F} := \operatorname{Grad} \mathbf{f}(\mathbf{X}, t) \equiv \frac{\partial f_k}{\partial X^K} \mathbf{e}_k \otimes \mathbf{e}_K, \tag{2.2}$$

$$\text{Grad} := \frac{\partial}{\partial \mathbf{X}} \equiv \frac{\partial}{\partial X^K} \mathbf{e}_\alpha \equiv \left\{ \frac{\partial}{\partial X^1}, \frac{\partial}{\partial X^2}, \frac{\partial}{\partial X^3} \right\}.$$

The deformation gradient is the most important quantity for the modeling of deformations. In order to describe the local change of the geometry of a body in the Euclidean space we investigate a material point $\mathbf{X}$ and the corresponding infinitesimal material vector $d\mathbf{X}$ belonging to the reference configuration $\mathcal{B}_0$. In the current configuration $\mathcal{B}_t$ the length of this vector is determined by the following relation

$$d\mathbf{x}{\cdot}d\mathbf{x} = (\mathbf{F}d\mathbf{X}) \cdot (\mathbf{F}d\mathbf{X}) = d\mathbf{X}{\cdot} \left(\mathbf{F}^T\mathbf{F}\right) d\mathbf{X}. \tag{2.3}$$

The change of length of the vector is thus

$$d\mathbf{x}{\cdot}d\mathbf{x} - d\mathbf{X}{\cdot}d\mathbf{X} = d\mathbf{X}{\cdot} \left(\mathbf{F}^T\mathbf{F} - \mathbf{1}\right) d\mathbf{X}, \tag{2.4}$$

and the length remains unchanged if

$$\mathbf{F}^T\mathbf{F} - \mathbf{1} = 0 \quad \Rightarrow \quad \mathbf{F}^T = \mathbf{F}^{-1}, \tag{2.5}$$

this means, if the deformation gradient is orthogonal. This is only the case for a local rigid rotation of the body.

2.1.3 Measures of deformation

Relation (2.3) shows that not the deformation gradient $\mathbf{F}$ itself measures the deformation but the symmetric combination $\mathbf{C} := \mathbf{F}^T\mathbf{F}$. This tensor is called the *right Cauchy-Green deformation tensor*. This is a measure of deformation whose reference configuration coincides with the configuration $\mathcal{B}_0$ for which $\mathbf{C} = \mathbf{1}$. There exist further deformation measures, as for example the *left Cauchy-Green deformation tensor* $\mathbf{B}$ or the *Almansi-Hamel tensor* $\mathbf{e}$, whose reference situation is the current configuration $\mathcal{B}_t$. We have

$$d\mathbf{X} = \mathbf{F}^{-1}d\mathbf{x} \quad \Rightarrow \quad d\mathbf{X}{\cdot}d\mathbf{X} = d\mathbf{x}{\cdot} \left(\mathbf{F}^{-T}\mathbf{F}^{-1}\right) d\mathbf{x} = d\mathbf{x}{\cdot} \left(\mathbf{F}\mathbf{F}^T\right)^{-1} d\mathbf{x}, \tag{2.6}$$

and

$$d\mathbf{X} \cdot d\mathbf{X} - d\mathbf{x}{\cdot}d\mathbf{x} = d\mathbf{x}{\cdot} \left[\mathbf{1} - \left(\mathbf{F}\mathbf{F}^T\right)^{-1}\right] d\mathbf{x}, \tag{2.7}$$

and define

$$\mathbf{B} := \mathbf{F}\mathbf{F}^T, \qquad \mathbf{e} := \tfrac{1}{2}\left(\mathbf{1} - \mathbf{c}\right) \quad \text{with} \quad \mathbf{c} := \mathbf{B}^{-1}. \tag{2.8}$$

These and also other measures of deformation (see Section 2.2 of Part I) for other configurations are equivalent because the matrix of the deformation gradient is invertible. Some of them are summarized in Table 2.1. They are matter of the following illustration.

B. R. Seth describes in his contribution to the IUTAM Symposium on Second Order Effects in Elasticity, Plasticity and Fluid Mechanics in 1962 [351] that in the classical theory of elasticity mostly the Cauchy measure is employed when the strain is small and that the theory of finite deformation usually uses the Almansi-Hamel or Green-St. Venant measure. The Hencky measure is widely established in plasticity. K. H. Swainger (see, for instance, [371, 372]) uses linear displacement gradients referred to the strained state. Seth perceived that it would be of interest to suggest a generalized strain measure which includes the others as particular cases. In [351] he introduced such a generalized measure

$$\mathbf{e}^{(m)} = \frac{1}{m}\left(\mathbf{1} - \mathbf{V}^{-m}\right). \tag{2.9}$$

Table 2.1: *Some measures of deformation.*

right Cauchy-Green	$\mathbf{C} = \mathbf{F}^T \mathbf{F}$
left Cauchy-Green (Finger tensor)	$\mathbf{B} = \mathbf{F}\mathbf{F}^T$
Cauchy	$\mathbf{c} = \mathbf{B}^{-1}$
Almansi-Hamel (Euler)	$\mathbf{e} = \frac{1-\mathbf{c}}{2}$
Green-St. Venant (Lagrange)	$\mathbf{E} = \frac{\mathbf{C}-1}{2}$
Swainger	$\mathbf{S} = 1 - \mathbf{c}^{-1/2}$
Hencky	$\mathbf{H} = \frac{1}{2}\ln \mathbf{C}$
generalized (Seth)	$\mathbf{e}^{(m)} = \frac{1}{m}\left(1 - \mathbf{V}^{-m}\right)$

He found it in the form $\mathbf{V}^{(m)} = (1 - m\mathbf{e})^{-1/m}$, where $\mathbf{V} = \sqrt{\mathbf{B}}$ is the left stretch tensor and by the parameter m all the known measures are included:

$$
\begin{aligned}
m = -2 : \quad & \text{Green-St. Venant,} \\
m = -1 : \quad & \text{Cauchy,} \\
m = 0 : \quad & \text{Hencky,} \\
m = 1 : \quad & \text{Swainger,} \\
m = 2 : \quad & \text{Almansi-Hamel.}
\end{aligned}
\tag{2.10}
$$

For the simple example of uniaxial deformation where l/l_0 denotes the ratio of the current to the initial length there is only one component of the strain tensor, denoted by ε. For the above mentioned values of m follows then from (2.9)

$$
\begin{aligned}
m = -2 : \quad & \varepsilon = \frac{1}{2}\left[\left(\frac{l}{l_0}\right)^2 - 1\right], \\
m = -1 : \quad & \varepsilon = \frac{l}{l_0} - 1, \\
m = 0 : \quad & \varepsilon = \log\left(\frac{l}{l_0}\right), \\
m = 1 : \quad & \varepsilon = 1 - \left(\frac{l}{l_0}\right)^{-1}, \\
m = 2 : \quad & \varepsilon = \frac{1}{2}\left[1 - \left(\frac{l}{l_0}\right)^{-2}\right].
\end{aligned}
\tag{2.11}
$$

These functions are illustrated in Figure 2.1. It becomes obvious that for small deformations all measures are equivalent but for large deformations the behavior is very different.

2.1.4 Polar decomposition

The essential property of the deformation tensors is their tensorial symmetry. In contrast to the deformation gradient which possesses nine independent components, this symmetry

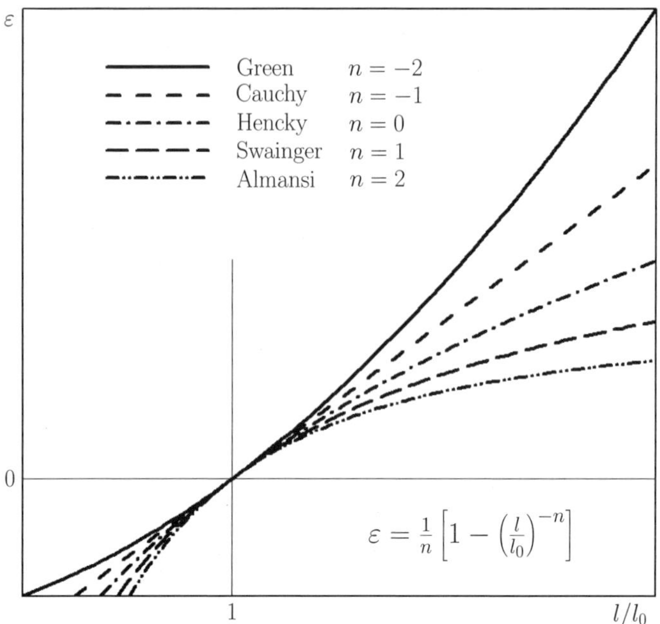

Figure 2.1: *Strain measures for uniaxial deformation.*

results in only six independent components of the deformation tensors. This follows from the *polar decomposition*[1] of the deformation gradient (for an example see: Appendix A.1):
For each tensor $\mathbf{F}$ with $\det\mathbf{F} \neq 0$ unique tensors $\mathbf{R}$ and $\mathbf{U}$ exist with

$$\mathbf{F} = \mathbf{R}\mathbf{U}, \qquad \mathbf{R}^T = \mathbf{R}^{-1}, \quad \mathbf{U} = \mathbf{U}^T, \qquad \det\mathbf{U} \neq 0. \qquad (2.12)$$

$\mathbf{R}$ is the so-called *rotation matrix*, an orthogonal tensor with three independent components. The symmetric tensor $\mathbf{U}$ is the *right stretch tensor*. Using the definition of $\mathbf{C}$, the following relation follows

$$\mathbf{C} = (\mathbf{R}\mathbf{U})^T (\mathbf{R}\mathbf{U}) = \mathbf{U}^2. \qquad (2.13)$$

For the tensor $\mathbf{C}$ mainly those material tensors are important which by action of $\mathbf{C}$ change only their length but not their direction. They determine the properties concerning invariance of the tensor $\mathbf{C}$ and follow from the three-dimensional eigenvalue problem

$$(\mathbf{C}-\lambda\mathbf{1})\,\mathbf{r} = 0 \quad \Rightarrow \quad \begin{aligned} &\mathbf{C} = \sum_{\alpha=1}^{3} \lambda_C^{(\alpha)}\mathbf{K}_C^\alpha \otimes \mathbf{K}_C^\alpha, \\[1em] &\det\left(\mathbf{C}-\lambda_C^{(\alpha)}\mathbf{1}\right) = 0, \end{aligned} \qquad (2.14)$$

---

[1]Procedure of the polar decomposition of a given deformation gradient $\mathbf{F}$:
1. $\mathbf{C} = \mathbf{F}^T\mathbf{F}$,
2. Solution of the eigenvalue problem for the tensor $\mathbf{C}$ (see (2.14)),
3. $\mathbf{U} = \sum_{\alpha=1}^{3} \sqrt{\lambda_C^{(\alpha)}}\mathbf{K}_C^\alpha \otimes \mathbf{K}_C^\alpha$,
4. $\mathbf{R} = \mathbf{F}\mathbf{U}^{-1}$.

where $\lambda_C^{(\alpha)}$ are the eigenvalues and $\mathbf{K}_C^\alpha$ the eigenvectors of tensor $\mathbf{C}$. The principal stretches $\lambda_C^{(\alpha)}$ are the solutions of the equation

$$\lambda_C^3 - I\lambda_C^2 + II\lambda_C - III = 0, \tag{2.15}$$

with

$$
\begin{aligned}
I &= \operatorname{tr} \mathbf{C} = \lambda_C^{(1)} + \lambda_C^{(2)} + \lambda_C^{(3)}, \\
II &= \frac{1}{2}\left(I^2 - \operatorname{tr} \mathbf{C}^2\right) = \lambda_C^{(1)}\lambda_C^{(2)} + \lambda_C^{(1)}\lambda_C^{(3)} + \lambda_C^{(2)}\lambda_C^{(3)}, \\
III &= J^2 = \det \mathbf{C} = \lambda_C^{(1)}\lambda_C^{(2)}\lambda_C^{(3)}.
\end{aligned}
\tag{2.16}
$$

These *principal invariants* of tensor $\mathbf{C}$ agree with those of tensor $\mathbf{B}$. J denotes the *Jacobi determinant*. The scalars I, II and III are called invariants because a change of the base system, indeed, causes a change of the eigenvectors in the spectral representation of tensor $\mathbf{C}$, but not a change of the eigenvalues. Thus, the invariants do not change during an arbitrary coordinate transformation of the given tensor $\mathbf{C}$.

2.1.5 Homogeneous deformations

In general, the deformation gradient $\mathbf{F}$ varies from point to point in $\mathcal{B}_0$. Such a deformation is called inhomogeneous. In special cases, when $\mathbf{F}$ is independent of $\mathbf{X}$ we say that the deformation of the body is homogeneous. The function of motion for such deformations has the general form

$$\mathbf{x} = \mathbf{F}\mathbf{X} + \mathbf{c}, \qquad \mathbf{X} \in \mathcal{B}_0, \tag{2.17}$$

where the deformation gradient $\mathbf{F}$ and the vector of rigid translation $\mathbf{c}$ are independent of $\mathbf{X}$.

2.2 Universal solutions

Many authors reported on a very important special class of solutions: the so-called *universal solutions* (e.g. C.-C. Wang & C. A. Truesdell [415](1973), C. A. Truesdell [391] (1977) and later on K. Wilmanski [437] (1998)). Universal solutions are the foundation of experimental verification of constitutive relations for various materials under static conditions. These families of problems have been discussed in details by R. S. Rivlin in a series of ten papers (partly together with other authors) starting from 1948 [313, 314, 315, 316, 317, 318, 319, 320, 2, 3] and by J. L. Ericksen [114](1954) and by J. L. Ericksen & R. S. Rivlin [116] (1954).

While the ten papers by Rivlin with the main title "Large elastic deformations of isotropic materials" are related to

I.	Fundamental concepts,
II.	Some uniqueness theorems for pure, homogeneous deformation,
III.	Some simple problems in cylindrical polar coordinates,
IV.	Further developments of the general theory,
V.	The problem of flexture,
VI.	Further results in the theory of torsion, shear and flexture,
VII.	Experiments on the deformation of rubber,
VIII.	Strain distribution around a hole in a sheet,
IX.	The deformation of thin shells,
X.	Reinforcement by inextensible cords.

C. A. Truesdell, and also K. Wilmanski, group the problems in the following five (plus one) families. The supplementation of the additional zeroth group which summarizes the simplest class of deformations, namely homogeneous plane deformations, traces back to C.-C. Wang [414] (1968).

Family 0:   Homogeneous plane deformations,
Family 1:   Pure bending, stretching, and shearing of a rectangular block,
Family 2:   Straithening, stretching, and shearing of a section of a hollow cylinder,
Family 3:   Inflation, eversion, bending, torsion, extension, and shearing of an annular wedge,
Family 4:   Inflation and eversion of a sector of a spherical shell,
Family 5:   Inflation, azimuthal bending and shearing, and extension of an annular wedge.

Several authors give attention to the topic "universal solutions" or "universal deformations" in different extent. Amongst others these are: A. E. Green & W. Zerna (1968) [143] or R. W. Ogden (1984) [281]. More recent treatises e.g. by M. F. Beatty (2001) [36], G. Saccomandi (2001) [333], R. Batra (2006) [32] and A. Romano & A. Marasco (2010) [324] pick up the matter again. While in the first book chapter examples for different materials, models and experiments are shown, the second not only addresses some families of universal solutions but also non-universal solutions. The latter two are books on continuum mechanics in which this important topic is also discussed. A PhD-thesis covering this area of research has been published recently by R. de Pascalis [103].

2.2.1   Families of universal solutions and corresponding geometric quantities

Capital letters denote Lagrangian coordinates: (X, Y, Z) are rectangular Cartesian, (R, Θ, Z) are cylindrical polar and (R, Θ, Φ) are spherical polar. Small letters denote Eulerian coordinates: $(x, y, z), (r, \theta, z), (r, \theta, \phi)$ with a similar meaning as before. The letters $A, ..., E$ are constants.

2.2.1.1   Family 0: Homogeneous plane deformations

In this case both for the reference and for the current configuration Cartesian coordinates are used. Deformations of this family are described by the function of motion

$$x = AX, \quad y = BY + CZ, \quad z = DY + EZ, \tag{2.18}$$

the strain tensor **B** is then

$$\left(B^{kl}\right) = \begin{pmatrix} A^2 & 0 & 0 \\ 0 & B^2 + C^2 & BD + CE \\ 0 & BD + CE & D^2 + E^2 \end{pmatrix}, \tag{2.19}$$

its inverse

$$\left(B^{-1}_{kl}\right) = \begin{pmatrix} \dfrac{1}{A^2} & 0 & 0 \\ 0 & \dfrac{D^2 + E^2}{B^2 E^2 + C^2 D^2 - 2BCDE} & -\dfrac{BD + CE}{B^2 E^2 + C^2 D^2 - 2BCDE} \\ 0 & -\dfrac{BD + CE}{B^2 E^2 + C^2 D^2 - 2BCDE} & \dfrac{B^2 + C^2}{B^2 E^2 + C^2 D^2 - 2BCDE} \end{pmatrix}, \tag{2.20}$$

and the invariants of $\mathbf{B}$

$$I = \operatorname{tr} \mathbf{B} = A^2 + B^2 + C^2 + D^2 + E^2,$$

$$II = \frac{1}{2}\left(I^2 - \operatorname{tr}\mathbf{B}^2\right) = A^2\left(B^2 + C^2 + D^2 + E^2\right) - 2BCDE + C^2D^2 + B^2E^2, \quad (2.21)$$

$$III = \det \mathbf{B} = A^2(B^2E^2 + C^2D^2 - 2BCDE).$$

2.2.1.2 Family 1: Bending, stretching and shearing of a rectangular block

For this family it is convenient to choose cylindrical coordinates (r, θ, z) (see Appendix A.2) for the current configuration

$$r = \sqrt{2AX}, \quad \theta = BY, \quad z = \frac{Z}{AB} - BCY, \qquad AB \neq 0. \quad (2.22)$$

$$\left(B^{kl}\right) = \begin{pmatrix} B^{rr} & B^{r\theta} & B^{rz} \\ B^{r\theta} & B^{\theta\theta} & B^{\theta z} \\ B^{rz} & B^{\theta z} & B^{zz} \end{pmatrix} = \begin{pmatrix} \dfrac{A^2}{r^2} & 0 & 0 \\ 0 & B^2 & -B^2C \\ 0 & -B^2C & B^2C^2 + \dfrac{1}{A^2B^2} \end{pmatrix}, \quad (2.23)$$

$$\left(B_{kl}^{-1}\right) = \begin{pmatrix} \dfrac{r^2}{A^2} & 0 & 0 \\ 0 & \dfrac{1}{B^2} + A^2B^2C^2 & A^2B^2C \\ 0 & A^2B^2C & A^2B^2 \end{pmatrix}, \quad (2.24)$$

$$I = B_k^k = \frac{A^2}{r^2} + B^2r^2 + B^2C^2 + \frac{1}{A^2B^2},$$

$$II = \frac{1}{2}\left[\left(B_k^k\right)^2 - B_l^k B_k^l\right] = \frac{r^2}{A^2} + \frac{1}{r^2}\left(\frac{1}{B^2} + A^2B^2C^2\right) + A^2B^2, \quad (2.25)$$

$$III = 1.$$

2.2.1.3 Family 2: Straightening, stretching and shearing of a sector of a circular-cylindrical tube

In this case for the current configuration Cartesian coordinates are used and for the reference configuration cylindrical coordinates (R, Θ, Z)

$$x = \frac{1}{2}AB^2R^2, \quad y = \frac{\Theta}{AB}, \quad z = \frac{Z}{B} + \frac{C\Theta}{AB}, \qquad AB \neq 0. \quad (2.26)$$

$$\left(B^{kl}\right) = \begin{pmatrix} 2AB^2x & 0 & 0 \\ 0 & \dfrac{1}{2Ax} & \dfrac{C}{2Ax} \\ 0 & \dfrac{C}{2Ax} & \dfrac{1}{B^2} + \dfrac{C^2}{2Ax} \end{pmatrix}, \quad (2.27)$$

$$\left(B_{kl}^{-1}\right) = \begin{pmatrix} \dfrac{1}{2AB^2x} & 0 & 0 \\ 0 & 2Ax + B^2C^2 & -B^2C \\ 0 & -B^2C & B^2 \end{pmatrix}, \tag{2.28}$$

$$I = \frac{1}{B^2} + 2AB^2x + \frac{1}{2Ax}\left(1 + C^2\right),$$

$$II = B^2\left(1 + C^2\right) + 2Ax + \frac{1}{2AB^2x}, \tag{2.29}$$

$$III = 1.$$

2.2.1.4   Family 3: Inflation, eversion, bending, torsion, extension, and shearing of an annular wedge

Here, both reference coordinates (R, Θ, Z) and current coordinates (r, θ, z) are cylindrical.

$$r = \sqrt{AR^2 + B}, \quad \theta = C\Theta + DZ, \quad z = E\Theta + FZ, \quad A(CF - DE) = 1. \tag{2.30}$$

$$\left(B^{kl}\right) = \begin{pmatrix} \dfrac{A^2R^2}{r^2} & 0 & 0 \\ 0 & \dfrac{C}{R^2} + D^2 & \dfrac{CE}{R^2} + DF \\ 0 & \dfrac{CE}{R^2} + DF & \dfrac{E^2}{R^2} + F^2 \end{pmatrix}, \tag{2.31}$$

$$\left(B_{kl}^{-1}\right) = \begin{pmatrix} \dfrac{r^2}{A^2R^2} & 0 & 0 \\ 0 & A^2\left(E^2 + F^2R^2\right) & -A^2\left(CE + DFR^2\right) \\ 0 & -A^2\left(CE + DFR^2\right) & A^2\left(C^2 + D^2R^2\right) \end{pmatrix}, \tag{2.32}$$

$$I = \frac{A^2R^2}{r^2} + r^2\left(\frac{C^2}{R^2} + D^2\right) + F^2 + \frac{E^2}{R^2},$$

$$II = \frac{r^2}{A^2R^2} + \frac{A^2}{r^2}\left(E^2 + F^2R^2\right) + A^2\left(C^2 + D^2R^2\right), \tag{2.33}$$

$$III = 1.$$

2.2.1.5   Family 4: Inflation or eversion of a sector of a spherical shell

In this case spherical coordinates (R, Θ, Φ) in the reference configuration and (r, θ, ϕ) in the current configuration are used.

$$r = \left(\pm R^3 + A\right)^{\frac{1}{3}}, \quad \theta = \pm\Theta, \quad \varphi = \Phi. \tag{2.34}$$

$$\left(B^{kl}\right) = \begin{pmatrix} \dfrac{R^4}{r^4} & 0 & 0 \\ 0 & \dfrac{r^2}{R^2} & 0 \\ 0 & 0 & \dfrac{r^2}{R^2} \end{pmatrix}, \quad \left(B_{kl}^{-1}\right) = \begin{pmatrix} \dfrac{r^4}{R^4} & 0 & 0 \\ 0 & \dfrac{R^2}{r^2} & 0 \\ 0 & 0 & \dfrac{R^2}{r^2} \end{pmatrix}, \tag{2.35}$$

$$I = \frac{R^4}{r^4} + 2\frac{r^2}{R^2},$$
$$II = \frac{r^4}{R^4} + 2\frac{R^2}{r^2}, \tag{2.36}$$
$$III = 1.$$

2.2.1.6 Family 5: Inflation, stretching and shearing of a circular cylinder

Again, for both configurations cylindrical coordinates are used.

$$r = AR, \quad \theta = B \log R + C\Theta, \quad z = DZ, \quad A^2CD = 1. \tag{2.37}$$

$$\left(B^{kl}\right) = \begin{pmatrix} A^2 & \dfrac{AB}{R} & 0 \\ \dfrac{AB}{R} & \dfrac{B^2 + C^2}{R^2} & 0 \\ 0 & 0 & D^2 \end{pmatrix}, \tag{2.38}$$

$$\left(B_{kl}^{-1}\right) = \begin{pmatrix} \dfrac{B^2 + C^2}{A^2C^2} & -\dfrac{rB}{A^2C^2} & 0 \\ -\dfrac{rB}{A^2C^2} & \dfrac{r^2}{A^2C^2} & 0 \\ 0 & 0 & \dfrac{1}{D^2} \end{pmatrix}, \tag{2.39}$$

$$I = A^2 + D^2 + A^2\left(B^2 + C^2\right),$$
$$II = A^2\left[D^2 + A^2C^2 + B^2D^2 + B^2C^2\right], \tag{2.40}$$
$$III = 1.$$

In the next section a few special problems belonging to the above introduced families are discussed. While in this chapter the geometric behavior is presented, in Chapter 5 also properties arising due to forces and stresses are discussed. There are several books in which further examples for universal deformations can be found, e.g. in Green & Zerna [143], Ogden [281], Romano & Marasco [324], Wang & Truesdell [415], de Pascalis [103].

2.3 A few examples of universal deformations

2.3.1 Isochoric extension

During isochoric deformation the volume of a body is kept constant. If, again, Capital letters (X, Y, Z) denote Lagrangian coordinates belonging to the reference configuration $\mathcal{B}_0$ and small letters (x, y, z) Eulerian ones which pertain to the current configuration $\mathcal{B}_t$ then the function of motion is given by

$$x = \frac{1}{\sqrt{\lambda}}X, \quad y = \frac{1}{\sqrt{\lambda}}Y, \quad z = \lambda Z, \tag{2.41}$$

Table 2.2: *Summary of families of universal solutions.*

Capital letters denote Lagrangian coordinates: (X, Y, Z) are rectangular Cartesian, (R, Θ, Z) are cylindrical polar and (R, Θ, Φ) are spherical polar. Small letters denote Eulerian coordinates: $(x, y, z), (r, \theta, z), (r, \theta, \phi)$ with a similar meaning as before. The letters $A, ..., E$ are constants.

Family 0: Homogeneous plane deformations

$$x = AX, \quad y = BY + CZ, \quad z = DY + EZ.$$

Family 1: Bending, stretching and shearing of a rectangular block

$$r = \sqrt{2AX}, \quad \theta = BY, \quad z = \frac{Z}{AB} - BCY, \quad AB \neq 0.$$

Family 2: Straightening, stretching and shearing of a sector of a circular-cylindrical tube

$$x = \frac{1}{2}AB^2R^2, \quad y = \frac{\Theta}{AB}, \quad z = \frac{Z}{B} + \frac{C\Theta}{AB}, \quad AB \neq 0.$$

Family 3: Inflation or eversion, bending, torsion, extension and shearing of a sector of a circular-cylindrical tube

$$r = \sqrt{AR^2 + B}, \quad \theta = C\Theta + DZ, \quad z = E\Theta + FZ, \quad A(CF - DE) = 1.$$

Family 4: Inflation or eversion of a sector of a spherical shell

$$r = (\pm R^3 + A)^{\frac{1}{3}}, \quad \theta = \pm\Theta, \quad \varphi = \Phi.$$

Family 5: Inflation, stretching and shearing of a circular cylinder

$$r = AR, \quad \theta = B \log R + C\Theta, \quad z = DZ, \quad A^2CD = 1.$$

---

where λ is a constant bigger than zero. This means that for the present example of isochoric (volume preserving) extension (illustrated in Figure 2.2) in the description of homogeneous plane deformations of family zero (2.18) the constants have the values $A = \frac{1}{\sqrt{\lambda}}, B = \frac{1}{\sqrt{\lambda}}, C = 0, D = 0$ and $E = \lambda$.

From the definition of the deformation gradient (2.2) and the definitions of the deformation measures directly follows

$$\mathbf{F} = \begin{pmatrix} \frac{1}{\sqrt{\lambda}} & 0 & 0 \\ 0 & \frac{1}{\sqrt{\lambda}} & 0 \\ 0 & 0 & \lambda \end{pmatrix}, \qquad J = \det \mathbf{F} = 1, \qquad \mathbf{F}^{-1} = \begin{pmatrix} \sqrt{\lambda} & 0 & 0 \\ 0 & \sqrt{\lambda} & 0 \\ 0 & 0 & \frac{1}{\lambda} \end{pmatrix}, \qquad (2.42)$$

$$\mathbf{R} = 1, \qquad \mathbf{U} = \mathbf{F}, \qquad \mathbf{V} = \mathbf{F}^{-1}, \qquad (2.43)$$

$$\mathbf{C} = \begin{pmatrix} \frac{1}{\lambda} & 0 & 0 \\ 0 & \frac{1}{\lambda} & 0 \\ 0 & 0 & \lambda^2 \end{pmatrix}, \qquad \mathbf{B} = \mathbf{C}, \qquad \mathbf{c} = \mathbf{B}^{-1} = \begin{pmatrix} \lambda & 0 & 0 \\ 0 & \lambda & 0 \\ 0 & 0 & \frac{1}{\lambda^2} \end{pmatrix}, \qquad (2.44)$$

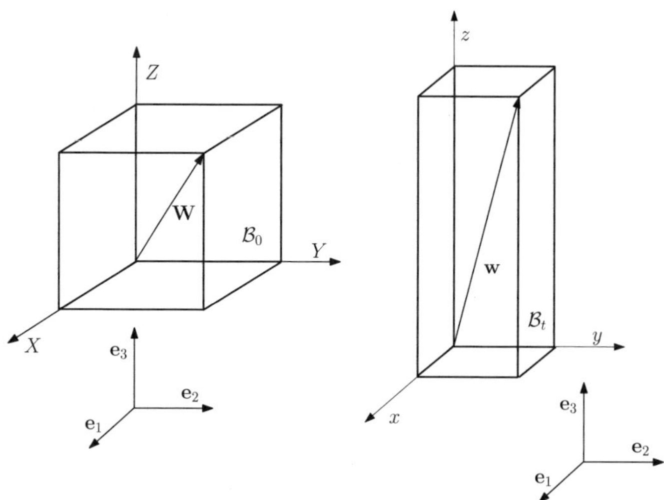

Figure 2.2: *Extension of a prism.*

$$\mathbf{E} = \frac{1}{2}\begin{pmatrix} \frac{1}{\lambda} - 1 & 0 & 0 \\ 0 & \frac{1}{\lambda} - 1 & 0 \\ 0 & 0 & \lambda^2 - 1 \end{pmatrix}, \qquad \mathbf{e} = \frac{1}{2}\begin{pmatrix} 1 - \lambda & 0 & 0 \\ 0 & 1 - \lambda & 0 \\ 0 & 0 & 1 - \frac{1}{\lambda^2} \end{pmatrix}. \tag{2.45}$$

The principal invariants are

$$I - \operatorname{tr}\mathbf{C} = \lambda_C^{(1)} + \lambda_C^{(2)} + \lambda_C^{(3)} = \frac{1}{\lambda} + \frac{1}{\lambda} + \lambda^2 = \frac{\lambda^3 + 2}{\lambda},$$

$$II = \frac{1}{2}\left(I^2 - \operatorname{tr}\mathbf{C}^2\right) = \lambda_C^{(1)}\lambda_C^{(2)} + \lambda_C^{(1)}\lambda_C^{(3)} + \lambda_C^{(2)}\lambda_C^{(3)} = \frac{1}{\lambda^2} + \lambda + \lambda = \frac{2\lambda^3 + 1}{\lambda^2}, \tag{2.46}$$

$$III = J^2 = \det\mathbf{C} = \lambda_C^{(1)}\lambda_C^{(2)}\lambda_C^{(3)} = \frac{1}{\lambda}\frac{1}{\lambda}\lambda^2 = 1.$$

It is worth mentioning that none of the strain tensors reacts symmetrically to a transformation tension $\leftrightarrow$ compression. The notion strain, which stems from the linear beam theory, describes the relative change of the length

$$\varepsilon = \frac{z - Z}{Z} = \lambda - 1. \tag{2.47}$$

If $\varepsilon > 0$ then $\infty > \lambda > 1$ (tension) and for $\varepsilon < 0$ is $0 < \lambda < 1$ (compression). In the present particular case we could symmetrize the measure of deformation. This means that λ could be replaced by a function $d(\lambda)$ which fulfils the condition

$$d\left(\frac{1}{\lambda}\right) = -d(\lambda), \tag{2.48}$$

e.g.

$$\begin{aligned} &\text{a)} \quad d(\lambda) = \log\lambda && \Rightarrow -\infty < d(\lambda) < \infty, \\ &\text{b)} \quad d(\lambda) = \frac{4}{\pi}\arctan\lambda - 1 && \Rightarrow -1 < d(\lambda) < 1. \end{aligned} \tag{2.49}$$

These functions (compare (2.49) a) and the Hencky strain measure) are sometimes used for the interpretation of experimental results. However, in the nonlinear theory there is no possibility to define such symmetric functions for arbitrary deformations.

Furthermore, the lack of rotations in this example ($\mathbf{R} = \mathbf{1}$) does not mean that all material vectors preserve the same directions during such a deformation. The eigenvalue problem (2.14) can be easily formulated also for the right stretch tensor $\mathbf{U}$. If we denote the eigenvalues by $\lambda_U^{(A)}$, $A = 1, 2, 3$, and the normalized (unit) eigenvectors by $\mathbf{r}^{(A)}$, $A = 1, 2, 3$, we see easily from

$$\left(\mathbf{U} - \lambda_U^{(A)} \mathbf{1} \right) \mathbf{r}^{(A)} = 0, \qquad A = 1, 2, 3, \qquad \mathbf{r}^{(A)} \cdot \mathbf{r}^{(B)} = \delta^{AB}, \tag{2.50}$$

that only the eigenvectors of the stretch tensor $\mathbf{U}$, and consequently, the eigenvectors of the tensor $\mathbf{C}$ rotate around the eigenvector of the orthogonal tensor $\mathbf{R}$ (corresponding to the single real eigenvalue $\lambda_R = 1$) on the angle determined by the complex eigenvalues of $\mathbf{R}$. Namely, for the eigenvectors $\mathbf{r}^{(A)}$ of $\mathbf{U}$ we have

$$\mathbf{F}\mathbf{r}^{(A)} = \mathbf{R}\mathbf{U}\mathbf{r}^{(A)} = \lambda_U^{(A)} \mathbf{R}\mathbf{r}^{(A)}, \tag{2.51}$$

with

$$\left(\mathbf{F}\mathbf{r}^{(A)} \right) \cdot \left(\mathbf{F}\mathbf{r}^{(A)} \right) = \left(\lambda_U^{(A)} \right)^2 \left(\mathbf{R}\mathbf{r}^{(A)} \right) \cdot \left(\mathbf{R}\mathbf{r}^{(A)} \right) = \left(\lambda_U^{(A)} \right)^2 \mathbf{r}^{(A)} \cdot \mathbf{r}^{(A)}. \tag{2.52}$$

Hence, for $\mathbf{R} = \mathbf{1}$ the vectors $\mathbf{r}^{(A)}$ change the length but not the direction. In the example, the eigenvectors $\mathbf{r}^{(A)}$ coincide with the edges of the prism and they, indeed, do not rotate. However, any other material vector changes its direction in spite of the "lack of rotations". For instance, the vector $\mathbf{W}$ (see Figure 2.2)

$$\mathbf{W} = \mathbf{e}_1 + \mathbf{e}_2 + \mathbf{e}_3, \tag{2.53}$$

transforms according to the formula

$$\mathbf{w} = \mathbf{F}\mathbf{W} = \mathbf{U}\mathbf{W} = \frac{1}{\sqrt{\lambda}}\mathbf{e}_1 + \frac{1}{\sqrt{\lambda}}\mathbf{e}_2 + \lambda\mathbf{e}_3, \tag{2.54}$$

and its cosines of angles with the axes of coordinates change in the following way:

$$\left(\frac{1}{\sqrt{3}}, \frac{1}{\sqrt{3}}, \frac{1}{\sqrt{3}} \right) \quad \rightarrow \quad \left(\frac{1}{\sqrt{\lambda + \frac{2}{\lambda^2}}}, \frac{1}{\sqrt{\lambda + \frac{2}{\lambda^2}}}, \frac{\sqrt{\lambda}}{\sqrt{\lambda + \frac{2}{\lambda^2}}} \right). \tag{2.55}$$

2.3.2   Simple shear

Next we consider the geometry of a somewhat more sophisticated deformation of a prism: a simple shearing in the plane perpendicular to the basis vector $\mathbf{e}_1$ described by

$$x = X, \qquad y = Y + Z \tan\varphi, \qquad z = Z, \tag{2.56}$$

(see Figure 2.3). It is a deformation in which points of planes parallel to the xy-plane move parallel to the y-axis. In this case the constants of family 0 in (2.18) have the values $A = 1$, $B = 1$, $C = \tan\varphi$, $D = 0$ and $E = 1$.

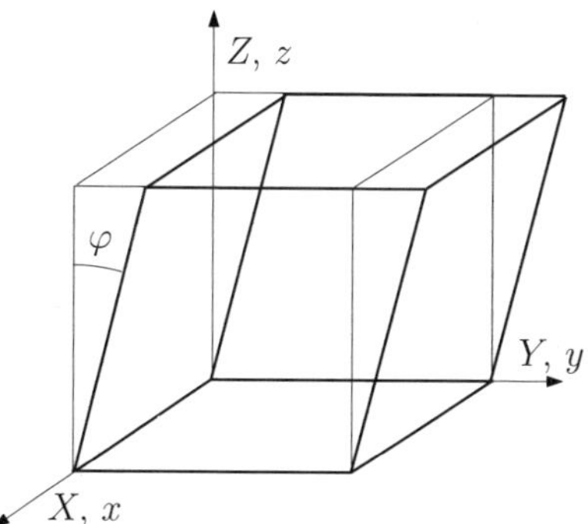

Figure 2.3: *Geometry of a simple shear.*

The corresponding deformation gradient and the corresponding quantities are

$$\mathbf{F} = \begin{pmatrix} 1 & 0 & 0 \\ 0 & 1 & \tan\varphi \\ 0 & 0 & 1 \end{pmatrix}, \qquad \det\mathbf{F} = 1, \qquad \mathbf{F}^{-1} = \begin{pmatrix} 1 & 0 & 0 \\ 0 & 1 & -\tan\varphi \\ 0 & 0 & 1 \end{pmatrix}. \tag{2.57}$$

The polar decomposition for this problem can be described by

$$\mathbf{R} = \begin{pmatrix} 1 & 0 & 0 \\ 0 & R & \sqrt{1-R^2} \\ 0 & -\sqrt{1-R^2} & R \end{pmatrix}, \qquad \mathbf{U} = \begin{pmatrix} 1 & 0 & 0 \\ 0 & R & \sqrt{1-R^2} \\ 0 & \sqrt{1-R^2} & \frac{2}{R}-R \end{pmatrix},$$

$$\mathbf{V} = \begin{pmatrix} 1 & 0 & 0 \\ 0 & R+\frac{2}{R}(1-R^2) & \sqrt{1-R^2} \\ 0 & \sqrt{1-R^2} & R \end{pmatrix}, \tag{2.58}$$

with

$$R := \frac{2}{\sqrt{4+\tan^2\varphi}} \leq 1. \tag{2.59}$$

Thus,

$$R = \cos\alpha \quad \Rightarrow \quad \frac{1}{\sqrt{1+\frac{1}{4}\tan^2\varphi}} = \cos\alpha \quad \Rightarrow \quad \tan\alpha = \frac{1}{2}\tan\varphi, \tag{2.60}$$

and (2.58) can be written in another form

$$\mathbf{R} = \begin{pmatrix} 1 & 0 & 0 \\ 0 & \cos\alpha & \sin\alpha \\ 0 & -\sin\alpha & \cos\alpha \end{pmatrix}, \qquad \mathbf{U} = \begin{pmatrix} 1 & 0 & 0 \\ 0 & \cos\alpha & \sin\alpha \\ 0 & \sin\alpha & \frac{2}{\cos\alpha}-\cos\alpha \end{pmatrix},$$

$$\mathbf{V} = \begin{pmatrix} 1 & 0 & 0 \\ 0 & \cos\alpha + 2\frac{\sin^2\alpha}{\cos\alpha} & \sin\alpha \\ 0 & \sin\alpha & \cos\alpha \end{pmatrix}. \tag{2.61}$$

We use again the original notation to specify further deformation measures:

- right Cauchy-Green:

$$\mathbf{C} = \begin{pmatrix} 1 & 0 & 0 \\ 0 & 1 & \tan\varphi \\ 0 & \tan\varphi & 1+\tan^2\varphi \end{pmatrix}, \tag{2.62}$$

- Green-St. Venant:

$$\mathbf{E} = \frac{1}{2}\begin{pmatrix} 0 & 0 & 0 \\ 0 & 0 & \tan\varphi \\ 0 & \tan\varphi & \tan^2\varphi \end{pmatrix}, \tag{2.63}$$

- left Cauchy-Green:

$$\mathbf{B} = \begin{pmatrix} 1 & 0 & 0 \\ 0 & 1+\tan^2\varphi & \tan\varphi \\ 0 & \tan\varphi & 1 \end{pmatrix}, \tag{2.64}$$

- Almansi-Hamel:

$$\mathbf{e} = \frac{1}{2}\begin{pmatrix} 0 & 0 & 0 \\ 0 & 0 & \tan\varphi \\ 0 & \tan\varphi & -\tan^2\varphi \end{pmatrix}. \tag{2.65}$$

Obviously, for $\varphi \ll 1$ these results agree with the classical results of materials sciences. For

$$\varphi \ll 1 \quad \Rightarrow \quad \mathbf{E} = \mathbf{e} = \frac{1}{2}\begin{pmatrix} 0 & 0 & 0 \\ 0 & 0 & \frac{1}{2}\varphi \\ 0 & \frac{1}{2}\varphi & 0 \end{pmatrix} \quad \text{and} \quad \mathbf{R} \approx \begin{pmatrix} 1 & 0 & 0 \\ 0 & 1 & \frac{1}{2}\varphi \\ 0 & -\frac{1}{2}\varphi & 0 \end{pmatrix}. \tag{2.66}$$

The eigenvalues of $\mathbf{C}$ are

$$\lambda_C^{(1)} = 1, \qquad \begin{matrix} \lambda_C^{(2)} \\ \lambda_C^{(3)} \end{matrix} = 1 + \frac{1}{2}\tan^2\varphi \pm \frac{1}{2}\sqrt{4\tan^2\varphi + \tan^4\varphi} = \left(\frac{1 \mp \sin\alpha}{\cos\alpha}\right)^2, \tag{2.67}$$

thus, the principal invariants follow in the form

$$\begin{aligned} I &= \lambda_C^{(1)} + \lambda_C^{(2)} + \lambda_C^{(3)} = 3 + \tan^2\varphi, \\ II &= \lambda_C^{(1)}\lambda_C^{(2)} + \lambda_C^{(1)}\lambda_C^{(3)} + \lambda_C^{(2)}\lambda_C^{(3)} = 3 + \tan^2\varphi, \\ III &= \lambda_C^{(1)}\lambda_C^{(2)}\lambda_C^{(3)} = 1, \end{aligned} \tag{2.68}$$

and the corresponding eigenvectors are

$$\begin{aligned} \mathbf{r}^{(1)} &= \mathbf{e}_1, \qquad \mathbf{r}^{(2)} = -\frac{1}{\sqrt{2}}\frac{\cos\alpha}{\sqrt{1-\sin\alpha}}\mathbf{e}_2 + \frac{1}{\sqrt{2}}\sqrt{1-\sin\alpha}\,\mathbf{e}_3, \\ \mathbf{r}^{(3)} &= \frac{1}{\sqrt{2}}\frac{\cos\alpha}{\sqrt{1+\sin\alpha}}\mathbf{e}_2 + \frac{1}{\sqrt{2}}\sqrt{1+\sin\alpha}\,\mathbf{e}_3. \end{aligned} \tag{2.69}$$

It is obvious from (2.61) and (2.69) that the real eigenvector of $\mathbf{R}$ coincides with the direction $\mathbf{e}_1$ (x-axis) and that the angle α, defining the complex eigenvalues $\exp(\pm i\alpha)$ of $\mathbf{R}$, describes the angle of rotation of the eigenvectors of $\mathbf{U}$ around this axis. However, these eigenvectors do not have such a simple geometrical interpretation anymore as it was the case in the example of extension (Subsection 2.3.1). It is easy to see from (2.69) that

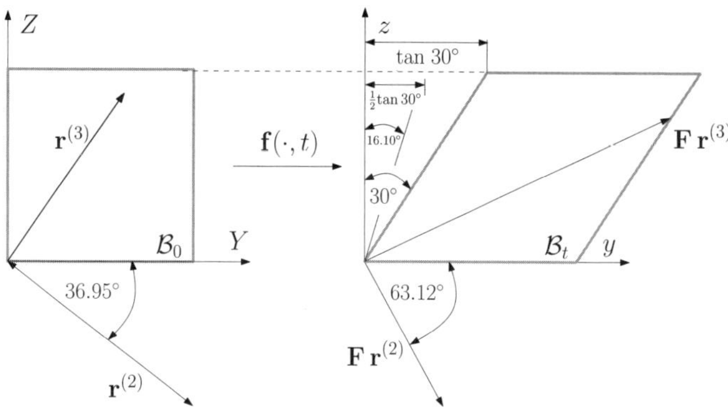

Figure 2.4: *Numerical example of simple shear.*

the eigenvectors do not coincide with the directions of the edges of the prism. In Figure 2.4 we illustrate these properties using the following numerical example

$$\varphi = 30° \quad \Rightarrow \quad \tan\alpha = \tfrac{1}{2}\tan\varphi = 0.2887 \quad \Rightarrow \quad \alpha = 16.1021°,$$

$$\lambda_C^{(1)} = 1, \quad \lambda_C^{(2)} = 0.5657, \quad \lambda_C^{(3)} = 1.7676. \tag{2.70}$$

The transformations of the eigenvectors by operation of the deformation gradient $\mathbf{F}$ have the following form

$$\mathbf{F}\,\mathbf{r}^{(2)} = \begin{pmatrix} 1 & 0 & 0 \\ 0 & 1 & 0.5774 \\ 0 & 0 & 1 \end{pmatrix} \begin{pmatrix} 0 \\ 0.7992 \\ -0,.6011 \end{pmatrix} = \begin{pmatrix} 0 \\ 0.4521 \\ -0.6011 \end{pmatrix},$$

$$\mathbf{F}\,\mathbf{r}^{(3)} = \begin{pmatrix} 0 \\ 1.0625 \\ 0.79917 \end{pmatrix}, \tag{2.71}$$

and the new vectors have the lengths

$$|\mathbf{F}\,\mathbf{r}^{(2)}| = \sqrt{0.45212^2 + 0.6011^2} = 0.7521 \quad \text{and} \quad |\mathbf{F}\cdot\mathbf{r}^{(3)}| = 1.3295. \tag{2.72}$$

In this case the eigenvectors of the stretch tensor do not coincide with any characteristic material vector. In contrast to the linear theory of shearing ($\tan\varphi \approx \varphi$, $\alpha \approx \tfrac{1}{2}\varphi$) the inverse rotation $\mathbf{R}^{-1}$ does not produce the current configuration symmetrical with respect to the diagonal of the y- and z-axes.

2.3.3 Pure torsion of a circular cylinder

We consider the torsional deformation of a circular cylinder of radius R due to twisting moments at its ends. This problem belongs to family 3 and in (2.30) the values of the constants are

$$A = 1,\ B = 0,\ C = 1,\ E = 0,\ F = 1, \tag{2.73}$$

thus from (2.30) remains

$$r = R,\ \theta = \Theta + DZ,\ z = Z. \tag{2.74}$$

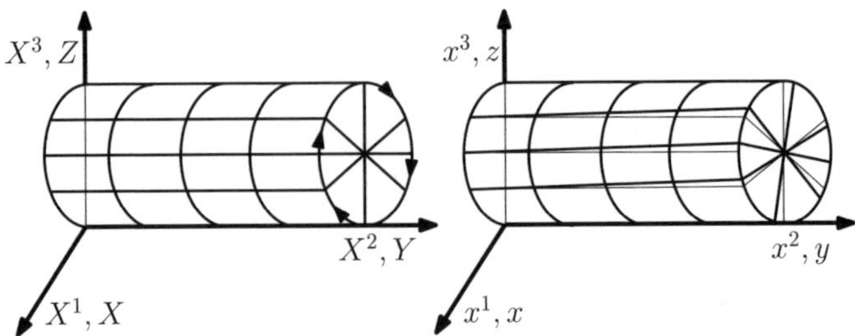

Figure 2.5: *Pure torsion of a circular cylinder.*

Here, again, (R, Θ, Z) are cylindrical coordinates in the reference configuration while (r, θ, z) belong to the deformed state. Obviously, the constant D describes the twist of the cylinder (for an illustration of the problem, see Figure 2.5). The coefficients of the left Cauchy-Green deformation tensor are

$$
\left(B^{kl}\right) = \begin{pmatrix} \dfrac{R^2}{r^2} & 0 & 0 \\ 0 & \dfrac{1}{R^2} + D^2 & D \\ 0 & D & 1 \end{pmatrix}. \tag{2.75}
$$

In this chapter only geometrical aspects are of interest. We refer to Section 5.4 for an inspection of the forces and stresses appearing in this and in the preceding examples.

Chapter 3

Kinematics of continua in different descriptions

3.1 Summary: Kinematics of one-component media

Again, for further reading on continuum mechanics we refer to the books of Wilmanski (e.g. Part I or [437]) Liu [231] or Marsden/Hughes [242].

Now we account also for the time dependence of the motion. To this aim the motion $\mathbf{f}(\mathbf{X}, \cdot)$ of a special material point $\mathbf{X} \in \mathcal{B}_0$ is considered. Provided that $\mathbf{f}(\mathbf{X}, \cdot)$ is twice differentiable the velocity of the material point can be expressed by

$$\forall t : \quad \mathbf{v} - \dot{\mathbf{x}}(\mathbf{X}, t) = \frac{\partial \mathbf{f}}{\partial t}(\mathbf{X}, t), \tag{3.1}$$

and its acceleration by

$$\forall t : \quad \mathbf{a} - \ddot{\mathbf{x}}(\mathbf{X}, t) - \frac{\partial^2 \mathbf{f}}{\partial t^2}(\mathbf{X}, t). \tag{3.2}$$

These two fields are defined on the reference configuration $\mathcal{B}_0$. The configuration of each time t is a possible reference configuration, i.e., there are countless possibilities. However, if possible, the reference configuration will be chosen in such a way that the stresses in the reference configuration are zero.

This type of description, where the motion is referred to a prescribed reference configuration, is called *material* or *Lagrangian description*. Traditionally, it is used to describe the motion of solids.

In fluid mechanics another type of description is commonly chosen, namely the *spatial* or *Eulerian description*. In this formulation the motion at any time t is related to the current configuration. I.e., it is observed in which way the body moves in time through a fixed point $\mathbf{x}$ in space. The motion is described by an instantaneous picture of the velocities $\mathbf{v}$ and the acceleration $\mathbf{a}$ of all points:

$$\mathbf{v} = \mathbf{v}(\mathbf{x},t), \qquad \mathbf{a} = \mathbf{a}(\mathbf{x},t) := \frac{\partial \mathbf{v}}{\partial t} + (\operatorname{grad} \mathbf{v})\,\mathbf{v} \equiv \dot{\mathbf{v}}(\mathbf{x},t). \tag{3.3}$$

The definition of the acceleration $(3.3)_2$ contains the so-called material time derivative. It follows from the assumption that the change of velocity is calculated along a trajectory. This corresponds to the partial time derivative in the Lagrangian description.

Physically, the Lagrangian description has the disadvantage that it is not directly verifiable by observations. Examples are velocity measurements which are undertaken in

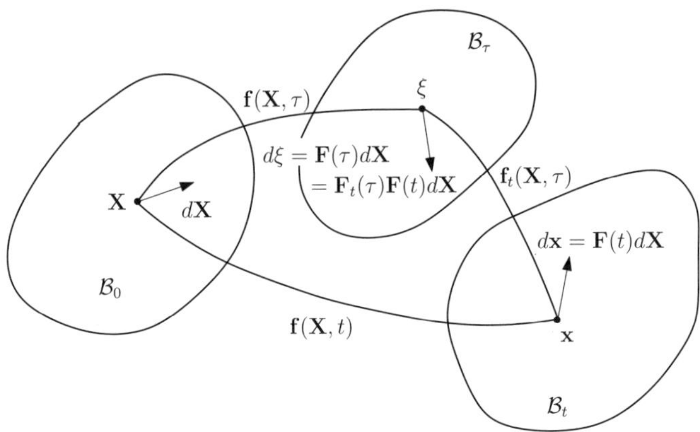

Figure 3.1: *Reference configurations.*

the space of motion. Two important questions arise in this connection: 1) How does the description of motion react on a transformation of the reference configuration, especially if the new configuration is identical with the current configuration, and 2) How does the description of motion react on a transformation of the observer in the space of motion.

In order to clarify these questions, we look at the following two configurations

$$\xi = \mathbf{f}\left(\mathbf{X},\tau\right), \quad \mathbf{x} = \mathbf{f}\left(\mathbf{X},t\right), \qquad \begin{aligned} \mathbf{X} &\in \mathcal{B}_0, \\ \xi &\in \mathcal{B}_\tau := \mathbf{f}\left(\mathcal{B}_0,\tau\right), \\ \mathbf{x} &\in \mathcal{B}_t := \mathbf{f}\left(\mathcal{B}_0,t\right), \end{aligned} \tag{3.4}$$

where ξ is the position of material point $\mathbf{X}$ at the moment τ and $\mathbf{x}$ is the position of the same material point at the moment t. Obviously, the position ξ can be described in reference to $\mathbf{x}$ instead of $\mathbf{X}$. Then, we have

$$\xi = \mathbf{f}\left[\mathbf{f}^{-1}(\mathbf{x},t),\tau\right]\mathbf{f}_t\left(\mathbf{x},\tau\right), \tag{3.5}$$

where the instant t is understood to be chosen and fixed (see Figure 3.1).

This type of description of the motion is especially advantageous if the velocity is available as a given function $\mathbf{v}$ of the position $\mathbf{x}$ and the time t

$$\mathbf{v} \equiv \frac{\partial \mathbf{x}}{\partial t}\left(\mathbf{X},t\right) = \frac{\partial \mathbf{x}}{\partial t}\left[\mathbf{f}^{-1}(\mathbf{x},t),t\right] = \mathbf{v}\left(\mathbf{x},t\right). \tag{3.6}$$

In this case, according to (3.5)

$$\frac{\partial \xi}{\partial \tau} = \mathbf{v}\left(\xi,\tau\right), \tag{3.7}$$

with the condition

$$\xi\left(\mathbf{x},\tau\right)\big|_{\tau=t} = \mathbf{x}. \tag{3.8}$$

This system is a differential equation with the initial condition (3.8) whose solution describes the trajectory of the particle backwards.

Simultaneously, Relation (3.5) allows to determine the deformation gradient in the configuration $\mathcal{B}_0$ with respect to configuration $\mathcal{B}_t$. Namely (Figure 3.1)

$$\mathbf{F}\left(\tau\right) = \frac{\partial \mathbf{f}}{\partial \mathbf{X}}\left(\mathbf{X}, \tau\right) = \frac{\partial \mathbf{f}_t}{\partial \mathbf{x}}\left(\mathbf{x}, \tau\right) \frac{\partial \mathbf{f}}{\partial \mathbf{X}}\left(\mathbf{X}, t\right). \tag{3.9}$$

It follows that

$$\mathbf{F}\left(\tau\right) = \mathbf{F}_t\left(\tau\right)\mathbf{F}\left(t\right), \qquad \mathbf{F}_t\left(\tau\right) := \frac{\partial \mathbf{f}_t}{\partial \mathbf{x}}\left(\mathbf{x}, \tau\right). \tag{3.10}$$

Of course

$$\mathbf{F}_t\left(\tau\right)\big|_{\tau=t} = \mathbf{1} \tag{3.11}$$

holds.

Of particular importance is the time change of the relative deformation gradient $\mathbf{F}_t\left(\tau\right)$ in the current configuration $\mathcal{B}_t$. We define

$$\dot{\mathbf{F}}_t\left(t\right) := \frac{\partial \mathbf{F}_t\left(\tau\right)}{\partial \tau}\bigg|_{\tau=t}. \tag{3.12}$$

With definition (3.10) we obtain

$$\dot{\mathbf{F}}_t\left(t\right) = \frac{\partial^2 \mathbf{f}_t}{\partial \tau \partial \mathbf{x}}\left(\mathbf{x}, \tau\right)\bigg|_{\tau=t} = \frac{\partial \mathbf{v}}{\partial \mathbf{x}}\left(\mathbf{x}, t\right) = \operatorname{grad}\mathbf{v}. \tag{3.13}$$

This means that the time derivative of the relative deformation gradient in relation to the current configuration $\mathcal{B}_t$ gives an account of the spatial gradient of the velocity. According to (3.10) furthermore

$$\mathbf{L} := \operatorname{grad}\mathbf{v} = \dot{\mathbf{F}}\left(t\right)\mathbf{F}^{-1}\left(t\right) \tag{3.14}$$

holds. Usually, the velocity gradient $\mathbf{L}$ is split into the symmetric part $\mathbf{D}$ and the skew-symmetric part $\mathbf{W}$ (Euler-Cauchy-Stokes decomposition)

$$\mathbf{L} = \mathbf{D} + \mathbf{W}, \qquad \mathbf{D} := \frac{1}{2}\left(\mathbf{L} + \mathbf{L}^T\right), \quad \mathbf{W} := \frac{1}{2}\left(\mathbf{L} - \mathbf{L}^T\right). \tag{3.15}$$

The tensor $\mathbf{D}$ is called the stretching tensor or the rate of deformation tensor and $\mathbf{W}$ the spin tensor. In order to interpret these tensors we consider an example. However, also directly from the polar decomposition we can illustrate the kinematic importance of these two tensors. Namely,

$$\mathbf{L} = \left(\mathbf{RU}\right)^{\cdot}\left(\mathbf{RU}\right)^{-1} = \left(\dot{\mathbf{R}}\mathbf{U} + \mathbf{R}\dot{\mathbf{U}}\right)\mathbf{U}^{-1}\mathbf{R}^T = \dot{\mathbf{R}}\mathbf{R}^T + \mathbf{R}\dot{\mathbf{U}}\mathbf{U}^{-1}\mathbf{R}^T, \tag{3.16}$$

and moreover

$$\left(\mathbf{RR}^T\right)^{\cdot} = \dot{\mathbf{R}}\mathbf{R}^T + \mathbf{R}\dot{\mathbf{R}}^T \quad \Rightarrow \quad \left(\dot{\mathbf{R}}\mathbf{R}^T\right)^T = \mathbf{R}\dot{\mathbf{R}}^T = -\dot{\mathbf{R}}\mathbf{R}^T. \tag{3.17}$$

From (3.15) we then obtain

$$\mathbf{D} = \tfrac{1}{2}\mathbf{R}\left(\dot{\mathbf{U}}\mathbf{U}^{-1} + \mathbf{U}^{-1}\dot{\mathbf{U}}\right)\mathbf{R}^T,$$

$$\mathbf{W} = \mathbf{\Omega} + \tfrac{1}{2}\mathbf{R}\left(\dot{\mathbf{U}}\mathbf{U}^{-1} + \mathbf{U}^{-1}\dot{\mathbf{U}}\right)\mathbf{R}^T, \quad \mathbf{\Omega} := \dot{\mathbf{R}}\mathbf{R}^T = -\mathbf{\Omega}^T. \tag{3.18}$$

This means that the deformation velocity $\mathbf{D}$ is only related to the time change of the stretch tensor $\mathbf{U}$, and the spin $\mathbf{W}$ arises as a superposition of the rigid rotation velocity $\boldsymbol{\Omega} = \dot{\mathbf{R}}\mathbf{R}^T$ and the rotation due to deformation (the second term of the right-hand side of $(3.18)_2$).

The spin, as it is a skew-symmetric tensor, can also be described by a vector $\boldsymbol{\omega}$. In Cartesian coordinates we have then

$$\omega_k := \frac{1}{2}\epsilon_{klm}W_{lm} \quad \Rightarrow \quad W_{kl} = \epsilon_{klm}\omega_m. \tag{3.19}$$

The vector $\boldsymbol{\omega}$ is called vorticity and is often used in fluid mechanics (compare Chapter 5). Since $\mathbf{W}$ is the skew-symmetric part of the velocity gradient, follows from (3.9)

$$\boldsymbol{\omega} = \frac{1}{2}\nabla \times \mathbf{v} = \frac{1}{2}\operatorname{curl}\mathbf{v}, \qquad \nabla := \operatorname{grad}. \tag{3.20}$$

As an example, the above introduced quantities are calculated for the case of simple shear (see Subsection 2.3.2). The angle of rotation φ is considered here as a given function of time. Then,

$$\mathbf{L} = \dot{\mathbf{F}}\mathbf{F}^{-1} = \begin{pmatrix} 0 & 0 & 0 \\ 0 & 0 & \frac{\dot{\varphi}}{\cos^2\varphi} \\ 0 & 0 & 0 \end{pmatrix}\begin{pmatrix} 1 & 0 & 0 \\ 0 & 1 & -\tan\varphi \\ 0 & 0 & 1 \end{pmatrix} = \begin{pmatrix} 0 & 0 & 0 \\ 0 & 0 & \frac{\dot{\varphi}}{\cos^2\varphi} \\ 0 & 0 & 0 \end{pmatrix}, \tag{3.21}$$

where

$$\dot{\varphi} := \frac{d\varphi}{dt}. \tag{3.22}$$

After simple calculation we obtain

$$\mathbf{D} = \frac{1}{2}\frac{\dot{\varphi}}{\cos^2\varphi}\begin{pmatrix} 0 & 0 & 0 \\ 0 & 0 & 1 \\ 0 & 1 & 0 \end{pmatrix}, \qquad \mathbf{W} = \frac{1}{2}\frac{\dot{\varphi}}{\cos^2\varphi}\begin{pmatrix} 0 & 0 & 0 \\ 0 & 0 & 1 \\ 0 & -1 & 0 \end{pmatrix},$$

$$\boldsymbol{\Omega} = 2\frac{\dot{\varphi}}{1 + 3\cos^2\varphi}\begin{pmatrix} 0 & 0 & 0 \\ 0 & 0 & 1 \\ 0 & -1 & 0 \end{pmatrix}, \tag{3.23}$$

$$\boldsymbol{\omega} = \left(\frac{1}{2}\frac{\dot{\varphi}}{\cos^2\varphi}, 0, 0\right).$$

On the other hand, the time derivative of the right Cauchy-Green deformation tensor is given by

$$\frac{d\mathbf{C}}{dt} \equiv \dot{\mathbf{C}} = 2\mathbf{F}^T\mathbf{D}\mathbf{F} = \begin{pmatrix} 0 & 0 & 0 \\ 0 & 0 & \frac{\dot{\varphi}}{\cos^2\varphi} \\ 0 & \frac{\dot{\varphi}}{\cos^2\varphi} & 2\frac{\dot{\varphi}\sin\varphi}{\cos^3\varphi} \end{pmatrix}. \tag{3.24}$$

Obviously, neither the derivative $\dot{\mathbf{C}}$ nor the derivative of any other deformation tensor is identical with the deformation velocity tensor $\mathbf{D}$. In this sense, the denotation of tensor $\mathbf{D}$ is misleading. Only in the case of small deformations ($\varphi \approx \sin\varphi$) the tensors $\mathbf{D}$ and $\dot{\mathbf{C}}$ are the same. Then, we have

$$\mathbf{D} \approx \frac{1}{2}\dot{\varphi}\begin{pmatrix} 0 & 0 & 0 \\ 0 & 0 & 1 \\ 0 & 1 & 0 \end{pmatrix}, \qquad \mathbf{W} \approx \boldsymbol{\Omega} \approx \frac{1}{2}\dot{\varphi}\begin{pmatrix} 0 & 0 & 0 \\ 0 & 0 & 1 \\ 0 & -1 & 0 \end{pmatrix}, \qquad \boldsymbol{\omega} = \left(\frac{1}{2}\dot{\varphi}, 0, 0\right). \tag{3.25}$$

The kinematic considerations will be completed by a few remarks on the transformation properties of the above mentioned tensors and vectors. The basic equations of continuum mechanics are formulated in an inertial coordinate system. However, it is not always favorable to solve problems in such systems. Thus, we need transformation rules for the change of the global observer in the space of motion. His observations of motion satisfy the following rules

$$\mathbf{x}^* = \mathbf{O}^*\mathbf{x} + \mathbf{c}^*, \qquad \mathbf{O}^* = \mathbf{O}^*\left(t\right), \qquad \mathbf{O}^{*T} = \mathbf{O}^{*-1}, \qquad \mathbf{c}^* = \mathbf{c}^*\left(t\right). \tag{3.26}$$

Similarly to the classical dynamics of rigid bodies, the transformation (3.26) can be understood either as the additional rigid motion of a body in a chosen coordinate system or as the description of the same motion in two different coordinate systems. We use the second interpretation.

The orthogonal matrix $\mathbf{O}^*$ is called the rotation matrix. From the orthogonality, it follows that

$$\det \mathbf{O}^* = 1. \tag{3.27}$$

The vector $\mathbf{c}^*$ indicates the relative translational motion of the two coordinate systems. After easy calculations we achieve the following results for the transformation rules of the deformation gradient and some deformation tensors

$$\mathbf{F}^* = \mathbf{O}^*\mathbf{F}, \qquad \mathbf{R}^* = \mathbf{O}^*\mathbf{R}, \qquad \mathbf{U}^* = \mathbf{U}, \qquad \mathbf{V}^* = \mathbf{O}^*\mathbf{V}\mathbf{O}^{*T},$$

$$\mathbf{C}^* = \mathbf{C}, \qquad \mathbf{E}^* = \mathbf{E}, \qquad \mathbf{B}^* = \mathbf{O}^*\mathbf{B}\mathbf{O}^{*T}, \qquad \mathbf{e}^* = \mathbf{O}^*\mathbf{e}\mathbf{O}^{*T}, \qquad \text{etc.} \tag{3.28}$$

More interesting is the transformation of time derivatives. For the velocity and the acceleration we obtain

$$\mathbf{v}^* = \mathbf{O}^*\mathbf{v} + \dot{\mathbf{O}}^*\mathbf{x} + \dot{\mathbf{c}}^*,$$

$$\mathbf{a}^* = \mathbf{O}^*\mathbf{a} + 2\dot{\mathbf{O}}^*\mathbf{v} + \ddot{\mathbf{O}}^*\mathbf{x} + \ddot{\mathbf{c}}^*, \tag{3.29}$$

or in relation to the new system

$$\mathbf{O}^*\mathbf{v} = \mathbf{v}^* - \mathbf{\Omega}^*\left(\mathbf{x}^* - \mathbf{c}^*\right) - \dot{\mathbf{c}}^*,$$

$$\mathbf{O}^*\mathbf{a} = \mathbf{a}^* - 2\mathbf{\Omega}^*\left(\mathbf{v}^* - \dot{\mathbf{c}}^*\right) + \mathbf{\Omega}^{*2}\left(\mathbf{x}^* - \mathbf{c}^*\right) - \dot{\mathbf{\Omega}}^*\left(\mathbf{x}^* - \mathbf{c}^*\right) - \ddot{\mathbf{c}}^*, \tag{3.30}$$

$$\mathbf{\Omega}^* := \dot{\mathbf{O}}^*\mathbf{O}^{*T} = -\mathbf{\Omega}^{*T}.$$

The second term on the right-hand side of the transformation rule for the velocity is called the relative rotation velocity, the third is the relative translation velocity. The rule for the acceleration contains the additional terms

$$-2\mathbf{\Omega}^*\left(\mathbf{v}^* - \dot{\mathbf{c}}^*\right) \quad - \quad \text{Coriolis acceleration,}$$

$$\mathbf{\Omega}^{*2}\left(\mathbf{x}^* - \mathbf{c}^*\right) \quad - \quad \text{centrifugal acceleration,}$$

$$\dot{\mathbf{\Omega}}^*\left(\mathbf{x}^* - \mathbf{c}^*\right) \quad - \quad \text{Euler acceleration,}$$

$$\ddot{\mathbf{c}}^* \quad\qquad\qquad - \quad \text{relative translational acceleration.}$$

The matrix $\boldsymbol{\Omega}^*$ denotes the angular velocity of the relative motion of the systems.

The most important property of the transformation rules (3.30) is the non-homogeneous dependence of the velocity and of the acceleration on the system.

Similarly, the rule for the velocity gradient can be described. We obtain

$$\mathbf{O}^*\mathbf{L}\mathbf{O}^{*T} = \mathbf{L}^* - \boldsymbol{\Omega}^*, \tag{3.31}$$

and, finally,

$$\mathbf{O}^*\mathbf{D}\mathbf{O}^{*T} = \mathbf{D}^*,$$

$$\mathbf{O}^*\mathbf{W}\mathbf{O}^{*T} = \mathbf{W}^* - \boldsymbol{\Omega}^*. \tag{3.32}$$

The occurrence of the angular velocity $\boldsymbol{\Omega}^*$ in the transformation rules has serious consequences for the formulation of constitutive laws. However, in this chapter, constitutive properties are not discussed. It remains only to note that the scalar temperature θ does not change by the change of the systems, so that $\theta = \theta^*$. If a temperature gradient appears, this is influenced by a change of coordinate systems because $\mathrm{grad}^*\theta^* = \mathbf{O}^*\mathrm{grad}\,\theta$.

3.2   Two-component materials with the skeleton as reference

We have mentioned above that for solids the Lagrangian description is common while the motion of fluids is usually described in the Eulerian frame. In systems containing both solid and fluid components, often the reference to the skeleton is useful since it usually undergoes smaller deformations than the fluid and serves as a kind of confinement for the pore fluid(s). The Lagrangian description of motion of porous materials with the skeleton as reference has been introduced in Chapter 13 of Part I and in the original papers by K. Wilmanski [433], [436]. The fields describing a multi-component porous medium are transformed to the common space-time domain defined by the reference configuration of the skeleton. We come back to such a formulation in Sections 12.1 and 12.2. In the present section we consider a simple example of two bars in extension in order to illustrate the geometrical meaning of the Lagrangian description of relative motion.

3.2.1   Example clarifying the Lagrangian description of relative motion

We consider the simple example of two bars in extension shown in Figure 3.2. The motions of the two bars are given by the following relations

$$\text{bar } \boxed{1}: \quad x = X\left(1 + \frac{V_1}{L}t\right),$$

$$V_2 > V_1 > 0. \tag{3.33}$$

$$\text{bar } \boxed{2}: \quad x = X\left(1 + \frac{V_2}{L}t\right),$$

V_1 and V_2 denote constants of the dimension of velocity, L is the initial length of both bars, x and X are Eulerian and Lagrangian coordinates, respectively, and t denotes the time.

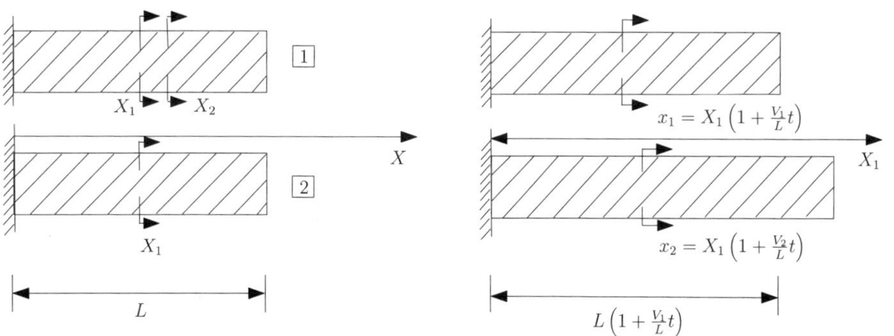

Figure 3.2: *Two bars in reference and current configurations.*

In Figure 3.2 also the current images x_1 and x_2 of an arbitrarily chosen initial cross-section X_1 for both bars are shown.

It is quite obvious that in this example with explicit Lagrangian description the image x_2 of the cross-section in the bar $\boxed{2}$ can be directly projected backwards on the reference configuration of bar $\boxed{1}$. This is due to Relation (3.33) and the result is as follows

$$X_2 = \frac{x_2}{1 + \dfrac{V_1}{L}t} = X_1\frac{1 + \dfrac{V_2}{L}t}{1 + \dfrac{V_1}{L}t}. \tag{3.34}$$

This backward projection ("pull-back" relative to the bar $\boxed{1}$) can certainly be performed solely for cross-sections of the bar $\boxed{2}$ satisfying the condition

$$0 < x < L\left(1 + \frac{V_1}{L}t\right). \tag{3.35}$$

The rest of the bar $\boxed{2}$ has "flown out" of the bar $\boxed{1}$ as shown in Figure 3.2.

In the real case of the skeleton and of a fluid we do not have at the disposal equations of the sort (3.33) because the motion of the fluid is given in Eulerian description. To transform our example to such a mixed description we need the velocities of both bars. According to (3.33), they are

$$\acute{x}_1 = X\frac{V_1}{L}, \qquad \acute{x}_2 = X\frac{V_2}{L}. \tag{3.36}$$

Hence, the Eulerian description of the second bar is given by the following velocity field

$$v_2 = x\frac{\dfrac{V_2}{L}}{1 + \dfrac{V_2}{L}t}, \tag{3.37}$$

which follows easily from $(3.36)_2$ by replacing the Lagrangian coordinates of bar $\boxed{2}$ by Eulerian coordinates according to Relation $(3.33)_2$.

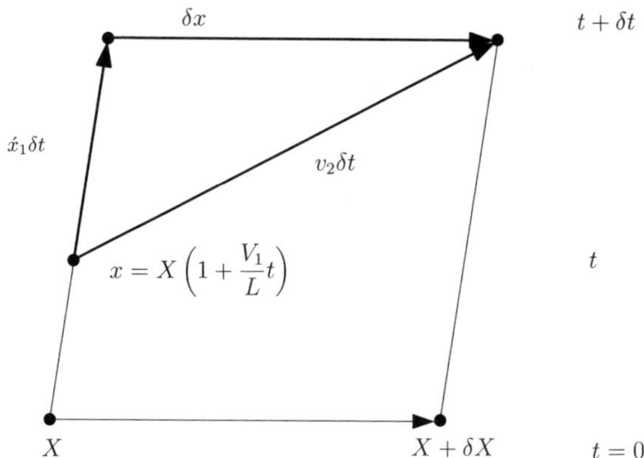

Figure 3.3: *Relative motion of the cross-section of bar 2.*

If we now ask the same question, how the backward projections of the cross-section of bar $\boxed{2}$ change in the reference configuration of bar $\boxed{1}$, this cannot be directly answered because the function of motion $(3.33)_2$ of bar $\boxed{2}$ is now unknown. We have to rely only on Relation (3.37). To solve the problem we proceed by calculating the velocity of these backward projections.

We choose an arbitrary point x of the configuration space. At this point the cross-section X of bar $\boxed{1}$ is located at the instant of time t. This cross-section moves to a new position (see Figure 3.3)

$$x + \dot{x}_1 \delta t = X\left(1 + \frac{V_1}{L}t\right) + X\frac{V_1}{L}\delta t, \tag{3.38}$$

as the time changes on the amount δt. We assume the time increments to be infinitesimal.

Simultaneously, the unknown cross-section of bar $\boxed{2}$ which has been located in the point x at the instant of time t, moves to the new position

$$x + v_2 \delta t = x\left(1 + \frac{\dfrac{V_2}{L}\delta t}{1 + \dfrac{V_2}{L}t}\right). \tag{3.39}$$

These two new positions do not coincide as we show in Figure 3.3. Their difference is

$$\delta x = X\left(1 + \frac{V_1}{L}t\right)\left(1 + \frac{\dfrac{V_2}{L}\delta t}{1 + \dfrac{V_2}{L}t}\right) - X\left(1 + \frac{V_1}{L}(t + \delta t)\right)$$

$$= X\frac{\dfrac{V_2}{L} - \dfrac{V_1}{L}}{1 + \dfrac{V_2}{L}t}\delta t. \tag{3.40}$$

The corresponding change of the cross-section of bar $\boxed{1}$ in the reference configuration follows from the relation

$$\delta x = (X + \delta X)\left(1 + \frac{V_1}{L}(t + \delta t)\right) - X\left(1 + \frac{V_1}{L}(t + \delta t)\right) \quad \Rightarrow$$

$$\delta X = \frac{\delta x}{1 + \frac{V_1}{L}t}. \tag{3.41}$$

Consequently,

$$\frac{\delta X}{\delta t} = X \frac{\frac{V_2}{L} - \frac{V_1}{L}}{\left(1 + \frac{V_1}{L}t\right)\left(1 + \frac{V_2}{L}t\right)}. \tag{3.42}$$

In the limit $\delta t \to 0$ this relation describes the velocity of the backward projection of cross-sections of bar $\boxed{2}$ on the reference configuration of bar $\boxed{1}$. It corresponds to the general condition for the Lagrangian velocity of the fluid F relative to the skeleton S (compare Section 12.1.2)

$$\acute{\mathbf{X}}^F\left(\mathbf{X}^S, t\right) := \lim_{\delta t \to 0} \frac{\delta \mathbf{X}^S}{\delta t} = \mathbf{F}^{S-1}\left(\acute{\mathbf{x}}^F - \acute{\mathbf{x}}^S\right),$$

$$\acute{\mathbf{x}}^F := \mathbf{v}^F\left(\mathbf{f}^S\left(\mathbf{X}^S, t\right), t\right) = \acute{\mathbf{x}}^F\left(\mathbf{X}^S, t\right),$$
$$\acute{\mathbf{x}}^S := \mathbf{v}^S\left(\mathbf{f}^S\left(\mathbf{X}^S, t\right), t\right) = \acute{\mathbf{x}}^S\left(\mathbf{X}^S, t\right), \tag{3.43}$$

due to the following formulae for our particular case (compare (3.36))

$$v_1(x, t) = x\frac{V_1}{L}\left(1 + \frac{V_1}{L}\right)^{-1}, \qquad v_2(x, t) = x\frac{V_2}{L}\left(1 + \frac{V_2}{L}\right)^{-1},$$

$$F = \left(1 + \frac{V_1}{L}\right). \tag{3.44}$$

Hence

$$\acute{X} = F^{-1}(v_2 - v_1) = \left(1 + \frac{V_1}{L}t\right)^{-1} X\left(1 + \frac{V_1}{L}t\right)\left[\frac{V_2}{L}\left(1 + \frac{V_2}{L}t\right)^{-1}\right.$$

$$\left. - \frac{V_1}{L}\left(1 + \frac{V_1}{L}\right)^{-1}\right] = X\frac{\frac{V_2}{L} - \frac{V_1}{L}}{\left(1 + \frac{V_1}{L}t\right)\left(1 + \frac{V_2}{L}t\right)}, \tag{3.45}$$

i.e., we obtain Relation (3.42).

It remains to check if the above considerations give the result (3.34). In order to do so we have to solve Equation (3.42) for X with the initial condition

$$X(t = 0) = X_1. \tag{3.46}$$

We have

$$\frac{dX}{X} = \left(\frac{V_2}{L} - \frac{V_1}{L}\right)\frac{dt}{\left(1 + \frac{V_1}{L}t\right)\left(1 + \frac{V_2}{L}t\right)} \equiv \left(\frac{\frac{V_2}{L}}{1 + \frac{V_2}{L}t} - \frac{\frac{V_1}{L}}{1 + \frac{V_1}{L}t}\right)dt. \tag{3.47}$$

Integration yields the following current position of the backward projection of the X_1-intersection of the bar $\boxed{2}$

$$X = X_1 \frac{1 + \dfrac{V_2}{L}t}{1 + \dfrac{V_1}{L}t}. \tag{3.48}$$

This is, indeed, identical with (3.34).

Chapter 4

Balance equations

The quantities introduced in the last chapter could be deduced from the function of motion $\mathbf{f}$. The form of this function has to be found for a chosen system as a solution of the governing equations, i.e., the unknown fields must be determined as functions of $\mathbf{X}$ and t (Lagrangian description). To this aim one has to find a set of field equations which describes the desired problem and which is fulfilled by the set of fields. The field equations are based on balance equations which in contrast to the field equations not yet include material properties. By insertion of constitutive relations, the system of balance equations changes into a system of field equations which has to be solved using appropriate initial and boundary conditions.

Balance equations consist of three parts

1. The change in time of the unknown quantity — a property of the subbody $\mathcal{P} \subset \mathcal{B}_0$,
2. the flux which is driven by extraneous causes to $\mathcal{P}$ and
3. an inner production.

If during the process the balanced quantity remains conserved, then the production vanishes and we refer to the laws as *conservation laws*.

We begin with the balance equations in integral form as a global statement for subbodies $\mathcal{P} \in \mathcal{B}_0$. Afterward they are converted into a local form. Local balance laws correspond to single points of the body. For regular points they have another form than for singular points (e.g. on the surface of the body).

Global balances. It is assumed that any scalar field $\varphi(\cdot, t)$ fulfills the global balance law

$$\forall \mathcal{P} \subset \mathcal{B}_0: \quad \frac{d}{dt} \int_{\mathcal{P}} \varphi(\mathbf{X}, t) dV = \oint_{\partial \mathcal{P}} \boldsymbol{\Psi}(\mathbf{X}, t) \mathbf{N}(\mathbf{X}) dS$$

$$+ \int_{\mathcal{P}} \left[\Psi_V(\mathbf{X}, t) + \varphi^*(\mathbf{X}, t) \right] dV. \tag{4.1}$$

In this equation $\boldsymbol{\Psi}(\cdot, t)$ is the flux density of φ, $\Psi_V(\cdot, t)$ is the supply of φ and $\varphi^*(\cdot, t)$ denotes the production of φ. $\mathbf{N}$ is the normal vector on the boundary $\partial \mathcal{P}$, V denotes the volume measure of the body and S is the surface measure of the body.

Local balances. Due to the assumption that the balance equation (4.1) is valid for *all* subbodies, it expresses also local properties of the field φ. In order to investigate these properties we look at a single point $\mathbf{X}$ and embed it into a family of subbodies $\mathcal{P}_i$ whose members are decreasing so that $\bigcap_{i=1}^{\infty} \mathcal{P}_i = \{\mathbf{X}\}$. Then, using Stokes' law which converts the surface integral into a volume integral, the global balance equation (4.1) can be transformed as follows

$$\forall \mathcal{P}_i : \frac{1}{V(\mathcal{P}_i)} \int_{\mathcal{P}_i} \left[\frac{\partial \varphi}{\partial t} - \mathrm{Div}\,\boldsymbol{\Psi} - \Psi_V - \varphi^* \right] dV. \tag{4.2}$$

The divergence with respect to $\mathbf{X}$ is denoted by Div. As a limit value for $\mathcal{P}_i \to \mathbf{X}$ the *local balance equation* for the field φ *in regular points* follows:

$$\frac{\partial \varphi}{\partial t} = \mathrm{Div}\,\boldsymbol{\Psi} + \Psi_V + \varphi^*. \tag{4.3}$$

Regular points are such points where $\frac{\partial \varphi}{\partial t}$ and Div $\boldsymbol{\Psi}$ exist.

If this is not the case, i.e., if the subbody contains discontinuities of these quantities and therefore singular surfaces as, for example, the boundaries of the body, the above balance equation becomes the following *dynamic compatibility condition* or *jump condition* (for the derivation for porous media see: Wilmanski [433]):

$$[\![\varphi]\!]\,U + [\![\boldsymbol{\Psi}]\!]\,\mathbf{N} = 0. \tag{4.4}$$

This is the *local balance equation* for points *on a singular surface*. Therein, U denotes the velocity in direction of the normal vector, $\mathbf{N}$, of the singular surface, the so-called *propagation velocity*. Double brackets represent jumps of a quantity: $[\![...]\!] = (...)^+ - (...)^-$. $(-)$- and $(+)$-signs correspond to the limits of the respective function on the two sides of the surface, once inside the body and once outside.

Three quantities are of particular importance for thermomechanical processes, namely: mass, momentum and energy. All of them satisfy conservation laws, i.e., balance laws without production terms. These conservation laws are given in Table 4.1. The left-hand side of the energy balance, the total energy, is the sum of the kinetic energy and the inner energy.

The mass density in the reference configuration is denoted by ρ_0, i.e.,

$$\rho_0 = \rho\,(\mathbf{X},t=0), \qquad \rho_0 = \rho J, \qquad J = \det \mathbf{F}, \tag{4.5}$$

where $\mathbf{F}$ is the deformation gradient and J the Jacobi determinant. Furthermore, $\mathbf{P}$ is the first Piola-Kirchhoff stress tensor, $\mathbf{b}$ denotes the body forces, ε the inner energy, $\mathbf{Q}$ the heat flux and r the supply of energy.

Often, particularly in fluid mechanics, the Eulerian description of the balance laws is used. They are obtained from the balance laws in Lagrangian formulation by a simple transformation and are shown in Table 4.2.

Analogously to the propagation velocity U in the Lagrangian description which corresponds to the kinematics of the singular surface in the reference configuration, in the Eulerian formulation the propagation velocity c is used which corresponds to the kinematics of the singular surface in the space of motion. Furthermore, $\mathbf{T}$ denotes the usual stress notation in the Eulerian description, the Cauchy stress tensor, and $\mathbf{q}$ is the Eulerian formulation of the heat flux.

Table 4.1: *Lagrangian description of the classical balance laws for a one-component body.*

	global
mass	$\dfrac{d}{dt}\int_{\mathcal{P}}\rho_0 dV = 0,$
momentum	$\dfrac{d}{dt}\int_{\mathcal{P}}\rho_0\mathbf{v}dV = \oint_{\partial\mathcal{P}}\mathbf{PN}dS + \int_{\mathcal{P}}\rho_0\mathbf{b}dV,$
energy	$\dfrac{d}{dt}\int_{\mathcal{P}}\rho_0\left(\tfrac{1}{2}v^2 + \varepsilon\right)dV = \oint_{\partial\mathcal{P}}\left(\mathbf{v}\cdot\mathbf{PN} - \mathbf{Q}\cdot\mathbf{N}\right)dS + \int_{\mathcal{P}}\rho_0\left(\mathbf{v}\cdot\mathbf{b}+r\right)dV,$

	local	
	regular	singular
mass	$\dfrac{\partial\rho_0}{\partial t} = 0,$	$U\,[\![\rho_0]\!] = 0,$
momentum	$\rho_0\mathbf{a} = \operatorname{Div}\mathbf{P}+\rho_0\mathbf{b},$	$U\,[\![\rho_0\mathbf{v}]\!] + [\![\mathbf{P}]\!]\,\mathbf{N} = 0,$
energy	$\rho_0\dfrac{\partial}{\partial t}\left(\tfrac{1}{2}v^2 + \varepsilon\right)$ $= \operatorname{Div}\left(\mathbf{P}^T\mathbf{v} - \mathbf{Q}\right) + \rho_0\mathbf{b}\cdot\mathbf{v}+\rho_0 r,$	$U\,[\![\rho_0\left(\tfrac{1}{2}v^2 + \varepsilon\right)]\!]$ $+ [\![\mathbf{P}^T\mathbf{v} - \mathbf{Q}]\!]\cdot\mathbf{N} = 0.$

Table 4.2: *Eulerian description of the classical balance laws for a one-component body.*

	global
mass	$\dfrac{d}{dt}\int_{\mathcal{P}_t}\rho dV = 0,\qquad \rho = \rho\left(\mathbf{x},t\right),\qquad \mathcal{P}_t := f\left(\mathcal{P},t\right),$
momentum	$\dfrac{d}{dt}\int_{\mathcal{P}_t}\rho\mathbf{v}dV = \oint_{\partial\mathcal{P}_t}\mathbf{Tn}dS + \int_{\mathcal{P}_t}\rho\mathbf{b}dV,\qquad \mathbf{T} = J^{-1}\mathbf{PF}^T,$
energy	$\dfrac{d}{dt}\int_{\mathcal{P}_t}\rho\left(\tfrac{1}{2}v^2 + \varepsilon\right)dV = \oint_{\partial\mathcal{P}_t}\left(\mathbf{vT} - \mathbf{q}\right)\cdot\mathbf{n}dS + \int_{\mathcal{P}_t}\rho\left(\mathbf{v}\cdot\mathbf{b}+r\right)dV,\quad \mathbf{q} = \mathbf{FQ},$

	local	
	regular	singular
mass	$\dfrac{\partial\rho}{\partial t} + \operatorname{div}\left(\rho\mathbf{v}\right) = 0,$	$[\![\rho\left(c - \mathbf{v}\cdot\mathbf{n}\right)]\!] = 0,$
momentum	$\dfrac{\partial\rho\mathbf{v}}{\partial t} + \operatorname{div}\left(\rho\mathbf{v}\otimes\mathbf{v} - \mathbf{T}\right) = \rho\mathbf{b},$	$[\![\rho\mathbf{v}\left(c - \mathbf{v}\cdot\mathbf{n}\right)]\!] + [\![\mathbf{T}]\!]\,\mathbf{n} = 0,$
energy	$\dfrac{\partial}{\partial t}\left[\rho\left(\tfrac{1}{2}v^2 + \varepsilon\right)\right]$ $+ \operatorname{div}\left[\rho\left(\tfrac{1}{2}v^2 + \varepsilon\right)\mathbf{v} - \mathbf{Tv} + \mathbf{q}\right]$ $= \rho\left(\mathbf{v}\cdot\mathbf{b}+r\right),$	$[\![\rho\left(\tfrac{1}{2}v^2 + \varepsilon\right)\left(c - \mathbf{v}\cdot\mathbf{n}\right)]\!]$ $= [\![\left(\mathbf{q} - \mathbf{Tv}\right)]\!]\cdot\mathbf{n}.$

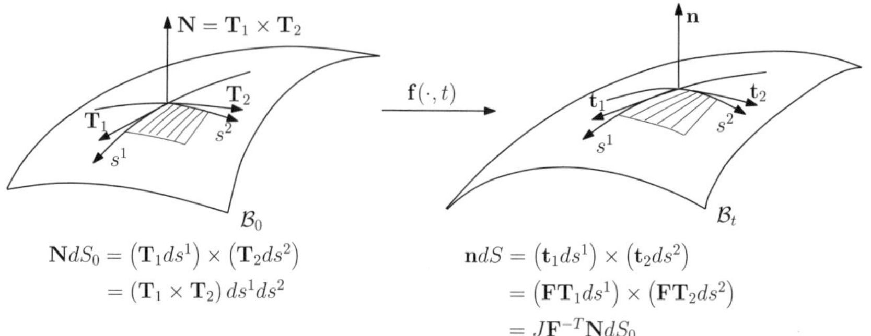

$$\mathbf{N}dS_0 = \left(\mathbf{T}_1 ds^1\right) \times \left(\mathbf{T}_2 ds^2\right) \qquad\qquad \mathbf{n}dS = \left(\mathbf{t}_1 ds^1\right) \times \left(\mathbf{t}_2 ds^2\right)$$
$$= \left(\mathbf{T}_1 \times \mathbf{T}_2\right) ds^1 ds^2 \qquad\qquad = \left(\mathbf{F}\mathbf{T}_1 ds^1\right) \times \left(\mathbf{F}\mathbf{T}_2 ds^2\right)$$
$$= J\mathbf{F}^{-T}\mathbf{N}dS_0$$

Figure 4.1: *Transformation of a surface element.*

We examine the relation between the Cauchy stress $\mathbf{T}$ and the Piola-Kirchhoff stress $\mathbf{P}$. Therefore, we repeat the global form of the momentum balance equation in Eulerian form. The stress tensor has been replaced according to Cauchy's lemma

$$\mathbf{t}_n\left(\mathbf{x}, t; \mathbf{n}\right) = \mathbf{T}\left(\mathbf{x}, t\right)\mathbf{n}, \tag{4.6}$$

by the stress vector $\mathbf{t}_n$

$$\forall \mathcal{P}_t: \quad \frac{d}{dt}\int_{\mathcal{P}_t} \rho\mathbf{v}\left(\mathbf{x}, t\right) dV = \oint_{\partial\mathcal{P}_t} \mathbf{t}_n\left(\mathbf{x}, t; \mathbf{n}\right) dS + \int_{\mathcal{P}_t} \rho\mathbf{b}\left(\mathbf{x}, t\right) dV, \qquad \mathbf{x} \in \mathcal{B}_t. \tag{4.7}$$

In this equation the coordinates of the surface integrals are transformed

$$\oint_{\partial\mathcal{P}_t} \mathbf{t}_n dS = \oint_{\partial\mathcal{P}_t} \mathbf{T}\mathbf{n} dS = \oint_{\partial\mathcal{P}_t} \mathbf{T}\left(\mathbf{t}_1 ds^1 \times \mathbf{t}_2 ds^2\right)$$

$$= \oint_{\partial\mathcal{P}_0} \mathbf{T}\left(\mathbf{F}\mathbf{P}_1 ds^1\right) \times \left(\mathbf{F}\mathbf{P}_2 ds^2\right) = \oint_{\partial\mathcal{P}_0} J\mathbf{T}\mathbf{F}^{-T}\mathbf{N} dS_0, \tag{4.8}$$

where the relation between the normal vectors in the reference configuration and in the current configuration

$$\mathbf{n} = J\mathbf{F}^{-T}\mathbf{N}, \tag{4.9}$$

and Figure 4.1 have been accounted for. As mentioned above, J denotes the Jacobi determinant.

If this result is used in the balance law, we obtain the following local relation

$$\mathbf{P} = J\mathbf{T}\mathbf{F}^{-T}. \tag{4.10}$$

The difference between $\mathbf{P}$ and $\mathbf{T}$ gets more obvious if we look at an example. Let us assume that the measurement of the force which is necessary to realize the isochoric extension (2.42) gives the following result

$$\mathbf{K} = \left(K_1, 0, 0\right), \qquad K_1 = K_0\varepsilon, \qquad \varepsilon := \log\lambda, \qquad K_0 := \text{const.} \tag{4.11}$$

The Cauchy stress on the area $x = \text{const.}$ is then

$$T_{11} = \frac{K_1}{A} = \lambda\frac{K_1}{A_0} = \mathrm{e}^\varepsilon\frac{K_1}{A_0} = \frac{K_0}{A_0}\left(\varepsilon\mathrm{e}^\varepsilon\right), \tag{4.12}$$

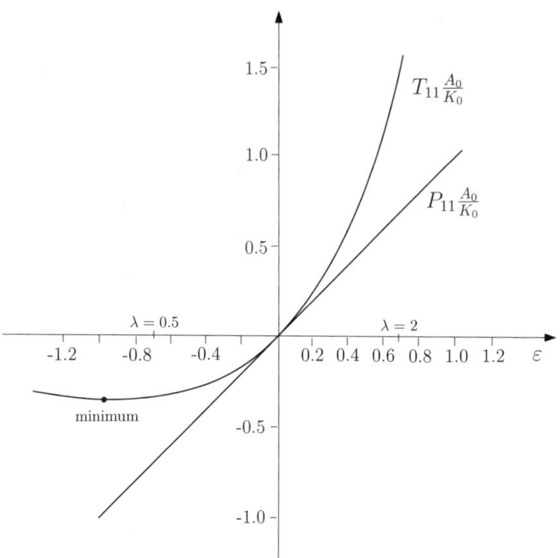

Figure 4.2: *Example for the difference between Cauchy and Piola-Kirchhoff tensile stress.*

and the Piola-Kirchhoff stress on the same area

$$P_{11} = \frac{K_1}{A_0} = \frac{K_0}{A_0}\varepsilon, \tag{4.13}$$

where A_0 denotes the initial cross section and A the current one. Roughly speaking, the stress P_{11} reflects the behavior of the external force, while T_{11} represents the local properties of the medium. For small deformations ($\varepsilon \ll 1 \Rightarrow e^\varepsilon \approx 1$) the difference between the two stresses vanishes (compare Figure 4.2). However, this is not the case for large deformations – the stresses differ not only in their values but also qualitatively: in the present example T_{11} exhibits a minimum which does not appear for P_{11} (see Figure 4.2). The Cauchy stress tensor $\mathbf{T}$ and the Piola-Kirchhoff stress tensor $\mathbf{P}$ have also different transformation rules (3.26):

$$\mathbf{x}^* = \mathbf{O}^*\mathbf{x} + \mathbf{c}^* \quad \Rightarrow \quad \mathbf{T}^* = \mathbf{O}^*\mathbf{T}\mathbf{O}^{*T}, \quad \mathbf{P}^* = \mathbf{O}^*\mathbf{P}, \tag{4.14}$$

and

$$J^* = J, \quad \mathbf{F}^{*-T} = (\mathbf{O}^*\mathbf{F})^{-T} = \mathbf{O}^*\mathbf{F}^{-T}$$

$$\Rightarrow \quad \mathbf{T}^* = J\mathbf{O}^*\mathbf{T}\mathbf{O}^{*T}\mathbf{O}^*\mathbf{F}^{-T} = \mathbf{O}^*\left(J\mathbf{T}\mathbf{F}^{-T}\right). \tag{4.15}$$

In contrast to the above mentioned quantities, the entropy is not conserved independently of whether the flux through the surface $\partial\mathcal{P}_t$ and the supply vanish (isolation) or not. The entropy satisfies the global balance equation with a production term on the right-hand side

$$\forall \mathcal{P}_t : \quad \frac{d}{dt}\int_{\mathcal{P}_t} \rho\eta\, dV + \oint_{\partial\mathcal{P}_t} \mathbf{h}\cdot\mathbf{n}\, dS - \int_{\mathcal{P}_t} \rho s\, dV = \int_{\mathcal{P}_t} \hat{\eta}\, dV, \tag{4.16}$$

where η denotes the entropy density, $\mathbf{h}$ is the entropy flux, s is the entropy supply and $\hat{\eta}$ the entropy production. The form of these quantities depends on the considered material.

As it was the case for the other quantities, from (4.16) we obtain a local balance equation for the entropy. For regular points it has the form

$$\frac{\partial \rho \eta}{\partial t} + \operatorname{div}\left(\rho \eta \mathbf{v} + \mathbf{h}\right) - \rho s - \hat{\eta}$$

$$\equiv \rho \dot{\eta} + \operatorname{div} \mathbf{h} - \rho s - \hat{\eta} = 0, \qquad \dot{\eta} := \frac{\partial \eta}{\partial t} + \mathbf{v} \cdot \operatorname{grad} \eta, \tag{4.17}$$

and on the singular surface, provided that the supply s and the production $\hat{\eta}$ are continuous

$$m \left[\!\left[\eta \right]\!\right] - \left[\!\left[\mathbf{h} \cdot \mathbf{n} \right]\!\right] = 0. \tag{4.18}$$

These equations in Eulerian description have the following form in Lagrangian notation

$$\rho_0 \frac{\partial \eta}{\partial t} + \operatorname{Div} \mathbf{H} = \rho_0 s + \hat{\eta}, \qquad \left[\!\left[\rho_0 \eta \right]\!\right] U - \left[\!\left[\mathbf{H} \right]\!\right] \cdot \mathbf{N} = 0. \tag{4.19}$$

A detailed discussion of entropy and of the formulation of the second law of thermodynamics by means of the inequality for this quantity can be found in Chapters 4 and 5 of Part I.

Chapter 5

Some solutions for fluids and solids

5.1 Preliminaries

In this chapter we present some equilibrium and stationary solutions of field equations for purely mechanical problems of fluids and nonlinear elastic solids. These solutions form the basis for static and stationary experiments in which constitutive functions and parameters of particular models are measured. Even though we concentrate on mechanical problems there is an important thermodynamical aspect of this investigation. As we see in Chapter 6 on stability, thermodynamics gives rise to restrictions which eliminate some models from practical applications in spite of their pertinence in simple experiments.

We begin with the presentation of a certain general property of the momentum balance equation for infinite media which yields the so-called *Euler-d'Alembert paradox*. The result is independent of the particular form of the constitutive strain-stress relation. Then we demonstrate on the example of the simplest flow problem of an ideal fluid how this paradox influences fluid mechanics. In particular, we introduce the notion of the so-called *added mass*. This notion was used to justify certain contributions to linear models of porous materials. This subject is discussed e.g. in Chapter 13 of Part I. The following two sections contain some solutions for viscous and viscoelastic fluids. For viscous Newtonian fluids we show also some properties of boundary layers. Non-Newtonian fluids discussed in the subsequent section are experimentally investigated in the so-called viscometric flows. We present some solutions for these flows. The second part of this chapter is devoted to the universal solutions for nonlinear elastic solids which were the subject of geometrical considerations in Chapter 2. Here we discuss some properties of the strain-stress relations for these universal solutions. We indicate the possibility of extension of this class of solutions on the so-called simple materials.

The equilibrium and stationary solutions deliver basic relations for the experimental verification of the models but, of course, there also exist some other experimental methods of validation of thermomechanical models. One of the most important methods is the analysis of acoustic waves in such systems. This is the subject of Chapter 10.

5.2 D'Alembert paradox

One of the issues which we would like to expose in the discussion of motions of ideal fluids is the so-called d'Alembert paradox. It follows from certain general properties of the momentum balance equation and it gives rise to various speculations concerning constitutive relations for continua with microstructure. In this section we discuss its origin within the frame of the general momentum conservation law. We solve the following problem (compare Section 202 in the book of C. Truesdell and R. A. Toupin [395]). Let us consider a rigid body $\mathcal{P}_{\text{inclusion}}$ with the surface $\mathcal{S}_{\text{inclusion}}$ and let it be submerged and fixed in the infinite continuous medium. This rigid inclusion is assumed to be immobile in the chosen reference system while the surrounding continuum is moving. We calculate the force exerted on the continuum by this inclusion in the case of steady motion of the continuum. Let $\mathcal{S}$ be an arbitrary closed control surface sufficiently large as to include the whole inclusion $\mathcal{P}_{\text{inclusion}}$. For simplicity, we neglect body forces. Then the momentum balance equation has the form (see Formula (4.36) in Part I)

$$\int_{\mathcal{S}_{\text{inclusion}} \cup \mathcal{S}} (\rho \mathbf{v} \otimes \mathbf{v} - \mathbf{T}) \, \mathbf{n} dS = 0, \tag{5.1}$$

where $\mathbf{T}$ is the Cauchy stress tensor and $\mathbf{v}$ is the velocity field of the medium. Hence, according to our choice of reference, it is zero in the inclusion. Simultaneously, the force exerted by the inclusion is defined by the relation

$$\mathbf{F} = - \oint_{\mathcal{S}_{\text{inclusion}}} \mathbf{T} \mathbf{n} dS, \tag{5.2}$$

where the minus sign appears due to the orientation of this surface – it is positive when the normal vector is pointing in the direction of the inclusion and not outwards as in Relation (5.1). In order to evaluate the momentum conservation we use the identity

$$\oint_{\mathcal{S}} \rho(\mathbf{v} \otimes \mathbf{v}) \mathbf{n} dS = \oint_{\mathcal{S}} [\rho(\mathbf{v} - \mathbf{V}) \otimes (\mathbf{v} - \mathbf{V})] \, \mathbf{n} dS + \mathbf{V} \cdot \mathbf{n} \oint_{\mathcal{S}} \rho(\mathbf{v} - \mathbf{V}) \, dS + \mathbf{V} \oint_{\mathcal{S}} \rho \mathbf{v} \cdot \mathbf{n} dS \tag{5.3}$$

for an arbitrary constant vector $\mathbf{V}$. In the steady motion the integral $\oint_{\mathcal{S}} \rho \mathbf{v} \cdot \mathbf{n} dS$ vanishes due to the mass conservation. Consequently, bearing the conservation of momentum (5.1) in mind, we can write Relation (5.2) in the following form

$$\mathbf{F} = \oint_{\mathcal{S}} [\rho(\mathbf{v} - \mathbf{V}) \otimes (\mathbf{v} - \mathbf{V})] \, \mathbf{n} dS + \mathbf{V} \cdot \mathbf{n} \oint_{\mathcal{S}} \rho(\mathbf{v} - \mathbf{V}) \, dS - \oint_{\mathcal{S}} (\mathbf{T} + p\mathbf{1}) \, \mathbf{n} dS, \tag{5.4}$$

where p is an arbitrary constant (obviously $\oint_{\mathcal{S}} \mathbf{n} dS = 0$). If the constants $\mathbf{V}$ and p can be chosen in such a way that the following conditions are satisfied[1]

$$\rho(\mathbf{v} - \mathbf{V}) \otimes (\mathbf{v} - \mathbf{V}) = \bar{o}\left(\frac{1}{|\mathbf{x}|^2}\right), \quad \rho(\mathbf{v} - \mathbf{V}) = \bar{o}\left(\frac{1}{|\mathbf{x}|^2}\right), \quad (\mathbf{T} + p\mathbf{1})\mathbf{n} = \bar{o}\left(\frac{1}{|\mathbf{x}|^2}\right), \tag{5.5}$$

then the rigid inclusion exerts no drag force on the continuous medium. The order of magnitude $\bar{o}\left(\dfrac{1}{|\mathbf{x}|^2}\right)$ means that surface integrals vanish in the limit of the surface $\mathcal{S}$ going to infinity, i.e., for $|\mathbf{x}| \to \infty$.

Similar considerations based on the moment of momentum conservation law can be performed for the calculation of the torque. They yield the same result (compare [395]).

---

[1]The symbol $\bar{o}$ may be read "lower mean order than" and it has been introduced by C. Truesdell [387].

In the case of the ideal fluid which we consider in the next section, the following sufficient conditions, instead of (5.5), are fulfilled

$$\mathbf{v} - \mathbf{V} = o\left(\frac{1}{|\mathbf{x}|^2}\right), \quad \rho = O\left(1\right), \quad \left(\mathbf{T} + p\mathbf{1}\right)\mathbf{n} = o\left(\frac{1}{|\mathbf{x}|^2}\right). \tag{5.6}$$

This is a very general and very unexpected result: if the velocity of the medium approaches its constant limit $\mathbf{V}$ in infinity sufficiently quickly and if the stress tensor approaches a uniform pressure $-p\mathbf{1}$ in infinity sufficiently quickly then the inclusion exerts no force on the continuous medium. This observation of L. Euler, rediscovered and improved by J.-B. d'Alembert in 1752, was the reason for the criticism of the fluid mechanics, which at these times was considered as the mechanics of ideal fluids, not being suitable to support theoretically the engineering hydraulics. First the development of mechanics of viscous fluids and, in particular, contributions of Ludwig Prandtl at the end of XIX century, suggested the explanation that these are the adherence forces to the boundary of the inclusion which yield the existence of the drag. Surprisingly, this issue seems to be still controversial as recent publications on the subject indicate (e.g. [145]).

5.3 Fluids

5.3.1 Ideal fluids

There is a vast literature concerning modeling and solving problems for ideal fluids. The classical monographs of L. Prandtl and O. G. Tietjens [298], J. Serrin [350], L. D. Landau and E. M. Lifshitz [219] as well as more recent books of I. G. Currie [102] or K. Hutter [172] should be mentioned. In this section we present only one simple example related to the d'Alembert paradox and to the so-called added mass effect. This special attention is due to the role which this problem plays in the theory of boundary layers as well as in the linear modeling of porous materials which we discuss further in this book.

Ludwig Prandtl   *Lev D. Landau*   *Kolumban Hutter*
1875-1953   *1908-1968*   *1941-*

The ideal fluid model describes the fields of mass density $\rho\left(\mathbf{x}, t\right)$, velocity $\mathbf{v}\left(\mathbf{x}, t\right)$ and temperature $T\left(\mathbf{x}, t\right)$. The field equations follow from the balance equations of mass, momentum and internal energy

$$\frac{\partial \rho}{\partial t} + \operatorname{div}(\rho \mathbf{v}) = 0,$$

$$\frac{\partial \rho \mathbf{v}}{\partial t} + \operatorname{div}(\rho \mathbf{v} \otimes \mathbf{v} - \mathbf{T}) = \rho \mathbf{b}, \tag{5.7}$$

$$\frac{\partial \rho \varepsilon}{\partial t} + \operatorname{div}(\rho \varepsilon \mathbf{v} + \mathbf{q}) = \operatorname{tr}(\mathbf{T} \operatorname{grad} \mathbf{v}),$$

and constitutive relations (see Section 6.1 of Part I for details)

$$\mathbf{T} = -p\mathbf{1}, \quad p = p(\rho, T), \quad \varepsilon = \varepsilon(\rho, T), \quad \eta = \eta(\rho, T), \quad \mathbf{q} = -K \operatorname{grad} T, \tag{5.8}$$

with the following consequences of the second law of thermodynamics

$$d\eta = \frac{1}{T}\left(d\varepsilon - \frac{1}{\rho^2} p d\rho\right), \quad K \geq 0. \tag{5.9}$$

In these relations, p is the pressure, ε denotes the specific internal energy, η is the specific entropy, $\mathbf{q}$ is the heat flux vector, K the thermal conductivity. Equation $(5.9)_1$ is the Gibbs equation for ideal fluids. Relations (5.8) are assumed to be invertible.

For the purpose of this section it is convenient to choose $\{p, \eta\}$ as constitutive variables instead of $\{\rho, T\}$. Then the Gibbs equation has the form (after the Legendre transformation; see Part I)

$$dh = T d\eta + \frac{1}{\rho} dp, \quad h = \varepsilon + \frac{p}{\rho}, \tag{5.10}$$

where h denotes the enthalpy.

An important particular case is the class of the so-called adiabatic processes. In these processes the heat flux vector is zero. However, such processes in ideal fluids are reversible (see Section 6.1 in Part I) and then the change of the entropy $d\eta$ must be zero as well (i.e., these processes are isentropic). Hence, according to (5.10) we have

$$\operatorname{grad} h = \frac{1}{\rho} \operatorname{grad} p. \tag{5.11}$$

Consequently, we can write the momentum balance equation in the form[2]

$$\frac{\partial \mathbf{v}}{\partial t} - \mathbf{v} \times (\operatorname{rot} \mathbf{v}) = \operatorname{grad}\left(\frac{v^2}{2} + h + \varphi\right), \quad \mathbf{b} = \operatorname{grad} \varphi, \tag{5.12}$$

or, equivalently,

$$\dot{\mathbf{v}} = \operatorname{grad}(h + \varphi), \quad \dot{\mathbf{v}} = \frac{\partial \mathbf{v}}{\partial t} + (\mathbf{v} \cdot \operatorname{grad}) \mathbf{v}. \tag{5.13}$$

---

[2]We use the following identity

$$\frac{1}{2} \operatorname{grad} v^2 = v_k \frac{\partial v_k}{\partial x_i} \mathbf{e}_i$$

$$= v_k \frac{\partial v_i}{\partial x_k} \mathbf{e}_i + v_k \left(\frac{\partial v_k}{\partial x_i} - \frac{\partial v_i}{\partial x_k}\right) \mathbf{e}_i = v_k \frac{\partial v_i}{\partial x_k} \mathbf{e}_i + \epsilon_{ikj}\epsilon_{jpq}\frac{\partial v_p}{\partial x_q} \mathbf{e}_i = (\mathbf{v} \cdot \operatorname{grad} \mathbf{v} + \mathbf{v} \times (\operatorname{rot} \mathbf{v})).$$

This is the so-called Euler equation. It means that for adiabatic processes the acceleration $\dot{\mathbf{v}}$ is derivable from the potential $(h + \varphi)$.

In the particular case of stationary flow the scalar multiplication of Equation (5.12) by a unit vector $\mathbf{m}$ parallel to the velocity $\mathbf{v}$ yields

$$\frac{\partial \mathbf{v}}{\partial t} = 0 \quad \Rightarrow \quad \mathbf{m} \cdot \operatorname{grad}\left(\frac{v^2}{2} + h + \varphi\right) = 0, \tag{5.14}$$

i.e.,

$$\frac{d}{ds}\left(\frac{v^2}{2} + h + \varphi\right) = 0, \tag{5.15}$$

where s is the parametrization of the curve to which the vector $\mathbf{m}$ is tangent. These are the so-called streamlines. The above equation is called the Bernoulli equation.

If we perform the operation of rotation on Equation (5.12), then the right-hand side becomes zero and we are left with the following equation

$$\frac{\partial \omega}{\partial t} - (\operatorname{rot}(\mathbf{v} \times \omega)) = 0, \quad \omega = \operatorname{rot}\mathbf{v}, \tag{5.16}$$

where the quantity ω is called the vorticity[3]. It is the characteristic feature of the model of ideal fluids that the equation for vorticity is independent of material properties. This is, of course, not the case for the velocity, described by Equation (5.12) where material properties are hidden in the constitutive relation for the enthalpy. As we see further the vorticity is dependent on material properties within the model of the viscous fluid.

An important notion related to the vorticity is the circulation. It is defined by the line integral of the velocity along a closed curve $\mathcal{C}$

$$\Gamma = \oint_{\mathcal{C}} \mathbf{v} \cdot d\mathbf{l}, \tag{5.18}$$

where $d\mathbf{l}$ is the line element multiplied by the unit vector tangent to the curve $\mathcal{C}$. This notion was introduced to fluid mechanics by Frederick Lanchester, Wilhelm Kutta, and Nikolai Zhukovsky in connection with their works on the motion of airfoils and it plays an important role in the numerical analysis of boundary layers. According to the Kelvin (Stokes) Theorem (Appendix A of Part I) it can be written in the form

$$\Gamma = \int_{\mathcal{S}} \omega \cdot \mathbf{n} dS, \quad \partial \mathcal{S} = \mathcal{C}, \tag{5.19}$$

where $\mathcal{S}$ is an arbitrary surface spanned by the curve $\mathcal{C}$. If the curve is material, i.e., if it is moving with the particle velocity $\mathbf{v}$ then we have

$$\frac{d\Gamma}{dt} = \oint_{\mathcal{C}} \dot{\mathbf{v}} \cdot d\mathbf{l} + \oint_{\mathcal{C}} \mathbf{v} \cdot d\frac{d\mathbf{l}}{dt} = \int_{\mathcal{S}} (\operatorname{rot}\dot{\mathbf{v}}) \cdot \mathbf{n} dS$$

$$\tag{5.20}$$

$$= \int_{\mathcal{S}} \operatorname{rot}(\operatorname{grad}(h + \varphi)) \cdot \mathbf{n} dS = 0,$$

---

[3]This definition of the vorticity is, of course, in the one-to-one relation to the notion of the spin tensor $\mathbf{W}$ defined by Relation (3.7) of Part I or of (3.19) of this part. Namely, we have in the Cartesian coordinates

$$W_{ij} = \frac{1}{2}\epsilon_{ijk}\omega_k \quad \Leftrightarrow \quad \omega_k = \epsilon_{kij}W_{ji}, \quad \mathbf{W} = W_{ij}\mathbf{e}_i \otimes \mathbf{e}_j, \quad \omega = \omega_k\mathbf{e}_k. \tag{5.17}$$

where Equation (5.13) was used and the second contribution vanishes because it is the integral of the full differential $d\dfrac{v^2}{2}$ along the closed curve. This result is called the Thomson[4] Circulation Theorem.

Even though the circulation of the fluid is defined by the integration of the vorticity, these two notions describe different properties of the flow as pointed out by Prandtl and Tietjens (p. 154 of the book [298]). The vorticity ω describes the rotation of individual particles and, for instance for potential flows for which $\mathbf{v} = \operatorname{grad}\phi$ it is identically zero. We say that the flow is irrotational. On the other hand, the circulation even in the potential flow may be different from zero, for instance for two-dimensional flows in which the curve $\mathcal{C}$ surrounds a singular point.

Thomson's Theorem indicates an important property of the circulation. Let us consider a small closed curve $\delta\mathcal{C}$. Then

$$\oint_{\delta\mathcal{C}} \mathbf{v} \cdot d\mathbf{l} \approx \delta S \mathbf{n} \cdot \operatorname{rot} \mathbf{v} = \delta S \mathbf{n} \cdot \omega = \text{const.} \tag{5.21}$$

Hence, the vorticity is transported along the trajectories (pathlines) of the fluid. Thus, if the circulation is zero in one point of the trajectory it remains zero on this trajectory for all times. Two conclusions follow from this property:

1. In the stationary flow around an obstacle submerged in an infinite medium, the flow in infinity must be homogeneous, i.e., the velocity is constant. Consequently, the vorticity is zero in infinity and this means that it must be zero in the whole space.

2. If the flow is potential (see above) at one instant of time then it remains potential for all times. The vorticity cannot be created in the potential flow.

These properties are sometimes called the Helmholtz Theorems (compare: J. Serrin [350], Section 25).

We proceed to investigate a class of fluids which is of great importance in practical applications. Namely, we assume that the fluid is incompressible, i.e., $\rho = \text{const.}$ Then the equations for the velocity can be written in the form

$$\begin{aligned} \operatorname{div} \mathbf{v} &= 0, \\ \frac{\partial \omega}{\partial t} - (\operatorname{rot}(\mathbf{v} \times \omega)) &= 0, \end{aligned} \tag{5.22}$$

where, obviously, the first equation follows from the conservation of mass and the second one is the equation for the vorticity (5.16). We investigate only potential flows and, hence, the second equation is fulfilled identically while the first one yields

$$\mathbf{v} = \operatorname{grad}\phi \quad \Rightarrow \quad \Delta^2\phi = 0, \quad \Delta^2(\dots) = \operatorname{div}\operatorname{grad}(\dots). \tag{5.23}$$

This means that the potential of velocity ϕ is in the case of incompressible fluids a harmonic function – it fulfills the Laplace equation. In this particular case Equation (5.12) leads immediately to the relation

$$\operatorname{rot}\left(\frac{\partial\phi}{\partial t} + \frac{v^2}{2} + h\right) = 0 \quad \Rightarrow \quad \frac{\partial\phi}{\partial t} + \frac{v^2}{2} + h = f(t), \tag{5.24}$$

---

[4]William Thomson is known as the 1st Baron Kelvin (1824-1907). His most prominent scientific contributions belong to thermodynamics. The Circulation Theorem was proved by him in 1869.

where $f(t)$ is an arbitrary function of time.

Let us demonstrate the solution of the flow problem on two simple examples.

1. A sphere of radius R moves with the velocity $\mathbf{v}_0$ in an incompressible fluid. We seek the solution for the motion of the fluid which vanishes in infinity. For this purpose we choose the solution of the Laplace equation in the form

$$\phi = \mathbf{A} \cdot \operatorname{grad} \frac{1}{r} \equiv -\frac{\mathbf{A} \cdot \mathbf{n}}{r^2}, \tag{5.25}$$

where the vector $\mathbf{n}r$ has the origin in the center of the sphere and $\mathbf{n}$ is the unit vector. The vector $\mathbf{A}$ should be found from the boundary condition which has the form

$$\mathbf{v} \cdot \mathbf{n} = \mathbf{v}_0 \cdot \mathbf{n} \quad \text{for } r = R. \tag{5.26}$$

It follows

$$\mathbf{A} = \mathbf{v}_0 \frac{R^3}{2} \quad \Rightarrow \quad \phi = -\frac{R^3}{2r^2} \mathbf{v_0} \cdot \mathbf{n}. \tag{5.27}$$

Consequently, for the velocity and the pressure (we use Relation $(5.24)_2$) we obtain the following relations

$$\mathbf{v} = \frac{R^3}{2r^3} \left(3\mathbf{n}\mathbf{v}_0 \cdot \mathbf{n} - \mathbf{v}_0 \right), \quad p = p_\infty - \frac{\rho v^2}{2} - \rho \frac{\partial \phi}{\partial t}. \tag{5.28}$$

2. An infinite cylinder of radius R moves with the velocity $\mathbf{v}_0$ perpendicular to its axis. Then the potential has the form

$$\phi = \mathbf{A} \cdot \operatorname{grad} \ln r = \frac{\mathbf{A} \cdot \mathbf{n}}{r}. \tag{5.29}$$

Again the boundary condition for the surface of the cylinder yields

$$\mathbf{A} = -R^2 \mathbf{v}_0 \quad \Rightarrow \quad \phi = -\frac{R^2}{r} \mathbf{v}_0 \cdot \mathbf{n}. \tag{5.30}$$

The corresponding velocity is

$$\mathbf{v} = \frac{R^2}{r^2} \left(2\mathbf{n}\mathbf{v_0} \cdot \mathbf{n} - \mathbf{v}_0 \right), \tag{5.31}$$

and the pressure follows from Relation $(5.28)_2$.

The above examples demonstrate the typical dependence of the potential on the distance in the infinite medium: for the three-dimensional case it is $1/r^2$ and for the two-dimensional case it is $1/r$. Such a contribution cannot appear in three-dimensional flows due to the incompressibility. The potential $-a/r$ would yield a constant mass flow through the surface of an arbitrary sphere of radius, say, R: $4\pi\rho a$, where a is a constant. This is, obviously, impossible and, consequently, $a = 0$.

It is easy to see that in both examples the independence of the velocity $\mathbf{v}_0$ from time yields solutions which satisfy the conditions (5.6). This means that the reaction force on the motion of the inclusion – the drag is zero and we have the d'Alembert paradox.

In the case of the velocity $\mathbf{v}_0$ dependent on time there is a drag which follows from the inertial forces. We consider an arbitrary inclusion of the volume V_0 which moves with the velocity $\mathbf{v}_0$ in the infinite medium. Sufficiently far from the inclusion, the velocity potential can be chosen to be $\phi = \mathbf{A} \cdot \mathrm{grad}\,(1/r)$. The easiest way to find the action of the fluid on the inclusion is to estimate the energy. The full energy accumulated in the system is given by the integral

$$E = \frac{\rho}{2} \int v^2 dV, \quad v^2 = \mathbf{v} \cdot \mathbf{v}, \tag{5.32}$$

where the volume integration is carried out over the whole space. It is convenient to separate the contribution of the kinetic energy of the inclusion. We perform the integration over the large sphere of the radius R and the volume $V = 4\pi R^3/3$ and then take the limit $R \to \infty$. We have

$$\int_V v^2 dV = \int_V v_0^2 dV + \int_V (\mathbf{v} + \mathbf{v}_0)(\mathbf{v} - \mathbf{v}_0)\, dV$$

$$= v_0^2 (V - V_0) + \int_V \mathrm{div}\,[(\phi + \mathbf{v}_0{\cdot}\mathbf{x})(\mathbf{v} - \mathbf{v}_0)]\, dV, \tag{5.33}$$

where the following relations were used

$$\mathbf{v} + \mathbf{v}_0 = \mathrm{grad}\,(\phi + \mathbf{v}_0 \cdot \mathbf{x}), \quad \mathrm{div}\,\mathbf{v} = 0. \tag{5.34}$$

Hence, using the Stokes Theorem, we obtain

$$\int_V v^2 dV = v_0^2 (V - V_0) + \oint_{\partial V} (\phi + \mathbf{v}_0{\cdot}\mathbf{x})(\mathbf{v} - \mathbf{v}_0) \cdot \mathbf{n}dS. \tag{5.35}$$

Obviously, the integration over the sphere of radius R means that the surface element has the form $dS = R^2 do$, where do is the surface element of the unit sphere.

Bearing in mind the identity

$$\oint_{\partial V} \mathbf{n} \otimes \mathbf{n}dS = \oint_{\partial V} \mathbf{n} \otimes \mathbf{n}R^2 do = \frac{4}{3}\pi R^2 \mathbf{1}, \tag{5.36}$$

and integrating in Relation (5.35) we obtain after substitution in (5.33)

$$E = \frac{\rho}{2}\left(4\pi \mathbf{A} \cdot \mathbf{v}_0 - V_0 v_0^2\right), \quad v_0^2 = \mathbf{v}_0 \cdot \mathbf{v}_0, \tag{5.37}$$

where we have taken the limit $R \to \infty$.

The vector $\mathbf{A}$ follows from the boundary condition on the surface of the inclusion. However, as the whole problem is linear, it must be linear with respect to the velocity $\mathbf{v}_0$. Hence, both contributions of this velocity to the energy are quadratic and we can write in the general case

$$E = \frac{1}{2}\mathbf{m} \cdot (\mathbf{v}_0 \otimes \mathbf{v}_0), \quad \mathbf{m} = m_{ij}\mathbf{e}_i \otimes \mathbf{e}_j. \tag{5.38}$$

The tensor $\mathbf{m}$ is called the tensor of the added mass. For example, in the case of the spherical inclusion which we have considered above, we have

$$\mathbf{A} = \mathbf{v}_0 \frac{R^3}{2} \quad \Rightarrow \quad m_{ij} = \frac{2}{3}\pi R^3 \rho \delta_{ij}. \tag{5.39}$$

We proceed to demonstrate the consequences of this result. First of all, the full energy E is, obviously, related to the full momentum, say $\mathbf{P}$, of the fluid as the increment of E must be balanced by the working of momentum on the increment of the velocity $\mathbf{v}_0$. This, in turn, yields infinitesimal changes of the reaction force, say $\mathbf{F}$, acting on the inclusion. Consequently, we have

$$\mathbf{P} = \frac{dE}{d\mathbf{v}_0} = 4\pi\rho\mathbf{A} - \rho V_0\mathbf{v}_0, \quad \mathbf{F} = -\frac{d\mathbf{P}}{dt}, \tag{5.40}$$

where the minus sign follows from the fact that it is a force with which the fluid acts on the inclusion. The component of the force in the direction of the velocity $\mathbf{v}_0$ is called, as we have already mentioned, the drag and the component of the force perpendicular to this direction is called the buoyancy force.

Bearing the above relations in mind, we can write the equation of motion for the inclusion in the following form

$$M\frac{d\mathbf{v}_0}{dt} + \frac{d\mathbf{P}}{dt} = \mathbf{f}, \quad \text{i.e.,} \quad (M\delta_{ij} + m_{ij})\frac{dv_{0j}}{dt} = f_i, \tag{5.41}$$

where M is the mass of the inclusion and $\mathbf{f}$ is the external force. For the sphere which we have considered above the result is as follows

$$\frac{4\pi R^3}{3}\left(\rho_0 + \frac{\rho}{2}\right)\frac{d\mathbf{v}_0}{dt} = \mathbf{f}, \tag{5.42}$$

where ρ_0 is the mass density of the material of the inclusion.

Summing up we see that the d'Alembert paradox is not directly connected with the appearance of additional inertial forces which yield the added mass effect. One can say that in the case of more sophisticated models, such as viscous fluids, two reaction forces on the moving inclusion should appear – one which is present in both stationary and nonstationary flows and which should be, as we have seen, related to the viscosity of the fluid and another one which is related to the inertia of the fluid and the inclusion which yields the added mass effect.

5.3.2 Viscous fluids

5.3.2.1 Navier-Stokes equation, thermodynamical properties, uniqueness of solutions

Mechanics of non-ideal fluids was most likely besides the theory of heat conduction one of the first nontrivial subjects of the constitutive description of matter in which dissipation processes play the essential role. The notion of viscosity characteristic for such fluids was already anticipated by I. Newton in his Book II of *Principia* (1687): "The resistance which arises from the lack of lubricity or slipperiness of the parts of a fluid is, other things being equal, proportional to the velocity with which the parts of the fluid are separated from one another." However, the first scientist who consciously dwelled on the notion of viscosity in the description of the properties of fluids was Claude Navier in 1827. Independently, this notion was introduced by George Gabriel Stokes in 1845. He also described essential ideas of fluidity which James Serrin [350] formulated as the following postulates

1. the Cauchy stress tensor $\mathbf{T}$ is a continuous function of the stretching tensor $\mathbf{D} = \frac{1}{2}\left[\operatorname{grad}\mathbf{v} + (\operatorname{grad}\mathbf{v})^T\right]$ and it is independent of all other kinematic quantities,

2. $\mathbf{T}$ does not depend explicitly on the position $\mathbf{x}$ (spatial homogeneity),

3. there is no preferred direction in space (isotropy),

4. when $\mathbf{D} = 0$, the stress tensor reduces to the pressure $\mathbf{T} = -p\mathbf{1}$.

Such materials are called Stokesian fluids. They are defined by the constitutive relation $\mathbf{T} = \mathbf{T}(\mathbf{D})$ where the constitutive function is isotropic. As it is independent of the mass density, the Stokesian fluid is incompressible. This is a property well confirmed in many experiments but there are, certainly, important exceptions. Therefore we begin with a more general situation in which a constitutive dependence on ρ may appear as well. Then

$$\mathbf{T} = -p\mathbf{1} + \lambda \left(\operatorname{div}\mathbf{v}\right)\mathbf{1} + 2\eta\mathbf{D} + \kappa\mathbf{D}^2, \tag{5.43}$$

where the pressure p and the coefficients λ, η, κ may depend on the mass density ρ and all three invariants of the stretching $\mathbf{D}$. According to the Cayley-Hamilton Theorem (Appendix A of Part I) this is the most general form of this constitutive relation. The coefficient $\zeta = \lambda + \frac{2}{3}\eta$ is called the volume viscosity and the coefficient η is the dynamic shear viscosity[5]. All these coefficients are also dependent on the temperature.

Claude L. M. H. Navier   *George Gabriel Stokes*   *Osborn Reynolds*
1785-1836   *1819-1903*   *1842-1912*

In this section we present a few facets of processes described by the so-called Navier-Stokes equation which follows under the assumption of the linearity of the constitutive law (5.43) with respect to the dependence on the stretching $\mathbf{D}$. However, it should be mentioned that the linear description of viscosity which forms the basis for the derivation of the Navier-Stokes equation is by far not always an acceptable assumption. In the first part of the XX century a new branch of mechanics – rheology, started to develop. Most likely Eugene Bingham was the first who suggested a nonlinear relation between the deviatoric stresses $\mathbf{T}^D$ and the stretching tensor $\mathbf{D}$ (e.g. [48]) similar to this mentioned above.

---

[5]There is a confusion in the notation related to viscosity coefficients. In this part of the book, η denotes the dynamic shear viscosity, $\nu = \eta/\rho$ is the kinematic viscosity while in Part I, we have denoted by ν the dynamic shear viscosity. The latter notation or, equivalently, μ as the dynamic shear viscosity are used in extended thermodynamics and some rheological papers. Simultaneously, the notation λ, μ is common for the Lamé constants in the linear elasticity. At the same time, we denote by η the specific entropy and this notation is so common that we would rather like to keep it. This means that a distinction between those various quantities can be made only by a cautious reading of the accompanying text. All this mess cannot be changed anymore and we have to advise the reader to check carefully the notation before drawing conclusions.

Extensive work of M. Reiner (e.g. [308, 309]) and then of R. S. Rivlin (see his collected works edited by Barenblatt and Joseph [29]) provided the basis for the vehement development of the thermodynamic theory of simple materials whose part became the rheology. We return to a few problems of this general model of fluids in the next subsection. In this subsection we limit the attention to the classical linear viscous fluids.

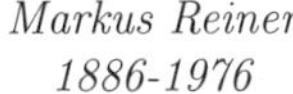

Markus Reiner   *Ronald S. Rivlin*   *Eugene C. Bingham*

1886-1976   *1915-2005*   *1878-1945*

In contrast to the ideal fluid (5.8) the linear Newtonian fluid is the substance defined by the constitutive relations

$$\mathbf{T} = -p\left(\rho, T\right)\mathbf{1} + \lambda\left(\rho, T\right)\left(\operatorname{div}\mathbf{v}\right)\mathbf{1} + 2\eta\left(\rho, T\right)\mathbf{D},$$

$$\varepsilon - \varepsilon\left(\rho, T\right) + \varepsilon_D\left(\rho, T\right)\operatorname{tr}\mathbf{D}, \quad \eta = \eta\left(\rho, T\right) + \eta_D\left(\rho, T\right)\operatorname{tr}\mathbf{D}, \tag{5.44}$$

$$\mathbf{q} = -\kappa\operatorname{grad}T, \quad \mathbf{h} = -H\operatorname{grad}T,$$

which should satisfy the entropy inequality

$$\frac{\partial\left(\rho\eta\right)}{\partial t} + \operatorname{div}\left(\rho\eta\mathbf{v} + \mathbf{h}\right) \geq 0, \tag{5.45}$$

for all solutions of the field equations. The latter follow by the substitution of constitutive relations in the balance laws (5.7). In Part I (e.g. Chapter 9) we have demonstrated the application of the method of Lagrange multipliers to solve the problem of the thermodynamical admissibility. Relations (9.44) of Part I which correspond to the below quoted results follow from the considerations of the extended thermodynamics after an appropriate truncation. We skip here the technical details which are rather easy for this particular case of the constitutive problem. Within the classical thermodynamical approach they can be found, for instance, in the book of I. Müller [261] (Section 1.3.2). The results of the exploitation of the entropy inequality (the second law of thermodynamics) have the following form

$$\varepsilon_D = 0, \quad \eta_D = 0, \quad \mathbf{h} = \frac{\mathbf{q}}{T},$$

$$\mathcal{D} = \frac{\kappa}{T^2}\left|\operatorname{grad}T\right|^2 + \frac{1}{T}\left(\lambda\left|\operatorname{tr}\left(\mathbf{D}\right)\right|^2 + 2\eta\operatorname{tr}\left(\mathbf{D}\mathbf{D}\right)\right) \geq 0, \tag{5.46}$$

where the quantity $\mathcal{D}$ is called the dissipation (entropy production). The dissipation inequality is satisfied if the following conditions hold

$$\eta \geq 0, \quad 3\lambda + 2\eta \geq 0, \quad \kappa \geq 0. \tag{5.47}$$

In the linear model under considerations they are necessary and sufficient for the dissipation to possess a minimum in the thermodynamical equilibrium state defined by the relations

$$\operatorname{grad} T|_E = 0, \quad \mathbf{D}|_E = 0. \tag{5.48}$$

These conditions are also called the stability conditions of the thermodynamical equilibrium.[6]

We should mention that the result $(5.46)_{1,2}$ means that the specific energy ε and the entropy η are the same in an arbitrary nonequilibrium state and in the thermodynamical equilibrium: $\varepsilon = \varepsilon|_E(\rho, T)$, $\eta = \eta|_E(\rho, T)$. Consequently, they possess the same thermodynamical properties as the corresponding quantities for ideal fluids. In particular, they satisfy the Gibbs equation (5.9) with the equilibrium pressure contribution $p(\rho, T)$.

Obviously, the constitutive relations (5.44) satisfy the conditions of the material objectivity and material symmetry. In contrast to similar relations of the linear elasticity in which, instead of the stretching tensor, the Almansi-Hamel deformation tensor appears, it is not an approximation for small deformation measures only.[7] We comment further on this issue again. The above linear model is often called Navier-Stokes-Fourier fluid.

In many cases of practical bearing one can assume additionally the incompressibility of the fluid, i.e., $\operatorname{div} \mathbf{v} = 0$. Then the contribution with the coefficient λ in (5.45) vanishes and the mass density is a constant. For isothermal processes in such materials (i.e., $\operatorname{grad} T = 0$) the dissipation $\mathfrak{D}(\mathcal{P}, t)$ in the domain $\mathcal{P}$ characterizing the irreversibility of processes is given by the so-called Bobylew-Forsyth formula [52, 127][8]

$$\mathfrak{D}(\mathcal{P}, t) = \int_{\mathcal{P}} \mathcal{D}(\mathbf{x}, t)\, dV = \eta \int_{\mathcal{P}} \omega^2 dV + 2\eta \oint_{\partial \mathcal{P}} \dot{\mathbf{v}} \cdot \mathbf{n} dS \geq 0. \tag{5.51}$$

Obviously, for the velocity equal identically zero on the boundary (i.e. $\partial \mathbf{v}/\partial t = 0$, $\mathbf{v} \cdot \operatorname{grad} \mathbf{v} = 0 \;\Rightarrow\; \dot{\mathbf{v}} = 0$ on $\partial \mathcal{P}$) the dissipation is determined solely by the vorticity ω.

For incompressible fluids the substitution of Relation (5.45) in the momentum conservation law yields the following equation

$$\rho \left[\frac{\partial \mathbf{v}}{\partial t} + (\mathbf{v} \cdot \operatorname{grad}) \mathbf{v} \right] = -\operatorname{grad} p + \eta \Delta^2 \mathbf{v} + \rho \mathbf{b}, \quad \Delta^2(\ldots) = \operatorname{div} \operatorname{grad}(\ldots) \tag{5.52}$$

---

[6]Some remarks on thermodynamics of the compressible Navier-Stokes model and its relations to variational formulations can be found in Section 2.3 of the book of A. M. Sonnet and E. G. Virga [361].

[7]I-Shih Liu states: "It needs not be regarded as the linear approximation of a Reiner-Rivlin fluid. Thus it is conceivable that there are some fluids which obey the constitutive equation (5.45) for arbitrary rate of deformation. Indeed, water and air are usually treated as Navier-Stokes fluids in most practical applications with very satisfactory results even under rapid flow conditions" [234].

[8]The entropy inequality (5.46) yields for the dissipation in incompressible viscous fluids the following inequality

$$\mathcal{D}(\mathbf{x}, t) = \frac{2\eta}{T} \operatorname{tr}(\mathbf{DD}) \geq 0. \tag{5.49}$$

Simultaneously, we have the identity

$$\begin{aligned}
\operatorname{div} \dot{\mathbf{v}} &= (D_{ij} + W_{ij})(D_{ji} + W_{ji}) = \operatorname{tr}(\mathbf{DD}) + \frac{1}{4}(\epsilon_{ijk}\omega_k \epsilon_{jil}\omega_l) \\
&= \operatorname{tr}(\mathbf{DD}) - \frac{1}{2}\omega^2,
\end{aligned} \tag{5.50}$$

where we have applied the relation $\epsilon_{ijk}\epsilon_{ijl} = 2\delta_{kl}$.

which is the Navier-Stokes equation. Obviously, $\Delta^2(\ldots)$ is the Laplace operator. Together the condition of incompressibility and this equation form the field equations for the fields $\{p, \mathbf{v}\}$. There exists a vast literature concerning these equations, their mathematical properties, analytical and numerical solutions and applications of particular cases.[9]

Horace Lamb   *Olga A. Olejnik*   *Roger Temam*
1849-1934   *1925-2001*   *1940-*

Let us mention that, as in the case of the Euler equation (5.17), we can construct the equation for the vorticity by taking the rotation of Equation (5.52). For incompressible fluids we do not have to assume that processes are isentropic because the mass density ρ is constant. It follows

$$\frac{\partial \omega}{\partial t} - (\operatorname{rot}(\mathbf{v} \times \omega)) = \nu \Delta^2 \omega, \quad \nu = \frac{\eta}{\rho}, \tag{5.53}$$

where ν is the so-called kinematic viscosity. Consequently, the Helmholtz Theorems proved above for the ideal fluids do not hold. The presence of the right-hand side yields a diffusion of the vorticities which cannot be generated in the interior of the viscous incompressible fluid but they diffuse inward from the boundaries. This was the motivation for the very extensive research of boundary layers initiated by Ludwig Prandtl in 1904 in his lecture during the III International Congress of Mathematicians in Heidelberg (compare Figure 5.1) [297].

The overwhelming majority of problems considered for the Navier-Stokes equation (5.52) relies on one of the two types of boundary conditions

1. At a free surface or at an interface between two dissimilar fluids the stress vector should be continuous

$$[\![\mathbf{Tn}]\!] = 0, \tag{5.54}$$

where $[\![\ldots]\!]$ is a jump, i.e., the difference of the limits on both sides. Whether this condition is equivalent for Navier-Stokes fluids to the continuity of the velocity is questionable but often claimed.

2. The adherence boundary conditions are required on the boundary with the solid, i.e.,

$$[\![\mathbf{v}]\!] = 0. \tag{5.55}$$

---

[9]e.g. R. Temam [378], O. A. Olejnik, V. N. Samochin [282] on mathematical properties, J. Serrin [350] on continuum mechanical foundations, H. Lamb [216], J. N. Hunt [169], C. Truesdell, K. R. Rajagopal [394] on analytical solutions, F. M. White [424] on analytical and numerical solutions, H. Schlichting [338], H. Schlichting, K. Gersten [339] on the boundary layer theory, to name a few.

Über Flüssigkeitsbewegung bei sehr kleiner Reibung.

Von

L. Prandtl aus Hannover.

(Hierzu eine Figurentafel.)

In der klassischen Hydrodynamik wird vorwiegend die Bewegung der *reibungslosen* Flüssigkeit behandelt. Von der *reibenden Flüssigkeit* besitzt man die Differentialgleichung der Bewegung, deren Ansatz durch physikalische Beobachtungen wohl bestätigt ist. An Lösungen dieser Differentialgleichung hat man außer eindimensionalen Problemen, wie sie u. a. von Lord Rayleigh*) gegeben wurden, nur solche, bei denen die Trägheit der Flüssigkeit vernachlässigt ist, oder wenigstens keine Rolle spielt. Das zwei- und dreidimensionale Problem mit Berücksichtigung von Reibung *und* Trägheit harrt noch der Lösung. Der Grund hierfür liegt wohl in den unangenehmen Eigenschaften der Differentialgleichung. Diese lautet in Gibbsscher Vektorsymbolik**)

$$\varrho \left(\frac{\partial v}{\partial t} + v \circ \nabla v \right) + \nabla \left(V + p \right) = k \nabla^2 v \tag{1}$$

(v Geschwindigkeit, ϱ Dichte, V Kräftefunktion, p Druck, k Reibungskonstante); dazu kommt noch die Kontinuitätsgleichung: für inkompressible Flüssigkeiten, die hier allein behandelt werden sollen, wird einfach

$$\operatorname{div} v = 0.$$

Figure 5.1: *An initial fragment of the lecture of Ludwig Prandtl on the III International Congress of Mathematicians in Heidelberg, 1904.*

The last condition is not satisfied on porous boundaries where the diffusion in the porous solids yields a tangential component of the velocity relative to the boundary different from zero. This is related to a special boundary layer problem which was considered by G. S. Beavers and D. D. Joseph [37]. Such a condition appears also in the theory of porous media which we consider further in this book.

For the second type of conditions one can easily prove the uniqueness of the solutions of the incompressible Navier-Stokes equation. Following J. Serrin [350] we present a sketch of this proof. We consider two flows in a bounded region $\mathcal{P}(t)$ whose boundary consists of finitely many closed rigid surfaces moving in some prescribed manner. Let us denote these two solutions $\mathbf{v}$ and $\mathbf{v}^*$ and let us assume that they satisfy the same initial conditions at $t = 0$. Let us define

$$\mathbf{w} = \mathbf{v}^* - \mathbf{v}, \quad K = \frac{1}{2} \int_{\mathcal{P}(t)} w^2 dV, \quad w^2 = \mathbf{w} \cdot \mathbf{w}. \tag{5.56}$$

By subtraction of the Navier-Stokes equation (5.52) for $\mathbf{v}^*$ and $\mathbf{v}$ we obtain

$$\begin{aligned}
&\frac{\partial \mathbf{w}}{\partial t} + (\mathbf{v}^* \cdot \mathrm{grad}) \, \mathbf{v}^* - (\mathbf{v} \cdot \mathrm{grad}) \, \mathbf{v} \\
&= -\mathrm{grad} \left(\frac{p^* - p}{\rho} \right) + \nu \Delta^2 \mathbf{w}.
\end{aligned} \tag{5.57}$$

Scalar product of this relation with $\mathbf{w}$ and the incompressibility conditions $\mathrm{div}\,\mathbf{v} = 0$, $\mathrm{div}\,\mathbf{v}^* = 0$, yield

$$\frac{\partial}{\partial t} \left(\frac{1}{2} w^2 \right) = -\nu \, \mathrm{tr} \left((\mathrm{grad}\,\mathbf{w})^T (\mathrm{grad}\,\mathbf{w}) \right)$$

$$-\mathbf{w} \cdot \mathbf{D}\mathbf{w} + \mathrm{div} \left(\nu \, \mathrm{grad} \left(\frac{1}{2} w^2 \right) - \frac{p^* - p}{\rho} \mathbf{w} - \frac{1}{2} w^2 \mathbf{v}^* \right), \tag{5.58}$$

where $\mathbf{D}$ denotes the stretching in the motion with the velocity field $\mathbf{v}$. Now the integration over the region $\mathcal{P}(t)$ provides the following relation

$$\frac{dK}{dt} = -\int_{\mathcal{P}(t)} \left[\nu \, \mathrm{tr} \left((\mathrm{grad}\,\mathbf{w})^T (\mathrm{grad}\,\mathbf{w}) \right) + \mathbf{w} \cdot \mathbf{D}\mathbf{w} \right] dV, \tag{5.59}$$

where we have used the Stokes Theorem and accounted for the boundary condition $\mathbf{w} = 0$ on the boundary $\partial \mathcal{P}(t)$.

Due to the incompressibility condition we have the following relations for the invariants of the stretching tensor

$$I_D = \mathrm{tr}\,\mathbf{D} = \sum_{\alpha=1}^{3} d^{(\alpha)} = \mathrm{div}\,\mathbf{v} = 0,$$

$$II_D = d^{(1)}d^{(2)} + d^{(1)}d^{(3)} + d^{(3)}d^{(2)} = \frac{1}{2} \left[I_D^2 - \left(\left(d^{(1)} \right)^2 + \left(d^{(2)} \right)^2 + \left(d^{(3)} \right)^2 \right) \right] \leq 0, \tag{5.60}$$

$$III_D = d^{(1)}d^{(2)}d^{(3)},$$

where $d^{(\alpha)}$, $\alpha = 1, 2, 3$ are eigenvalues of $\mathbf{D}$. It follows from the first and second relations that the eigenvalues cannot have the same sign. Let us denote by $-m$, $m > 0$, the lower bound of these eigenvalues. Then

$$
\mathbf{w} \cdot \mathbf{D}\mathbf{w} = \mathbf{w} \cdot \sum_{\alpha=1}^{3} d^{(\alpha)} \left(\mathbf{k}^{\alpha} \otimes \mathbf{k}^{\alpha}\right) \mathbf{w} \geq -m w^2, \quad \mathbf{w} = \sum_{\alpha=1}^{3} w_\alpha \mathbf{k}^{\alpha}, \tag{5.61}
$$

where $\mathbf{k}^{\alpha}$ are eigenvectors of $\mathbf{D}$. Hence, Relation (5.59) implies

$$
\frac{dK}{dt} \leq m \int_{\mathcal{P}(t)} w^2 dV = 2mK \quad \Rightarrow \quad \frac{d}{dt}\left(K e^{-2mt}\right) \leq 0. \tag{5.62}
$$

Integration with respect to time in the interval between 0 and, say, T yields

$$
K\left(T\right) e^{-2mT} \leq 0, \tag{5.63}
$$

which means, as T is arbitrary, that K must be identically zero. Hence, $\mathbf{w} = 0$ and this means that the flows $\mathbf{v}$ and $\mathbf{v}^*$ are identical.

The above theorem can be easily extended on infinite domains if we make appropriate assumptions on the asymptotic behavior of the solutions.

We are now in the position to discuss a few important particular solutions of the Navier-Stokes equations. Due to the nonlinearity of this equation only a few exact analytical solutions are known. Before we construct some of them let us write the governing equations in dimensionless form. To this aim we introduce a characteristic velocity V and a characteristic length of the system L. Then we define the following dimensionless quantities

$$
\widetilde{t} = \frac{Vt}{L}, \quad \widetilde{\mathbf{x}} = \frac{\mathbf{x}}{L}, \quad \widetilde{\mathbf{v}} = \frac{\mathbf{v}}{V}, \quad \widetilde{p} = \frac{p}{\rho V^2}. \tag{5.64}
$$

Consequently, the Navier-Stokes equation (5.52) has the form

$$
\frac{\partial \widetilde{\mathbf{v}}}{\partial t} + \left(\widetilde{\mathbf{v}} \cdot \widetilde{\mathrm{grad}}\right) \widetilde{\mathbf{v}} = -\widetilde{\mathrm{grad}}\, \widetilde{p} + \frac{1}{Re}\, \widetilde{\mathrm{div}}\, \widetilde{\mathrm{grad}}\, \widetilde{\mathbf{v}}, \tag{5.65}
$$

where

$$
Re = \frac{VL}{\nu} = \frac{\rho V^2}{\eta V/L}, \tag{5.66}
$$

is the so-called Reynolds number. It is obviously the ratio of the inertial force to the viscous force. Solutions of various boundary problems possess physically acceptable properties for Reynolds numbers smaller than approximately $Re_{CRIT} = 2300$. Above this range the flow loses the stability which yields the creation of vorticities called in this range eddies. They may be very small or extremely large. For example, eddies are common in the ocean, and range in diameter from centimeters, to hundreds of kilometers. The smallest scale eddies may last for a matter of seconds, while the larger features may persist for months to years. Those eddies which are between 10 and 500 km in diameter, and persist for periods of days to months are commonly referred to in oceanography as mesoscale eddies. The flow in the range of Reynolds numbers $Re < Re_{CRIT}$ is said to be laminar and above this number turbulent. In the range $Re > Re_{CRIT}$ most likely solutions of the Navier-Stokes equation would still give a reasonable description but the interactions of vorticities produce cascades of instabilities which make even the construction of numerical solutions practically impossible. These instabilities yield a stochastic character of macroscopic description. Therefore for turbulent flows independent macroscopic theories are developed which possess certain features of the continuum (e.g. the notion of stresses – the so-called Reynolds stresses growing from the stochastic fluctuation of the velocity which correct the viscous contribution to the Navier-Stokes equation; see [424]). However, some thermodynamical arguments seem to be not applicable in this theory. We skip further details in our short presentation and for details on this subject we refer to the book of K. Hutter and K. Jöhnk [173].

Let us return to boundary value problems for Reynolds numbers smaller than the above mentioned critical value Re_{CRIT}. As already mentioned one of the first works on this subject was the publication of L. Prandtl in the book of the III International Congress of Mathematicians in Heidelberg in 1904 where the author presented the notion of the boundary layer (Figure 5.1). This is the intermediate spacial range between the boundary on which the tangential component of the velocity is equal to the tangential component of the velocity of the boundary and the stationary flow range far from the boundary where the velocity in the same tangential direction has a different value. The theory of boundary layers determining the velocity in this intermediate range became one of the main research subjects for flows of viscous fluids. We proceed to present a few solutions illustrating this problem.

5.3.2.2 Lamellar flows

We begin with the simplest class of solutions in which, in spite of the viscosity, the boundary layer does not appear and the problem is linear. Let us assume that the flow is such that only one component of the velocity, say $v_1 \equiv u$ is different from zero. Such flows belong to the class of lamellar flows.[10] Then the incompressibility condition yields $\partial u/\partial x = 0$, $x = x^1$. The momentum balance equations for the direction of coordinates $y = x^2$, $z = x^3$ lead to the relations $\partial p/\partial y = \partial p/\partial z = 0$. Consequently,

$$\frac{\partial u}{\partial t} = -\frac{1}{\rho}\frac{\partial p}{\partial x} + \nu\left(\frac{\partial^2 u}{\partial y^2} + \frac{\partial^2 u}{\partial z^2}\right), \quad p = p(x,t), \quad u = u(y,z,t). \tag{5.67}$$

For a given pressure gradient this linear diffusion equation can be easily solved. In the particular case of the stationary motion between two planes which have velocities zero for $y = 0$ and V for $y = h$ in the direction of the z-axis we have

$$\frac{\partial p}{\partial x} = \eta\frac{\partial^2 u}{\partial y^2}, \quad u(y = 0) = 0, \quad u(y = h) = V. \tag{5.68}$$

Obviously, the solution has the following form

$$u = V\frac{y}{h} - \frac{\partial p}{\partial x}\frac{yh}{2\eta}\left(1 - \frac{y}{h}\right). \tag{5.69}$$

This is the so-called generalized Couette flow. For constant pressure the velocity profile is linear and the flow is called the Couette flow. On the other hand, for $U = 0$ the profile is symmetric parabolic and the flow is called the Poiseuille flow. It is driven by the external gradient of pressure.

---

[10]**lamellar flow**: "flow of a liquid in which layers glide over one another" (www.websters-online-dictionary.org). C. A. Truesdell and K. R. Rajagopal [394] (page 20) call any irrotational (potential) flow with $\mathbf{v} = \operatorname{grad}\phi$ lamellar as in the direction perpendicular to equipotential surfaces, $\phi = \text{const.}$ equal increments c divide the region of flow into laminae with $|\mathbf{v}| \approx c/d$, where $c = \delta\phi$ and d is the normal distance between equipotential surfaces.

A similar solution called the Hagen-Poiseuille flow we obtain for a circular tube of the radius R. The set of equations in cylindrical coordinates has in this case the following form

$$\frac{\partial p}{\partial r} = 0, \quad \frac{\partial p}{\partial z} = \eta \left(\frac{\partial^2 w}{\partial r^2} + \frac{1}{r} \frac{\partial w}{\partial r} \right), \quad w\,(r = R) = 0, \tag{5.70}$$

where $r = x^1$, $z = x^3$, $w = v_3$. Then the solution is as follows

$$w = -\frac{1}{4\eta} \frac{\partial p}{\partial z} R^2 \left(1 - \frac{r^2}{R^2} \right). \tag{5.71}$$

In spite of the lack of any restrictions on the Reynolds number Re in those solutions it has been found that they agree with observations only for laminar flows, i.e., for Reynolds numbers smaller than Re_{CRIT}.

5.3.2.3 Creeping flows

Another class of flows for which one can find analytical solutions are the so-called creeping flows. These are flows in which the Reynolds number is sufficiently small (i.e., viscous forces sufficiently large) for neglecting nonlinear contributions to the acceleration. These are flows of fluids with sufficiently large viscosity, or flows with very low velocities. The simplified Navier-Stokes equation (5.52) has then the following form

$$\frac{\partial \mathbf{v}}{\partial t} + \operatorname{grad} \frac{p}{\rho} = \nu \Delta^2 \mathbf{v}. \tag{5.72}$$

Bearing the incompressibility condition in mind, the divergence of this equation yields

$$\Delta^2 p = 0, \tag{5.73}$$

which means that the pressure in creeping flows is a harmonic function. Simultaneously, the rotation of the above equation gives rise to the equation for the vorticity

$$\frac{\partial \omega}{\partial t} = \nu \Delta^2 \omega, \tag{5.74}$$

which is, of course, a known diffusion equation.

Let us consider a two-dimensional flow for which the velocity possesses the potential ϕ. In Cartesian coordinates (compare (5.22)) this means

$$u = v_1 = \frac{\partial \phi}{\partial y}, \quad v = v_2 = -\frac{\partial \phi}{\partial x}, \quad \text{i.e.} \quad \omega_3 = -\Delta^2 \phi, \tag{5.75}$$

and Equation (5.74) has the form

$$\frac{\partial}{\partial t} \Delta^2 \phi = \nu \Delta^2 \left(\Delta^2 \phi \right). \tag{5.76}$$

For stationary processes the velocity potential is a biharmonic function. Solutions of such a problem can be considered to be the first approximation of the full solution for small Reynolds numbers.

An important example of the creeping flow is the flow in an infinite medium due to the motion of a sphere in, say, direction of the x-axis. On this example we demonstrate that the conditions for the d'Alembert paradox discussed earlier are not fulfilled for the viscous fluid. We consider the solution of Equation (5.73) in the following form

$$p = -\eta A \frac{x}{r^3}, \quad r^2 = x^2 + y^2 + z^2, \tag{5.77}$$

where A is a constant. We have chosen Cartesian coordinates in which the sphere is not moving. Obviously, this solution satisfies the condition that the pressure goes to zero in infinity. Substitution of this solution in Equation (5.72) yields the following equations for the components of the velocities in stationary flow

$$\Delta^2 u = A \left(\frac{3x}{r^5} - \frac{1}{r^3} \right), \quad u = v_1,$$

$$\Delta^2 v = A \frac{3xy}{r^5}, \quad v = v_2, \tag{5.78}$$

$$\Delta^2 w = A \frac{3xz}{r^5}, \quad w = v_3.$$

We have to satisfy the boundary conditions that all components of the velocity are zero on the boundary of the sphere $r = R$ (i.e., $u\,(r = R) = v\,(r = R) = w\,(r = R) = 0$) and the velocity of the fluid in infinity with respect to the immobile sphere is $u\,(r \to \infty) = u_\infty$. It follows immediately that

$$u = u_\infty - \frac{3}{4} u_\infty \frac{r^2 R}{r^3} \left(1 - \frac{R^2}{r^2} \right) - \frac{1}{4} u_\infty \frac{R}{r} \left(3 + \frac{R^2}{r^2} \right),$$

$$v = -\frac{3}{4} u_\infty \frac{xyR}{r^3} \left(1 - \frac{R^2}{r^2} \right), \tag{5.79}$$

$$w = -\frac{3}{4} u_\infty \frac{xzR}{r^3} \left(1 - \frac{R^2}{r^2} \right),$$

and this yields the relation for the pressure

$$p = -\frac{3}{2} \eta u_\infty \frac{xR}{r^3}. \tag{5.80}$$

Let us calculate the force exerted by the sphere on the fluid. Essential is here, of course, only the x-component (compare Relation (5.2))

$$F_1 = \oint_{\text{sphere}} t_{1i} n_i dS, \quad \mathbf{T} = T_{ij} \mathbf{e}_i \otimes \mathbf{e}_j, \quad \mathbf{n} = n_i \mathbf{e}_i,$$

$$\tag{5.81}$$

$$T_{11} = -p + 2\eta \frac{\partial u}{\partial x}, \quad T_{12} = \eta \left(\frac{\partial v}{\partial x} + \frac{\partial u}{\partial y} \right), \quad T_{13} = \eta \left(\frac{\partial w}{\partial x} + \frac{\partial u}{\partial z} \right),$$

where Relation (5.45) with $\lambda = 0$ was used. Easy integration yields then the following Stokes formula

$$F_1 = 6\pi \eta R u_\infty. \tag{5.82}$$

In the case of an object of arbitrary shape moving through the viscous fluid, a more general formula was proposed by John Strutt, 3rd Baron Rayleigh in the following form

$$F_D = \frac{1}{2}\rho A C_d u_\infty^2, \tag{5.83}$$

where A is the reference area – in the case of the Stokes formula $A = \pi R^2$, and C_d is the drag coefficient. For the sphere described by the Stokes formula (5.82) it is given by the relation ($Re = 2Ru_\infty\rho/\eta$)

$$C_d = \frac{24}{Re}. \tag{5.84}$$

Clearly, in contrast to the ideal fluid there is no paradox here.

5.3.2.4   Boundary layers

We proceed to present the basic ideas of the Ludwig Prandtl boundary layer theory [297]. As already indicated experimental observations and theoretical results reveal that the influence of the assumption on the "no slip" boundary conditions for flows of fluids with low viscosity is limited to a small region of the boundary layer. The theoretical description of such fluids was proposed by Prandtl who assumed that the fluid in the bulk moves as it were an ideal fluid and the influence of viscosity appears only in the boundary layer. We shall derive the equations for the boundary layer in an infinite medium for the flow around an inclusion in the form of an infinite strip $y = 0$ of the length L in the x-direction. The problem is two-dimensional and the Navier-Stokes equations in the Cartesian coordinates for the velocity $\mathbf{v}\,(x, y, t) = u\,(x, y, t)\,\mathbf{e}_1 + v\,(x, y, t)\,\mathbf{e}_2$ and the pressure $p\,(x, y, t)$ have the form

$$\frac{\partial u}{\partial t} + u\frac{\partial u}{\partial x} + v\frac{\partial u}{\partial y} = -\frac{1}{\rho}\frac{\partial p}{\partial x} + \nu\left(\frac{\partial^2 u}{\partial x^2} + \frac{\partial^2 u}{\partial y^2}\right),$$

$$\frac{\partial v}{\partial t} + u\frac{\partial v}{\partial x} + v\frac{\partial v}{\partial y} = -\frac{1}{\rho}\frac{\partial p}{\partial y} + \nu\left(\frac{\partial^2 v}{\partial x^2} + \frac{\partial^2 v}{\partial y^2}\right), \tag{5.85}$$

$$\frac{\partial u}{\partial x} + \frac{\partial v}{\partial y} = 0,$$

with the initial and boundary conditions

$$u\,(x, y, t = 0) = u_0\,(x, y)\,, \quad v\,(x, y, t = 0) = v_0\,(x, y)$$

$$u\,(x, y = 0, t) = 0, \quad v\,(x, y = 0, t) = 0 \quad \text{for} \quad 0 < x < L, \tag{5.86}$$

$$\mathbf{v}\,(x, y, t) \rightarrow \mathbf{v}_\infty = u_\infty\mathbf{e}_1 \quad \text{for} \quad x^2 + y^2 \rightarrow \infty,$$

where u_0, v_0 are given initial components of the velocity and u_∞ is the constant x-component of the velocity in infinity.

This problem is solved under the assumption that the viscosity is small which means that the Reynolds number $Re = u_\infty L/\nu$ is large (approximately larger than 1000). Then, the Prandtl hypothesis implies the existence of a layer of thickness H in the positive and negative y-directions in which the component of the velocity u changes from the zero value on the strip to its value $U\,(x)$ beyond the boundary layer, i.e., for $|y| > H$. The thickness of the layer H and the vertical component of the velocity v within the layer are assumed to satisfy the conditions which easily follow from the dimensional analysis of the problem

$$H = O(\varepsilon), \quad \frac{v}{u} = O(\varepsilon), \quad \varepsilon = 1/\sqrt{Re} \ll 1, \quad Re = \frac{u_\infty L}{\nu}. \tag{5.87}$$

In order to select the most important contributions to the Navier-Stokes equations under these assumptions, we change the variables for $0 < x < L$

$$t' = t, \quad x' = x, \quad y' = \varepsilon^{-1}y. \tag{5.88}$$

This rescaling, obviously, blows up the region of the boundary layer in the y-direction. Equations (5.85) now have the form

$$\frac{\partial u'}{\partial t'} + u'\frac{\partial u'}{\partial x'} + v'\frac{\partial u'}{\partial y'} = -\frac{1}{\rho}\frac{\partial p}{\partial x'} + \varepsilon^2 u_\infty L\frac{\partial^2 u'}{\partial x'^2} + u_\infty L\frac{\partial^2 u'}{\partial y'^2},$$

$$\varepsilon^2\frac{\partial v'}{\partial t'} + \varepsilon^2\left(u'\frac{\partial v'}{\partial x'} + v'\frac{\partial v'}{\partial y'}\right) = -\frac{1}{\rho}\frac{\partial p}{\partial y'} + \varepsilon^4 u_\infty L\frac{\partial^2 v'}{\partial x'^2} + \varepsilon^2 u_\infty L\frac{\partial^2 v'}{\partial y'^2}, \tag{5.89}$$

$$\frac{\partial u'}{\partial x'} + \frac{\partial v'}{\partial y'} = 0, \quad u'(x', y', t') = u(x', \varepsilon y', t'), \quad v'(x', y', t') = \frac{1}{\varepsilon}v(x', \varepsilon y', t').$$

In the limit $\varepsilon \to 0$ these equations yield the equations for the plain nonstationary laminar boundary layer or the Prandtl equations for the boundary layer. We return to the original variables and then the equations are as follows:

$$\frac{\partial u}{\partial t} + u\frac{\partial u}{\partial x} + v\frac{\partial u}{\partial y} = -\frac{1}{\rho}\frac{\partial p}{\partial x} + \nu\frac{\partial^2 u}{\partial y^2},$$

$$\frac{\partial u}{\partial x} + \frac{\partial v}{\partial y} = 0, \quad p = p(x, t). \tag{5.90}$$

Simultaneously, beyond the boundary layer the flow is described by the Euler equations for the ideal incompressible fluid which in the case under consideration reduce to the single relation

$$\frac{\partial U}{\partial t} + U\frac{\partial U}{\partial x} = -\frac{1}{\rho}\frac{\partial p}{\partial x}. \tag{5.91}$$

Consequently, the horizontal freestream velocity $U(x)$ outside the boundary layer can be found from the problem of flow of an ideal fluid around an obstacle and the pressure $p(x)$ within the layer is then a known function. This important observation is due to Prandtl.

Many analytical and numerical solutions have been obtained for the plain boundary as well as for some curvilinear systems (e.g. a sphere, a cylinder or a wedge). We quote here only three books. The classical books of H. Schlichting [338, 339] are the main reference on the subject. The book of O. Olejnik and V. N. Samochin is primarily devoted to mathematical problems of stationary and nonstationary plain boundary layers. On the other hand, the book of F. White [424] contains also, apart from some analytical and numerical solutions for the Prandtl boundary layer, an extensive discussion of the problem of coupling with thermal effects. We return to some of them further in this section.

An interesting feature of the boundary layer problem should be pointed out. The approximation for large Reynolds numbers which we have described above is from the mathematical point of view far from trivial. As the inspection of Equations (5.85) reveals, the small parameter ε for large values of $Re \sim 1/\nu$ appears in front of the operator of the highest order. This means that the limit $\varepsilon \to 0$ is singular. This is a characteristic feature of all problems in physics in which a boundary layer appears. From the mathematical standpoint a solution should be constructed by means of the singular perturbation method which consists of two contributions: a regular expansion with respect to the small parameter and the so-called second perturbation series which appears with a large parameter (rescaling of variables measuring the distance from the boundary; in our example it is y/ε). We shall not elaborate this problem in this book (compare a simple mathematical presentation in the book of J. G. Simmonds and J. E. Mann [357] and a modern mathematical approach to thermodynamical singular perturbation problems by E. V. Radkevich [301], in particular Chapter 8 of this book).

Let us mention one more issue of great practical importance in relation to boundary layers. It is the so-called flow separation. In spite of the assumption on large Reynolds numbers ($\varepsilon = 1/\sqrt{Re} \ll 1$) within the boundary layer there may appear lines (points in two-dimensional cases) on the boundaries in which the tangential component of the velocity changes the sign. Then the assumption on the order of the vertical component $v/u = O\left(\varepsilon\right)$ is not fulfilled and at such places a vorticity is created. This may move away from the boundary and finally it may enter the region of the external flow where Thomson's Circulation Theorem (see (5.20)) holds. Certainly, Prandtl's model of the boundary layer can be applied only in the range before such a point of separation $x < x_0$, where x_0 is the coordinate of the separation point on the two-dimensional boundary (compare Figure 5.2).

We can easily formulate conditions for the creation of the point of separation in the stationary flow. Namely, in such a point the following conditions must be satisfied ($u \geq 0$ in the range of the boundary layer!)

$$u\left(x_0,0\right) = 0, \quad \frac{\partial u}{\partial y}\left(x_0,0\right) = 0 \quad \frac{\partial^2 u}{\partial y^2} \geq 0. \tag{5.92}$$

Bearing Equation (5.90)$_1$ in mind we obtain

$$\nu \left.\frac{\partial^2 u}{\partial y^2}\right|_{y=0} = \frac{1}{\rho}\frac{\partial p}{\partial x} \geq 0. \tag{5.93}$$

Simultaneously, as the pressure p does not vary in the boundary layer in the direction of the y-coordinate, the flow in the exterior must fulfill the condition (compare (5.91) in the stationary case)

$$\frac{1}{\rho}\frac{\partial p}{\partial x} = -U\frac{\partial U}{\partial x} \quad \Rightarrow \quad \frac{\partial U}{\partial x} \leq 0. \tag{5.94}$$

Hence, the gradient of the velocity in the exterior dU/dx is changing the sign from positive to negative at the point of separation x_0. Consequently, the separation of the flow and the creation of vorticities on the boundary appear always for finite obstacles.

One of the important problems of the theory of boundary layers is the coupling of mechanical and thermal effects. Thermal boundary layers appear in many practical applications such as phase transformations (melting, evaporation, solidification, condensation, etc.) but also problems of heat transfer between civil engineering constructions and environment, air conditioning systems, etc. contain field equations whose boundary conditions concern transition regions in which thermal boundary layers appear. We shall discuss some of these problems in Chapter 7 of this book.

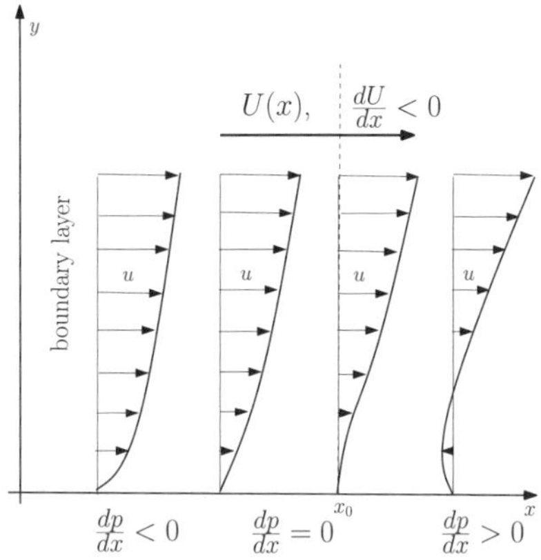

Figure 5.2: *Formation of the point of separation.*

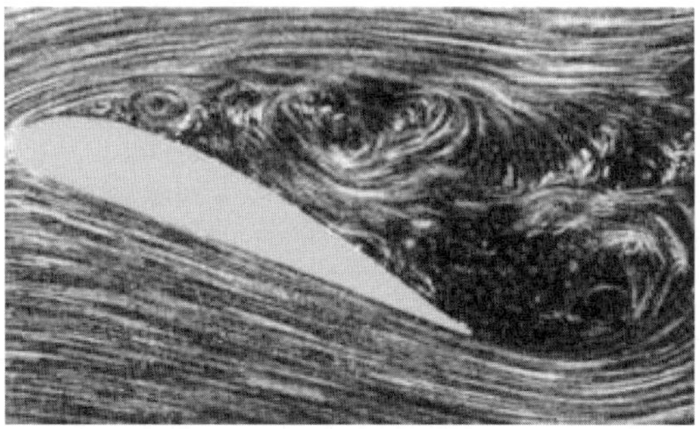

Figure 5.3: *An example of the formation of the flow separation.*

However, one should mention the problem which appears in some theoretical works on the subject of boundary layers in fluids. The extension of the set of equations for the boundary layer which account for the temperature distribution includes the energy balance law and additional contributions to the constitutive relations for the stresses. Under the assumption of a small deviation of the temperature T from the initial reference value T_0 the full set for the Navier-Stokes-Fourier fluid which follows from the balance equations (5.7) and constitutive equations (5.44) has the form

$$\frac{\partial \rho}{\partial t} + \mathrm{div}\,(\rho \mathbf{v}) = 0,$$

$$\frac{\partial \rho \mathbf{v}}{\partial t} + \mathrm{div}\,(\rho \mathbf{v} \otimes \mathbf{v} - \mathbf{T}) = \rho \mathbf{b},$$

$$\frac{\partial \rho \varepsilon}{\partial t} + \mathrm{div}\,(\rho \epsilon \mathbf{v} + \mathbf{q}) = \mathrm{tr}\,(\mathbf{T}\,\mathrm{grad}\,\mathbf{v}),$$

$$\mathbf{T} = -p\mathbf{1} + \lambda\,(\mathrm{div}\,\mathbf{v})\,\mathbf{1} + 2\eta \mathbf{D} - \gamma\,(T - T_0)\,\mathbf{1},$$

$$\mathbf{q} = -K\,\mathrm{grad}\,T, \quad \varepsilon = \varepsilon\,(\rho, T),$$

(5.95)

where γ is the thermal expansion coefficient $\alpha = -\left(\partial \rho / \partial T\right)/\rho$ divided by the compressibility $\kappa_T = \left(\partial \rho / \partial p\right)/\rho$ and all material parameters may be dependent on the mass density ρ. The dependence on the temperature falls out due to the small deviation from the reference temperature T_0. If we denote by c_p the specific heat at constant pressure then the following inequality follows from the stability condition of the thermodynamical equilibrium (see Section 6.1 of Part I and Section 1.4 of the book [261])

$$\alpha^2 \leq \frac{\rho}{T} c_p \kappa_T. \tag{5.96}$$

Consequently, the assumption on incompressibility of the fluid ($\kappa_T = 0$) yields the lack of the thermal expansion $\alpha = 0$.

This property has been overlooked in some works on extensions of the Prandtl theory of boundary layers by the heat conduction equation and the coupling in the form of thermal expansion to the stress tensor. For instance, in the book of F. M. White [424] Relations (4-35) are wrong as they violate the second law of thermodynamics in the above explained sense. However, in order to preserve the simplicity of the boundary layer theory for incompressible fluids we can try to construct the thermal boundary layer equations for compressible Navier-Stokes-Fourier fluids with an additional constraint. Such a constraint has been introduced in 1903 by J. Boussinesq [54] (p. 625 of vol. II). He made the assumption that for stationary flows changes of the mass density $(\rho - \rho_0)/\rho_0$ are sufficiently small to be neglected in the mass balance. Then, for the two-dimensional case the following equations for the thermal boundary layer follow

$$\frac{\partial u}{\partial x} + \frac{\partial v}{\partial y} \approx 0,$$

$$u\frac{\partial u}{\partial x} + v\frac{\partial u}{\partial y} = U\frac{\partial U}{\partial x} + \frac{\gamma}{\rho}\,(T - T_0) + \frac{\eta}{\rho}\frac{\partial^2 u}{\partial y^2}, \tag{5.97}$$

$$u\frac{\partial T}{\partial x} + v\frac{\partial T}{\partial y} = \frac{K}{\rho c_p}\frac{\partial^2 T}{\partial y^2} + \frac{\eta}{\rho c_p}\left(\frac{\partial u}{\partial y}\right)^2.$$

Except for the time dependence, this set is identical with the set of equations of F. M.White but the incompressibility is not assumed. The problem of the thermal boundary layer with the Boussinesq approximation has an extensive literature (compare the book of Schlichting and Gersten [339] where many further references can be found).

The Boussinesq approximation is applied even in such extreme cases as the motion of the atmosphere. Various generalizations are then proposed (e.g. [108]).

5.3.3 Maxwell and N-th grade (Rivlin-Ericksen) fluids; viscometric flows

5.3.3.1 Preliminaries

Observations of such substances as canada balsam, glue in water, almond oil, whale liver oil and, in particular, almost all polymer solutions and melts reveal that they possess elastic features which under normal circumstances are not observed in such fluids as water or glycerin. The elasticity of viscous fluids gives rise, in turn, to many secondary effects – such as a non-concave free surface of a rotating fluid, swelling at the exhaust of a pipe, etc. All of them carry the common name of the normal stress effect or the Weissenberg effect.

It is known to rheologists that, to account for this effect, viscous fluids must have certain memory. In such a case, we call the fluid non-Newtonian.

There are two main classes of models reflecting the memory of fluids, commonly used by rheologists:

1. Maxwell fluid, described either by the rate-type equation for the deviatoric part of the Cauchy stress tensor (compare Formulae (9.104) and (9.124) of Part I; the shear viscosity is in Part I denoted by ν rather than η)

$$\tau_t \overset{\triangle}{\mathbf{T}^D} + \mathbf{T}^D = 2\eta \mathbf{D}^D,$$

$$\mathbf{T}^D = \mathbf{T} - \frac{1}{3}\left(\operatorname{tr}\mathbf{T}\right)\mathbf{1}, \quad \mathbf{D}^D = \mathbf{D} - \frac{1}{3}\left(\operatorname{tr}\mathbf{D}\right)\mathbf{1},$$

(5.98)

where τ_t is the stress relaxation time, η the shear viscosity and $\overset{\triangle}{(\ldots)}$ an objective time derivative, defined by the relation[11]

$$\overset{\triangle}{\mathbf{T}^D} = \dot{\mathbf{T}}^D + \mathbf{T}^D\mathbf{W} - \mathbf{W}\mathbf{T}^D + \gamma_t\left(\mathbf{T}^D\mathbf{D} + \mathbf{D}\mathbf{T}^D - \frac{2}{3}\operatorname{tr}\left(\mathbf{T}^D\mathbf{D}\right)\mathbf{1}\right),$$

(5.99)

or by the memory functional in which the current value of the stresses is given by a functional dependence on the history of stretching; a linear example of such a constitutive relation is given by the following functional

$$\mathbf{T}\left(t\right) = \int_0^\infty G\left(s\right)\mathbf{D}^t\left(s\right)ds, \quad \mathbf{D}^t\left(s\right) = \mathbf{D}\left(t - s\right),$$

(5.100)

---

[11]Obviously, this time derivative is objective for any value of the constant γ_t. When $\gamma_t = 0$ we obtain the Jaumann-Zaremba (corotational) derivative. For $\gamma_t = -1$ the Oldroyd derivative follows.

where $G(s)$ is the linear stress relaxation modulus with the following properties

$$G(s=0) > 0, \quad \frac{dG}{ds}(s=0) < 0, \quad \lim_{s\to\infty} G(s) = 0. \tag{5.101}$$

2. Nth grade fluid

$$\mathbf{T} = -p\mathbf{1} + \eta\mathbf{A}_1 + \alpha_1\mathbf{A}_2 + \alpha_2\left(\mathbf{A}_1\right)^2 + \alpha_3\left(\mathbf{A}_2\right)^2 + \alpha_4\left(\mathbf{A}_1\mathbf{A}_2 + \mathbf{A}_2\mathbf{A}_1\right) + \ldots, \tag{5.102}$$

where

$$\mathbf{A}_1 = 2\mathbf{D} = \mathbf{L} + \mathbf{L}^T, \quad \mathbf{A}_2 = \frac{d\mathbf{A}_1}{dt} + \mathbf{A}_1\mathbf{L} + \mathbf{L}^T\mathbf{A}_1 \ldots, \tag{5.103}$$

are Rivlin-Ericksen tensors (see Section 3.2 and Section 9.4 of Part I). An extensive presentation of such fluids can be found in the book of C. A. Truesdell and K. R. Rajagopal [394].

Apart from the Nth grade fluids also the so-called Nth order fluids are considered. These models follow from the Maxwell model or from the memory functional by expansion of the history of the velocity gradient $\mathbf{L}(t-s)$ into a Taylor series around the present instant of time t and a subsequent truncation on the Nth term. In contrast to the constitutive relations of the form (5.102) such models must additionally satisfy certain convergence conditions.

An extensive analysis of fluids with memory was presented in numerous monographs and books (e.g. C. A. Truesdell and W. Noll [393], Chapter E: Fluidity, C. A. Truesdell, K. R. Rajagopal [394]). One of the important issues discussed in these books is the measurability of the material properties of such fluids in simple experiments. In particular, the notion of the monotonous motion is introduced. This is a motion in which the history of deformation is constant with respect to any chosen instant of time. For such a motion one can prove the Wang Theorem that such a history is uniquely determined by the first three Rivlin-Ericksen tensors. Then, for instance (Noll's Theorem, [394]),

$$\mathbf{L} = \kappa\mathbf{N}, \quad \mathbf{A}_1 = \kappa\left(\mathbf{N} + \mathbf{N}^T\right), \quad \mathbf{A}_2 = \kappa^2\left(2\mathbf{N}^T\mathbf{N} + \mathbf{N}^2 + \left(\mathbf{N}^T\right)^2\right), \quad \text{etc.,}$$
$$|\mathbf{N}| = \sqrt{\operatorname{tr}\left(\mathbf{N}\mathbf{N}^T\right)} = 1. \tag{5.104}$$

κ is the so-called shear rate and the matrix $\mathbf{N}$ has a very special form for the particular case of the viscometric flow in which $\operatorname{tr}\mathbf{N}^2 = 0$. Then there exists a basis in which

$$\mathbf{N} = \begin{pmatrix} 0 & 0 & 0 \\ 1 & 0 & 0 \\ 0 & 0 & 0 \end{pmatrix}. \tag{5.105}$$

In Part I we have extensively presented the construction of various models of non-Newtonian fluids on the basis of extended thermodynamics. The evolution equation for the stresses (Equation (9.103) of Part I) which follows from this procedure contains contributions which violate the principle of objectivity. This is not unusual in models of the extended thermodynamics. However, the first approximation which results from the so-called Maxwell iteration (the lowest order approximation with respect to the deviation from the thermodynamical equilibrium) yields the nonlinear evolution equation for stresses (9.104) of Part I which is objective. It yields as well the hyperbolicity of the set of field

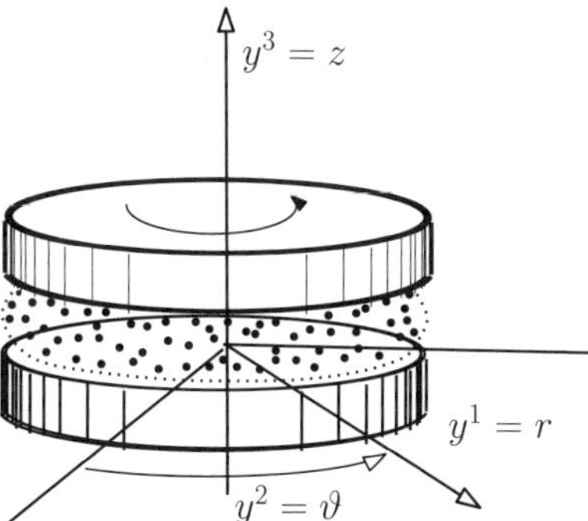

Figure 5.4: *Schematic of torsional (plate-and-plate) flow.*

equations. Simultaneously, the result of the next step of the Maxwell iteration which has the form of the above quoted equation for the Nth grade fluid for $N = 2$, i.e., the constitutive relation for the stresses (5.102) with the material constants η, α_1, α_2 yields already a set of parabolic equations. There appear various problems with the stability of such models which we discuss in Chapter 6 of this book. Some of them were also presented in Section 9.4.3 of Part I and in the book [437].

5.3.3.2 Viscometric flows

In order to demonstrate the way in which properties of non-Newtonian fluids are measured in steady-state flow experiments we present below typical examples of modeling such flows. A detailed discussion of these and some other steady-state flows can be found in the books of B. D. Coleman, H. Markovitz and W. Noll [87] as well as S. Zahorski [454]. The main purpose of viscometric flow experiments is to measure three quantities: $\tau = T_{(1)(2)}$ – the shear stress function, $\sigma_1 = T_{(1)(1)} - T_{(3)(3)}$, $\sigma_2 = T_{(2)(2)} - T_{(3)(3)}$ – the normal stress functions in dependence on the shear rate $\kappa = \omega'$. In these relations $t_{(\alpha)(\beta)}$ are physical components of the stress tensor, i.e., $\mathbf{T} = T_{(\alpha)(\beta)}\mathbf{e}_\alpha \otimes \mathbf{e}_\beta$, where $\mathbf{e}_\alpha$ are unit covariant basis vectors of an appropriate curvilinear coordinate system, $\mathbf{e}_\alpha \cdot \mathbf{e}_\beta = \delta_{\alpha\beta}$. The shear rate is defined by the derivative of the angular velocity ω with respect to the coordinate which is appropriate for changes of velocity in a chosen geometry of the device. We consider here examples in which this geometry is described either by cylindrical or by spherical coordinates.[12]

Torsional flow (plate-and-plate)

As indicated in Figure 5.4, the solution of this viscometric flow problem will be constructed in cylindrical coordinates. The general form of the momentum balance equations

---

[12]For details concerning the geometrical properties of these coordinate systems we refer to Appendix A.2 or to Appendix A (in particular Table 13) of Part I. The metric tensors needed for the transformation to physical components have the form:
cylindrical coordinates: $g_{11} = 1$, $g_{22} = r^2$, $g_{33} = 1$,
spherical coordinates: $g_{11} = 1$, $g_{22} = r^2 \sin^2 \vartheta$, $g_{33} = r^2$.
The remaining components are zero.

in these coordinates is as follows (for simplicity, we leave out the body forces)

$$\rho \left\{ \frac{\partial v^1}{\partial t} + v^1 \frac{\partial v^1}{\partial y^1} + v^2 \frac{\partial v^1}{\partial y^2} + v^3 \frac{\partial v^1}{\partial y^3} - y^1 \left(v^2\right)^2 \right\}$$

$$= \frac{\partial T^{11}}{\partial y^1} + \frac{\partial T^{12}}{\partial y^2} + \frac{\partial T^{13}}{\partial y^3} + \frac{1}{y^1} T^{11} - y^1 T^{22},$$

$$\rho \left\{ \frac{\partial v^2}{\partial t} + v^1 \frac{\partial v^2}{\partial y^1} + v^2 \frac{\partial v^2}{\partial y^2} + v^3 \frac{\partial v^2}{\partial y^3} + \frac{2}{y^1} v^1 v^2 \right\}$$

$$= \frac{\partial T^{12}}{\partial y^1} + \frac{\partial T^{22}}{\partial y^2} + \frac{\partial T^{23}}{\partial y^3} + \frac{3}{y^1} T^{12}, \tag{5.106}$$

$$\rho \left\{ \frac{\partial v^3}{\partial t} + v^1 \frac{\partial v^3}{\partial y^1} + v^2 \frac{\partial v^3}{\partial y^2} + v^3 \frac{\partial v^3}{\partial y^3} \right\}$$

$$= \frac{\partial T^{13}}{\partial y^1} + \frac{\partial T^{23}}{\partial y^2} + \frac{\partial T^{33}}{\partial y^3} + \frac{1}{y^1} T^{13},$$

where

$$\mathbf{v} = v^\alpha \mathbf{g}_\alpha, \quad \mathbf{T} = T^{\alpha\beta} \mathbf{g}_\alpha \otimes \mathbf{g}_\beta, \quad \mathbf{g}_\alpha = \frac{\partial \mathbf{x}}{\partial y^\alpha}. \tag{5.107}$$

The covariant basis vectors $\{\mathbf{g}_1, \mathbf{g}_2, \mathbf{g}_3\}$ point, of course, along the coordinate lines drawn in Figure 5.4 and the contravariant basis vectors satisfy the usual orthogonality relation $\mathbf{g}_\alpha \cdot \mathbf{g}^\beta = \delta_\alpha^\beta$.

For the plate-and-plate device the velocity field $\mathbf{v}$ and the shear rate κ in the steady-state motion have the form

$$\mathbf{v} = v^2 \mathbf{g}_2, \quad v^2 = \omega\left(z\right), \quad \kappa = r \frac{d\omega}{dz}, \tag{5.108}$$

which yield the velocity gradient

$$\mathbf{L} = L^\alpha_{.\beta} \mathbf{g}_\alpha \otimes \mathbf{g}^\beta \equiv v^\alpha_{.;\beta} \mathbf{g}_\alpha \otimes \mathbf{g}^\beta, \quad \left(L^\alpha_{.\beta}\right) = \begin{pmatrix} 0 & -r\omega & 0 \\ \omega/r & 0 & d\omega/dz \\ 0 & 0 & 0 \end{pmatrix},$$

$$\tag{5.109}$$

$$\mathbf{L} = L_{(\alpha)(\beta)} \mathbf{e}_\alpha \otimes \mathbf{e}_\beta, \quad \left(L_{(\alpha)(\beta)}\right) = \begin{pmatrix} 0 & -\omega & 0 \\ \omega & 0 & r\left(d\omega/dz\right) \\ 0 & 0 & 0 \end{pmatrix},$$

where $v^\alpha_{.;\beta}$ is the covariant derivative of the velocity and for the acceleration we obtain

$$\dot{\mathbf{v}} = \left(\frac{\partial v^\alpha}{\partial t} + v^\beta v^\alpha_{.;\beta} \right) \mathbf{g}_\alpha = -r\omega^2 \mathbf{g}_1. \tag{5.110}$$

Clearly, the physical components of the shearing tensor $\mathbf{D} = \frac{1}{2}\left(\mathbf{L} + \mathbf{L}^T\right)$ form the matrix

$$\left(D_{(\alpha)(\beta)}\right) = \frac{1}{2} \begin{pmatrix} 0 & 0 & 0 \\ 0 & 0 & r\left(d\omega/dz\right) \\ 0 & r\left(d\omega/dz\right) & 0 \end{pmatrix}, \tag{5.111}$$

which justifies the definition of the shear rate κ.

The symmetry conditions for this flow yield vanishing components of the shear stresses

$$T_{(1)(2)} = T_{(1)(3)} = 0. \tag{5.112}$$

The remaining physical components of the stresses are combined in the following way

$$\tau = T_{(2)(3)} \quad \sigma_1 = T_{(3)(3)} - T_{(1)(1)}, \quad \sigma_1 = T_{(2)(2)} - T_{(1)(1)}, \quad p = -\frac{1}{3}\left(T_{(1)(1)} + T_{(2)(2)} + T_{(3)(3)}\right). \tag{5.113}$$

The relation between the given shear rate κ and the measured τ, σ_1, σ_2 is the subject of the viscometric flow experiment.

It is easy to prove that the momentum balance equations (5.106) reduce in this case to the following form

$$\frac{\partial T_{(3)(3)}}{\partial z} = 0,$$

$$r\frac{\partial \tau}{\partial z} - \frac{\partial p}{\partial \varphi} = 0, \tag{5.114}$$

$$\frac{\partial T_{(1)(1)}}{\partial r} - \frac{\sigma_2}{r} = -\rho r \omega^2.$$

The solution of the problem follows if we use a constitutive relation for the stresses. It can be shown by an iteration procedure (compare [431]) that the evolution equation (5.99) for the Maxwellian fluid and the constitutive relations (5.102) for the second grade fluid ($N = 2$) yield the same result if we properly choose the parameters. Namely

$$\tau = \eta\kappa,$$

$$T_{(1)(1)} + p = -\frac{2}{3}\left(\alpha_1 + \alpha_2\right)\kappa^2,$$

$$T_{(2)(2)} + p = -\frac{1}{3}\left(2\alpha_1 - \alpha_2\right)\kappa^2, \tag{5.115}$$

$$T_{(2)(2)} + p = \frac{1}{3}\left(4\alpha_1 + \alpha_2\right)\kappa^2,$$

where

$$\alpha_1 = -\eta\tau_t, \quad \alpha_2 = -\alpha_1\left(1 - \gamma_t\right). \tag{5.116}$$

The same relations between the parameters hold for the remaining viscometric flows.

The normal stress functions are then given by the relations

$$\sigma_1 = \left(2\alpha_1 + \alpha_2\right)\kappa^2, \quad \sigma_2 = \alpha_2\kappa^2. \tag{5.117}$$

The relations between the coefficients $\{\alpha_1, \alpha_2\}$ and $\{\tau_t, \gamma_t\}$ are very important practical conclusions. They show that in viscometric flow experiments one cannot test the validity of any of these models. As we see in Chapter 6 on stability, the Maxwell model yields in general hyperbolicity of the field equations and it is stable while the Nth grade models are parabolic and unstable. These differences could not be observed in steady-state flows commonly investigated in the rheology.

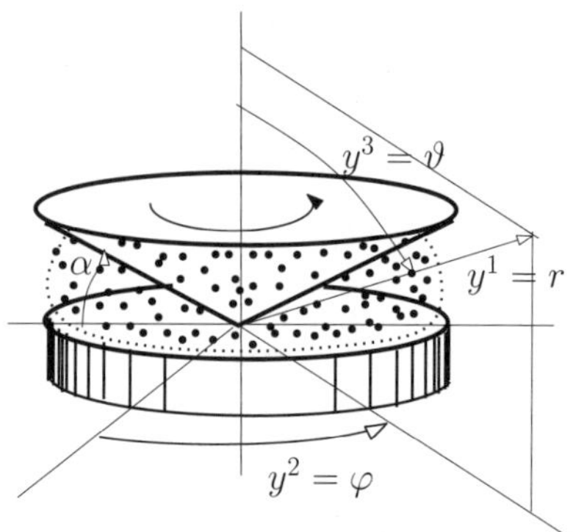

Figure 5.5: *Schematic of cone-and-plate flow* ($\alpha = 2\text{-}3°$).

Relations (5.115) demonstrate the symmetry of the constitutive relations in viscometric flows which is common for Nth grade models and functional models. Namely, the shear stress τ is an odd function of the shearing rate κ, in our simple case – a linear function, and the normal stresses even functions of the shearing κ, in our case – quadratic.

Cone-and-plate

We solve the next problem in spherical coordinates indicated in Figure 5.5. In the general case the momentum balance equations in spherical coordinates have the following form (again, we leave out the body forces)

$$\rho\left\{\frac{\partial v^1}{\partial t} + v^1\frac{\partial v^1}{\partial y^1} + v^2\frac{\partial v^1}{\partial y^2} + v^3\frac{\partial v^1}{\partial y^3} - y^1\sin^2 y^3 \left(v^2\right)^2 - y^1\left(v^3\right)^2\right\}$$

$$= \frac{\partial T^{11}}{\partial y^1} + \frac{\partial T^{12}}{\partial y^2} + \frac{\partial T^{13}}{\partial y^3} + \frac{2}{y^1}T^{11} - y^1\sin^2 y^3 T^{22} + \cot y^3 T^{13} - y^1 T^{33},$$

$$\rho\left\{\frac{\partial v^2}{\partial t} + v^1\frac{\partial v^2}{\partial y^1} + v^2\frac{\partial v^2}{\partial y^2} + v^3\frac{\partial v^2}{\partial y^3} - \frac{2}{y^1}v^1 v^2 + 2\cot y^3 v^2 v^3\right\}$$

$$= \frac{\partial T^{12}}{\partial y^1} + \frac{\partial T^{22}}{\partial y^2} + \frac{\partial T^{23}}{\partial y^3} + \frac{4}{y^1}T^{12} + 2\cot y^3 T^{23},$$

$$\rho\left\{\frac{\partial v^3}{\partial t} + v^1\frac{\partial v^3}{\partial y^1} + v^2\frac{\partial v^3}{\partial y^2} + v^3\frac{\partial v^3}{\partial y^3} + \frac{2}{y^1}v^1 v^3 - \sin y^3\cos y^3 \left(v^2\right)^2\right\}$$

$$= \frac{\partial T^{13}}{\partial y^1} + \frac{\partial T^{23}}{\partial y^2} + \frac{\partial T^{33}}{\partial y^3} + \frac{4}{y^1}T^{13} - \sin y^3\cos y^3 T^{22} + \cot y^3 T^{33}.$$

$$(5.118)$$

In the case of the cone-and-plate device the velocity field for steady-state motion is assumed to have the form (in practical applications $\sin\vartheta \approx 1$, $\cos\vartheta \approx 0$)

$$\mathbf{v} = v^2\mathbf{g}_2, \quad v^2 = \omega\left(\vartheta\right), \quad \kappa = d\omega/d\vartheta\sin\vartheta. \tag{5.119}$$

The velocity gradient is as follows

$$\mathbf{L} = L^{\alpha}_{.\beta}\mathbf{g}_{\alpha} \otimes \mathbf{g}^{\beta} \equiv v^{\alpha}_{.;\beta}\mathbf{g}_{\alpha} \otimes \mathbf{g}^{\beta}, \quad \left(L^{\alpha}_{.\beta}\right) = \begin{pmatrix} 0 & -r\sin^2\vartheta\,\omega & 0 \\ \dfrac{1}{r}\omega & 0 & d\omega/d\vartheta + \omega\cot\vartheta \\ 0 & -\cos\vartheta\sin\vartheta\,\omega & 0 \end{pmatrix},$$

$$\mathbf{L} = L_{(\alpha)(\beta)}\mathbf{e}_{\alpha} \otimes \mathbf{e}_{\beta}, \quad \left(L_{(\alpha)(\beta)}\right) = \begin{pmatrix} 0 & -\sin\vartheta\,\omega & 0 \\ \sin\vartheta\,\omega & 0 & d\omega/d\vartheta\sin\vartheta + \omega\cos\vartheta \\ 0 & -\omega\cos\vartheta & 0 \end{pmatrix},$$

$$(5.120)$$

where $v^{\alpha}_{.;\beta}$ is the covariant derivative of the velocity and for the acceleration we obtain

$$\dot{\mathbf{v}} = \left(\frac{\partial v^{\alpha}}{\partial t} + v^{\beta}v^{\alpha}_{.;\beta}\right)\mathbf{g}_{\alpha} = -\omega^2 r\sin^2\vartheta\,\mathbf{g}_1 - \omega^2\sin\vartheta\cos\vartheta\,\mathbf{g}_3. \tag{5.121}$$

Again the symmetry of the problem yields

$$T_{(1)(3)} = T_{(1)(2)} = 0. \tag{5.122}$$

The equations of motion (5.118) can now be written in the form

$$-\rho\omega^2 r\sin^2\vartheta = r\frac{\partial T_{(1)(1)}}{\partial r} - (\sigma_1 + \sigma_2),$$

$$0 - \frac{\partial\tau}{\partial\vartheta} + 2\cot\vartheta\,\tau, \tag{5.123}$$

$$-\rho\omega^2 r^2\sin\vartheta\cos\vartheta = \frac{\partial T_{(3)(3)}}{\partial\vartheta} + (\sigma_1 - \sigma_2)\cot\vartheta,$$

where

$$\tau = T_{(2)(3)}, \quad \sigma_1 = T_{(3)(3)} - T_{(1)(1)}, \quad \sigma_2 = T_{(2)(2)} - T_{(1)(1)}. \tag{5.124}$$

It remains to choose the constitutive model. In the two cases which we have used for the plate-and-plate model, i.e., the Maxwell equation (5.99) and the second grade fluid ($N = 2$) (5.102) we have again Relation (5.115)$_1$ for the shear stress function τ and (5.117) for the normal stress functions σ_1, σ_2.

Couette flow (cylindrical)

The device for the viscometric Couette flow is schematically shown in Figure 5.6. Obviously, we use again cylindrical coordinates. The motion is described by the angular velocity

$$\mathbf{v} = v^2\mathbf{g}_2, \quad v^2 = \omega\,(r), \quad \kappa = r\frac{d\omega}{dr}. \tag{5.125}$$

Then the viscometric functions are given by the constitutive relations

$$\begin{aligned} \tau &= T_{(1)(2)} = \eta\kappa, \\ \sigma_1 &= T_{(11)} - T_{(3)(3)} = (2\alpha_1 + \alpha_2)\,\kappa^2, \\ \sigma_2 &= T_{(2)(2)} - T_{(3)(3)} = \alpha_2\kappa^2. \end{aligned} \tag{5.126}$$

The symmetry of the problem implies

$$T_{(1)(3)} = T_{(2)(3)} = 0. \tag{5.127}$$

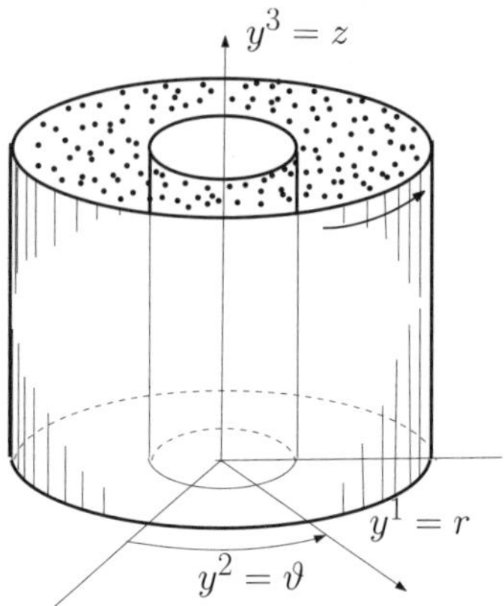

Figure 5.6: *Schematic of cylindrical Couette flow.*

The equations of motion follow in the form

$$-\rho r \omega^2 = \frac{\partial T_{(11)}}{\partial r} + \frac{\sigma_1 - \sigma_2}{r},$$

$$\frac{\partial \tau}{\partial r} + \frac{2}{r}\tau = 0, \tag{5.128}$$

$$0 = 0.$$

Poiseuille flow

This is the viscometric flow in a circular pipe with the velocity $\mathbf{v} = v\,(r)\,\mathbf{g}_2$, with $\mathbf{g}_2$ pointing along the pipe, the shear rate $\kappa = dv/dr$, and the adherence boundary conditions. With the application of the cylindrical coordinates $y^1 = r, y^2 = z, y^3 = \varphi$ the stress functions are defined by the relations

$$\tau = T_{(1)(3)}, \quad \sigma_1 = T_{(1)(1)} - T_{(3)(3)}, \quad \sigma_2 = T_{(2)(2)} - T_{(3)(3)}, \quad T_{(1)(2)} = T_{(2)(3)} = 0. \tag{5.129}$$

The equations of motion are as follows

$$\sigma_1 = \frac{\partial p}{\partial r},$$

$$\frac{\partial \tau}{\partial r} + \frac{1}{r}\tau = \frac{\partial p}{\partial z}, \tag{5.130}$$

$$0 = 0.$$

The constitutive relations for viscometric functions are the same as before.

In the table of Section 9.4.1 of Part I we have quoted experimental values of the material parameters obtained by R. F. Ginn and A. B. Metzner in the cone-and-plate experiment with the angle $\alpha = 2.01°$ (compare Figure 5.5) for the non-Newtonian fluid: 6.5 wt% solution of polyisobutylene (Vistanex L-100) in decaline [135]. We present some of these

Table 5.1: *Stress functions and material parameters of the solution of polyisobutylene in decaline (after [135]).*

$\kappa[1/\mathrm{s}]$	$\tau[\mathrm{Pa}]$	$\sigma_1[\mathrm{Pa}]$	$\sigma_2[\mathrm{Pa}]$	$\eta[\mathrm{Pas}]$	$\alpha_1[\mathrm{Pas}^2]$	$\alpha_2[\mathrm{Pas}^2]$	$\tau_t[\mathrm{s}]$	$\gamma_t[\text{-}]$
3	-	-1.516	7.861	-	-0.521	0.873	-	-0.676
5	32.405	-2.827	17.995	6.81	-0.416	0.720	0.061	-0.731
10	56.537	-10.549	38.191	5.654	-0.244	0.382	0.043	-0.566
20	91.700	-15.167	102.211	4.585	-0.147	0.256	0.032	-0.741

results again in Table 5.1 (compare Tables III and VIII in [135]) in order to illustrate the order of magnitude of the quantities appearing in steady-state viscometric flows.

Variations of the data for the viscosity η in this table show that the applicability of the models which we have presented as examples is rather limited. We return in Chapter 6 on stability to the values of normal stress coefficients and, in particular, to the property that α_1 is negative and the sum of α_1 and α_2 is different from zero.

5.4 Nonlinear elastic solids

The constitutive relations for the class of nonlinear elastic solids have already been presented in Section 5.4 of Part I. Several examples have been listed there. The maybe most actual application of nonlinear elasticity are biological tissues. They are not discussed in this section but we come back to several aspects concerning such materials in Chapter 12 on the mechanics of porous media. In this section we select only a single material belonging to the class of nonlinear elastic solids, namely rubber-like materials, in order to illustrate the reaction of such materials on universal solutions which were already the subject of Chapter 2.

5.4.1 Rubber-like materials

Only isothermal processes in an isotropic material are considered. The assumption of a constant temperature reflects rather badly what happens in rubber-like materials. Thus, the results of this section can be understood only as a rough approximation. Due to isotropy, the material properties have to be independent of the direction in the reference configuration. Not all rubber-like materials satisfy this condition. However, very often this assumption is appropriate.

Then, all material properties follow from the Helmholtz free energy

$$\psi = \psi(\mathbf{C}), \tag{5.131}$$

satisfying the isotropy condition

$$\forall \mathbf{H}, \ \mathbf{H}^T = \mathbf{H}^{-1} : \psi(\mathbf{C}) = \psi(\mathbf{HCH}^T). \tag{5.132}$$

In (5.132) $\mathbf{H}$ denotes an arbitrary orthogonal matrix corresponding to a rotation of the material directions. Equation (5.132) means that ψ can depend only on such components of $\mathbf{C}$ which do not change by such a rotation. Obviously, these are the invariants of $\mathbf{C}$

(compare (2.16))

$$I^C = \operatorname{tr} \mathbf{C}, \qquad II^C = \frac{1}{2}\left((I^C)^2 - \operatorname{tr}\mathbf{C}^2\right), \qquad III^C = \det\mathbf{C}. \tag{5.133}$$

Thus, it follows that

$$\psi = \psi(I^C, III^C, III^C). \tag{5.134}$$

The following considerations are easier if we use $\mathbf{B}$ instead of $\mathbf{C}$. In Chapter 2 we already noticed that the principal invariants of tensor $\mathbf{C}$ agree with those of tensor $\mathbf{B}$. Therefore they are written without an index in the following. We repeat the relation between $\mathbf{B}$ and $\mathbf{C}$

$$\mathbf{C} = \mathbf{F}^T\mathbf{F} = \mathbf{F}^{-1}\mathbf{B}\mathbf{F} \Rightarrow \mathbf{C}^{-1} = \mathbf{F}^{-1}\mathbf{B}^{-1}\mathbf{F}. \tag{5.135}$$

In thermodynamic equilibrium the second law of thermodynamics reduces the problem of constitutive equations for the stress tensor to a scalar constitutive equation for ψ. The Piola-Kirchhoff stress tensor can be written as

$$\begin{aligned}
\mathbf{P} &= 2\rho_0\mathbf{F}\frac{\partial\psi}{\partial\mathbf{C}} = 2\rho_0\left(\frac{\partial\psi}{\partial I}\frac{\partial I}{\partial\mathbf{C}} + \frac{\partial\psi}{\partial II}\frac{\partial II}{\partial\mathbf{C}} + \frac{\partial\psi}{\partial III}\frac{\partial III}{\partial\mathbf{C}}\right) \\
&= 2\rho_0\mathbf{F}\left(\frac{\partial\psi}{\partial I}\mathbf{I} + \frac{\partial\psi}{\partial II}(I\,\mathbf{I} - \mathbf{C}) + \frac{\partial\psi}{\partial III}III\mathbf{C}^{-1}\right) \\
&= 2\rho_0\left[\left(\frac{\partial\psi}{\partial I} + \frac{\partial\psi}{\partial II}I\right)\mathbf{B} - \frac{\partial\psi}{\partial II}\mathbf{B}^2 + \frac{\partial\psi}{\partial III}III\mathbf{I}\right]\mathbf{F}^{-T}.
\end{aligned} \tag{5.136}$$

Simultaneously, follows from the Cayley-Hamilton Theorem

$$\mathbf{B}^2 = I\,\mathbf{B} - II\,\mathbf{I} + III\,\mathbf{B}^{-1}. \tag{5.137}$$

Then, we obtain

$$\boldsymbol{T} = \frac{1}{J}\mathbf{P}\mathbf{F}^T = \Im_0\mathbf{I} + \Im_1\mathbf{B} + \Im_{-1}\mathbf{B}^{-1}, \tag{5.138}$$

where

$$\Im_0 := 2\rho\left(\frac{\partial\psi}{\partial II}II + \frac{\partial\psi}{\partial III}III\right) = \Im_0(I, II, III),$$

$$\Im_1 := 2\rho\frac{\partial\psi}{\partial I} = \Im_1(I, II, III), \tag{5.139}$$

$$\Im_{-1} := -2\rho\frac{\partial\psi}{\partial II}III = \Im_{-1}(I, II, III),$$

and Relation (4.10) between Piola-Kirchhoff and Cauchy stresses, $\mathbf{P}$ and $\mathbf{T}$, has been used.

Obviously, the Helmholtz free energy ψ, as a given function of the three invariants I, II, III, is sufficient to determine the material coefficients $\Im_0, \Im_1, \Im_{-1}$. If, though, direct measurements of these coefficients are available, it is useful to bear in mind the following properties. Let $\mathbf{r}^B$ be the eigenvector of tensor $\mathbf{B}$. Then,

$$\mathbf{T}\mathbf{r}^B = \Im_0\mathbf{r}^B + \Im_1(\lambda^B)\mathbf{r}^B + \Im_{-1}\left(\frac{1}{\lambda^B}\right)\mathbf{r}^B, \tag{5.140}$$

where λ^B denotes the corresponding eigenvalue of tensor $\mathbf{B}$. This relation implies that the eigenvectors of the deformation tensor $\mathbf{B}$ coincide with the eigenvectors of the Cauchy stress tensor $\mathbf{T}$ and the eigenvalues λ^T of the stress tensor are given by the following equations

$$
\begin{aligned}
\lambda^T_{(i)} &= \Im_0 + \Im_1 \lambda^B_{(i)} + \Im_{-1}\frac{1}{\lambda^B_{(i)}}, \quad i = 1,2,3, \\
\Im_r &= \Im_r(I,II,III), \quad r = 0,1,-1, \\
I &= \lambda^B_{(1)} + \lambda^B_{(2)} + \lambda^B_{(3)}, \\
II &= \lambda^B_{(1)}\lambda^B_{(2)} + \lambda^B_{(1)}\lambda^B_{(3)} + \lambda^B_{(2)}\lambda^B_{(3)}, \\
III &= \lambda^B_{(1)}\lambda^B_{(2)}\lambda^B_{(3)}.
\end{aligned}
\tag{5.141}
$$

As a matter of course, Relations (5.141) uniquely determine the constitutive equations of isotropic material.

Rubber and many rubber-like materials can be considered to be incompressible. In this case, only the isochoric motion is admitted, so that the deformation is restricted by the condition

$$
\mathcal{J}\left(\mathbf{C}\right) = \left(\det \mathbf{F}\right)^2 - 1 \equiv \det \mathbf{C} - 1.
\tag{5.142}
$$

This means

$$
\frac{\partial \mathcal{J}}{\partial \mathbf{C}} = \mathbf{C}^{-1}\det \mathbf{C} \quad \Rightarrow \quad \mathbf{n} = p\mathbf{F}\left(\mathbf{C}^{-1}\det \mathbf{C}\right)\mathbf{F}^T
\tag{5.143}
$$

$$
- p\mathbf{F}\left(\mathbf{F}^T\mathbf{F}\right)^{-1}\mathbf{F}^T - p\mathbf{F}\mathbf{F}^{-1}\mathbf{F}^{\ T}\mathbf{F}^T = p\mathbf{1}.
$$

Thus, $\mathbf{n} = -\mathrm{p}\mathbf{1}$, with $\mathrm{p} := -p$ being an arbitrary scalar. This reflects the well-known result of Poincaré, that the stress in an incompressible material follows from the motion, but in such a way, that an unknown scalar p remains. Then, the Cauchy stress is given by

$$
\mathbf{T} = -\mathrm{p}\mathbf{1} + 2\rho\mathbf{F}\frac{\partial \psi}{\partial \mathbf{C}}\mathbf{F}^T, \qquad \det \mathbf{F} = 1,
\tag{5.144}
$$

where p denotes the pressure which is not specified by the constitutive law. In this case, the free energy ψ for isotropic materials only depends on two invariants $(III = 1)$

$$
\psi = \psi(I,II) \quad \Rightarrow \quad \frac{\partial \psi}{\partial \mathbf{C}} = \frac{\partial \psi}{\partial I}\mathbf{1} + \frac{\partial \psi}{\partial II}(I\,\mathbf{1} - \mathbf{C}).
\tag{5.145}
$$

It follows

$$
\mathbf{T} = -\mathrm{p}\mathbf{1} + \Im_1\mathbf{B} + \Im_{-1}\mathbf{B}^{-1},
\tag{5.146}
$$

with

$$
\Im_1 := 2\rho\frac{\partial \psi}{\partial I} = \tilde{\Im}_1(I,II), \qquad \Im_{-1} := -2\rho\frac{\partial \psi}{\partial II} = \tilde{\Im}_{-1}(I,II).
\tag{5.147}
$$

One of the most important special cases can be obtained from (5.146) under the additional assumption of constant coefficients $\Im_1$ and $\Im_{-1}$. Then,

$$
\mathbf{T} = -\mathrm{p}\mathbf{1} + \mu(\frac{1}{2} + \beta)\mathbf{B} - \mu(\frac{1}{2} - \beta)\mathbf{B}^{-1}, \qquad \mu, \beta - \text{constant},
\tag{5.148}
$$

where

$$
\beta := \frac{1}{2}\frac{\Im_1 + \Im_{-1}}{\Im_1 - \Im_{-1}}, \qquad \mu := \Im_1 - \Im_{-1}.
\tag{5.149}
$$

The material described by (5.148) is called Mooney-Rivlin material (for the discussion of compressible and incompressible Mooney-Rivlin materials compare Section 5.3 – *Materials with constraints* of Part I). The corresponding Helmholtz free energy then has the form

$$\rho\psi = \frac{1}{2}\mu\left[(\frac{1}{2} + \beta)(I - 3) + (\frac{1}{2} - \beta)(II - 3)\right], \tag{5.150}$$

where the integration constant is chosen in such a way that ψ is zero in the undeformed state. If we, additionally, require that the undeformed state induces the minimum of the free energy, then

$$\mu > 0, \qquad -\frac{1}{2} \le \beta \le \frac{1}{2}. \tag{5.151}$$

In the special case $\beta = \frac{1}{2}$, Equation (5.148) defines neo-Hookean material.

The application of (5.148) for the description of rubber and rubber-like materials is widely discussed in the literature. We refer here only to the classical book by L. R. G. Treloar [386].

5.4.1.1   Homogeneous deformations

We study a few solutions of the equations of motion in the static case

$$\operatorname{div} \mathbf{T} = 0, \tag{5.152}$$

with the Cauchy stress (5.138) for compressible materials and (5.146) for incompressible materials.

Examples

1. *Isochoric extension.*

We recall the form of the left Cauchy-Green deformation tensor (compare (2.44))

$$\mathbf{B} = \begin{pmatrix} \lambda^2 & 0 & 0 \\ 0 & \frac{1}{\lambda} & 0 \\ 0 & 0 & \frac{1}{\lambda} \end{pmatrix} \quad \Rightarrow \quad \mathbf{B}^{-1}\begin{pmatrix} \frac{1}{\lambda^2} & 0 & 0 \\ 0 & \lambda & 0 \\ 0 & 0 & \lambda \end{pmatrix}. \tag{5.153}$$

The Cauchy stress for compressible material is given by

$$\mathbf{T} = \begin{pmatrix} \Im_0 + \lambda^2\Im_1 + \frac{1}{\lambda^2}\Im_{-1} & 0 & 0 \\ 0 & \Im_0 + \frac{1}{\lambda}\Im_1 + \lambda\Im_{-1} & 0 \\ 0 & 0 & \Im_0 + \frac{1}{\lambda}\Im_1 + \lambda\Im_{-1} \end{pmatrix}, \tag{5.154}$$

and with $-p$ instead of the $\Im_0$-function in the case of incompressible material. Let us assume that the side walls of the sample are not loaded. Then, we can eliminate either $\Im_0$ or p

$$\Im_0 = -\left(\frac{1}{\lambda}\Im_1 + \lambda\Im_{-1}\right), \tag{5.155}$$

and the stress component in the loading direction is obtained in the form

$$T_{11} = \frac{K}{A} = \lambda\frac{K}{A_0} = \left(\lambda^2 - \frac{1}{\lambda}\right)\left(\Im_1 - \frac{1}{\lambda}\Im_{-1}\right). \tag{5.156}$$

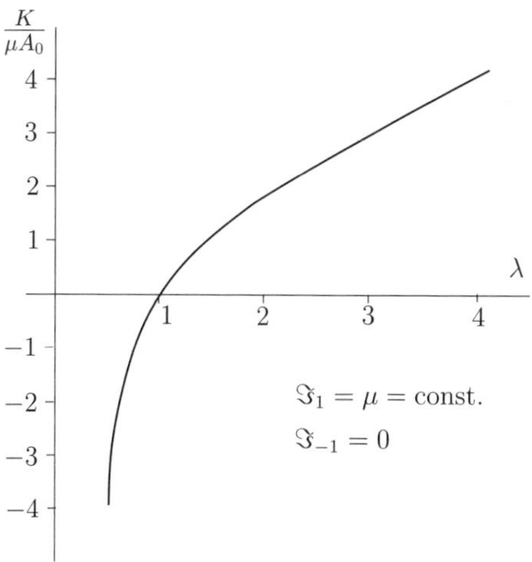

Figure 5.7: *Behavior of the force K for neo-Hookean material.*

Figure 5.7 shows the behavior of the force K for neo-Hookean material:

$$\Im_1 = \mu = \text{const.}, \qquad \Im_{-1} = 0 \quad \Rightarrow \quad \beta = \frac{1}{2}, \tag{5.157}$$

that means

$$\mathbf{T} = -p\mathbf{1} + \mu\mathbf{B}. \tag{5.158}$$

Obviously, the arbitrary constant stretch λ is the solution of Equation (5.152) with the following boundary conditions

$$T_{11}\big|_{X=0} = T_{11}\big|_{X=L} = \lambda\frac{K}{A_0},$$

$$T_{12} = \big|_{X=0} = T_{13}\big|_{X=0} = T_{12}\big|_{X=L} = T_{13}\big|_{X=L} = 0, \tag{5.159}$$

$$T_{22}\big|_{(Y,Z)\epsilon\Sigma} = T_{33}\big|_{(Y,Z)\epsilon\Sigma} = T_{12}\big|_{(Y,Z)\epsilon\Sigma} = T_{23}\big|_{(Y,Z)\epsilon\Sigma} = 0,$$

where Σ denotes the side wall of the sample.

2. *Simple shear.*

According to Equation (2.64) in the case of simple shear, the left-Cauchy-Green deformation tensor has the form

$$\mathbf{B} = \begin{pmatrix} 1 & 0 & 0 \\ 0 & 1+\tan^2\varphi & \tan\varphi \\ 0 & \tan\varphi & 1 \end{pmatrix} \quad \Rightarrow \quad \mathbf{B}^{-1} = \begin{pmatrix} 1 & 0 & 0 \\ 0 & 1 & -\tan\varphi \\ 0 & -\tan\varphi & 1+\tan^2\varphi \end{pmatrix}, \tag{5.160}$$

and for the stress we obtain

$$\mathbf{T} = \begin{pmatrix} \Im_0 + \Im_1 + \Im_{-1} & 0 & 0 \\ 0 & \Im_0 + \Im_1(1 + \tan^2\varphi) + \Im_{-1} & (\Im_1 - \Im_{-1})\tan\varphi \\ 0 & (\Im_1 - \Im_{-1})\tan\varphi & \Im_0 + \Im_1 + \Im_{-1}(1 + \tan^2\varphi) \end{pmatrix},$$

$$(5.161)$$

with

$$I = II = 3 + \tan^2\varphi, \qquad III = 1 \quad \Rightarrow \quad \Im_\Gamma = \Im_\Gamma(\tan^2\varphi), \tag{5.162}$$

with $\Gamma = 0, 1, -1$ in the case of compressible material and $\Im_0 \to -\mathrm{p}$, $\Gamma = 1, -1$ for incompressible material.

The shear stress

$$T_{12} = \mu(\kappa^2)\kappa, \qquad \mu(\kappa^2) := \Im_1(\kappa^2) - \Im_{-1}(\kappa^2), \qquad \kappa := \tan\varphi, \tag{5.163}$$

is, as expected, an odd function of the shear $\kappa = \tan\varphi$. If the shear modulus μ is a differentiable function of κ, then each deviation from the classical linear theory must be at least quadratic in κ

$$\mu(\kappa^2) = \mu(\kappa = 0) + o\kappa^2. \tag{5.164}$$

We can also write

$$\frac{1}{3}\mathrm{tr}\,\mathbf{T} = (\Im_0 + \Im_1 + \Im_{-1}) + \frac{1}{3}(\Im_1 - \Im_{-1})\kappa^2. \tag{5.165}$$

The condition

$$\lim_{\kappa \to 0} \frac{1}{3}\mathrm{tr}\,\mathbf{T} \equiv \lim_{\kappa \to 0}(\Im_0 + \Im_1 + \Im_{-1}) = 0, \tag{5.166}$$

yields therefore that the limit $\kappa \to 0$ corresponds to the stress-free undeformed state. For a differentiable function $(\Im_0 + \Im_1 + \Im_{-1})$ in the vicinity of $\kappa = 0$ we obtain

$$\left| \lim_{\kappa \to 0} \frac{\Im_0 + \Im_1 + \Im_{-1}}{\kappa^2} \right| < \infty, \tag{5.167}$$

and, depending on the sign of the medium normal stress

$$\frac{1}{3}\frac{\mathrm{tr}\,\mathbf{T}}{\kappa^2} = \frac{\Im_0 + \Im_1 + \Im_{-1}}{\kappa^2} + (\Im_1 + \Im_{-1}), \tag{5.168}$$

the sample under shear stress will either expand or contract. This property is called the *Kelvin effect*.[13]

Another attribute of large shear is the inevitable inequality of the normal stresses T_{22} and T_{33}. From (5.161) follows

$$T_{22} = T_{33} \quad \Rightarrow \quad \Im_1 - \Im_{-1} = 0 \quad \Rightarrow \quad \mu = 0, \tag{5.169}$$

which implies the absence of the shear stress. This difference between the normal stresses is known as the *Poynting effect*.

Obviously, both effects can be visualized only for large deformations because both $\mathrm{tr}\,\mathbf{T}$ and T_{22} and T_{33} behave quadratic in κ.

---

[13] "But it is possible that a distorting stress may produce, in a truly isotropic solid condensation or dilatation in proportion to the square of its value: and it is possible that such effects may be sensible in India rubber or cork, or other bodies susceptible of great deformations or compressions with persistent elasticity" (W. Thomson (Lord Kelvin) and P. G. Tait: Treatise on Natural Philosophy, Part I, Cambridge, footnote to §679, 1867).

The above two examples belong to the class of universal solutions (see Section 2.2). These are such deformations which are solutions of the static equations of motion without external forces (5.152) and, simultaneously, do not depend on material properties.

This definition leads for *compressible materials* (5.138) to the condition, that the coefficients of $\mathfrak{S}_0, \frac{\partial \mathfrak{S}_0}{\partial I}, \frac{\partial \mathfrak{S}_0}{\partial II}, \frac{\partial \mathfrak{S}_0}{\partial III}, ..., \frac{\partial \mathfrak{S}_{-1}}{\partial III}$ must vanish. We obtain 12 conditions for 6 components of the deformation tensor $\mathbf{B}$. Additionally, $\mathbf{B}$ must be a positive-definite and integrable matrix. Jerald Ericksen proved in [115], that only homogeneous deformations $\mathbf{B} = \text{const.}$ satisfy all these conditions. This means – except for homogeneous deformations – no controllable finite deformations exist in isotropic compressible elastic materials.

Ericksen's Theorem

As introduced above, $\mathbf{B}$ is the left Cauchy-Green tensor of the deformation gradient $\mathbf{F}$: $\mathbf{B} = \mathbf{F}\mathbf{F}^T$. The reference configuration $\mathcal{B}_0$ is undistorted. The constitutive equation of a homogeneous elastic body has the form

$$\mathbf{T} = \mathfrak{S}_0 \mathbf{1} + \mathfrak{S}_1 \mathbf{B} + \mathfrak{S}_{-1} \mathbf{B}^{-1},$$

where $\mathfrak{S}_0$, $\mathfrak{S}_1$ and $\mathfrak{S}_{-1}$ are functions of the three principal invariants I, II and III of $\mathbf{B}^{-1}$. The stress $\mathbf{T}$ required to effect a homogeneous deformation is a constant tensor and, thus, $\text{div}\mathbf{T} = 0$. Then,

$$\frac{\partial \mathfrak{S}_0}{\partial I_a} \operatorname{grad} I_a + \frac{\partial \mathfrak{S}_1}{\partial I_a} \mathbf{B} \operatorname{grad} I_a + \frac{\partial \mathfrak{S}_{-1}}{\partial I_a} \mathbf{B}^{-1} \operatorname{grad} I_a$$
$$+ \mathfrak{S}_1 \operatorname{div} \mathbf{B} + \mathfrak{S}_{-1} \operatorname{div} \mathbf{B}^{-1} = 0,$$

where the index a is summed from 1 to 3 and $I_1 = I$, $I_2 = II$, $I_3 = III$. Since $\mathfrak{S}_0$, $\mathfrak{S}_1$ and $\mathfrak{S}_{-1}$ are arbitrary functions, the governing equations for $\mathbf{B}$ are

$$\operatorname{grad} I_a = 0, \quad a = 1, 2, 3, \qquad \operatorname{div} \mathbf{B} = 0, \qquad \operatorname{div} \mathbf{B}^{-1} = 0.$$

Since not every tensor field $\mathbf{F}$ can be integrated to yield a deformation relative to $\mathcal{B}_0$, the integrability must be proven by the condition that the curvature tensor $\mathbf{R}$ based on $\mathbf{B}^{-1}$ is equal to zero for all $\mathbf{x}$ in the domain of $\mathbf{F}$

$$\mathbf{R}\left(\mathbf{x}\right) = 0.$$

It can be proven (see [415]) that both $\mathbf{B}$ and $\mathbf{F}$ are constant tensors.

These considerations lead to a weaker version of the ERICKSEN THEOREM:
All universal solutions for a hyperelastic homogeneous isotropic body are homogeneous deformations from an undistorted configuration.

Ericksen studied also incompressible materials [114]. However, for *incompressible materials* the situation is much more complicated than for compressible materials. In this case the stress is given by (5.146) and we obtain

$$-\operatorname{grad} p + \operatorname{div}(\mathfrak{S}_1 \mathbf{B} + \mathfrak{S}_{-1} \mathbf{B}^{-1}) = 0. \tag{5.170}$$

It follows that

$$\operatorname{rot} \operatorname{div} = (\mathfrak{S}_1 \mathbf{B} + \mathfrak{S}_{-1} \mathbf{B}^{-1}) = 0, \tag{5.171}$$

i.e.,

$$(\Im_1 B_{kl} + \Im_{-1} B_{kl}^{-1})_{,lm} = (\Im_1 B_{ml} + \Im_{-1} B_{ml}^{-1})_{,lk}. \tag{5.172}$$

This equation leads to 12 conditions for a universal solution with the additional constraint $\det \mathbf{B} = 1$. Rivlin showed in [313]-[320] that besides the homogeneous solution $\mathbf{B} = $ const. a series of heterogeneous and universal solutions exist which have practical bearing.

5.4.1.2   Heterogeneous deformations

In Section 2.2 we already showed the classification of these heterogeneous universal solutions into certain families. Ericksen tried to find all possible universal solutions for incompressible materials. The families 1 to 5 introduced above seem to exhaust the possibilities. One important example, which has already been tackled in Subsection 2.3.3 and belongs to family 3 is conclusively studied here.

Example   *Pure torsion of a circular cylinder*

We shortly repeat the problem (for an illustration see Subsection 2.3.3). In this case, in Equation (2.30) the parameters are

$$A = 1, \quad B = 0, \quad C = 1, \quad E = 0, \quad F = 1, \tag{5.173}$$

that is

$$r = R, \quad \vartheta = \Theta + DZ, \quad z = Z. \tag{5.174}$$

Constant D denotes the twist of the cylinder.

The physical coordinates of the stress tensor $\mathbf{T}$ have to fulfill the following equations

$$\frac{\partial}{\partial r} T_{(r)(r)} + \frac{1}{r} \left(T_{(r)(r)} - T_{(\vartheta)(\vartheta)} \right) = 0,$$
$$\frac{1}{r^2} \frac{\partial}{\partial r} \left(r^2 T_{(r)(\vartheta)} \right) - \frac{1}{r} \frac{\partial \mathrm{p}}{\partial \vartheta} = 0, \tag{5.175}$$
$$\frac{1}{r^2} \frac{\partial}{\partial r} \left(r T_{(r)(z)} \right) - \frac{\partial \mathrm{p}}{\partial z} = 0,$$

where

$$T_{(r)(r)} = T^{rr}, \quad T_{(r)(\vartheta)} = r T^{r\vartheta}, \quad T_{(\vartheta)(\vartheta)} = r^2 T^{\vartheta\vartheta}, \quad T_{(r)(z)} = T^{rz}. \tag{5.176}$$

The additional assumption

$$T_{(r)(\vartheta)} = 0, \quad T_{(r)(z)} = 0, \tag{5.177}$$

yields $\mathrm{p} = \mathrm{p}(r)$. It follows

$$T_{(r)(r)} = -\int \frac{T_{(r)(r)} - T_{(\vartheta)(\vartheta)}}{r} dr, \quad T_{(\vartheta)(\vartheta)} = \frac{d}{dr} \left(r T_{(r)(r)} \right). \tag{5.178}$$

Using (2.31) or (2.32) we can write the constitutive equation (5.146) in the following form

$$(T_{(k)(l)}) = \begin{pmatrix} -\mathrm{p} + \Im_1 + \Im_{-1} & 0 & 0 \\ 0 & -\mathrm{p} + \Im_1(1 + D^2 r^2) + \Im_{-1} & Dr(\Im_1 - \Im_{-1}) \\ 0 & Dr(\Im_1 - \Im_{-1}) & -\mathrm{p} + \Im_1 + \Im_{-1}(1 + D^2 r^2) \end{pmatrix}, \tag{5.179}$$

with the material coefficients

$$\Im_1 = \Im_1(D^2 r^2), \qquad \Im_{-1} = \Im_{-1}(D^2 r^2), \qquad I = II = (D^2 r^2). \tag{5.180}$$

The comparison of (5.179) and (5.161) shows that simple shear and the torsion of a cylinder locally are equivalent. Measurements of the functions $\Im_1$ and $\Im_{-1}$ under shear – for instance, by measurement of the shear stresses and the normal stress T_{11} as functions of $\kappa = \tan\varphi$ – determine Relations (5.179) completely.

One could also proceed vice versa and make measurements at the cylinder. In this case, it is favorable to use the following two notions.

The pair of shear forces

$$S := \int r T_{(\vartheta)(z)} 2\pi r \, dr, \tag{5.181}$$

describes the external load which is necessary for the twist of the cylinder. We obtain

$$S = 2\pi D \int r^3 \mu(D^2 r^2) dr, \quad \mu(D^2 r^2) = \Im_1(D^2 r^2) - \Im_{-1}(D^2 r^2). \tag{5.182}$$

The normal force

$$N := 2\pi \int T_{(z)(z)} r \, dr, \tag{5.183}$$

reflects the Poynting effect for torsion. We have

$$N = -\pi D^2 \int r^3 (\Im_1 - 2\Im_{-1}) dr \tag{5.184}$$

It is easy to show that $N < 0$. This means that the torsion of an incompressible cylinder with free surfaces produces an expansion proportional to D^2.

5.5   Viscoelastic solids

Materials which we were already discussing in this chapter belong to the broad class of simple materials in which the set constitutive variables consists of the deformation gradient $\mathbf{F}(t)$, temperature $T(t)$ and the gradient of temperature $\operatorname{Grad} T(t)$. Formally, these are materials whose response to all homogeneous thermokinetic processes (i.e. processes defined by the fields which we listed above) determines their response in all processes. This class of materials was introduced in 1968 in a memoir of M. E. Gurtin [148]. In this section we present only a few properties of such materials which yield the notion of the linear viscoelastic fluid and the linear viscoelastic solid. Further in this book we return to some nonlinear viscoelastic fluids – we consider some stability properties of these materials in the next chapter, as well as to linear viscoelastic solids – we present properties of some waves in these materials in Chapter 10.

As already indicated in Section 5.3.3 we expect the stress tensor to depend on the history of strain (i.e., the material possesses a memory). For small deformations the strain is described by the Almansi-Hamel deformation tensor $\mathbf{e} = e_{ij} \mathbf{e}_i \otimes \mathbf{e}_j$. We can formally postulate the following relation

$$T_{ij}(t) = \overset{\infty}{\underset{s=0}{\mathcal{S}_{ij}}} \left(e_{kl}(t-s), e_{kl}(t) \right), \tag{5.185}$$

where $\overset{\infty}{\underset{s=0}{\mathcal{S}_{ij}}}$ denotes a linear tensor valued functional mapping the strain history $e_{ij}(t)$, $-\infty \le t \le \infty$, into the stress history $T_{ij}(t)$. In addition, the functional possesses a parametric dependence upon the current value of the strain $e_{ij}(t)$ which describes the instantaneous elastic response mentioned in the above presented properties of viscoelastic materials. We do not include a dependence on the spatial variable $\mathbf{x}$ as the material is assumed to be homogeneous. The above functional has an integral representation for continuous histories of strain. It follows from the Riesz Representation Theorem [106]. Namely, it has the form of the Stieltjes integral

$$T_{ij} = \int\limits_0^\infty e_{kl}(t-s)\,dG_{ijkl}(s)\,, \tag{5.186}$$

where each component of the fourth order tensor G_{ijkl} is of bounded variation. The components of this tensor are called relaxation functions. The above convolution of the constitutive law implies that it is invariant with respect to arbitrary shifts in the time scale. This invariance is related to the energy conservation but we shall not discuss it any further.

Integral constitutive relations of this type are called Boltzmann integrals (Boltzmann superposition principle).

The tensor G_{ijkl} possesses obvious symmetries following from the symmetry of the stress and strain tensors

$$G_{ijkl}(t) = G_{jikl}(t) = G_{ijlk}(t)\,. \tag{5.187}$$

Assuming additionally the continuity of the first derivative of the tensor G_{ijkl} and $e_{ij}(t) = 0$ for $t < 0$ we can write Relation (5.186) in the form

$$T_{ij} = G_{ijkl}(0)\,e_{kl}(t) + \int\limits_0^t e_{kl}(t-s)\,\frac{dG_{ijkl}(s)}{ds}\,ds. \tag{5.188}$$

This form exposes the instantaneous elastic reaction of the material. Bearing the continuity of $e_{ij}(t)$ in mind, we can integrate (5.188) by parts. It follows

$$T_{ij} = \int\limits_0^t G_{ijkl}(t-s)\,\frac{de_{kl}(s)}{ds}\,ds. \tag{5.189}$$

Under the weaker assumption of a step discontinuity at $t = 0$ one can generalize the above relation [149] and obtain the following one

$$T_{ij}(t) = G_{ijkl}(t)\,e_{kl}(0) + \int\limits_0^t G_{ijkl}(t-s)\,\frac{de_{kl}(s)}{ds}\,ds. \tag{5.190}$$

In spite of the discontinuity at $t = 0$ the lower limit can be shifted from 0 to $-\infty$ due to the above mentioned shift invariance provided $e_{ij}(t \to -\infty) \to 0$. Integration by parts yields then

$$T_{ij}(t) = \int\limits_{-\infty}^t G_{ijkl}(t-s)\,\frac{de_{kl}(s)}{ds}\,ds. \tag{5.191}$$

An alternative to the above constitutive relation is the inverse

$$e_{ij}(t) = \int\limits_{-\infty}^{t} J_{ijkl}(t-s)\frac{dT_{kl}(s)}{ds}ds, \tag{5.192}$$

where

$$J_{ijkl}(t) = J_{jikl}(t) = J_{ijlk}(t), \quad J_{ijkl}(t) = 0 \quad \text{for} \quad -\infty < t < 0. \tag{5.193}$$

These functions are assumed to possess continuous first derivatives and they are called creep functions.

We limit all further considerations to isotropic materials. The most general isotropic representation of the fourth order tensor contains two independent parameters (compare the Hooke law for linear elastic isotropic materials). It is convenient to write it in the form

$$G_{ijkl}(t) = \frac{1}{3}\left[G_2(t) - G_1(t)\right]\delta_{ij}\delta_{kl} + \frac{1}{2}\left[G_1(t)\right]\left(\delta_{ik}\delta_{jl} + \delta_{il}\delta_{jk}\right), \tag{5.194}$$

where $G_1(t)$ and $G_2(t)$ are independent relaxation functions. If we separate spherical and deviatoric parts of the stress and strain tensors

$$T_{ij} = \frac{1}{3}T_{kk}\delta_{ij} + T_{ij}^{D}, \quad T_{kk}^{D} = 0, \tag{5.195}$$

$$e_{ij} = \frac{1}{3}e_{kk}\delta_{ij} + e_{ij}^{D}, \quad e_{kk}^{D} = 0,$$

then Relation (5.191) splits in the following way

$$T_{ij}^{D}(t) = \int\limits_{-\infty}^{t} G_1(t-s)\frac{de_{ij}^{D}(s)}{ds}ds, \tag{5.196}$$

$$T_{kk} = \int\limits_{-\infty}^{t} G_2(t-s)\frac{de_{kk}(s)}{ds}ds.$$

Similarly, the inverse relations obtain the form

$$e_{ij}^{D}(t) = \int\limits_{-\infty}^{t} J_1(t-s)\frac{dT_{ij}^{D}(s)}{ds}ds, \tag{5.197}$$

$$e_{kk} = \int\limits_{-\infty}^{t} J_2(t-s)\frac{dT_{kk}(s)}{ds}ds,$$

where $J_1(t)$ and $J_2(t)$ are two independent isotropic creep functions. Obviously, the functions G_1, J_1 are appropriate for shear processes and G_2, J_2 for dilatation processes.

The relaxation and creep functions are, of course, related to each other. The easiest way to find this relation is to perform the Laplace transformation on Relations (5.196) and (5.197).[14] We obtain

$$\bar{T}_{ij}^{D} = z\bar{G}_1 \bar{e}_{ij}^{D}, \quad \bar{T}_{kk} = z\bar{G}_2 \bar{e}_{kk},$$
$$\bar{e}_{ij}^{D} = z\bar{J}_1 \bar{T}_{ij}^{D}, \quad \bar{e}_{kk} = z\bar{J}_2 \bar{T}_{kk}. \tag{5.198}$$

These relations imply

$$J_\alpha = \left(z^2 G_\alpha\right)^{-1}, \quad \alpha = 1, 2. \tag{5.199}$$

The isotropic relations for linear elastic materials can be written in the form

$$T_{kk} = 3K e_{kk}, \quad T_{ij}^{D} = 2\mu e_{ij}^{D}, \tag{5.200}$$
$$e_{kk} = \frac{1}{3K} T_{kk}, \quad e_{ij}^{D} = \frac{1}{2\mu} T_{ij}^{D}.$$

They would suggest that $J_\alpha(t) = [G_\alpha(t)]^{-1}$. Relations (5.199) show that this is not correct. However, it can be shown using the properties of the Laplace transform that the limit values indeed satisfy such relations

$$\lim_{t\to 0} J_\alpha(t) = \lim_{t\to 0} [G_\alpha(t)]^{-1}, \quad \lim_{t\to\infty} J_\alpha(t) = \lim_{t\to\infty} [G_\alpha(t)]^{-1}. \tag{5.201}$$

Incidentally, to be consistent with the linear elasticity of isotropic materials Relations (5.200) imply the following notation for the relaxation and creep functions

$$\mu(t) = G_1(t)/2, \quad K(t) = G_2(t)/3. \tag{5.202}$$

The above presented results suggest a useful short-hand notation for Stieltjes convolution integrals which has been introduced by Gurtin and Sternberg in the earlier quoted paper. Namely, we write instead of (5.186) the following relation

$$T_{ij} = e_{ij} * dG_{ijkl}, \tag{5.203}$$

i.e., we write for two arbitrary functions f and g

$$f * dg = \int_{-\infty}^{t} f(t-s)\, dg(s), \quad g(t \to -\infty) = 0, \tag{5.204}$$

where $f(t)$ is continuous for $0 \le t \le \infty$. If $f(t) = 0$ for $t < 0$ then one can show the commutativity relation

$$f * dg = g * df. \tag{5.205}$$

Consequently,

$$T_{ij} = G_{ijkl} * de_{kl},$$

---

[14]It should be mentioned that, instead of the classical Laplace transform (compare Appendix B) it may be more convenient to use a modification which is called Laplace-Carson transform. It is defined by the relation

$$\bar{f}(z) = z \int_0^{\infty} f(t) \exp(-zt)\, dt.$$

It is applied in the presentation of viscoelasticity by Lemaitre and Chaboche [224]. We follow here rather the older approach of Christensen [75] and Pipkin [294].

which is the counterpart of (5.191).

The above notation leads to the following useful identities

$$f * d(g * dh) = (f * dg) * dh = f * dg * dh, \qquad (5.206)$$
$$f * d(g + h) = f * dg + f * dh.$$

We shall not enhance the subject of a general theory of materials with memory. This can be found in classical monographs on the subject [393], [390].

We proceed to investigate an example which helps to clear the distinction between the viscoelastic fluid and the viscoelastic solid. We consider the case of the simple shear defined in Section 2.3.2. The amount of shear $\kappa = \kappa_0 H(t)$, $\kappa_0 = \tan\varphi$ is assumed to be a step function in time and it yields the stretching in the linear theory to be identical with the time derivative of strain which is given by the rate of shearing $\dot{\kappa} = \kappa_0 \delta(t)$

$$D_{ij}\mathbf{e}_i \otimes \mathbf{e}_j = \frac{de_{ij}}{dt}\mathbf{e}_i \otimes \mathbf{e}_j = \frac{\dot{\kappa}}{2}(\mathbf{e}_1 \otimes \mathbf{e}_2 + \mathbf{e}_2 \otimes \mathbf{e}_1). \qquad (5.207)$$

Hence,

$$T_{12}^D = \frac{\kappa_0}{2}\int_0^t G_1(t-s)\,\delta(s)\,ds = \frac{\kappa_0}{2}G_1(t), \quad G_1(t) = 0 \text{ for } t < 0. \qquad (5.208)$$

It follows from the definition of a linear isotropic viscoelastic solid that the following condition must hold

$$\lim_{t\to\infty} G_1(t) \to \text{nonzero constant} \quad \Rightarrow \quad \text{solids.} \qquad (5.209)$$

On the other hand, for the viscoelastic (non-Newtonian) fluid

$$\lim_{t\to\infty} G_1(t) = 0 \quad \Rightarrow \quad \text{fluids.} \qquad (5.210)$$

The latter condition is necessary but not sufficient. In addition, the relaxation function G_1 must fulfill a condition for the steady state flow. Then the viscoelastic fluid at large values of time, where the steady state will be achieved, must behave like a viscous fluid of the viscosity η. Hence, the relaxation time must have the property

$$\eta = \frac{1}{2}\int_0^\infty G_1(s)\,ds \quad \text{(fluids).} \qquad (5.211)$$

5.5.1 Differential constitutive relations

Examples of simple rheological models indicate that the constitutive relations of these models may have the form of evolution equations (rate-type constitutive relations). This can be taken over to the description of a three-dimensional continuum. Let us consider the following differential operator

$$p_0 T_{ij}^D + p_1 \frac{dT_{ij}^D}{dt} + p_2 \frac{d^2 T_{ij}^D}{dt^2} + ... = q_0 e_{ij}^D + q_1 \frac{de_{ij}^D}{dt} + q_2 \frac{d^2 e_{ij}^D}{dt^2} + ..., \qquad (5.212)$$

or, in compact form,

$$P\left(D\right)T_{ij}^{D} = Q\left(D\right)e_{ij}^{D}, \quad P\left(D\right) = \sum_{k=0}^{N} p_k D^k, \quad Q\left(D\right) = \sum_{k=0}^{N} q_k D^k, \quad D^k = \frac{d^k}{dt^k}. \quad (5.213)$$

This type of operators appear, for instance, for rheological generalized Maxwell and Kelvin models. In order to see the significance of such models for viscoelasticity, we take the Laplace transform of (5.213)

$$\bar{P}\left(z\right)\bar{T}_{ij}^{D} - \frac{1}{z}\sum_{k=1}^{N} p_k \sum_{r=1}^{N} z^r \left[\frac{d^{k-r}T_{ij}^{D}}{dt^{k-r}}\left(t=0\right)\right]$$

$$= \bar{Q}\left(z\right)\bar{e}_{ij}^{D} - \frac{1}{z}\sum_{k=1}^{N} q_k \sum_{r=1}^{N} z^r \left[\frac{d^{k-r}e_{ij}^{D}}{dt^{k-r}}\left(t=0\right)\right], \quad (5.214)$$

where

$$\bar{P}\left(z\right) = \sum_{k=0}^{N} p_k z^k, \quad \bar{Q}\left(z\right) = \sum_{k=0}^{N} q_k z^k. \quad (5.215)$$

These relations follow easily by integration by parts. If we compare these relations with (5.198) then they specify the relaxation function by the formula

$$z\bar{G}_1 = \bar{Q}\left(z\right)/\bar{P}\left(z\right), \quad (5.216)$$

provided the initial conditions are constrained by the relations

$$\sum_{r=k}^{N} p_r \left[\frac{d^{k-r}T_{ij}^{D}}{dt^{k-r}}\left(t=0\right)\right] = \sum_{r=k}^{N} q_r \left[\frac{d^{k-r}e_{ij}^{D}}{dt^{k-r}}\left(t=0\right)\right], \quad k = 1, 2, ..., N. \quad (5.217)$$

Consequently, the relaxation function G_1 is specified in terms of $2\left(N+1\right)$ parameters $p_0, ..., p_N, q_0, ..., q_N$ which are related to a sequence of relaxation times. Similar relations can be introduced for the dilatational part of stress and strain

$$L\left(D\right)T_{kk}\left(t\right) = M\left(D\right)e_{kk}\left(t\right), \quad (5.218)$$

and these are again specified by a finite sequence of parameters.

For viscoelastic solids simple differential models are often based on the simplest differential equation describing the evolution of stresses. This is the generalization of the Maxwell model (row connection of a spring and a dashpot) constructed within the classical rheology. It is based on the equation for stresses

$$\tau\frac{dT_{ij}^{D}}{dt} + T_{ij}^{D} = 2\eta\frac{de_{ij}^{D}}{dt}, \quad (5.219)$$

where τ is the relaxation time and η is the viscosity. It is the so-called standard linear viscoelastic solid [437].

5.5.2 Steady state processes and elastic-viscoelastic correspondence principle

Now we investigate a class of problems appearing in the spectral analysis of waves in which we seek solutions in the form of monochromatic waves. This subject shall be discussed in Chapter 10. It is then important to know the behavior of the constitutive relation (5.196) if the time dependence of the strain is harmonic. We denote representatives of deviatoric and spherical strains and stresses by $\tilde{e}$ and $\tilde{\sigma}$, respectively, and assume

$$\tilde{e} = \tilde{e}_0 e^{i\omega t}, \tag{5.220}$$

where ω is the frequency and $\tilde{e}_0$ an amplitude. We write the typical contribution to (5.196) in the following form

$$\tilde{\sigma} = \int_{-\infty}^{t} G_\alpha (t - s) \frac{d\tilde{e}}{ds} ds, \tag{5.221}$$

where $\alpha = 1$ or 2 in dependence on the choice of $\tilde{\sigma}$. It is convenient to split the relaxation modulus into two parts: G_α^0 and G_α^1, where the first part is equal to the limit of the relaxation modulus for $t \to \infty$ and, consequently, $G_\alpha^1 (t \to \infty) = 0$. Both parts are zero for the time smaller than 0. Hence,

$$\tilde{\sigma} = G_\alpha^0 \tilde{e}_0 e^{i\omega t} + i\omega e_0 \int_{-\infty}^{t} G_\alpha^1 (t - s) e^{i\omega s} ds. \tag{5.222}$$

Changing the variables $\eta = t - s$ we obtain

$$\tilde{\sigma} = \left[G_\alpha^0 + \omega \int_0^\infty G_\alpha^1 (\eta) \sin \omega\eta d\eta + i\omega \int_0^\infty G_\alpha^1 (\eta) \cos \omega\eta d\eta \right] \tilde{e}_0 e^{i\omega t}. \tag{5.223}$$

Consequently, the stress $\tilde{\sigma}$ is given by the complex modulus G_α^*

$$\tilde{\sigma} = G_\alpha^* \tilde{e}_0 e^{i\omega t},$$

$$\operatorname{Re} G_\alpha^* = G_\alpha^0 + \omega \int_0^\infty G_\alpha^1 (\eta) \sin \omega\eta d\eta, \tag{5.224}$$

$$\operatorname{Im} G_\alpha^* = \omega \int_0^\infty G_\alpha^1 (\eta) \cos \omega\eta d\eta.$$

The real part is called the storage modulus and the imaginary part the loss modulus.

It is obvious that the real and imaginary components of the complex modulus are not independent. Their relation follows easily from (5.224)

$$\operatorname{Re} G_\alpha^* = G_\alpha^0 + \frac{2}{\pi} \int_0^\infty \frac{\omega^2 \operatorname{Im} G_\alpha^*}{\eta (\omega^2 - \eta^2)} d\eta. \tag{5.225}$$

This equation is called the Kramers-Kronig relation. It states that viscoelastic materials are inherently dispersive, i.e., the speed of propagation of the mechanical disturbance is frequency dependent. It can be shown as well that this relation is necessary and sufficient for the modulus to satisfy the principle of causality which states that in a physical system the reaction can never precede the action.

The integration by parts in (5.224) yields the frequency limit behavior of the above moduli. We have

$$\operatorname{Re} G_\alpha^* \left(\omega \right) = G_\alpha^0 + G_\alpha^1 \left(0 \right) + \int_0^\infty \frac{dG_\alpha^1 \left(\eta \right)}{d\eta} \cos \omega \eta d\eta, \tag{5.226}$$

$$\operatorname{Im} G_\alpha^* \left(\omega \right) = - \int_0^\infty \frac{dG_\alpha^1 \left(\eta \right)}{d\eta} \sin \omega \eta d\eta.$$

Hence,

$$\operatorname{Re} G_\alpha^* \left(\omega = 0 \right) = G_\alpha^0 = G_\alpha \left(t \right)|_{t \to \infty}, \quad \operatorname{Im} G_\alpha^* \left(\omega = 0 \right) = 0. \tag{5.227}$$

For the other limit, changing the variables $\omega \eta = \tau$ we easily obtain

$$\operatorname{Re} G_\alpha^* \left(\omega \to \infty \right) = G_\alpha^0 + G_\alpha^1 \left(t = 0 \right) = G_\alpha \left(t \right)|_{t \to 0}, \quad \operatorname{Im} G_\alpha^* \left(\omega \to \infty \right) = 0. \tag{5.228}$$

These relations show that for very high frequency the viscoelastic solid behaves as an elastic solid. The same concerns very low frequencies and this differs solids from viscous fluids.

The above considerations determine, obviously, the Fourier transforms of the constitutive relations. With the definitions for an arbitrary function f

$$\bar{f} \left(\omega \right) = \int_{-\infty}^\infty f \left(t \right) e^{-i\omega t} dt, \quad f \left(t \right) = \frac{1}{2\pi} \int_{-\infty}^\infty \bar{f} \left(\omega \right) e^{i\omega t} d\omega, \tag{5.229}$$

we have

$$\bar{T}_{ij}^D \left(\omega \right) = G_1^* \left(i\omega \right) \bar{e}_{ij}^D \left(\omega \right), \tag{5.230}$$

$$\bar{T}_{kk} = G_2^* \left(i\omega \right) \bar{e}_{kk} \left(\omega \right).$$

The similarity of these relations to the elastic relations (5.200) is called the elastic-viscoelastic correspondence principle. It has been first observed by W. T. Read in 1950 [305]. With respect to the important practical aspects of this principle we present it as well for the Laplace transform.

Let us consider the full set of governing equations describing the boundary value problem of a linear isotropic viscoelastic material. We have

$$e_{ij} = \frac{1}{2} \left(\frac{\partial u_i}{\partial x_j} + \frac{\partial u_j}{\partial x_i} \right), \tag{5.231}$$

$$\frac{\partial T_{ij}}{\partial x_j} + \rho b_i = 0, \quad \text{for } \mathbf{x} \in \mathcal{B}_t, \tag{5.232}$$

$$T_{ij}^D \left(t \right) = 2 \int_{-\infty}^t \mu \left(t - s \right) \frac{\partial e_{ij}^D}{\partial s} ds, \quad T_{kk} = 3 \int_{-\infty}^t K \left(t - s \right) \frac{\partial e_{kk}}{\partial s} ds, \tag{5.233}$$

$$T_{ij} = \frac{1}{3}T_{kk}\delta_{ij} + T_{ij}^D, \quad e_{ij} = \frac{1}{3}e_{kk}\delta_{ij} + e_{ij}^D, \tag{5.234}$$

$$T_{ij}(t)\,n_j = t_i^n \quad \text{for } \mathbf{x} \in \partial\mathcal{B}_t^\sigma, \quad u_i(t) = u_i^n \quad \text{for } \mathbf{x} \in \partial\mathcal{B}_t^u, \tag{5.235}$$

$$u_i(t) = e_{ij}(t) = T_{ij}(t) = 0 \quad \text{for } -\infty < t < 0, \tag{5.236}$$

where we have used the notation (5.202) for the relaxation moduli.

The Laplace transform of these equations has the form

$$\bar{e}_{ij} = \frac{1}{2}\left(\frac{\partial \bar{u}_i}{\partial x_j} + \frac{\partial \bar{u}_j}{\partial x_i}\right), \tag{5.237}$$

$$\frac{\partial \bar{T}_{ij}}{\partial x_j} + \rho \bar{b}_i = 0, \quad \text{for } \mathbf{x} \in \mathcal{B}_t, \tag{5.238}$$

$$\bar{T}_{ij}^D(t) = 2z\bar{\mu}(z)\,\bar{e}_{ij}^D \quad \bar{T}_{kk} = 3z\bar{K}(z)\,\bar{e}_{kk}, \tag{5.239}$$

$$\bar{T}_{ij} = \frac{1}{3}\bar{T}_{kk}\delta_{ij} + \bar{T}_{ij}^D, \quad \bar{e}_{ij} = \frac{1}{3}\bar{e}_{kk}\delta_{ij} + \bar{e}_{ij}^D, \tag{5.240}$$

$$\bar{T}_{ij}(z)\,n_j = \bar{t}_i^n \quad \text{for } \mathbf{x} \in \partial\mathcal{B}_t^\sigma, \quad \bar{u}_i(z) = \bar{u}_i^n \quad \text{for } \mathbf{x} \in \partial\mathcal{B}_t^u, \tag{5.241}$$

where z is the transformation variable and bars denote Laplace transforms.

Obviously the set (5.237)-(5.241) has a form identical with the equations of linear elasticity except of complex moduli $z\bar{K}(z)$ and $z\bar{\mu}(z)$ which replace real moduli K and μ of the elasticity theory. This correspondence reveals the possibility of converting numerous static solutions of elasticity into quasi-static solutions of viscoelasticity. The main problem is now the inversion of the Laplace transform

Let us mention that the above considerations can be easily extended on anisotropic materials.

One more general remark is appropriate for quasi-static problems of viscoelastic materials. Before we formulate it, let us collect the material functions corresponding to the material constants of elasticity. We already have Relations (5.202), i.e., $\mu(t) = G_1(t)/2$, $K(t) = G_2(t)/3$ and they yield in the transformed form

$$\bar{\lambda}(z) = \bar{K}(z) - \frac{2}{3}\bar{\mu}(z) = \frac{1}{3}\left(\bar{G}_2(z) - \bar{G}_1(z)\right),$$

$$\bar{E}(z) = \frac{3\bar{\mu}(z)\,\bar{K}(z)}{\bar{K}(z) + \frac{1}{3}\bar{\mu}(z)} = \frac{3\bar{G}_1(z)\,\bar{G}_2(z)}{2\bar{G}_2(z) + \bar{G}_1(z)}, \tag{5.242}$$

$$\bar{\nu}(z) = \frac{\bar{\lambda}(z)}{2\left(\bar{\lambda}(z) + \bar{\mu}(z)\right)} = \frac{\bar{G}_2(z) - \bar{G}_1(z)}{2\bar{G}_2(z) + \bar{G}_1(z)}.$$

The question arises whether we can apply the method of separation of variables in quasi-static problems of linear viscoelasticity. It means that, for instance, the displacement should have the form

$$u_i(\mathbf{x}, t) = \breve{u}_i(\mathbf{x})\,u(t), \tag{5.243}$$

where $u(t)$ is a common function for all components of the displacement. Hence, if we neglect the acceleration (quasi-static problem!) and body forces, the field equations for the displacements have the form

$$\frac{\partial^2 \breve{u}_i}{\partial x_k \partial x_k}\int_{-\infty}^{t} \mu(t-s)\frac{du(s)}{ds}ds + \frac{\partial^2 \breve{u}_k}{\partial x_i \partial x_k}\int_{-\infty}^{t}\left[\lambda(t-s) + \mu(t-s)\right]\frac{du(s)}{ds}ds = 0. \tag{5.244}$$

Obviously, the time contributions must be eliminated from this equation and this yields

$$\lambda(t) + \mu(t) = \beta\mu(t), \quad \beta = \text{const.} \tag{5.245}$$

This yields immediately that Poisson's ratio ν must be independent of time. Bearing the last Relation (5.242) in mind, we obtain the restriction

$$\frac{G_2(t)}{G_1(t)} = \frac{1+\nu}{1-2\nu} = \text{const.} \tag{5.246}$$

In a similar manner we can prove that the ratio of creep functions is a constant

$$\frac{J_2(t)}{J_1(t)} = \frac{1-2\nu}{1+\nu} = \text{const.} \tag{5.247}$$

These two conditions are necessary for the applicability of the method of separation of variables.

We demonstrate the application of the correspondence principle on a simple example. We consider the axial symmetric problem of a cylinder under the given radial loading on both lateral surfaces. The outer surface is pressurized by an elastic case [76]. The Laplace transform of the radial displacement $\bar{u}_r(r,z)$ must fulfill the equilibrium equation

$$\frac{\partial^2 \bar{u}_r}{\partial r^2} + \frac{1}{r}\frac{\partial \bar{u}_r}{\partial r} - \frac{\bar{u}}{r^2} = 0, \tag{5.248}$$

with the solution

$$\bar{u}_r = \bar{C}(z)r + \frac{\bar{D}(z)}{r}. \tag{5.249}$$

In order to apply the boundary conditions, we have to write the stress-strain relations in the transformed form. We have by the correspondence principle

$$\bar{T}_{rr} = 2z\bar{\mu}\left(\frac{\partial \bar{u}_r}{\partial r} + \frac{z\bar{\nu}}{1-2z\bar{\nu}}\bar{e}\right), \quad \bar{e} = \frac{\partial \bar{u}_r}{\partial r} + \frac{\bar{u}_r}{r},$$

$$\bar{T}_{\theta\theta} = 2z\bar{\mu}\left(\frac{\bar{u}_r}{r} + \frac{z\bar{\nu}}{1-2z\bar{\nu}}\bar{e}\right), \tag{5.250}$$

$$\bar{T}_{zz} = \frac{2z^2\bar{\nu}\bar{\mu}}{1-2z\bar{\nu}}.$$

The boundary conditions for the pressurized cylinder have the form

$$T_{rr}(r=a,t) = -p(t), \quad T_{rr}(r=b,t) = -q(t), \tag{5.251}$$

$$u(r=b,t) = q(t)\left[\frac{b^2(1-\nu_c^2)}{E_c h}\right],$$

where E_c, ν_c are elastic properties of the case, h is its thickness, b the outer radius and a the inner radius of the viscoelastic cylinder.

Easy calculations yield the following form of the transformed stresses

$$\bar{T}_{rr} = C_1 - \frac{b^2}{r^2}C_2, \quad \bar{T}_{\theta\theta} = C_1 + \frac{b^2}{r^2}C_2, \quad \bar{T}_{zz} = 2z\bar{\nu}C_1, \tag{5.252}$$

where

$$C_1 = \frac{-\bar{p}\left(S - z\bar{\mu}\right)}{\left(S - z\bar{\mu}\right) + \left(b^2/a^2\right)\left[S\left(1 - 2z\bar{\nu}\right) + z\bar{\mu}\right]},$$

$$C_2 = \frac{\bar{p}\left[S\left(1 - 2z\bar{\nu}\right) + z\bar{\mu}\right]}{\left(S - z\bar{\mu}\right) + \left(b^2/a^2\right)\left[S\left(1 - 2z\bar{\nu}\right) + z\bar{\mu}\right]}, \tag{5.253}$$

$$S = \frac{E_c h}{2h\left(1 - \nu_c^2\right)}.$$

The inverse transformations performed in the quoted work of Christensen and Schreiner were made under the assumption that the modulus $\bar{\mu}$ is a polynomial in z and the Poisson number ν is constant (see above). The polynomial form of $\bar{\mu}$ follows from the following considerations suggested by rheological models. It is assumed that the modulus $\mu\left(t\right)$ has the form

$$\mu\left(t\right) = G_0 + \sum_{n=1}^{N} G_n e^{-t/\tau_n}, \tag{5.254}$$

where $G_0, G_n, \tau_n, n = 1, ..., N$ are constants. Obviously, τ_n have the interpretation of relaxation times. The values of these parameters are obtained by fitting to experimental data (e.g. for harmonic torsional loading experiments of cylindrical samples). Laplace transformation of (5.254) yields

$$z\bar{\mu}\left(z\right) = \frac{A\left(z\right)}{\displaystyle\prod_{n=1}^{N}\left(z + 1/\tau_n\right)}, \tag{5.255}$$

where $A\left(z\right)$ is an Nth grade polynomial in z determined by the coefficients G_n.

Now the inverse transformation can be made by the technique of integration of the function $\bar{f}\left(z\right) = P\left(z\right)/Q\left(z\right)$ whose denominator yields simple pole singularities (i.e., zeros of $Q\left(z\right)$) in the complex domain. We shall not quote the rather complex final results.

Chapter 6

Stability

6.1 Preliminaries

The main concern of the stability analysis of thermomechanical systems is with applications and particular engineering problems. However, the general theory has not only this practical bearing. It yields as well an important information on the fundamental properties of thermodynamical models. In particular, it prescribes the ranges of material parameters in which such models are physically acceptable and relevant. In this chapter we indicate some general features of the stability analysis of continua but the majority of space is devoted to a few characteristic examples. We show stability properties of some flows of fluids, large static deformations of solids and thermodynamic equilibrium states of some continuous systems. We return to the stability analysis also in further chapters of this book when discussing special systems. This concerns in particular the stability analysis of ponderable systems interacting with electromagnetic fields (Chapter 11) and of a steady state 1D flow in a porous medium which is superposed by a small disturbance with adsorption (Section 13.3).

There exists an extensive literature on the subject. Fundamental definitions formulated in the language of topological properties of state spaces of continua are extensively discussed by R. J. Knops and E. W. Wilkes [201]. Even though the main part of the book concerns elastic solids general notions are common for all continuous systems. Knops and Wilkes emphasize particularly the definition introduced by Lyapunov as well as the method of stability analysis based on the so-called Lyapunov function. Among other general properties they indicate a non-uniqueness of the definition of stability. They show, for instance, that the same process may be stable in one norm of the space (of course, the norm yields a corresponding definition of the distance of states in the space of states) such as the supremum norm but it may not be stable in another norm, for instance the energy norm. For the purpose of this book we do not need any details of these considerations. The notion of stability which we demonstrate further in this chapter is based on the assumption that, in some sense, a state under investigation, in most cases – the state of equilibrium, will be spontaneously recovered after its small disturbance. In particular, we ask whether the disturbance of the system gradually dies down or whether it will grow in the amplitude in such a way that the system progressively departs from the initial state. In the former case we say that the system is stable with respect to the particular disturbance. In the latter case we say that it is unstable.

Aleksandr M. Lyapunov 1857-1918

Subrahmanyan Chandrasekhar 1910-1995

Daniel D. Joseph 1929-2011

Obviously, the system is considered to be unstable if there exists one special mode of disturbance yielding instability. On the other hand, the system is stable if it is stable with respect to every possible disturbance. Hence, the stability must imply that there exists no mode of disturbance for which it is unstable.

It should be stressed that we consider in this chapter only the class of small disturbances. In many cases of practical bearing a system may be unstable with respect to small disturbances but we would still consider it as stable with respect to disturbances of large amplitude. A typical example is furnished by materials with martensitic phase transformations. In these materials the transition between two phases – say martensite to austenite – is connected with an instability with respect to a small disturbance of temperature or deformation. However, when the process of transformation is completed which is related to large deformations of the system, the system behaves again in the stable manner. Moreover, the unstable behavior of such systems is strongly dependent on the boundary conditions. For instance, a purely mechanical loading of such a material may yield stable behavior in the so-called hard loading devices (external control of deformation) and unstable behavior in soft loading devices (external control of loading). We skip these problems here and return in Section 7.4 to a special problem concerning martensitic-austenitic phase transformations: the bending of a composite beam with embedded shape memory alloy.

The property of a state being stable or unstable depends on a set of parameters characterizing the system. These are material properties such as the elastic properties or the viscosity or external, for instance, boundary conditions. The space of these parameters can be divided into the part in which the states are stable and another part in which the states are unstable. The locus of parameters which separates these two parts corresponds to the states which are called marginally stable. They are also called the states of neutral stability.

The analysis of stability with respect to small disturbances proceeds in the following way. In the general field equations we assume the solution to consist of two parts. The first nonlinear part satisfies the general field equations simplified to the description of, say, stationary motion or an equilibrium state. The second linear part is described by the linearized general field equations. The solution of this linear set is sought in the form of the so-called normal mode expansion which must be complete. These normal modes yield equations for the evolution of the amplitude of the disturbance. If one of them grows in time the nonlinear solution under consideration is unstable. We demonstrate this procedure on a few examples.

6.2   Stability of the torsional Couette flow

As a first example we investigate the stability of the torsional Couette flow of an ideal fluid. The full analysis of this problem, as well as numerous other problems of stability of fluid motion, can be found in the classical monograph of S. Chandrasekhar [71]. A modern version of the analysis is contained in the book of D. D. Joseph [186]. The velocity field consists of the components $v_{(1)} = v_r$, $v_{(2)} = v_\vartheta = r\omega$, $v_{(3)} = v_z$ and the momentum balance equations (5.106) can be written in the form

$$\frac{\partial v_r}{\partial t} + (\mathbf{v} \cdot \mathrm{grad})\, v_r - \frac{v_\vartheta^2}{r} = -\frac{\partial}{\partial r}\left(\frac{p}{\rho}\right),$$

$$\frac{\partial v_\vartheta}{\partial t} + (\mathbf{v} \cdot \mathrm{grad})\, v_\vartheta + \frac{v_r v_\vartheta}{r} = -\frac{1}{r}\frac{\partial}{\partial \vartheta}\left(\frac{p}{\rho}\right), \tag{6.1}$$

$$\frac{\partial v_z}{\partial t} + (\mathbf{v} \cdot \mathrm{grad})\, v_z = -\frac{\partial}{\partial z}\left(\frac{p}{\rho}\right),$$

where

$$(\mathbf{v} \cdot \mathrm{grad}) = v_r\frac{\partial}{\partial r} + \frac{v_\vartheta}{r}\frac{\partial}{\partial \vartheta} + v_z\frac{\partial}{\partial z}, \tag{6.2}$$

and additionally the following incompressibility condition must be satisfied

$$\frac{\partial v_r}{\partial r} + \frac{v_r}{r} + \frac{1}{r}\frac{\partial v_\vartheta}{\partial \vartheta} + \frac{\partial v_z}{\partial z} - 0. \tag{6.3}$$

We investigate the stability of the following stationary solution of this set of equations

$$v_r = 0, \quad v_\vartheta = r\omega\,(r), \quad v_z = 0 \quad \Rightarrow \quad p = \rho\int r\omega^2 dr. \tag{6.4}$$

This corresponds to the problem of the velocity and pressure distribution in the vessel of two concentric cylinders rotating with constant angular velocities. The fluid is assumed to adhere to the cylinders. We consider an infinitesimal perturbation of this solution which we denote by a prime, i.e., the perturbed fields should have the form

$$v_r', \quad r\omega + v_\vartheta', \quad v_z', \quad p + p'. \tag{6.5}$$

They satisfy the set of linear equations

$$\frac{\partial v_r'}{\partial t} + \omega\frac{\partial v_r'}{\partial \vartheta} - 2\omega v_\vartheta' = -\frac{\partial}{\partial r}\left(\frac{p'}{\rho}\right),$$

$$\frac{\partial v_\vartheta'}{\partial t} + \omega\frac{\partial v_\vartheta'}{\partial \vartheta} + \left(2\omega + r\frac{d\omega}{dr}\right) v_r' = -\frac{1}{r}\frac{\partial}{\partial \vartheta}\left(\frac{p'}{\rho}\right),$$

$$\frac{\partial v_z'}{\partial t} + \omega\frac{\partial v_z'}{\partial \vartheta} = -\frac{\partial}{\partial z}\left(\frac{p}{\rho}\right), \tag{6.6}$$

$$\frac{\partial u_r'}{\partial r} + \frac{v_r'}{r} + \frac{1}{r}\frac{\partial v_\vartheta'}{\partial \vartheta} + \frac{\partial v_z'}{\partial z} = 0.$$

The analysis of these disturbances is performed in terms of normal modes which means that they are products of some function of the radius r, as the stationary solution and the function of the form

$$\exp\left(i\left(\lambda t + m\vartheta + kz\right)\right),\tag{6.7}$$

where λ is constant, maybe complex, m is integer (positive, zero or negative) and k is the wave number in the z-direction. Hence, the solution of the set (6.6) is assumed to have the form

$$v_r' = V_r\left(r\right)e^{i\left(\lambda t + m\vartheta + kz\right)}, \quad v_\vartheta' = V_\vartheta\left(r\right)e^{i\left(\lambda t + m\vartheta + kz\right)},$$

$$v_z' = V_z\left(r\right)e^{i\left(\lambda t + m\vartheta + kz\right)}, \quad p' = P\left(r\right)e^{i\left(\lambda t + m\vartheta + kz\right)}.\tag{6.8}$$

Substitution in Equations (6.6) yields

$$i\sigma_m V_r - 2\omega V_\vartheta = -\frac{d}{dr}\left(\frac{P}{\rho}\right)$$

$$i\sigma_m V_\vartheta + \left(2\omega + r\frac{d\omega}{dr}\right)V_r = -\frac{im}{r}\frac{P}{\rho}$$

$$i\sigma_m V_z = -ik\frac{P}{\rho},$$

$$\frac{dV_r}{dr} + \frac{V_r}{r} + \frac{im}{r}V_\vartheta + ikV_z = 0,\tag{6.9}$$

where

$$\sigma_m = \lambda + m\omega.\tag{6.10}$$

The analysis of this system can be simplified by the change of variables. If we introduce the so-called Lagrangian displacement vector ξ[1]

$$\xi_r = U_r\left(r\right)e^{i\left(\lambda t + m\vartheta + kz\right)}, \quad \xi_\vartheta = U_\vartheta\left(r\right)e^{i\left(\lambda t + m\vartheta + kz\right)}, \quad \xi_z = U_z\left(r\right)e^{i\left(\lambda t + m\vartheta + kz\right)},\tag{6.11}$$

whose time derivatives specify the velocities of the disturbance in the following way

$$i\sigma_m U_r = V_r, \quad i\sigma_m U_\vartheta = V_\vartheta + r\frac{d\omega}{dr}U_r, \quad i\sigma_m U_z = V_z,\tag{6.12}$$

then the set of equations $(6.9)_{1-3}$ has the form

$$\left(\sigma_m^2 - 2r\omega\frac{d\omega}{dr}\right)U_r + 2i\omega\sigma_m U_\vartheta = \frac{d}{dr}\left(\frac{P}{\rho}\right),$$

$$\sigma_m^2 U_\vartheta - 2i\omega\sigma_m U_r = \frac{im}{r}\frac{P}{\rho},\tag{6.13}$$

$$\sigma_m^2 U_z = ik\frac{P}{\rho}.$$

---

[1]This notion should be taken with a grain of salt as the time derivative of the displacement yields indeed the velocity if we assume $m = 0$. Otherwise the relations below define just the change of variables which simplifies the analysis.

Simultaneously, the incompressibility condition yields

$$\frac{dU_r}{dr} + \frac{U_r}{r} + \frac{im}{r}U_\vartheta + ikU_z = 0. \tag{6.14}$$

We can easily eliminate the components U_ϑ, U_z from these equations and finally the following set of two self-adjoint equations results

$$(\sigma_m^2 - \Phi(r))U_r = \frac{d}{dr}\left(\frac{P}{\rho}\right) + \frac{2m\omega}{\sigma_m r}\frac{P}{\rho},$$

$$\frac{1}{r}\frac{d}{dr}(rU_r) - \frac{2m\omega}{\sigma_m r}U_r = \frac{1}{\sigma_m^2}\left(\frac{m^2}{r^2} + k^2\right)\frac{P}{\rho}, \tag{6.15}$$

where

$$\Phi(r) = \frac{2\omega}{r}\frac{d}{dr}\left(r^2\omega\right) \tag{6.16}$$

is the so-called Rayleigh discriminant.

The stability analysis of the system (6.15) will be made for two cases separately, namely for $m = 0$ and for $m \neq 0$.

For $m = 0$ we can easily eliminate the pressure from the set (6.15). Consequently, it follows the equation for the U_r-component of the displacement

$$\frac{d}{dr}\left(\frac{1}{r}\frac{d}{dr}rU_r\right) - k^2 U_r = -\frac{k^2}{\lambda^2}\Phi(r)U_r. \tag{6.17}$$

In the case of two coaxial cylinders of radii R_I and R_E rotating with different angular velocities ω_I and ω_E we have to impose the boundary conditions of vanishing displacement $U_r = 0$ on both cylinders (the adherence!). Then, Equation (6.17) constitutes the Sturm-Liouville eigenvalue problem (e.g. see E. C. Titchmarsch [380], C. C. Lin, L. A. Segel [229], Section 5.2 and L. A. Segel [347], Appendix 12.1). This yields the following conclusion: the characteristic values k^2/λ^2 are all positive if $\Phi(r)$ is everywhere positive and they are all negative if $\Phi(r)$ is everywhere negative

$$\forall_{R_I \leq r \leq R_E}\left[\frac{k^2}{\lambda^2}\right]\Phi(r) > 0. \tag{6.18}$$

If $\Phi(r)$ should change the sign in the interval $[R_I, R_E]$ then there are two sets of real characteristic values which have the limit points $+\infty$ and $-\infty$.

Obviously, a negative square of the time coefficient λ in (6.11) means instability. This yields Rayleigh's criterion of stability

$$\Phi(r) > 0. \tag{6.19}$$

For $m \neq 0$ the treatment is technically more involved. It is more convenient to investigate the set of equations (6.15) with the above described boundary conditions as a characteristic value problem for k^2 for a given λ rather than the other way around. By integration, we can derive from (6.15) a relation for k^2. This relation indicates that if $\Phi(r)$ is everywhere negative k^2 cannot admit a positive characteristic value. For real λ the characteristic values of k are necessarily imaginary. Therefore for real k, λ must necessarily be complex and this means instability. Consequently, the Rayleigh criterion is valid also in this case and, hence, it is the stability criterion for the Couette flow of an incompressible inviscid fluid.

This analysis can be based also on a variational principle which we do not discuss in this presentation. It can be found in the book of S. Chandrasekhar [71].

We proceed to the problem of a viscous fluid. The momentum balance equations in cylindrical coordinates (5.106) together with the constitutive relations for the incompressible viscous fluid yield the following Navier-Stokes equations in these coordinates

$$\frac{\partial v_r}{\partial t} + (\mathbf{v} \cdot \mathrm{grad})\, v_r - \frac{v_\vartheta^2}{r} = -\frac{\partial}{\partial r}\left(\frac{p}{\rho}\right)$$
$$+ \eta\left(\Delta^2 v_r - \frac{2}{r^2}\frac{\partial v_\vartheta}{\partial \vartheta} - \frac{v_r}{r^2}\right),$$

$$\frac{\partial v_\vartheta}{\partial t} + (\mathbf{v} \cdot \mathrm{grad})\, v_\vartheta + \frac{v_r v_\vartheta}{r} = -\frac{\partial}{\partial r}\left(\frac{p}{\rho}\right) \tag{6.20}$$
$$+ \eta\left(\Delta^2 v_\vartheta + \frac{2}{r^2}\frac{\partial v_r}{\partial \vartheta} - \frac{v_\vartheta}{r^2}\right),$$

$$\frac{\partial v_z}{\partial t} + (\mathbf{v} \cdot \mathrm{grad})\, v_z = -\frac{\partial}{\partial r}\left(\frac{p}{\rho}\right) + \eta\Delta^2 v_z,$$

where the operator on the left-hand side is defined by (6.2) and the Laplace operator Δ^2 has the form

$$\Delta^2 = \frac{\partial^2}{\partial r^2} + \frac{1}{r}\frac{\partial}{\partial r} + \frac{1}{r^2}\frac{\partial^2}{\partial \vartheta^2} + \frac{\partial^2}{\partial z^2}. \tag{6.21}$$

The incompressibility condition is as follows

$$\frac{\partial v_r}{\partial r} + \frac{v_r}{r} + \frac{1}{r}\frac{\partial v_\vartheta}{\partial \vartheta} + \frac{\partial v_z}{\partial z} = 0. \tag{6.22}$$

It is easy to check that these equations are fulfilled by the stationary solution

$$v_r = v_z = 0, \quad v_\vartheta = r\omega\,(r)\,, \tag{6.23}$$

provided the following conditions are satisfied

$$\frac{d}{dr}\left(\frac{p}{\rho}\right) = r\omega^2, \quad \frac{d^2\,(r\omega)}{dr^2} = 0. \tag{6.24}$$

Obviously, the last equation can be immediately integrated. If we account for the above formulated boundary conditions on the inner and outer cylinders, we obtain

$$\omega = \omega_I\left(\frac{1-\varsigma}{1-\xi^2}\frac{r_I^2}{r^2} - \xi^2\frac{1-\varsigma/\xi^2}{1-\xi^2}\right), \quad \varsigma = \frac{\omega_E}{\omega_I}, \quad \xi = \frac{R_I}{R_E} < 1. \tag{6.25}$$

If the fluid were inviscid, the stability condition would follow from the Rayleigh condition (6.19): $\Phi\,(r) > 0$, where the function $\Phi\,(r)$ is defined by Relation (6.16). This would yield the condition

$$\varsigma > \xi^2 \quad \Rightarrow \quad \frac{\omega_E}{\omega_I} > \frac{R_I}{R_E}. \tag{6.26}$$

On physical grounds, we expect that the viscosity would postpone the onset of instability. This has been investigated by G. I. Taylor [376]. The corresponding perturbation problem follows from the set of field equations (6.20) by the requirement that the solution has the form (6.5). Then

$$\frac{\partial v'_r}{\partial t} - 2\omega v'_\vartheta = -\frac{\partial}{\partial r}\left(\frac{p}{\rho}\right) + \eta\left(\Delta^2 v'_r - \frac{v'_r}{r^2}\right),$$

$$\frac{\partial v'_\vartheta}{\partial t} + \left\{r\frac{d\omega}{dr} + 2\omega\right\}v'_r = \eta\left(\Delta^2 v'_\vartheta - \frac{v'_\vartheta}{r^2}\right),$$

$$\frac{\partial v'_z}{\partial t} = -\frac{\partial}{\partial z}\left(\frac{p}{\rho}\right) + \eta\Delta^2 v'_z,$$

$$\frac{\partial v'_r}{\partial r} + \frac{v'}{r} + \frac{\partial v'_z}{\partial z} = 0. \tag{6.27}$$

The solution is sought in the form

$$v'_r = e^{\lambda t}V_r\cos kz, \quad v'_z = e^{\lambda t}V_z\sin kz,$$
$$v'_\vartheta = e^{\lambda t}V_\vartheta\cos kz, \quad p = e^{\lambda t}P\cos kz, \tag{6.28}$$

and the boundary conditions are as follows

$$V_r = V_\vartheta = \left(\frac{d}{dr} + \frac{1}{r}\right)V_r = 0, \quad \text{for} \quad r = R_I \text{ and } R_E. \tag{6.29}$$

It can be shown analytically (see S. Chandrasekhar [71], Section 70) that the condition (6.26) is sufficient for the stability, i.e., $\mathrm{Re}\,(\lambda) < 0$. Otherwise, either the system must fulfill additional conditions, for instance $(R_E - R_I)/R_E \ll 1$ (small gap) or the analysis is carried out numerically.

6.3 Thermal instability of a layer of fluid heated from below – Rayleigh-Bénard problem

Coupled problems of mechanics and various nonmechanical fields such as interactions with thermal or magnetic fields both have a very important practical bearing as well as illustrate conflicting tendencies of the stability of the fluid motion influenced by those fields. The problem of a layer of fluid heated from below possesses all important stability features of such cases. For this reason we present this standard example in this book. The earliest experiments on the loss of stability under the circumstances of the fluid layer heated from below were performed by H. Bénard [39]-[41] in 1900 and then the analytical description was designed by Lord Rayleigh [304] in 1916.

Due to the thermal expansion the fluid at the bottom will be lighter than the fluid at the top. Such a top-heavy arrangement is potentially unstable. However, the natural tendency to rearrange this weakness will be inhibited by the viscosity of the fluid. Consequently, it is to be expected that the maintained adverse temperature gradient must exceed a certain value for the instability to appear.

The governing equations follow for this problem from the general mass, momentum and energy conservation laws with constitutive laws describing the linear viscous heat conducting fluid. Hence, for the fields $\{p, v_i, T\}$ we have in the Cartesian reference frame

$$\frac{\partial \rho}{\partial t} + \frac{\partial}{\partial x_i} (\rho v_i) = 0,$$

$$\rho \frac{\partial v_i}{\partial t} + \rho v_j \frac{\partial v_i}{\partial x_j} = -\frac{\partial p}{\partial x_i} + \eta \frac{\partial}{\partial x_j} \left[\left(\frac{\partial v_i}{\partial x_j} + \frac{\partial v_j}{\partial x_i} \right) - \frac{2}{3} \frac{\partial v_k}{\partial x_k} \delta_{ij} \right] + \rho b_i, \tag{6.30}$$

$$\rho \frac{\partial}{\partial t} (c_v T) + \rho v_i \frac{\partial}{\partial x_i} (c_v T) = K \frac{\partial^2 T}{\partial x_i \partial x_i} - p D_{kk} + 2\eta \left(D_{ij} D_{ij} - \frac{1}{3} (D_{kk})^2 \right),$$

where

$$\frac{\rho_0 - \rho}{\rho_0} \approx \alpha (T - T_0), \quad D_{ij} = \frac{1}{2} \left(\frac{\partial v_i}{\partial x_j} + \frac{\partial v_j}{\partial x_i} \right), \quad \alpha = \gamma \kappa_T, \tag{6.31}$$

and ρ_0, T_0 are the mass density and the temperature at the lower boundary of the layer. Obviously, α is the thermal expansion coefficient and κ_T is the compressibility.

For the infinite layer of the thickness d in the z-direction we investigate the stability of the solution

$$\mathbf{v} = 0, \quad T = T(z), \tag{6.32}$$

where the temperature T and the pressure p satisfy the equations (the momentum and energy balance equations)

$$\frac{dp}{dz} = -g\rho, \quad \frac{d^2 T}{dz^2} = 0. \tag{6.33}$$

Obviously, the solution has the form

$$T = T_0 - Gz, \quad \rho = \rho_0 (1 + \alpha G z), \quad p = p_0 - g\rho_0 \left(z + \frac{1}{2}\alpha G z^2 \right). \tag{6.34}$$

In these relations G denotes the adverse temperature gradient which is maintained (e.g. by a given heat flux $\mathbf{q} \cdot \mathbf{n} = -KG$ in the normal direction to the upper surface of the layer).

Let this state be disturbed in the following manner

$$\mathbf{v}'(x_i, t), \quad T + T'(x_i, t), \quad p + p'(x_i, t), \tag{6.35}$$

where $x_1 = x$, $x_2 = y$, $x_3 = z$ and the perturbations denoted by primes are sufficiently small to be described by the linear equations. These equations have the following form in Cartesian coordinates

$$\frac{\partial v_i'}{\partial x_i} = 0,$$

$$\frac{\partial v_i'}{\partial t} = -\frac{\partial}{\partial x_i} \left(\frac{p'}{\rho_0} \right) + g\alpha T' \delta_{3i} + \nu \Delta^2 v_i', \quad \nu = \frac{\eta}{\rho_0}, \tag{6.36}$$

$$\frac{\partial T'}{\partial t} = G v_3' + \kappa \Delta^2 T', \quad \kappa = \frac{K}{\rho_0 c_v},$$

where κ is called the coefficient of thermometric conductivity. The first equation – the solenoidal character of the perturbation of velocity – results from the mass conservation equation in which (6.34) is used for small values of the thermal expansion coefficient α.

From these equations we can easily eliminate the perturbation of pressure. If we perform the rotation on Equation $(6.36)_2$ we obtain the equation for the perturbation of the vorticity $\omega' = \mathrm{rot}\,\mathbf{v}'$. After simple manipulations we arrive at the following set of equations for the perturbations $\{\omega'_3, v'_3, T'\}$

$$\frac{\partial \omega'_3}{\partial t} = \nu \Delta^2 \omega,$$

$$\frac{\partial}{\partial t}\Delta^2 v'_3 = G\alpha \left(\frac{\partial^2 T'}{\partial x^2} + \frac{\partial^2 T'}{\partial y^2} \right) + \nu \Delta^2 \Delta^2 v'_3, \tag{6.37}$$

$$\frac{\partial T'}{\partial t} = G v'_3 + \kappa \Delta^2 T'.$$

The remaining two components of the velocity can be found by quadratures.

It remains to formulate boundary conditions.

The following four conditions are natural regardless of the kind of bounding surfaces

$$T'\left(z=0\right) = T'\left(z=d\right) = 0, \quad v'_3\left(z=0\right) = v'_3\left(z=d\right) = 0. \tag{6.38}$$

In addition, two kinds of surfaces are considered:

- the rigid surface on which all components of velocity vanish: $v'_i = 0$, $i = 1, 2, 3$;

- the free surface on which the shear stresses vanish: $t_{13} = t_{23} = 0$.

Then it can be easily shown that the z-component of the vorticity must satisfy the condition

$$\begin{cases} \omega'_z = 0 & \text{on a rigid surface,} \\ \dfrac{\partial \omega'_z}{\partial z} = 0 & \text{on a free surface.} \end{cases} \tag{6.39}$$

We are now in the position to analyze the normal modes of perturbation. We seek solutions of the system (6.37) in the following form

$$v'_3 = V\left(z\right) \exp\left[i\left(k_x x + k_y y\right) + \lambda t\right],$$

$$T' = \Theta\left(z\right) \exp\left[i\left(k_x x + k_y y\right) + \lambda t\right], \tag{6.40}$$

$$\omega'_3 = \Omega\left(z\right) \exp\left[i\left(k_x x + k_y y\right) + \lambda t\right].$$

Substitution in Equations (6.37) yields

$$\lambda \left(\frac{d^2}{dz^2} - k^2 \right) V = -g\alpha k^2 \Theta + \nu \left(\frac{d^2}{dz^2} - k^2 \right)^2 V,$$

$$\lambda \Theta = GV + \kappa \left(\frac{d^2}{dz^2} - k^2 \right) \Theta, \tag{6.41}$$

$$\lambda \Omega = \nu \left(\frac{d^2}{dz^2} - k^2 \right) \Omega, \quad k^2 = k_x^2 + k_y^2.$$

Simultaneously, the boundary conditions require

$$\Theta\,(z=0)=0,\quad \Theta\,(z=d)=0,\quad V\,(z=0)=0,\quad V\,(z=d)=0,$$

$$\begin{cases} \Omega=0 \quad\text{and}\quad \dfrac{dV}{dz}=0 \quad\text{on a rigid surface,} \\[3mm] \dfrac{d\Omega}{dz}=0 \quad\text{and}\quad \dfrac{d^2V}{dz^2}=0 \quad\text{on a free surface.} \end{cases} \tag{6.42}$$

It is convenient to introduce the following notation

$$a=kd,\quad \sigma=\frac{\lambda d^2}{\nu},\quad Pr=\frac{\nu}{\kappa}, \tag{6.43}$$

where Pr is called the Prandtl number. Then the equation for V which follows from $(6.41)_{1,2}$ has the form

$$\left(\frac{d^2}{d\tilde{z}^2}-a^2-Pr\,\sigma\right)\mathcal{F}=-Ra^2V,\quad \tilde{z}=\frac{z}{d}, \tag{6.44}$$

where

$$\mathcal{F}=\left(\frac{d^2}{d\tilde{z}^2}-a^2-\sigma\right)\mathcal{G},\quad \mathcal{G}=\left(\frac{d^2}{d\tilde{z}^2}-a^2\right)V,\quad R=\frac{g\alpha G}{\kappa\nu}d^4, \tag{6.45}$$

and R is the Rayleigh number. An identical equation governs Θ.

According to the boundary conditions the function $\mathcal{F}$ defined in Relations (6.45) satisfies the relations $\mathcal{F}=0$ for $\tilde{z}=0$ and $\tilde{z}=1$. Following the considerations of S. Chandrasekhar [71] we prove now that this yields the property that σ is real $(\mathrm{Im}\,(\sigma)=0)$ for all positive Rayleigh numbers R.

Multiplying (6.44) by $\mathcal{F}^*$ – the complex conjugate of $\mathcal{F}$ – we obtain

$$\int_0^1 \mathcal{F}^*\left(\frac{d^2}{d\tilde{z}^2}-a^2-Pr\,\sigma\right)\mathcal{F}\,d\tilde{z}=-Ra^2\int_0^1 \mathcal{F}^*V\,d\tilde{z} \tag{6.46}$$

and

$$\int_0^1 \mathcal{F}^*\frac{d^2\mathcal{F}}{d\tilde{z}^2}\,d\tilde{z}=-\int_0^1 \left|\frac{d\mathcal{F}}{d\tilde{z}}\right|^2 d\tilde{z}, \tag{6.47}$$

which easily follows from the boundary conditions and integration by parts. Hence,

$$\int_0^1 \left(\left|\frac{d\mathcal{F}}{d\tilde{z}}\right|^2+(a^2+Pr\,\sigma)\,|\mathcal{F}|^2\right)d\tilde{z}=Ra^2\int_0^1 V\mathcal{F}^*d\tilde{z}. \tag{6.48}$$

Now the right-hand side can be transformed in the following way

$$\int_0^1 V\mathcal{F}^*d\tilde{z}=\int_0^1 V\left(\frac{d^2}{d\tilde{z}^2}-a^2-\sigma^*\right)\mathcal{G}^*d\tilde{z}=\int_0^1 V\frac{d^2\mathcal{G}^*}{d\tilde{z}^2}d\tilde{z}-(a^2+\sigma^*)\int_0^1 V\mathcal{G}^*d\tilde{z}. \tag{6.49}$$

Again integrating by parts and using boundary conditions either on a rigid or free surface we arrive at the relation

$$\int_0^1 V\mathcal{F}^*d\tilde{z}=\int_0^1 |\mathcal{G}|^2\,d\tilde{z}+\sigma^*\int_0^1\left[\left|\frac{dV}{d\tilde{z}}\right|^2+a^2\,|V|^2\right]d\tilde{z}. \tag{6.50}$$

Combining these results we obtain

$$\int_0^1 \left(\left|\frac{d\mathcal{F}}{d\widetilde{z}}\right|^2 + \left(a^2 + Pr\,\sigma\right)|\mathcal{F}|^2 \right) d\widetilde{z} = Ra^2 \left\{ \int_0^1 |\mathcal{G}|^2\, d\widetilde{z} + \sigma^* \int_0^1 \left[\left|\frac{dV}{d\widetilde{z}}\right|^2 + a^2\,|V|^2 \right] d\widetilde{z} \right\}. \tag{6.51}$$

The imaginary part of this relation is as follows

$$\left\{ Pr \int_0^1 |\mathcal{F}|^2\, d\widetilde{z} + Ra^2 \int_0^1 \left[\left|\frac{dV}{d\widetilde{z}}\right|^2 + a^2\,|V|^2 \right] d\widetilde{z} \right\} \operatorname{Im}(\sigma) = 0. \tag{6.52}$$

The expression in curly brackets is positive definite for $R > 0$. Consequently, σ must be real for positive Rayleigh numbers.

In order to use the above property in the description of stability we have to introduce two notions. Obviously, the system is unstable when the disturbance departs from the initial state. Otherwise it is stable. The locus of states which separates those two domains defines the states of marginal stability. According to the property proved above the states of the layer for which $\sigma = 0$ form the range of marginal stability of this problem. Hence, the transition from stability to instability must cross over a stationary state. Substitution of $\sigma = 0$ in Equations (6.44), (6.45) yields the equation for V of such a stationary process

$$\left(\frac{d^2}{d\widetilde{z}^2} - a^2\right)^3 V = -Ra^2 V. \tag{6.53}$$

Similarly, we obtain the equation for Θ. Bearing the boundary conditions in mind we can easily see that, for a given a^2, i.e., for a given wave number k, only for some values of R the problem possesses a nontrivial solution. Hence, it is a characteristic value problem for R. For instance, in the case of two free boundaries the stationary solution has the form

$$V = A \sin\left(n\pi\widetilde{z}\right), \tag{6.54}$$

and the characteristic equation for R is as follows

$$R = \frac{\left(n^2\pi^2 + a^2\right)^3}{a^2}. \tag{6.55}$$

Consequently, the critical value of the temperature gradient for waves with the wave number k is given by the relation following from the above formula for $n = 1$, i.e.,

$$G = \frac{\left(\pi^2 + a^2\right)^3}{a^2}\,\frac{\kappa\nu}{g\alpha d^4}, \qquad a = kd. \tag{6.56}$$

If for a given wave number k, the temperature gradient G exceeds the value given by (6.56) then these disturbances are unstable. Obviously, the inverse of the wave number corresponds to the wavelength, i.e., to the size of the cells. As shown in Figure 6.1, for any given temperature gradient there exists a range of those sizes in which the cells are stable.

The other sets of boundary conditions yield analogous results.

If at the onset of instability a stationary motion prevails, then we say that the principle of the exchange of stabilities is valid and that the instability sets in as stationary cellular convection. These cellular structure was indeed observed experimentally by H. Bénard.

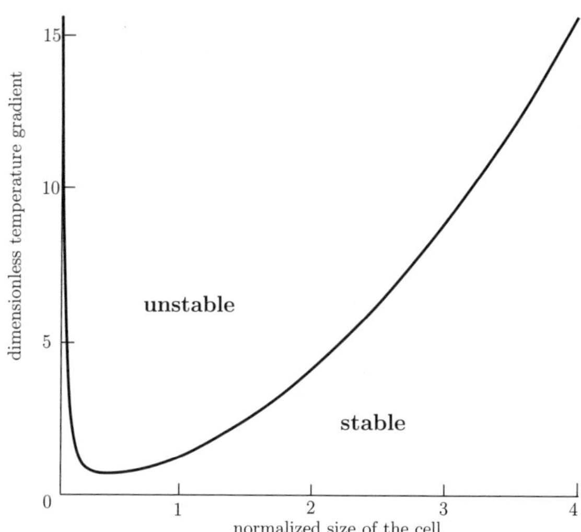

Figure 6.1: *The normalized temperature gradient $GR/1000$ as a function of the normalized size of the cell $1/a$.*

6.4 Stability of a nonlinear elastic strip

The linear stability analysis in solids proceeds in the similar manner as for fluids. However, in contrast to fluids one usually investigates equilibrium static cases rather than stationary flows. In practical applications these are states of large deformations, not necessarily elastic. We describe a typical and rather simple example of a homogeneous deformation of an elastic strip. We follow here the work of Z. Wesołowski [422] (see also the book of A. N. Guz [150]). Technicalities in this case are not very involved in contrast to almost any other system.

We consider a strip $\mathcal{B}_0$ of the initial thickness in the X^2-direction $2b_0$ and of the initial length in the X^1-direction l_0 (see Figure 6.2). In the X^3-direction the strip is infinitely long. This strip is loaded through two rigid plates on the edges $X^1 = 0$ and $X^1 = l_0$ in the direction perpendicular to the strip. The surfaces $X^2 = \pm b_0$ are free of stresses. The problem is assumed to be in the state of plain strains, i.e., the stretch in the X^3-direction is zero. Hence, the function of motion $\mathbf{x} = \mathbf{f}^0\left(\mathbf{X}, t\right)$ mapping the initial configuration $\mathcal{B}_0$ onto the final configuration $\mathcal{B}$ (the time t is here only a parameter) and the deformation gradient of this deformation $\mathbf{F}^0 = \mathrm{Grad}\,\mathbf{f}^0$ have in Cartesian coordinates the following form

$$x^1 = \sqrt{\lambda^{(1)}}X^1, \quad x^2 = \sqrt{\lambda^{(2)}}X^2, \quad x^3 = \sqrt{\lambda^{(3)}}X^3, \quad \lambda^{(1)}\lambda^{(2)}\lambda^{(3)} = 1, \quad \lambda^{(3)} = 1,$$

$$\mathbf{F}^0 = F^{0kK}\mathbf{e}_k \otimes \mathbf{e}_K, \quad \left(F^{0kK}\right) = \begin{pmatrix} \sqrt{\lambda^{(1)}} & 0 & 0 \\ 0 & \sqrt{\lambda^{(2)}} & 0 \\ 0 & 0 & \sqrt{\lambda^{(3)}} \end{pmatrix}, \tag{6.57}$$

where $\lambda^{(\alpha)}, \alpha = 1, \ldots, 3$, are constants. The upper index "0" distinguishes this base state of the system from the disturbed system which we consider later in order to find out if

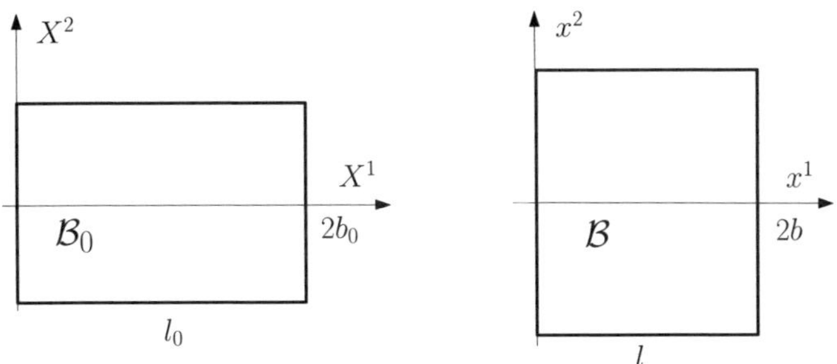

Figure 6.2: *Homogeneous stretch of a strip.*

the base state is stable.

We assume as well that the material is incompressible and satisfies the Mooney-Rivlin constitutive relations for Cauchy stresses (see Part I, Relation (5.149) or (5.148) of the present part)

$$\mathbf{T}^0 = -p^*\mathbf{1} + \Im_1\mathbf{B}^0 + \Im_{-1}\mathbf{B}^{0-1} = -p^0\mathbf{1} + \Im_1\mathbf{B}^0 - \left(I^0\mathbf{B}^0 - \mathbf{B}^{02}\right)\Im_{-1}, \quad p^0 = p^* - II^0\Im_{-1}, \tag{6.58}$$

where $\Im_1, \Im_{-1}$ are the material parameters (elasticities), $\mathbf{B}^0$ is the left Cauchy-Green deformation tensor and $I^0 = \operatorname{tr}\mathbf{B}^0$, $II^0 = \frac{1}{2}\left(I^{02} - \operatorname{tr}\mathbf{B}^{02}\right)$ are invariants of $\mathbf{B}^0$. Obviously, due to the incompressibility $III^0 = \det\mathbf{B}^0 = 1$. p^0 and p^* are reaction pressures, both undetermined from constitutive relations. They follow, in general, from the field equations. In our simple case, they are calculated from the equilibrium condition of the homogeneous state of deformation. The second form of the above relation follows from the first one by application of the Cayley-Hamilton Theorem (Relation (A.29) in Part I or (5.137) of the present part).

In the case which we consider the deformation tensor $\mathbf{B}^0 = \mathbf{F}^0\mathbf{F}^{0T}$ has the following form

$$\left(B^{0ij}\right) = \begin{pmatrix} \lambda & 0 & 0 \\ 0 & 1/\lambda & 0 \\ 0 & 0 & 1 \end{pmatrix}, \quad \lambda = \lambda^{(1)}, \tag{6.59}$$

where

$$l = \sqrt{\lambda}\,l_0, \quad b = b_0/\sqrt{\lambda}. \tag{6.60}$$

Hence, it describes the stretch deformation of the strip. The corresponding state of stresses is as follows[2]

$$\mathbf{T}^0 = T^{0ij}\mathbf{e}_i \otimes \mathbf{e}_j, \quad \begin{cases} T^{011} = -p^0 + \lambda^{(1)}\Im_1 - \lambda^{(1)}\left(\lambda^{(2)} + \lambda^{(3)}\right)\Im_{-1}, \\[2mm] T^{022} = -p^0 + \lambda^{(2)}\Im_1 - \lambda^{(2)}\left(\lambda^{(1)} + \lambda^{(3)}\right)\Im_{-1}, \\[2mm] T^{033} = -p^0 + \lambda^{(3)}\Im_1 - \lambda^{(3)}\left(\lambda^{(1)} + \lambda^{(2)}\right)\Im_{-1}, \end{cases} \tag{6.61}$$

---

[2]In the relations which we present further in this section we often leave the eigenvalues $\lambda^{(1)}, \lambda^{(2)}, \lambda^{(3)}$ not identified by the simple relations (6.57). This is convenient because in the derivation of some relations we have to differentiate first with respect to them as they were independent of each other. This is the case, for instance, in the derivation of equations for small disturbances.

and the remaining components are zero: $T^{012} = T^{013} = T^{023} = 0$.

Bearing the condition on the upper and lower surface in mind, and this means that $T^{022} = 0$, we can eliminate the pressure from the above relations

$$p^0 = \lambda^{(2)}\Im_1 - \lambda^{(2)}\left(\lambda^{(1)} + \lambda^{(3)}\right)\Im_{-1}. \tag{6.62}$$

If we account for the form of the deformation tensor $\mathbf{B}^0$ given by (6.59) we obtain the two normal components of stresses in the form

$$T^{011} = \left(\lambda - \frac{1}{\lambda}\right)(\Im_1 - \Im_{-1}),$$

$$T^{033} = \left(1 - \frac{1}{\lambda}\right)\Im_1 + (1 - \lambda)\Im_{-1}. \tag{6.63}$$

We are now in the position to formulate the linear problem of the stability. It means that on the above solution we have to superpose a small disturbance. We proceed to derive equations for this disturbance. In general, this means that we add a small displacement to the function of motion and a small additional pressure to the reaction pressure

$$\mathbf{x}' = \mathbf{f}\left(\mathbf{X}, t\right) = \mathbf{f}^0\left(\mathbf{X}, t\right) + \mathbf{u}'\left(\mathbf{x}, t\right), \quad \mathbf{x} = \mathbf{f}^0\left(\mathbf{X}, t\right)$$

$$\Rightarrow \quad \mathbf{F} = \mathbf{F}^0 + \operatorname{Grad}\mathbf{u}' = \mathbf{F}^0 + \left(\operatorname{grad}\mathbf{u}'\right)^T\mathbf{F}^0, \tag{6.64}$$

$$p\left(\mathbf{X}, t\right) = p^0\left(\mathbf{X}, t\right) + p'\left(\mathbf{x}, t\right).$$

This disturbance is small in the sense that all changes in the equations describing the disturbed system differ from the equations of the ground system by linear terms. The components of the stresses can then be written as follows

$$\mathbf{T} = \mathbf{T}^0 + \frac{\partial \mathbf{T}^0}{\partial \mathbf{F}^0}\mathbf{F}^{0T}\cdot\left(\operatorname{grad}\mathbf{u}'\right), \quad \text{i.e.,}$$

$$T^{ij} = T^{0ij} + A^{ijkl}\frac{\partial u_k}{\partial x_l}, \quad A^{ijkl} = \frac{\partial T^{0ij}}{\partial F^{0kK}}F^{0lK}. \tag{6.65}$$

For instance, bearing Relations (6.61) in mind, we immediately have

$$A^{1111} = 2\lambda^{(1)}\left(\Im_1 - \left(\lambda^{(2)} + \lambda^{(3)}\right)\Im_{-1}\right), \dots \text{etc.} \tag{6.66}$$

The equations for the disturbance $\mathbf{u}'$ follow from the momentum balance equation which in the static case under consideration has the form

$$\operatorname{div}\mathbf{T} = 0. \tag{6.67}$$

Obviously, an analogous equation holds true for the stress $\mathbf{T}^0$. After easy calculations we obtain in our case the following set

$$2\left[\lambda^{(1)}\Im_1 - \lambda^{(1)}\left(1 + \lambda^{(2)}\right)\Im_{-1}\right]\frac{\partial^2 u_1'}{\partial x^1 \partial x^1} + 2\lambda^{(2)}\left(\Im_1 - \Im_{-1}\right)\frac{\partial^2 u_1'}{\partial x^2 \partial x^2}$$

$$+ 2\left[\lambda^{(1)}\Im_1 - \lambda^{(1)}\left(1 + 2\lambda^{(2)}\right)\Im_{-1}\right]\frac{\partial^2 u_2'}{\partial x^1 \partial x^2} - \frac{\partial p'}{\partial x^1} = 0,$$

$$2 \left[\lambda^{(1)} \mathfrak{I}_1 - \lambda^{(2)} \left(1 + 2\lambda^{(2)} \right) \mathfrak{I}_{-1} \right] \frac{\partial^2 u_1'}{\partial x^1 \partial x^2} + 2\lambda^{(1)} \left(\mathfrak{I}_1 - \mathfrak{I}_{-1} \right) \frac{\partial^2 u_2'}{\partial x^1 \partial x^1}$$

$$+ 2 \left[\lambda^{(2)} \mathfrak{I}_1 - \lambda^{(2)} \left(1 + \lambda^{(1)} \right) \mathfrak{I}_{-1} \right] \frac{\partial^2 u_2'}{\partial x^2 \partial x^2} - \frac{\partial p'}{\partial x^2} = 0,$$

$$\frac{\partial u_1'}{\partial x^1} + \frac{\partial u_2'}{\partial x^2} = 0. \tag{6.68}$$

We have used here the assumption on the plain deformation $u_3' = 0$. The last equation is then the condition following from the incompressibility assumption.

We have to formulate the boundary conditions for the above set of partial differential equations. They follow from the linearization of the stress tensor in the disturbed state with respect to the disturbance. We skip here the details of the derivation (compare [422, 423]) and quote the results. They have the form

$$u_1' = 0 \quad \text{on} \quad x^1 = 0, l,$$

$$\frac{\partial u_1'}{\partial x^2} + \frac{\partial u_2'}{\partial x^1} = 0 \quad \text{on} \quad x^1 = 0, l \quad \text{and} \quad x^2 = \pm b, \tag{6.69}$$

$$-4\mathfrak{I}_{-1} \frac{\partial u_1'}{\partial x^1} + 2 \left[\lambda^{(2)} \mathfrak{I}_1 - \lambda^{(2)} \left(1 + \lambda^{(1)} \right) \mathfrak{I}_{-1} \right] \frac{\partial u_1'}{\partial x^2} - p' = 0 \quad \text{on} \quad x^2 = \pm b.$$

Obviously, the first group of conditions follows from the assumption that the loading is applied by rigid plates. The second group of conditions follows from the assumption that there is no friction on the surfaces of the strip and, consequently, the shear stresses are zero. Finally, the last group of conditions results from the lack of external loading on the upper and lower surface.

We seek the solution in the form of the following normal modes

$$u_1' = U_1 \left(x^2 \right) \sin \left(n\pi \frac{x^1}{l} \right), \qquad u_2' = U_2 \left(x^2 \right) \cos \left(n\pi \frac{x^1}{l} \right),$$

$$p' = P \left(x^2 \right) \sin \left(n\pi \frac{x^1}{l} \right). \tag{6.70}$$

In contrast to the problems which we have considered for the motion of fluids the present case is static and, consequently, we do not have a time contribution in the perturbation problem. This means that the problem of stability of the ground state does not reduce to the investigation of the time exponent. It is related to the homogeneity of the boundary conditions for the perturbation (6.69). These conditions yield an algebraic relation for the external loading at which the system loses the stability.

Substitution in (6.68) leads to ordinary differential equations for U_1, U_2, P. We can reduce them to a single equation for one of those functions. After some manipulations we obtain for U_2 the following boundary value problem

$$b^4 \frac{d^4 U_2}{d \left(x^2 \right)^4} - 2k^2 \beta \lambda^{(1)} b^2 \frac{d^2 U_2}{d \left(x^2 \right)^2} + k^4 \left(\lambda^{(2)} \right)^2 U_2 = 0,$$

$$\beta = \frac{1}{2} \left(\lambda^{(1)} + \frac{1}{\lambda^{(1)}} \right), \qquad k = n\pi \frac{b}{l}, \tag{6.71}$$

with

$$b^2 \frac{d^2 U_2}{d\left(x^2\right)^2} + k^2 U_2 = 0 \quad \text{on } x^2 = \pm b,$$

$$b^3 \frac{d^3 U_2}{d\left(x^2\right)^3} - 2k^2 \lambda^{(1)} \left[\beta + \frac{1}{2\lambda^{(1)}}\right] \frac{dU_2}{dx^2} = 0 \quad \text{on } x^2 = \pm b.$$

$$(6.72)$$

The linear equation (6.71) has a solution which contains four arbitrary constants. For those constants we have four conditions (6.72) which, obviously, yield a homogeneous set of linear algebraic equations. Hence, a nontrivial solution exists if the determinant of this system is equal to zero. After simple but tedious calculations we arrive at the following condition of stability ($\lambda \equiv \lambda^{(1)}$)

$$S_1 S_2 = 0,$$

$$S_1 = \frac{\left[\lambda\left(\beta + \sqrt{\beta^2 - 1}\right) + 1\right]^2 \sqrt{\beta - \sqrt{\beta^2 - 1}} \tanh\left(k\sqrt{\lambda}\sqrt{\beta - \sqrt{\beta^2 - 1}}\right)}{\left[\lambda\left(\beta - \sqrt{\beta^2 - 1}\right) + 1\right]^2 \sqrt{\beta + \sqrt{\beta^2 - 1}} \tanh\left(k\sqrt{\lambda}\sqrt{\beta + \sqrt{\beta^2 - 1}}\right)} - 1,$$

$$S_2 = \frac{\left[\lambda\left(\beta + \sqrt{\beta^2 - 1}\right) + 1\right]^2 \sqrt{\beta - \sqrt{\beta^2 - 1}} \tanh\left(k\sqrt{\lambda}\sqrt{\beta + \sqrt{\beta^2 - 1}}\right)}{\left[\lambda\left(\beta - \sqrt{\beta^2 - 1}\right) + 1\right]^2 \sqrt{\beta + \sqrt{\beta^2 - 1}} \tanh\left(k\sqrt{\lambda}\sqrt{\beta - \sqrt{\beta^2 - 1}}\right)} - 1.$$

$$(6.73)$$

For any given elongation λ this equation determines the ratio b/l for which the system is losing the stability. Hence, the variables λ and b/l for which this condition is satisfied determine the curves of marginal stability.

In Figure 6.3 we have plotted the relation between the dimensionless parameter k, describing the ratio of the dimensions b/l and the eigenvalue λ of the deformation tensor **B**. Let us recall that the latter describes the changes of the length of the edges of the strip $l = \sqrt{\lambda}l_0$, $b = b_0/\sqrt{\lambda}$. The region between the curves of marginal stability is unstable.

6.5　Stability of the thermodynamical equilibrium of second-grade fluids

In Section 9.4 of Part I we have presented extensively the problem of the thermodynamical description of non-Newtonian fluids, i.e., fluids possessing some elastic properties which yield the normal stress effect (Weissenberg effect). We have pointed out that the so-called second-grade model frequently used by rheologists possesses unpleasant mathematical properties. In Chapter 5 of this book we have also pointed out that viscometric flows are described by the same equations within the Maxwell model with an evolution equation for stresses and within the second-grade model. Consequently, by the results of such experiments one cannot select which of these models is physically more adequate to describe non-Newtonian fluids.

Since many years it is known that the second-grade model with negative α_1-coefficient (compare Relation (5.102) and Table 5.1) yields instability, non-existence and breakdown of pure shearing motions for a second-grade fluid. This has been shown by B. D. Coleman,

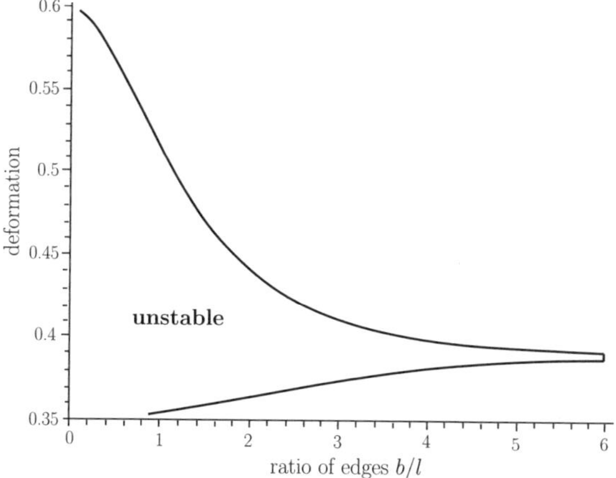

Figure 6.3: *Stability limits for the strip; variables: ratio of edges $k = n\pi b/l$ and deformation λ.*

R. J. Duffin and V. J. Mizel already in 1965 [80], [88]. We present below the main part of their results. In spite of a very special flow considered in those papers these results have a very general character. In the work of J. E. Dunn and R. L. Fosdick published in 1974 [107] it is shown that thermodynamical stability requires the coefficient α_1 of the second-grade fluid to be positive in contrast to all experimental results. Finally, D. D. Joseph has proven that the second-grade model yields unstable equilibrium states for negative α_1.

All these results should be sufficient to rule out Nth-grade models from any application in engineering. This is not the case. There still exist speculations that these substances which were investigated in experiments – in particular solutions of polymers – yield negative α_1 but it may well be the case that there exist not yet discovered materials for which the Nth-grade models possess positive α_1 and, as these models yield nice mathematical properties, such models should be developed and applied. Such statements made some 40 years ago are not supported by any experimental evidence until today.

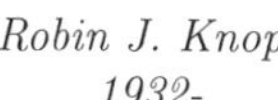

Robin J. Knops   *Roger L. Fosdick*   *Zbigniew*
1932-   *1939-*   *Wesołowski*
                                           1933-

We present in this section some details of stability considerations because such an analysis for second-grade fluid seems to be the most prominent example in which stability considerations in addition to the classical second law of thermodynamics constitute an important admissibility criterion for continuum models.

Let us begin with the analysis of the slow simple shearing flow in which we follow the work [80]. According to B. D. Coleman and W. Noll [89] the condition of slow flows yields a good approximation of the general memory functional for stresses by the constitutive relation (5.102) with $N = 2$, i.e.,

$$\mathbf{T} = -p\mathbf{1} + \eta \mathbf{A}_1 + \alpha_1 \mathbf{A}_2 + \alpha_2 \left(\mathbf{A}_1\right)^2,$$

$$(6.74)$$

$$\mathbf{A}_1 = \mathbf{L} + \mathbf{L}^T, \quad \mathbf{A}_2 = \dot{\mathbf{A}}_1 + \mathbf{L}^T \mathbf{A}_1 + \mathbf{A}_1 \mathbf{L}, \quad \mathbf{L} = \operatorname{grad} \mathbf{v}, \quad \operatorname{grad} \mathbf{L} = 0.$$

Assuming that the body forces possess a potential (compare (5.12)) we introduce the modified pressure ψ and the extra-stress $\mathbf{S}$ satisfying the momentum balance equation

$$\rho \dot{\mathbf{v}} + \operatorname{grad} \psi = \operatorname{div} \mathbf{S},$$
$$\psi = p + \rho\varphi, \quad \mathbf{T}^D = \mathbf{T} + p\mathbf{1}.$$

$$(6.75)$$

Relations (6.74) and (6.75) form the field equations of the second-grade fluid. We investigate this set for the flow satisfying in Cartesian coordinates the conditions

$$v_x = v_z = 0, \quad v_y = v\left(x, t\right).$$

$$(6.76)$$

Then, the combination of the field equations yields

$$\rho \frac{\partial v}{\partial t} + \frac{\partial \psi}{\partial y} = \eta \frac{\partial^2 v}{\partial x^2} + \alpha_1 \frac{\partial^3 v}{\partial t \partial x^2},$$

$$\frac{\partial \psi}{\partial x} = (2\alpha_1 + \alpha_2) \frac{\partial}{\partial x} \left(\frac{\partial v}{\partial x}\right)^2,$$

$$(6.77)$$

$$\frac{\partial \psi}{\partial z} = 0,$$

i.e., for this particular flow the field equations for the second-grade fluid are linear. It follows immediately

$$\psi = (2\alpha_1 + \alpha_2) \left(\frac{\partial v}{\partial x}\right)^2 + yg\left(t\right) + h\left(t\right).$$

$$(6.78)$$

Hence, the equation for the velocity has the form

$$\rho \frac{\partial v}{\partial t} + g\left(t\right) = \eta \frac{\partial^2 v}{\partial x^2} + \alpha_1 \frac{\partial^3 v}{\partial t \partial x^2}.$$

$$(6.79)$$

Due to the linearity of this equation it is sufficient to investigate the stability of the homogeneous problem: $g\left(t\right) = 0$.

Let us assume that $\alpha_1 < 0$ and let us introduce the notation

$$\widetilde{x} = x\sqrt{\frac{\rho}{-\alpha_1}}, \quad \widetilde{t} = t\frac{\eta}{-\alpha_1}, \quad \widetilde{v}\left(\widetilde{x}, \widetilde{t}\right) = v\left(x, t\right).$$

$$(6.80)$$

Then, erasing the tilde, we obtain the equation

$$\frac{\partial v}{\partial t} = \frac{\partial^2 v}{\partial x^2} - \frac{\partial^3 v}{\partial t \partial x^2},$$

$$(6.81)$$

which is the subject of the analysis in the work [80].

Two types of boundary conditions are considered:

$$\begin{cases} \text{Type I:} \quad \widetilde{v}\left(0,\widetilde{t}\right) = 0, \quad \widetilde{v}\left(\widetilde{d},t\right) = \widetilde{v}^d\left(\widetilde{t}\right), \quad \widetilde{d} = d\sqrt{\dfrac{\rho}{-\alpha_1}}, \\[3mm] \text{Type II:} \quad \widetilde{v}\left(0,\widetilde{t}\right) = 0, \quad \widetilde{T}_{12}\left(\widetilde{d},\widetilde{t}\right) = \widetilde{t}^d\left(\widetilde{t}\right). \end{cases} \tag{6.82}$$

Obviously, for the shear stress we have the relation

$$T_{12} = \eta\frac{\partial v}{\partial x} + \alpha_1\frac{\partial^2 v}{\partial t\partial x} \quad \Rightarrow \quad \widetilde{T}_{12} = T_{12}\frac{1}{\eta}\sqrt{\frac{-\alpha_1}{\rho}} = \frac{\partial v}{\partial \widetilde{x}} - \frac{\partial^2 v}{\partial t\partial \widetilde{x}}. \tag{6.83}$$

Further we erase the tilde also in these relations.

The initial data are assumed to have the form

$$v\left(x,0\right) = v^0\left(x\right). \tag{6.84}$$

We construct the solution of this problem in two steps. Firstly, it is clear that the function

$$v = Ae^{\lambda t}\sin kx, \tag{6.85}$$

satisfies Equation (6.81) provided

$$\lambda = \frac{k^2}{k^2 - 1}. \tag{6.86}$$

This function satisfies as well the homogeneous boundary condition of Type I $\left(v^d = 0\right)$ whenever

$$k = \frac{m\pi}{d}, \quad m = 0, 2, \ldots. \tag{6.87}$$

Clearly, such solutions are bounded, i.e., $\lambda < 0$ $(k^2 < 1)$, if and only if

$$m < \frac{d}{\pi}. \tag{6.88}$$

Hence, if $d \leq \pi$ every nontrivial solution (6.85) of the homogeneous boundary value problem is unbounded and there are many such solutions. Otherwise, it may be bounded as well as unbounded.

Now, in the case of the nonhomogeneous boundary value problem due to the linearity of Equation (6.81) we can write the solution in the form

$$v = v^* + \varepsilon\exp\left(\frac{k^2 t}{k^2 - 1}\right)\sin kx, \tag{6.89}$$

where v^* is a bounded solution which satisfies the boundary data of Type I and the initial condition $v^*\left(x,0\right) = v^0\left(x\right)$. Then there exists an m such that for any $\varepsilon > 0$ we have

$$\lim_{t\to\infty} v\left(x,t\right) = \infty, \tag{6.90}$$

for some $x \in [0,d]$. Similar considerations with the same result can be carried out for the boundary conditions of Type II. Consequently, we obtain the following

Coleman-Duffin-Mizel Theorem: *For fixed boundary data of Type I or Type II, all bounded solutions of (6.81) are unstable against arbitrary small perturbations of the initial data; i.e., if a solution $v^*(x,t)$ with initial data $v^{*0}(x)$ is bounded, then for every ε there exists an unbounded solution $v(x,t)$, with $|v^0(x) - v^{*0}(x)| < \varepsilon$.*

We do not need to go into any further details of this problem such as uniqueness or nonexistence which have been presented in the papers [80, 88]. We should only mention that attempts to avoid this disastrous result for the second-grade fluid by claiming that maybe better properties would follow if we had chosen a more general form of the velocity field than this of the form (6.76) have not been made. They would not resolve the problem anyway as we see from the general considerations which follow next.

After this simple example of the flow we proceed to discuss the stability condition following from the second law of thermodynamics for second-grade fluids. These considerations were published in 1974 by J. E. Dunn and R. L. Fosdick [107].

We recall first the results of the analysis of the thermodynamical admissibility as described in Chapter 5 of Part I. We shall do so for an incompressible, homogeneous fluid of complexity 2 (see Sections 27.1 and 35 of the book of C. A. Truesdell and W. Noll [393]). This is a bit more general than the second-grade fluid. Such a material is defined by the following constitutive relations

$$\mathbf{T} = -p\mathbf{1} + \mathbf{T}^D\left(\mathbf{L}, \dot{\mathbf{L}}, T, \mathbf{g}\right), \quad \mathbf{g} = \mathrm{grad}\, T, \quad \mathrm{tr}\,\mathbf{L} = 0,$$

$$\mathbf{q} = \mathbf{q}\left(\mathbf{L}, \dot{\mathbf{L}}, T, \mathbf{g}\right), \quad \varepsilon = \varepsilon\left(\mathbf{L}, \dot{\mathbf{L}}, T, \mathbf{g}\right), \tag{6.91}$$

$$\mathbf{h} = \mathbf{h}\left(\mathbf{L}, \dot{\mathbf{L}}, T, \mathbf{g}\right) \quad \eta = \eta\left(\mathbf{L}, \dot{\mathbf{L}}, T, \mathbf{g}\right),$$

where, obviously, $\mathbf{T}$ is the Cauchy stress tensor, p is the reaction pressure, ε is the specific internal energy, η denotes the specific entropy, $\mathbf{q}, \mathbf{h}$ are the heat flux and the entropy flux, respectively. The field equations for the fields $\{p, \mathbf{v}, T\}$ follow from the incompressibility condition $(6.91)_3$, the momentum conservation law and the energy conservation law. These constitutive relations must also satisfy the following entropy inequality

$$\rho\frac{\partial\eta}{\partial t} + \mathrm{div}\,(\rho\eta\mathbf{v} + \mathbf{h}) \geq 0, \tag{6.92}$$

for all solutions of the field equations. Obviously, the mass density ρ is constant.

The evaluation of this entropy principle can be made, for instance, by application of Lagrange multipliers (compare Part I). We skip here all technical details which are standard and present the final results. They are as follows

$$\varepsilon = \psi - T\frac{\partial\psi}{\partial T}, \quad \eta = -\frac{\partial\psi}{\partial T}, \quad \psi = \psi(T, \mathbf{L}),$$

$$\mathbf{T}^D = \mathbf{T}^D\left(\mathbf{L}, \dot{\mathbf{L}}, T, \mathbf{g}\right), \quad \mathbf{h} = \frac{\mathbf{q}}{T}, \tag{6.93}$$

$$\mathcal{D} = \frac{1}{T}\,\mathrm{tr}\left(\mathbf{SL} - \rho\frac{\partial\psi}{\partial\mathbf{L}}\dot{\mathbf{L}}^T\right) - \frac{1}{T^2}\mathbf{q}\cdot\mathbf{g} \geq 0,$$

for all thermodynamical processes. Certainly, the quantity $\mathcal{D}$ is the dissipation. In contrast to the Navier-Stokes-Fourier viscous heat conducting fluid (compare (5.46)) it still contains the contribution of the Helmholtz free energy function ψ. This is due to the fact that, in contrast to usual viscous fluids the function ψ may still be dependent on the velocity gradient $\mathbf{L}$. This is a nonequilibrium contribution as the thermodynamical equilibrium, defined by the vanishing dissipation $\mathcal{D}$, satisfies in the present case the following conditions

$$\mathbf{g}|_E = 0, \quad \mathbf{L}|_E = 0, \quad \dot{\mathbf{L}}\Big|_E = 0. \tag{6.94}$$

Let us return to the second-grade fluids for which the stress is assumed to satisfy Relation (6.74). The material parameters η, α_1, α_2 may be in general functions of the temperature T. This is immaterial for the argument which follows. Essential for further exploitation of the dissipation inequality $(6.93)_6$ is the linearity of this inequality with respect to $\dot{\mathbf{L}}$ because

$$\mathbf{T}^D = \eta \left(\mathbf{L} + \mathbf{L}^T \right) + \alpha_1 \left(\dot{\mathbf{L}} + \dot{\mathbf{L}}^T + \mathbf{L}^2 + \mathbf{L}^{T2} + 2\mathbf{L}^T\mathbf{L} \right) \\ + \alpha_2 \left(\mathbf{L}^2 + \mathbf{L}^{T2} + \mathbf{L}\mathbf{L}^T + \mathbf{L}^T\mathbf{L} \right). \tag{6.95}$$

As the linear contribution in $\dot{\mathbf{L}}$ would yield the violation of the dissipation inequality for some choices of $\dot{\mathbf{L}}$ we have to require

$$\rho \frac{\partial \psi}{\partial \mathbf{L}} \dot{\mathbf{L}}^T = \alpha_1 \operatorname{tr} \left[\left(\dot{\mathbf{L}} + \dot{\mathbf{L}}^T \right) \mathbf{L} \right]. \tag{6.96}$$

In the same way we have to require that cubic terms in $\mathbf{A}$ must vanish from the inequality. This yields

$$(\alpha_1 + \alpha_2) \operatorname{tr} \left[\left(\mathbf{A}^2 + \mathbf{A}^{T2} + 2\mathbf{A}^T\mathbf{A} \right) \mathbf{A} \right] = 0, \tag{6.97}$$

for arbitrary traceless $\mathbf{A}$.

It is obvious that the above considerations lead to the following conclusions

$$\mathbf{q} \cdot \mathbf{g} \geq 0, \quad \eta \geq 0, \quad \alpha_1 + \alpha_2 = 0,$$

$$\psi = \psi_E + \frac{\alpha_1}{4\rho} |\mathbf{A}|^2. \tag{6.98}$$

The first two conditions are standard. This is the heat conduction inequality which yields the positive definiteness of the heat conductivity. The second one is the non-negativeness of the viscosity η. However, the remaining two conditions concern the properties of the second-grade fluid and they are disastrous. Inspection of Table 5.1 shows that $\alpha_1 \neq \alpha_2$ for this particular material and, in fact, there are no reported experimental data for any material in which the above condition on those constants would be satisfied.

In addition, the structure of the relation for the free energy function ψ yields an additional restriction. It is required by the second law of thermodynamics that the thermodynamical equilibrium should appear at states which minimize the Helmholtz free energy function ψ (the convexity condition, e.g. compare the book of M. Šilhavý [355]). This is the so-called condition of stability of the thermodynamical equilibrium. Hence, in our case follows

$$\alpha_1 \geq 0. \tag{6.99}$$

This result rules out the physical applicability of the model of the second-grade fluid. J. E. Dunn and R. L. Fosdick [107] further investigated the model and found out many pleasant mathematical properties following from the above conditions. They expressed the hope that a material may be found which indeed fulfills those conditions. This is not the case until now.

The situation is even worse. Some years later, in 1981, D. D. Joseph [187] has shown that similar instabilities of the thermodynamical equilibrium appear in Nth grade fluids for an arbitrary N. We sketch here his proof.

D. D. Joseph considers asymptotic forms of the memory functional

$$\mathbf{T} = -p\mathbf{1} + \mathbf{T}^D,$$

$$\mathbf{T}^D = \mathbf{S} = \mathfrak{F}\left[\mathop{\mathbf{G}(s)}_{s=0}^{\infty}\right], \quad \mathbf{G}(s) = \mathbf{F}_t^T(t-s)\,\mathbf{F}_t(t-s) - \mathbf{1}, \tag{6.100}$$

where $\mathbf{F}_t$ is the relative deformation gradient with respect to the present configuration (compare Relation (3.9), Part I) and $\mathfrak{F}[\ldots]$ is the memory functional. B. D. Coleman and W. Noll proved in the paper [89] that this functional has the following sequence of approximations $\mathbf{T}^{D(N)} \sim \mathbf{T}^D$ for the so-called retarded motion, i.e., slow motions in slow times[3]

$$\mathbf{T}^{D(1)} = \eta\mathbf{A}_1,$$

$$\mathbf{T}^{D(2)} = \mathbf{T}^{D(1)} + \alpha_1\mathbf{A}_2 + \alpha_2\left(\mathbf{A}_1\right)^2,$$

$$\mathbf{T}^{D(3)} = \mathbf{T}^{D(2)} + \beta_1\mathbf{A}_3 + \beta_2\left(\mathbf{A}_1\mathbf{A}_2 + \mathbf{A}_2\mathbf{A}_1\right) + \beta_3\left(\mathrm{tr}\mathbf{A}_2\right)\mathbf{A}_1, \ldots \tag{6.101}$$

$$\mathbf{T}^{D(n)} = \mathbf{T}^{D(n-1)} + \widehat{\phi}_n\mathbf{A}_n + \text{nonlinear terms.}$$

These relations describe, obviously, Nth grade fluids (compare (5.102)). For motions which are not in slow times, the constitutive relations can be written in the form of multiple integrals whose example was shown in Chapter 5 (Relation (5.100)). This was shown for the first time by A. E. Green and R. Rivlin [142]. In particular, for $n = 1$ we obtain

$$\mathbf{T}^{D(1)} = \sum_{m=1}^{\infty}\frac{(-1)^m}{(m-1)!}\phi_m\mathbf{A}_m(t), \quad \phi_m = \int_0^{\infty} s^{m-1}G(s)\,ds, \tag{6.102}$$

and $G(s)$ is the linear stress relaxation modulus (compare Chapter 5).

Now setting $\widetilde{t} = \varepsilon t$, $\widetilde{\mathbf{v}} = \varepsilon\mathbf{v}$, $\varepsilon \ll 1$ (slow motion) we can compare the coefficients in the above relations and it follows

$$\eta = \int_0^{\infty} G(s)\,ds = \phi_1, \quad -\alpha_1 = \int_0^{\infty} sG(s)\,ds = \phi_2,$$

$$\beta_1 = \frac{1}{2}\int s^2 G(s)\,ds = \frac{1}{2}\phi_3, \quad \alpha_2 = \int_0^{\infty}\int_0^{\infty}\gamma(s_1, s_2)\,ds_1 ds_2, \tag{6.103}$$

where γ describes the contribution of the double integral to the constitutive relation. Joseph made certain general observations based solely on some integral inequalities. For

---

[3]According to the definition (3.13) of Part I Rivlin-Ericksen tensors are not deviatoric. Relations (6.101) should be made deviatoric on the right-hand side. This technicality has no influence on further considerations and, therefore, we skip the corresponding corrections.

instance, Schwarz's inequality implies

$$-\alpha_1 \leq \sqrt{\eta}\sqrt{2\beta_1}. \tag{6.104}$$

The proof of Joseph's Theorem is based on the observation that all moments ϕ_n of $G(s)$ are positive. This holds when $G(s) > 0$ for finite s and, simultaneously,

$$\eta > 0, \quad -\alpha_1 > 0, \quad \beta_1 > 0, \quad \text{etc.} \tag{6.105}$$

In order to check the stability one has to find the sign of the real part of the principal spectral values associated with the linearized field equations. Linearization yields

$$\mathbf{T}^{D(1)} = \eta\mathbf{A}_1, \quad \mathbf{T}^{D(2)} = \mathbf{T}^{D(1)} + \alpha_1\frac{\partial \mathbf{A}_1}{\partial t}, \quad \mathbf{T}^{D(3)} = \mathbf{T}^{D(2)} + \beta_1\frac{\partial^2 \mathbf{A}_1}{\partial t^2}, \quad \text{etc.} \tag{6.106}$$

The corresponding equation for the velocity of slow motion $\widetilde{\mathbf{v}}$ has the form

$$\rho\frac{\partial \widetilde{\mathbf{v}}}{\partial t} = -\operatorname{grad}\widetilde{p} + \operatorname{Lin}\left(\operatorname{div}\operatorname{grad}\widetilde{\mathbf{v}}\right), \quad \operatorname{div}\widetilde{\mathbf{v}} = 0, \tag{6.107}$$

with zero boundary conditions. Obviously, Lin is a linear operator whose form easily follows from the constitutive relations (6.106). The spectral problem follows by the assumption

$$\{\widetilde{\mathbf{v}}, \widetilde{p}\} = e^{-\lambda t}\{\mathbf{V}, P\}. \tag{6.108}$$

Substitution in the equation of motion yields

$$-\rho\lambda\mathbf{V} = -\operatorname{grad}P + k(\lambda)\operatorname{div}\operatorname{grad}\mathbf{V}, \tag{6.109}$$

where

$$k(\lambda) - \eta \quad \alpha_1\lambda,$$

$$k(\lambda) = \eta - \alpha_1\lambda + \beta_1\lambda^2,$$

$$k(\lambda) = \sum_{m=1}^{n} \frac{\phi_m\lambda^{m-1}}{(m-1)!}, \quad n > 1. \tag{6.110}$$

These equations show that $k(\lambda) = 0$ is a polynomial of degree $n - 1$ in λ with strictly positive coefficients. The coefficient of λ^{n-2} in this polynomial is minus the sum of all eigenvalues and it is positive because the coefficient of λ^{n-2} is positive. Hence, the roots satisfy the inequality

$$-(\lambda_1 + \lambda_2 + \cdots + \lambda_{n-1}) = (n-1)\frac{\phi_{n-1}}{\phi_n} > 0. \tag{6.111}$$

Hence, at least one root has a negative real part leading to instability. It follows

Joseph's Theorem. *The rest state of each and every fluid of grade $n \neq 1$ is unstable in the sense of the spectral problem of linearized theory if the ratio of the coefficients of $\mathbf{A}_n$ to $\mathbf{A}_{n-1}$ is negative.*

The conditions which must be satisfied for this theorem to hold require that the constitutive relations are formulated for slow motions. This rules out high frequency processes and, consequently, the applicability of the Nth grade fluid model could be very limited.

Chapter 7

Thermodynamical problems

In correspondence to Chapters 5 and 6 of Part I in which thermodynamical foundations are presented, in this chapter we discuss a few applications of the theory.

7.1 Some heat conduction problems

The contents of this section is oriented towards some examples of heat conduction problems presented by H. S. Carslaw and J. C. Jaeger in [66]. In this early book on Heat Conduction in Solids classical methods for the solution of various problems described in Cartesian, cylindrical and spherical coordinates (see Appendix A.2) are presented. Some of them are reproduced here. While for radial flow in cylinders and spheres the solution involves only one space variable and the time, problems on regions such as rectangular parallelepipeds and finite cylinders two or more space variables are involved. Carslaw and Jaeger mention different methods to obtain solutions: (i) The simplest and most important problems are those whose solutions can be expressed as a product of solutions of one-variable problems. The fundamental case is that of unit initial temperature and zero surface temperature (or radiation into the medium at zero temperature); from this the solution for zero initial temperature and surface temperature follows, and hence, by Duhamel's Theorem, that for surface temperature $\phi(t)$. (ii) Arbitrary initial and surface temperatures can be discussed by the use of multiple Fourier series or their generalizations (see Appendix B1). (iii) The use of Green's function (see Appendix C) also gives a complete solution of the general problem of arbitrary initial and surface temperature. For the simple cases referred to in (i) the same result is obtained after some reduction. It also reveals immediately solutions for heat production in the solid at a rate which is a given function of position and time. (iv) The direct application of the Laplace transformation method (see Appendix B.2) is particularly useful when some of the boundary surfaces are maintained at temperatures which are simple functions of the time. In such cases the results obtained by (ii) and (iii) (and (i), unless the surface temperature is constant), in the form of integrals, only give the simplest form of the solution after certain series have been summed.

7.1.1 Steady temperature

Here, we consider problems in which certain faces of a solid are maintained at constant temperature, while others are at zero temperature or radiate into a medium at zero.

7.1.1.1   Cartesian coordinates – heat flow in a rectangular parallelepiped

1. *The solid $0 < x < a$, $0 < y < b$, $0 < z < c$. Surface temperature v_1, constant, on $x = 0$; v_2, constant, on $x = a$; the other faces at zero.*

The differential equation for the temperature is

$$\frac{\partial^2 v}{\partial x^2} + \frac{\partial^2 v}{\partial y^2} + \frac{\partial^2 v}{\partial z^2} = 0. \tag{7.1}$$

The expression

$$\frac{v_1 \sinh l\,(a - x) + v_2 \sinh lx}{\sinh la} \sin \frac{m\pi y}{b} \sin \frac{n\pi z}{c}, \tag{7.2}$$

satisfies (7.1) provided

$$l^2 = \left(\frac{m^2}{b^2} + \frac{n^2}{c^2} \right) \pi^2. \tag{7.3}$$

Also, if m and n are integers, it vanishes when $y = 0$ and $y = b$, and when $z = 0$ and $z = c$. The expression

$$v = \sum_{m=1}^{\infty} \sum_{n=1}^{\infty} A_{m,n} \frac{v_1 \sinh l\,(a - x) + v_2 \sinh lx}{\sinh la} \sin \frac{m\pi y}{b} \sin \frac{n\pi z}{c}, \tag{7.4}$$

has the same properties, and also has the value v_1 when $x = 0$, and the value v_2 when $x = a$, provided that

$$\sum_{m=1}^{\infty} \sum_{n=1}^{\infty} A_{m,n} \sin \frac{m\pi y}{b} \sin \frac{n\pi z}{c} = 1. \tag{7.5}$$

Now expanding unity in a sine series in $(0, b)$ and $(0, c)$ gives

$$1 = \frac{4}{\pi} \sum_{p=0}^{\infty} \frac{\sin\left[(2p + 1)\frac{\pi y}{b}\right]}{2p + 1}, \qquad 1 = \frac{4}{\pi} \sum_{q=0}^{\infty} \frac{\sin\left[(2q + 1)\frac{\pi z}{c}\right]}{2q + 1}. \tag{7.6}$$

Multiplying the two parts of (7.6) gives

$$1 = \frac{16}{\pi^2} \sum_{p=0}^{\infty} \sum_{q=0}^{\infty} \frac{\sin\left[(2p + 1)\frac{\pi y}{b}\right] \sin\left[(2q + 1)\frac{\pi z}{c}\right]}{(2p + 1)\,(2q + 1)}. \tag{7.7}$$

Comparing (7.7) with (7.5) we see that $A_{m,n}$ must be zero unless m and n are both odd, and in the case it is equal to $16/\pi^2 mn$. Thus, finally from (7.4)

$$v = \frac{16}{\pi^2} \sum_{p=0}^{\infty} \sum_{q=0}^{\infty} \frac{[v_1 \sinh l\,(a - x) + v_2 \sinh lx] \sin\left[(2p + 1)\frac{\pi y}{b}\right] \sin\left[(2q + 1)\frac{\pi z}{c}\right]}{(2p + 1)\,(2q + 1)\sinh la}, \tag{7.8}$$

where

$$l^2 = \frac{(2p + 1)^2 \pi^2}{b^2} + \frac{(2q + 1)^2 \pi^2}{c^2}. \tag{7.9}$$

2. *Steady temperature in the solid $0 < x < a$, $-b < y < b$, $-c < z < c$ when the faces $x = 0$ and $x = a$ are maintained at temperatures v_1 and v_2, respectively, and there is radiation at the other faces into the medium at zero.*

Here, the surface conditions are

$$v = v_1, \quad \text{when} \quad x = 0, \quad \text{and} \quad v = v_2, \quad \text{when} \quad x = a, \tag{7.10}$$

$$\frac{\partial v}{\partial y} - hv = 0, \quad \text{when} \quad y = -b, \quad \text{and} \quad \frac{\partial v}{\partial y} + hv = 0, \quad \text{when} \quad y = b, \tag{7.11}$$

$$\frac{\partial v}{\partial z} - hv = 0, \quad \text{when} \quad z = -c, \quad \text{and} \quad \frac{\partial v}{\partial z} + hv = 0, \quad \text{when} \quad z = c. \tag{7.12}$$

The expression

$$\frac{v_1 \sinh l\,(a - x) + v_2 \sinh lx}{\sinh la} \cos \alpha_r y \cos \beta_s z, \tag{7.13}$$

satisfies the equation of conduction, provided

$$l^2 = \alpha_r^2 + \beta_s^2. \tag{7.14}$$

Also, it satisfies the surface conditions (7.11) and (7.12), provided α_r is a root of

$$\alpha \tan \alpha b = h, \tag{7.15}$$

and β_s is a root of

$$\beta \tan \beta c = h. \tag{7.16}$$

Thus, the solution of the problem is given by

$$\sum_{r=1}^{\infty} \sum_{s=1}^{\infty} A_{rs} \frac{v_1 \sinh l\,(a - x) + v_2 \sinh lx}{\sinh la} \cos \alpha_r y \cos \beta_s z, \tag{7.17}$$

provided that the constants A_{rs} are chosen so that

$$\sum_{r=1}^{\infty} \sum_{s=1}^{\infty} A_{rs} \cos \alpha_r y \cos \beta_s z = 1. \tag{7.18}$$

If α_r and β_s are positive roots of (7.15) and (7.16), respectively,

$$2h \sum_{r=1}^{\infty} \frac{\cos \alpha_r y}{\left[(\alpha_r^2 + h^2)\, b + h \right] \cos \alpha_r b} = 1, \qquad 2h \sum_{s=1}^{\infty} \frac{\cos \beta_s z}{\left[(\beta_s^2 + h^2)\, c + h \right] \cos \beta_s c} = 1. \tag{7.19}$$

Multiplying the latter two expressions gives

$$\sum_{r=1}^{\infty} \sum_{s=1}^{\infty} \frac{4h^2 \cos \alpha_r y \cos \beta_s z}{\cos \alpha_r b \cos \beta_s c \left[(\alpha_r^2 + h^2)\, b + h \right] \left[(\beta_s^2 + h^2)\, c + h \right]} = 1. \tag{7.20}$$

As before A_{rs} in (7.17) is found by comparing (7.18) and (7.20), and (7.17) becomes finally

$$v = \sum_{r=1}^{\infty} \sum_{s=1}^{\infty} \frac{4h^2 \left[v_1 \sinh l\,(a - x) + v_2 \sinh lx \right] \cos \alpha_r y \cos \beta_s z}{\cos \alpha_r b \cos \beta_s c \left[(\alpha_r^2 + h^2)\, b + h \right] \left[(\beta_s^2 + h^2)\, c + h \right] \sinh la}, \tag{7.21}$$

where l is defined in (7.14).

7.1.1.2　Cylindrical coordinates – radial heat flow in an infinite circular cylinder

The equation of conduction, when expressed in cylindrical coordinates (see Appendix A.2.2), becomes

$$\frac{1}{r}\frac{\partial}{\partial r}\left(r\frac{\partial v}{\partial r}\right) + \frac{1}{r^2}\frac{\partial^2 v}{\partial \theta^2} + \frac{\partial^2 v}{\partial z^2} = \frac{1}{\kappa}\frac{\partial v}{\partial t} \tag{7.22}$$

or

$$\frac{\partial v}{\partial t} = \kappa\left(\frac{\partial^2 v}{\partial r^2} + \frac{1}{r}\frac{\partial v}{\partial r} + \frac{1}{r^2}\frac{\partial^2 v}{\partial \theta^2} + \frac{\partial^2 v}{\partial z^2}\right), \tag{7.23}$$

where the position of a point (x, y, z) is determined by its distance from the origin r and its latitude θ, so that $x = r\cos\theta$ and $y = r\sin\theta$.

If a circular cylinder whose axis coincides with the axis of z is heated, and the initial and boundary conditions are independent of the coordinates θ and z, the temperature will be a function of r and t only, and this equation reduces to

$$\frac{\partial v}{\partial t} = \kappa\left(\frac{\partial^2 v}{\partial r^2} + \frac{1}{r}\frac{\partial v}{\partial r}\right). \tag{7.24}$$

In this case the flow of heat takes place in planes perpendicular to the axis, and the lines of flow are radial.

If the solid is a hollow cylinder, whose inner and outer radii are a and b, Equation (7.22) for the temperature becomes

$$\frac{d}{dr}\left(r\frac{dv}{dr}\right) = 0, \qquad a < r < b. \tag{7.25}$$

The general solution of this is

$$v = A + B\ln r, \tag{7.26}$$

where A and B are constants to be determined from the boundary conditions at $r = a$ and $r = b$.

1. $r = a$ *kept at temperature v_1 and $r = b$ at temperature v_2.*

   Here,

   $$v = \frac{v_1 \ln\frac{b}{r} + v_2 \ln\frac{r}{a}}{\ln\frac{b}{a}}. \tag{7.27}$$

   The rate of flow of heat per unit length is

   $$-2\pi r K\frac{dv}{dr} = \frac{2\pi K\left(v_1 - v_2\right)}{\ln\frac{b}{a}}, \tag{7.28}$$

   K being the thermal conductivity.

2. $r = a$ *kept at temperature v_1. At $r = b$ there is radiation into a medium at v_2, the boundary condition there being $\dfrac{dv}{dr} + h\left(v - v_2\right) = 0$, $r = b$.*

   Here

   $$v = \frac{v_1\left[1 + hb\ln\frac{b}{r}\right] + hbv_2 \ln\frac{r}{a}}{1 + hb\ln\frac{b}{a}}. \tag{7.29}$$

The outward rate of flow of heat per unit length of the cylinder is

$$2\pi K \left(v_1 - v_2\right) \frac{hb}{1 + hb \ln \frac{b}{a}}. \tag{7.30}$$

If $ah > 1$ Expression (7.30) decreases steadily as b increases from a, but if $ah < 1$ it has a maximum at $b = 1/h$. This implies that, under certain circumstances, it is possible to increase the heat loss from a pipe by surrounding it with insulating material.

3. *Heat supplied at a constant rate F_0, per unit length of the inner cylinder.*

Here, since it follows from (7.25) that $r(dv/dr)$ is constant, the flow of heat over any cylinder is independent of its radius, and

$$F_0 = -2\pi r K \frac{dv}{dr}, \qquad a < r < b. \tag{7.31}$$

Then, if v_1 and v_2 are the temperatures at radii r_1 and r_2 respectively, integrating (7.31) we have

$$2\pi K \left(v_1 - v_2\right) = F_0 \ln \frac{r_2}{r_1}. \tag{7.32}$$

This relation is independent of how the heat is supplied, and of the boundary conditions at the cylindrical surfaces. If the heat is supplied by a wire along the axis of the cylinder, of resistance R ohms per unit length and carrying current I amperes, we have

$$F_0 = jI^2 R, \tag{7.33}$$

where j is the number of calories in a joule.
If the thermal conductivity K depends on the temperature, a relation of type (7.32) still holds. Also (7.31) is still true, and if we introduce

$$K_m = \frac{1}{v_2 - v_1} \int_{v_1}^{v_2} K \, dv, \tag{7.34}$$

the mean conductivity over the range of temperature from v_1 to v_2, integrating (7.31) gives

$$2\pi K_m \left(v_1 - v_2\right) = F_0 \ln \frac{r_2}{r_1}. \tag{7.35}$$

4. *The composite hollow cylinder of n regions $(a_1, a_2), (a_2, a_3), ..., (a_n, a_{n+1})$ of conductivities $K_1, ..., K_n$.*

If $v_1, v_2, ..., v_{n+1}$ are the temperatures at $a_1, a_2, ..., a_{n+1}$, repeated application of (7.28) shows that the rate of flow of heat per unit length of the system, F, is

$$F = \frac{2\pi K_1 \left(v_1 - v_2\right)}{\ln \frac{a_2}{a_1}} = ... = \frac{2\pi K_n \left(v_n - v_{n+1}\right)}{\ln \frac{a_{n+1}}{a_n}}. \tag{7.36}$$

Therefore

$$v_1 - v_{n+1} = \frac{F}{2\pi} \sum_{r=1}^{n} \frac{\ln \frac{a_{r+1}}{a_r}}{K_r}. \tag{7.37}$$

If, in addition, there are contact resistances $R_1, R_2, ..., R_n, R_{n+1}$ per unit area over the surfaces $a_1, a_2, ..., a_n, a_{n+1}$, and v_0 and v_{n+2} are the temperatures inside and outside the composite cylinder,

$$v_0 - v_{n+2} = \frac{F}{2\pi} \left\{ \sum_{r=1}^{n} \frac{\ln \frac{a_{r+1}}{a_r}}{K_r} + \sum_{r=1}^{n+1} \frac{R_r}{a_r} \right\}. \tag{7.38}$$

Equation (7.32) is a simple special case of this.

7.1.1.3   Spherical polar coordinates – radial heat flow in a sphere

The conduction equation, when expressed in spherical polar coordinates (see Appendix A.2.3), becomes

$$\frac{\partial v}{\partial t} = \kappa \left\{ \frac{1}{r^2} \frac{\partial}{\partial r} \left(r^2 \frac{\partial v}{\partial r} \right) + \frac{1}{r^2 \sin \theta} \frac{\partial}{\partial \theta} \left(\sin \theta \frac{\partial v}{\partial \theta} \right) + \frac{1}{r^2 \sin^2 \theta} \frac{\partial^2 v}{\partial \phi^2} \right\}. \tag{7.39}$$

Here, $x = r \sin \theta \cos \phi$, $y = r \sin \theta \sin \phi$ and $z = r \cos \theta$, with r being the distance of the position of a point from the origin, θ its latitude and ϕ its azimuth.

In the case of flow of heat in a sphere, when the initial and surface conditions are such that the isothermal surfaces are concentric spheres, and the temperature, thus, depends only upon the coordinates r and t, this equation becomes

$$\frac{\partial v}{\partial t} = \kappa \left\{ \frac{\partial v^2}{\partial r^2} + \frac{2}{r} \frac{\partial v}{\partial r} \right\}. \tag{7.40}$$

On putting $u = vr$, we have

$$\frac{\partial u}{\partial t} = \kappa \frac{\partial u^2}{\partial r^2}. \tag{7.41}$$

In the case of steady temperature, the radial flow is described by

$$\frac{d}{dr} \left(r^2 \frac{dv}{dr} \right) = 0. \tag{7.42}$$

The general solution of this differential equation is

$$v = \frac{A}{r} + B, \tag{7.43}$$

where A and B are constants to be determined from the boundary conditions.

1. *The hollow sphere $a < r < b$. $r = a$ at v_1, and $r = b$ at v_2.*

$$v = \frac{av_1 (b - r) + bv_2 (r - a)}{r (b - a)}. \tag{7.44}$$

2. *The hollow sphere $a < r < b$. $r = a$ at v_1, and at $r = b$ radiation into medium at v_2.*

$$v = \frac{av_1 [hb^2 + r (1 - hb)] + hb^2 v_2 (r - a)}{r [hb^2 + a (1 - hb)]}. \tag{7.45}$$

3. *The hollow sphere $a < r < b$. At $r = a$ radiation from medium at v_1, and at $r = b$ radiation into medium at v_2.*

If the boundary conditions are

$$\frac{\partial v}{\partial r} + h_1 \left(v_1 - v \right) = 0, \quad r = a, \qquad \frac{\partial v}{\partial r} + h_2 \left(v - v_2 \right) = 0, \quad r = b, \tag{7.46}$$

the solution is

$$v = \frac{v_1 a^2 h_1 \left[b^2 h_2 - r \left(b h_2 - 1 \right) \right] + v_2 b^2 h_2 \left[r \left(a h_1 + 1 \right) - a^2 h_1 \right]}{r \left[b^2 h_2 \left(a h_1 + 1 \right) - a^2 h_1 \left(b h_2 - 1 \right) \right]}. \tag{7.47}$$

4. *Constant flux $Q_0/4\pi a^2$ at the inner surface $r = a$ of the hollow sphere $a < r < b$.*

Since $-4\pi r^2 K \frac{dv}{dr}$ is the rate of heat flow over any spherical surface of radius r, and by (7.42) this is constant, we have

$$Q_0 = -4\pi r^2 K \frac{dv}{dr}, \qquad a < r < b. \tag{7.48}$$

If v_1 and v_2 are the temperatures at $r = a$ and $r = b$, respectively, it follows on integrating again that

$$Q_0 = \frac{4\pi K \left(v_1 - v_2 \right) ab}{b - a}. \tag{7.49}$$

If the conductivity K is a function of the temperature, (7.48) is still true, and integrating gives

$$Q_0 \left(\frac{1}{a} - \frac{1}{b} \right) = 4\pi \int_{v_2}^{v_1} K \, dv = 4\pi K_m \left(v_1 - v_2 \right), \tag{7.50}$$

where K_m is the mean conductivity over the range of temperature from a to b. Thus, (7.49) remains true with K replaced by K_m.

5. *The composite hollow sphere of n regions $(a_1, a_2), (a_2, a_3), ..., (a_n, a_{n+1})$ of conductivities $K_1, ..., K_n$.*

If $v_1, v_2, ..., v_{n+1}$ are the temperatures at $a_1, a_2, ..., a_{n+1}$, repeated application of (7.49) gives

$$Q_0 = \frac{4\pi K_1 \left(v_1 - v_2 \right) a_1 a_2}{a_2 - a_1} = ... = \frac{4\pi K_n \left(v_n - v_{n+1} \right) a_{n+1} a_n}{a_{n+1} - a_n}. \tag{7.51}$$

Therefore,

$$v_1 - v_{n+1} = \frac{Q_0}{4\pi} \sum_{r=1}^{n} \frac{1}{K_r} \left(\frac{1}{a_r} - \frac{1}{a_{r+1}} \right). \tag{7.52}$$

If, in addition, there are contact resistances $R_1, R_2, ..., R_n, R_{n+1}$ per unit area over the surfaces $a_1, a_2, ..., a_n, a_{n+1}$, and v_0 and v_{n+2} are the temperatures inside and outside the composite sphere,

$$v_0 - v_{n+2} = \frac{Q_0}{4\pi} \left\{ \sum_{r=1}^{n} \frac{1}{K_r} \left(\frac{1}{a_r} - \frac{1}{a_{r+1}} \right) + \sum_{r=1}^{n+1} \frac{R_r}{a_r^2} \right\}. \tag{7.53}$$

7.1.2   Variable temperature

For certain types of initial and boundary conditions, the solution of problems in several space variables can be written down as the product of solutions of one-variable problems.

7.1.2.1   Cartesian coordinates

Consider the equation of heat conduction

$$\frac{\partial^2 v}{\partial x_1^2} + \frac{\partial^2 v}{\partial x_2^2} + \frac{\partial^2 v}{\partial x_3^2} = \frac{1}{\kappa}\frac{\partial v}{\partial t}, \quad t > 0, \tag{7.54}$$

in the rectangular parallelepiped

$$a_1 < x_1 < b_1, \qquad a_2 < x_2 < b_2, \qquad a_3 < x_3 < b_3. \tag{7.55}$$

For certain important types of initial and boundary conditions, the solution is the product of the solutions of three one-variable problems, and thus can be written down immediately if these are known.

Suppose $v_r(x_r, t)$, $r = 1, 2, 3$, is the solution of

$$\frac{\partial^2 v_r}{\partial x_r^2} = \frac{1}{\kappa}\frac{\partial v}{\partial t}, \qquad a_r < x_r < b_r, \quad t > 0, \tag{7.56}$$

with the boundary conditions

$$\alpha_r \frac{\partial v_r}{\partial x_r} - \beta_r v_r = 0, \qquad x_r = a_r, \qquad t > 0,$$

$$\alpha_r' \frac{\partial v_r}{\partial x_r} + \beta_r' v_r = 0, \qquad x_r = b_r, \qquad t > 0, \tag{7.57}$$

(where α_r, β_r, etc. are constants, either of which may be zero, so that the cases of zero surface temperature and of no flow of heat at the surface are included) and with initial conditions

$$v_r(x_r, t) = V_r(x_r), \qquad t = 0, \qquad a_r < x_r < b_r. \tag{7.58}$$

Then the solution of (7.54) in the region (7.55), with

$$v = V_1(x_1)V_2(x_2)V_3(x_3), \qquad \text{when} \quad t = 0, \tag{7.59}$$

and with the boundary conditions

$$\alpha_r \frac{\partial v_r}{\partial x_r} - \beta_r v_r = 0, \qquad x_r = a_r, \qquad t > 0, \quad r = 1, 2, 3,$$

$$\alpha_r' \frac{\partial v_r}{\partial x_r} + \beta_r' v_r = 0, \qquad x_r = b_r, \qquad t > 0, \quad r = 1, 2, 3, \tag{7.60}$$

is

$$v = v_1(x_1, t)v_2(x_2, t)v_3(x_3, t). \tag{7.61}$$

Substituting the latter equation in (7.54) gives, using (7.56),

$$v_2 v_3 \frac{\partial^2 v_1}{\partial x_1^2} + v_3 v_1 \frac{\partial^2 v_2}{\partial x_2^2} + v_1 v_2 \frac{\partial^2 v_3}{\partial x_3^2} - \frac{1}{\kappa}\left(v_2 v_3 \frac{\partial v_1}{\partial t} + v_3 v_1 \frac{\partial v_2}{\partial t} + v_1 v_2 \frac{\partial v_3}{\partial t} \right) = 0. \tag{7.62}$$

Obviously, the initial and boundary conditions (7.59) and (7.60) are satisfied.

Here, we give results for unit initial temperature and zero surface temperature (or radiation into a medium at zero). As are the solutions for steady temperature, they are a reproduction of the presentations in [66]. For further examples and their solutions see the original. It should be mentioned that solutions for zero initial temperature and unit surface temperature (or radiation into the medium at unity) are obtained by subtracting the results given below from unity. Solutions for arbitrary surface temperatures then follow by Duhamel's Theorem (see [66], §1.14). If the solid is anisotropic with thermal axes parallel to the coordinate planes, and if the surface conductances at the faces are different the method is still applicable.

1. *The region $x > 0$, $y > 0$, $z > 0$, with unit initial temperature and zero surface temperature.*

$$v = \operatorname{erf} \frac{x}{2\sqrt{\kappa t}} \operatorname{erf} \frac{y}{2\sqrt{\kappa t}} \operatorname{erf} \frac{z}{2\sqrt{\kappa t}}. \tag{7.63}$$

2. *The same region with unit initial temperature and radiation at the surface into the medium at zero.*

$$v = \phi\left(h, x\right) \phi\left(h, y\right) \phi\left(h, z\right), \tag{7.64}$$

where

$$\phi\left(h, x\right) = \operatorname{erf} \frac{x}{2\sqrt{\kappa t}} + e^{hx + h^2 \kappa t} \operatorname{erfc} \left\{ \frac{x}{2\sqrt{\kappa t}} + h\sqrt{\kappa t} \right\}. \tag{7.65}$$

3. *The region $-a < x < a$, $-b < y < b$, $z > 0$, with unit initial temperature and zero surface temperature.*

$$v = \psi\left(x, a\right) \psi\left(y, b\right) \operatorname{erf} \frac{x}{2\sqrt{\kappa t}}, \tag{7.66}$$

where

$$\psi\left(x, a\right) = \frac{4}{\pi} \sum_{n=0}^{\infty} \frac{(-1)^n}{2n + 1} e^{-\kappa(2n+1)^2 \pi^2 t / 4a^2} \cos \frac{(2n + 1)\pi x}{2a}. \tag{7.67}$$

4. *The region $-a < x < a$, $-b < y < b$, $-c < z < c$, with unit initial temperature and zero surface temperature.*

$$v = \psi\left(x, a\right) \psi\left(y, b\right) \psi\left(z, c\right)$$

$$= \frac{64}{\pi^3} \sum_{l=0}^{\infty} \sum_{m=0}^{\infty} \sum_{n=0}^{\infty} \frac{(-1)^{l+m+n}}{(2l + 1)(2m + 1)(2n + 1)} \tag{7.68}$$

$$\times \cos \frac{(2l + 1)\pi x}{2a} \cos \frac{(2m + 1)\pi y}{2b} \cos \frac{(2n + 1)\pi z}{2c} e^{-\alpha_{l,m,n} t},$$

where

$$\alpha_{l,m,n} = \frac{\kappa \pi^2}{4} \left[\frac{(2l + 1)^2}{a^2} + \frac{(2m + 1)^2}{b^2} + \frac{(2n + 1)^2}{c^2} \right]. \tag{7.69}$$

5. *For the region $-a < x < a$, $-b < y < b$, $-c < z < c$, with unit initial temperature and radiation at the surface into the medium at zero.*

$$v = \psi(x, a, h)\, \psi(y, b, h)\, \psi(z, c, h),\tag{7.70}$$

where

$$\psi(x, a, h) = \sum_{n=1}^{\infty} \frac{2h \cos \alpha_n x}{[(h^2 + \alpha_n^2) a + h] \cos \alpha_n a}\, e^{-\kappa \alpha_n^2 t},\tag{7.71}$$

and the α_n are the positive roots of

$$\alpha \tan \alpha a = h.\tag{7.72}$$

6. *The region $-a < x < a$, $-b < y < b$, $-c < z < c$, with zero initial temperature and surface temperature $\phi(t)$.*

If $\phi(t) = V$, constant, the solution, which follows from (7.68), is

$$
\begin{aligned}
v = V - \frac{64V}{\pi^3} \sum_{l=0}^{\infty} \sum_{m=0}^{\infty} \sum_{n=0}^{\infty} \frac{(-1)^{l+m+n}}{(2l+1)(2m+1)(2n+1)} \\
\times \cos \frac{(2l+1)\pi x}{2a} \cos \frac{(2m+1)\pi y}{2b} \cos \frac{(2n+1)\pi z}{2c} e^{-\alpha_{l,m,n} t}.
\end{aligned}
\tag{7.73}
$$

For surface temperature $\phi(t)$, by Duhamel's Theorem,

$$
\begin{aligned}
v = \frac{64}{\pi^3} \sum_{l=0}^{\infty} \sum_{m=0}^{\infty} \sum_{n=0}^{\infty} \frac{\alpha_{l,m,n}\,(-1)^{l+m+n}}{(2l+1)(2m+1)(2n+1)} \\
\times \cos \frac{(2l+1)\pi x}{2a} \cos \frac{(2m+1)\pi y}{2b} \cos \frac{(2n+1)\pi z}{2c} e^{-\alpha_{l,m,n} t} \int_0^t e^{\alpha_{l,m,n}\lambda} \phi(\lambda)\, d\lambda.
\end{aligned}
\tag{7.74}
$$

If $\phi(t) = kt$, this gives

$$
\begin{aligned}
v = kt - \frac{64k}{\pi^3} \sum_{l=0}^{\infty} \sum_{m=0}^{\infty} \sum_{n=0}^{\infty} \frac{(-1)^{l+m+n}}{(2l+1)(2m+1)(2n+1)} \left(1 - e^{-\alpha_{l,m,n} t}\right) \\
\times \cos \frac{(2l+1)\pi x}{2a} \cos \frac{(2m+1)\pi y}{2b} \cos \frac{(2n+1)\pi z}{2c},
\end{aligned}
\tag{7.75}
$$

where the value of (7.73) with $t = 0$ has been used to reduce one of the series.

7.1.2.2   Cylindrical coordinates

A similar result to (7.61) holds for combined radial and axial flow in a solid or hollow cylinder. Here, the differential equation (7.23) becomes, since we are assuming that all quantities are independent of θ,

$$\frac{1}{r}\frac{\partial}{\partial r}\left(r\frac{\partial v}{\partial r}\right) + \frac{\partial^2 v}{\partial z^2} = \frac{1}{\kappa}\frac{\partial v}{\partial t}.\tag{7.76}$$

Suppose it has to be solved in the region

$$a < r < b, \qquad z_1 < z < z_2. \tag{7.77}$$

Let $v_1(r, t)$ be the solution of

$$\frac{1}{r}\frac{\partial}{\partial r}\left(r\frac{\partial v_1}{\partial r}\right) = \frac{1}{\kappa}\frac{\partial v_1}{\partial t}, \qquad t > 0, \quad a < r < b, \tag{7.78}$$

with

$$\alpha_1\frac{\partial v_1}{\partial r} - \beta_1 v_1 = 0, \qquad r = a, \qquad t > 0,$$

$$\alpha_1'\frac{\partial v_1}{\partial r} + \beta_1' v_1 = 0, \qquad r = b, \qquad t > 0, \tag{7.79}$$

and $v_1 = V_1(r)$ when $t = 0$.

Also let $v_2(z, t)$ be the solution of

$$\frac{\partial^2 v_2}{\partial z^2} = \frac{1}{\kappa}\frac{\partial v_2}{\partial t}, \qquad t > 0, \quad z_1 < z < z_2, \tag{7.80}$$

with

$$\alpha_2\frac{\partial v_2}{\partial z} - \beta_2 v_2 = 0, \qquad z = z_1, \qquad t > 0,$$

$$\alpha_2'\frac{\partial v_2}{\partial z} + \beta_2' v_2 = 0, \qquad z = z_2, \qquad t > 0, \tag{7.81}$$

and $v_2 = V_2(z)$ when $t = 0$.

Then, $v = v_1(r, t)v_2(z, t)$ is the solution of (7.76) in the region (7.77) with the boundary conditions

$$\alpha_1\frac{\partial v}{\partial r} - \beta_1 v = 0, \qquad r = a, \qquad z_1 < z < z_2, \qquad t > 0,$$

$$\alpha_1'\frac{\partial v}{\partial r} + \beta_1' v = 0, \qquad r = b, \qquad z_1 < z < z_2, \qquad t > 0,$$

$$\alpha_2\frac{\partial v}{\partial z} - \beta_2 v = 0, \qquad z = z_1, \qquad a < r < b, \qquad t > 0, \tag{7.82}$$

$$\alpha_2'\frac{\partial v}{\partial z} + \beta_2' v = 0, \qquad z = z_2, \qquad a < r < b, \qquad t > 0,$$

and with the initial condition $v = V_1(r)V_2(z)$ when $t = 0$.

We consider now the radial flow in an infinite cylinder at variable temperature.

Let the initial temperature be given by $v = f(r)$ and let the surface $r = a$ be kept at a constant temperature, which may be taken as zero[1].

The equations for v are as follows:

$$\frac{\partial v}{\partial t} = \kappa\left(\frac{\partial^2 v}{\partial r^2} + \frac{1}{r}\frac{\partial v}{\partial r}\right), \qquad 0 < r < a, \tag{7.83}$$

$$v = 0, \quad \text{when} \quad r = a, \qquad \text{and} \qquad v = f(r), \quad \text{when} \quad t = 0.$$

---

[1]If the constant surface temperature is v_0, we may reduce this to the case of zero temperature by putting $v = v_0 + w$.

If we put $v = \mathrm{e}^{-\kappa\alpha^2 t}u$, where u is a function of r only, then we must have

$$\frac{\partial^2 u}{\partial r^2} + \frac{1}{r}\frac{\partial u}{\partial r} + \alpha^2 u = 0, \tag{7.84}$$

which is Bessel's equation of order zero (see Appendix D).

As the solution of the second kind is infinite at $r = 0$, the particular integral of the temperature equation suitable for our problem is

$$v = AJ_0\left(\alpha r\right)\mathrm{e}^{-\kappa\alpha^2 t}, \tag{7.85}$$

where $J_0(x)$ is the Bessel function of order zero of the first kind (see Appendix D).

To satisfy the boundary condition α must be a root of

$$J_0(a\alpha) = 0. \tag{7.86}$$

It is known [419] that this equation has no complex roots or repeated roots, and that it has an infinite number of real positive roots α_1, α_2, α_3, ... To each positive root α there corresponds a negative root $-\alpha$. The first few roots are given in [66].

If $f(r)$ can be expanded in the series

$$A_1 J_0(\alpha_1 r) + A_2 J_0(\alpha_2 r) + ..., \tag{7.87}$$

the conditions of the problem will be satisfied by

$$v = \sum_{n=1}^{\infty} A_n J_0(\alpha_n r)\mathrm{e}^{-\kappa\alpha_n^2 t}. \tag{7.88}$$

For further problems and their analytic solutions concerning cylinders (comparable to the different cases studied in Subsection 7.1.2.1) see [66].

7.1.2.3   Spherical polar coordinates

We consider here the sphere $0 \le r < a$ with initial temperature $f(r)$ and surface temperature $\phi(t)$. As remarked in Subsection 7.1.1.3 we substitute

$$u = vr, \tag{7.89}$$

and the equations for u are

$$\frac{\partial u}{\partial t} = \kappa\frac{\partial u^2}{\partial r^2}, \qquad 0 \le r < a, \tag{7.90}$$

with

$$\begin{aligned}
u &= 0, &\text{when} \quad r &= 0,\\
u &= a\phi(t), &\text{when} \quad r &= a,\\
u &= rf(r), &\text{when} \quad t &= 0.
\end{aligned} \tag{7.91}$$

These equations are identical to those of heat flow in a slab of thickness a, with its ends $r = 0$ and $r = a$ kept at zero and $a\phi(t)$ respectively, and with initial temperature $rf(r)$. The solution to these problems is

$$v = \frac{2}{ar}\sum_{n=1}^{\infty}\mathrm{e}^{-\kappa n^2\pi^2 t/a^2}\sin\frac{n\pi r}{a}$$

$$\times\left\{\int_0^a r'f(r')\sin\frac{n\pi r'}{a}\,dr' - n\pi\kappa(-1)^n\int_0^t \mathrm{e}^{\kappa n^2\pi^2\lambda/a^2}\phi(\lambda)\,d\lambda\right\}. \tag{7.92}$$

Results for a few special cases can be found below. For further examples see [66].

1. *Zero initial temperature, surface temperature V, constant.*

$$v = V + \frac{2aV}{\pi r} \sum_{n=1}^{\infty} \frac{(-1)^n}{n} \sin \frac{n\pi r}{a} e^{-\kappa n^2 \pi^2 t/a^2}$$

$$= \frac{2aV}{r} \sum_{n=0}^{\infty} \left\{ \operatorname{erfc} \frac{(2n+1)\,a - r}{2\,(\kappa t)^{\frac{1}{2}}} - \operatorname{erfc} \frac{(2n+1)\,a + r}{2\,(\kappa t)^{\frac{1}{2}}} \right\}. \tag{7.93}$$

The temperature v_c at the center, given by the limit as $r \to 0$ in (7.93), is

$$v_c = V + 2V \sum_{n=1}^{\infty} (-1)^n e^{-\kappa n^2 \pi^2 t/a^2}$$

$$= \frac{aV}{(\pi \kappa t)^{\frac{1}{2}}} \sum_{n=0}^{\infty} e^{-(2n+1)^2 a^2/4\kappa t}. \tag{7.94}$$

The average temperature v_{av} of the sphere at any time is

$$v_{av} = V - \frac{6V}{\pi^2} \sum_{n=1}^{\infty} \frac{1}{n^2} e^{-\kappa n^2 \pi^2 t/a^2}$$

$$= \frac{6V\,(\kappa t)^{\frac{1}{2}}}{a\pi^{\frac{1}{2}}} - \frac{3\kappa V t}{a^2} + \frac{12V\,(\kappa t)^{\frac{1}{2}}}{a} \sum_{n=1}^{\infty} \operatorname{ierfc} \frac{na}{(\kappa t)^{\frac{1}{2}}}. \tag{7.95}$$

The heat content of the sphere at any time is $4\pi a^3 \rho c v_{av}/3$.
In Figure 7.1 values of v/V calculated from (7.93) for various values of $T = \kappa t/a^2$ are plotted as a function of r/a.

2. *Zero initial temperature, surface temperature kt.*

$$v = k \left[t - \frac{a^2 - r^2}{6\kappa} \right] - \frac{2ka^3}{\kappa \pi^3 r} \sum_{n=1}^{\infty} \frac{(-1)^n}{n^3} \sin \frac{n\pi r}{a} e^{-\kappa n^2 \pi^2 t/a^2}. \tag{7.96}$$

The average temperature is

$$v_{av} = k \left\{ t - \frac{a^2}{15\kappa} \right\} + \frac{6ka^2}{\kappa \pi^4} \sum_{n=1}^{\infty} \frac{1}{n^4} e^{-\kappa n^2 \pi^2 t/a^2}. \tag{7.97}$$

3. *Initial temperature $V\,(a - r)\,/a$, zero surface temperature.*

$$v = \frac{8aV}{\pi^3 r} \sum_{n=0}^{\infty} \frac{1}{(2n+1)^3} \sin \frac{(2n+1)\,\pi r}{a} e^{-\kappa(2n+1)^2 \pi^2 t/a^2}$$

$$= \frac{V\,(a - r)}{a} - \frac{2V\kappa t}{ar} + \frac{8V\kappa t}{ar} \sum_{n=0}^{\infty} (-1)^n \left\{ i^2 \operatorname{erfc} \frac{na + r}{2\,(\kappa t)^{\frac{1}{2}}} + i^2 \operatorname{erfc} \frac{(n+1)\,a - r}{2\,(\kappa t)^{\frac{1}{2}}} \right\}. \tag{7.98}$$

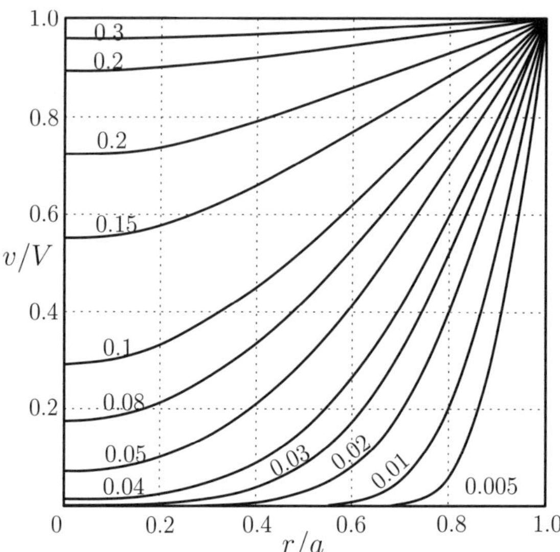

Figure 7.1: *Temperature distribution at various times in a sphere of radius a with zero initial temperature and surface temperature V. The numbers on the curves are the values of $\kappa t/a^2$ (after [66]).*

4. *Initial temperature V, constant, in $0 < r < b$, zero in $b < r < a$, zero surface temperature.*

$$v = \frac{2V}{r} \sum_{n=1}^{\infty} \left\{ \frac{a}{n^2\pi^2} \sin \frac{n\pi b}{a} - \frac{b}{n\pi} \cos \frac{n\pi b}{a} \right\} \sin \frac{n\pi r}{a} e^{-\kappa n^2 \pi^2 t/a^2}. \tag{7.99}$$

Several other examples and their solutions can be found in the source of the above mentioned cases, the book by Carslaw & Jaeger [66].

7.2 Heat conduction in anisotropic solids

Anisotropic media are of considerable importance in practice. Some examples are crystals, some sedimentary rocks, wood or laminated materials. Their behavior is described in [66] and this introduction is repeated here.

In contrast to isotropic media in which the direction of the heat flux vector at a point is normal to the isothermal through the point, for anisotropic media this is usually not true. Thus, instead of the fluxes across the three planes parallel to the axes of coordinates for isotropic media

$$f_x = -K\frac{\partial v}{\partial x}, \qquad f_y = -K\frac{\partial v}{\partial y}, \qquad f_z = -K\frac{\partial v}{\partial z}, \tag{7.100}$$

with the thermal conductivity K, we generalize for anisotropic media that each component of the flux vector at a point is a linear function of the components of the temperature gradient at the point, that is

$$
\left.
\begin{aligned}
-f_x &= K_{11}\frac{\partial v}{\partial x} + K_{12}\frac{\partial v}{\partial y} + K_{13}\frac{\partial v}{\partial z} \\[2mm]
-f_y &= K_{21}\frac{\partial v}{\partial x} + K_{22}\frac{\partial v}{\partial y} + K_{23}\frac{\partial v}{\partial z} \\[2mm]
-f_z &= K_{31}\frac{\partial v}{\partial x} + K_{32}\frac{\partial v}{\partial y} + K_{33}\frac{\partial v}{\partial z}
\end{aligned}
\right\},
\tag{7.101}
$$

where the quantities K_{rs} are the conductivity coefficients. This expression frequently relates two vectors in anisotropic media. Having regard to certain types of symmetry, the general form may be simplified to different extent.

We proceed to develop the theory on the general assumption (7.101). Substituting this into the equation

$$
\rho c\frac{\partial v}{\partial t} + \left(\frac{\partial f_x}{\partial x} + \frac{\partial f_y}{\partial y} + \frac{\partial f_z}{\partial z}\right) = 0,
\tag{7.102}
$$

which holds at any point of the solid, provided no heat is supplied at the point, and where ρ is the density and c the specific heat (at temperature v) of the solid, gives the equation of heat conduction

$$
\rho c\frac{\partial v}{\partial t} = K_{11}\frac{\partial^2 v}{\partial x^2} + K_{22}\frac{\partial^2 v}{\partial y^2} + K_{33}\frac{\partial^2 v}{\partial z^2} + (K_{23} + K_{32})\frac{\partial^2 v}{\partial y\partial z}
$$
$$
+ (K_{31} + K_{13})\frac{\partial^2 v}{\partial z\partial x} + (K_{12} + K_{21})\frac{\partial^2 v}{\partial x\partial y}.
\tag{7.103}
$$

This equation holds provided that the medium is homogeneous and that no heat is generated in it.

It will be used further that a quadric

$$
K_{11}x^2 + K_{22}y^2 + K_{33}z^2 + (K_{23} + K_{32})\,yz + (K_{31} + K_{13})\,zx + (K_{12} + K_{21})\,xy = \text{const.}
\tag{7.104}
$$

can be transformed to a new system of rectangular coordinates ξ, η, ζ, so that the left-hand side is reduced to a sum of squares

$$
K_1\xi^2 + K_2\eta^2 + K_3\zeta^2.
\tag{7.105}
$$

In terms of these variables, (7.103) becomes

$$
\rho c\frac{\partial v}{\partial t} = K_1\frac{\partial^2 v}{\partial \xi^2} + K_2\frac{\partial^2 v}{\partial \eta^2} + K_3\frac{\partial^2 v}{\partial \zeta^2}.
\tag{7.106}
$$

The new axes are the principal axes of conductivity and the coefficients K_1, K_2, K_3 are the principal conductivities.

If we make the additional transformation

$$
\xi_1 = \xi\,(K/K_1)^{\frac{1}{2}}, \qquad \eta_1 = \eta\,(K/K_2)^{\frac{1}{2}}, \qquad \zeta_1 = \zeta\,(K/K_3)^{\frac{1}{2}},
\tag{7.107}
$$

where K may be chosen arbitrarily, (7.106) becomes

$$
\frac{\partial v}{\partial t} = \frac{K}{\rho c}\left(\frac{\partial^2 v}{\partial \xi_1^2} + \frac{\partial^2 v}{\partial \eta_1^2} + \frac{\partial^2 v}{\partial \zeta_1^2}\right).
\tag{7.108}
$$

This equation has the same form as the equation for the isotropic solid

$$\frac{\partial^2 v}{\partial x^2} + \frac{\partial^2 v}{\partial y^2} + \frac{\partial^2 v}{\partial z^2} - \frac{1}{\kappa}\frac{\partial v}{\partial t} = 0, \tag{7.109}$$

with $\kappa = \frac{K}{\rho c}$, the diffusivity or thermometric conductivity. Thus, this transformation reduces problems on the anisotropic solid to the solution of corresponding problems on the isotropic solid when the solid is infinite, or when it is bounded by planes perpendicular to the principal axes of conductivity, or, in the case $K_2 = K_3$ by planes perpendicular to the axis of ξ and by circular cylinders with this as axis. For cases with a certain kind of symmetry the equation can further be simplified. For the homogeneous orthotropic solid, for example, the equation of conduction of heat is

$$K_1\frac{\partial^2 v}{\partial x^2} + K_2\frac{\partial^2 v}{\partial y^2} + K_3\frac{\partial^2 v}{\partial z^2} - \rho c\frac{\partial v}{\partial t} = 0, \tag{7.110}$$

and for a solid with cylindrical symmetry it is

$$\frac{K_1}{r}\frac{\partial}{\partial r}\left(r\frac{\partial v}{\partial r}\right) + \frac{K_2}{r^2}\frac{\partial^2 v}{\partial \theta^2} + K_3\frac{\partial^2 v}{\partial z^2} - \rho c\frac{\partial v}{\partial t} = 0. \tag{7.111}$$

Various important special cases in which the differential equation contains only one or two space variables may be reproduced from [66]:

1. *The temperature is a function of x only.*

   By symmetry, this is the case of flow into a semi-infinite solid or slab with faces perpendicular to the x-axis and surface conditions independent of y and z. In this case $\partial v/\partial y = \partial v/\partial z = 0$ and (7.101) gives

   $$-f_x = K_{11}\frac{\partial v}{\partial x}, \qquad -f_y = K_{21}\frac{\partial v}{\partial x}, \qquad -f_z = -K_{31}\frac{\partial v}{\partial x}, \tag{7.112}$$

   and the differential equation (7.103) becomes

   $$K_{11}\frac{\partial^2 v}{\partial x^2} - \rho c\frac{\partial v}{\partial t} = 0. \tag{7.113}$$

   Thus, the theory for isotropic media holds for the anisotropic solid with

   $$\kappa = \frac{K_{11}}{\rho c}. \tag{7.114}$$

   If the x-axis has direction cosines l, m, n relative to the principal axes of conductivity, then $K_{11} = l^2 K_1 + m^2 K_2 + n^2 K_3$. When v has been found, the fluxes f_x, f_y, f_z follow from (7.112) and it appears that the direction of the flux vector is not normal to the isothermals.

2. *Flow of heat in the x-direction only.*

   This is the case of a thin rod in the direction of the x-axis. We now have $f_y = f_z = 0$ and, from Equation (7.101) solved for $\partial v/\partial x$,

$$-\frac{\partial v}{\partial x} = R_{11}f_x + R_{12}f_y + R_{13}f_z$$

$$-\frac{\partial v}{\partial y} = R_{21}f_x + R_{22}f_y + R_{23}f_z \Bigg\} \,, \qquad (7.115)$$

$$-\frac{\partial v}{\partial z} = R_{31}f_x + R_{32}f_y + R_{33}f_z$$

we get

$$-\frac{\partial v}{\partial x} = R_{11}f_x, \qquad -\frac{\partial v}{\partial y} = R_{21}f_x, \qquad -\frac{\partial v}{\partial z} = R_{31}f_x, \qquad (7.116)$$

where the R_{rs} are called resistivity coefficients. The differential equation (7.103) now becomes

$$\frac{1}{R_{11}}\frac{\partial^2 v}{\partial x^2} - \rho c\frac{\partial v}{\partial t} = 0 \qquad (7.117)$$

and so it is the equation of linear flow of heat with

$$\kappa = \frac{1}{R_{11}\rho c}. \qquad (7.118)$$

It should be noted that R_{11} is not equal to $1/K_{11}$ but is given by

$$R_{11} = \frac{K_{22}K_{33} - K_{23}K_{32}}{\Delta}, \qquad \Delta = \begin{vmatrix} K_{11} & K_{12} & K_{13} \\ K_{21} & K_{22} & K_{23} \\ K_{31} & K_{32} & K_{33} \end{vmatrix}. \qquad (7.119)$$

If the x-axis has direction cosines l, m, n relative to the principal axes of conductivity, then $R_{11} = l^2/K_1 + m^2/K_2 + n^2/K_3$.

3. *The temperature is a function of x and y only.*

This is the case of flow into an infinite cylinder parallel to the z-axis with surface conditions independent of z. Since $\partial v/\partial z = 0$, (7.103) becomes

$$K_{11}\frac{\partial^2 v}{\partial x^2} + (K_{12} + K_{21})\frac{\partial^2 v}{\partial x \partial y} + K_{22}\frac{\partial^2 v}{\partial y^2} - \rho c\frac{\partial v}{\partial t} = 0, \qquad (7.120)$$

and the fluxes are given by (7.101) with $\partial v/\partial z = 0$. It will be noticed that $f_z \neq 0$.

7.2.1 Conduction in a thin crystal plate

One example covering conduction in two dimensions and illustrating the main features of the general case is reproduced from [66]. The general theory of flow in a thin crystal plate, without any assumptions of symmetry, will be developed.

Taking x- and y-axes in the plane of the plate, we assume that there is no flow of heat in the z-direction so that $f_z = 0$ and from (7.115) follows

$$-\frac{\partial v}{\partial x} = R_{11}f_x + R_{12}f_y, \qquad -\frac{\partial v}{\partial y} = R_{21}f_x + R_{22}f_y. \qquad (7.121)$$

Solving for f_x and f_y, and using (7.119) and analogously

$$R_{12} = \frac{K_{13}K_{32} - K_{12}K_{33}}{\Delta}, \qquad R_{21} = \frac{K_{31}K_{23} - K_{21}K_{33}}{\Delta}, \qquad R_{22} = \frac{K_{11}K_{33} - K_{31}K_{13}}{\Delta}, \tag{7.122}$$

and

$$R_{11}R_{22} - R_{12}R_{21} = \frac{K_{33}}{\Delta}, \tag{7.123}$$

gives

$$-f_x = K'_{11}\frac{\partial v}{\partial x} + K'_{12}\frac{\partial v}{\partial y}, \qquad -f_y = K'_{21}\frac{\partial v}{\partial x} + K'_{22}\frac{\partial v}{\partial y}, \tag{7.124}$$

where

$$\begin{aligned} K'_{11} &= (K_{11}K_{33} - K_{31}K_{13})/K_{33}, & K'_{12} &= (K_{12}K_{33} - K_{13}K_{32})/K_{33}, \\ K'_{21} &= (K_{21}K_{33} - K_{31}K_{23})/K_{33}, & K'_{22} &= (K_{22}K_{33} - K_{23}K_{32})/K_{33}. \end{aligned} \tag{7.125}$$

The four quantities K'_{11} etc., may be called the conductivity coefficients for a thin plate in the xy-plane: they reduce to K_{11}, etc., only if one of K_{13} and K_{31} and also one of K_{23} and K_{32} vanish.

Substituting (7.124) into (7.102) gives the equation of heat conduction in the steady state

$$K'_{11}\frac{\partial^2 v}{\partial x^2} + (K'_{12} + K'_{21})\frac{\partial^2 v}{\partial x \partial y} + K'_{22}\frac{\partial^2 v}{\partial y^2} = 0, \tag{7.126}$$

or in alternative form

$$R_{22}\frac{\partial^2 v}{\partial x^2} - (R_{12} + R_{21})\frac{\partial^2 v}{\partial x \partial y} + R_{11}\frac{\partial^2 v}{\partial y^2} = 0. \tag{7.127}$$

Referred to the principal axes ξ, η, (7.126) becomes

$$K'_1\frac{\partial^2 v}{\partial \xi^2} + K'_2\frac{\partial^2 v}{\partial \eta^2} = 0, \tag{7.128}$$

where K'_1 and K'_2 may be called the principal conductivities in the plane (the prime is used to distinguish them from the principal conductivities in three dimensions). Referred to these principal axes, the components of the flux, f_ξ, f_η, must take the form

$$-f_\xi = K'_1\frac{\partial v}{\partial \xi} + A\frac{\partial v}{\partial \eta}, \qquad -f_\eta = -A\frac{\partial v}{\partial \xi} + K'_2\frac{\partial v}{\partial \eta}, \tag{7.129}$$

since the equation of heat conduction (7.128) in these coordinates contains no term in $\partial^2 v/\partial\xi\partial\eta$, and, by comparison with (7.126), this requires that the sum of the coefficients of $\partial v/\partial\eta$ in the first equation of (7.129) and $\partial v/\partial\xi$ in the second equation of (7.129) must vanish. K'_1, K'_2, and A can, in principle, be determined from the coefficients in (7.121) and (7.124).

As a specific example of considerable practical importance, we determine the isothermals and lines of heat flow for the case of steady heat supply to the plate at the origin. Writing, as in (7.107),

$$\xi_1 = \xi\,(K/K'_1)^{\frac{1}{2}}, \qquad \eta_1 = \eta\,(K/K'_2)^{\frac{1}{2}}, \tag{7.130}$$

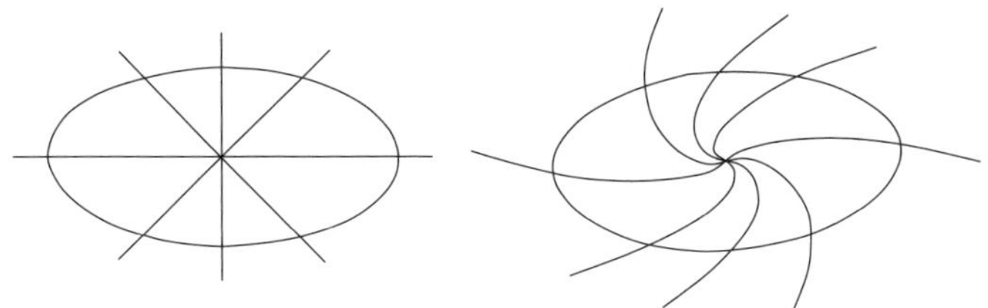

Figure 7.2: *Directions of heat flow from a point source in an infinite plate for $A = 0$ (left) and $A \neq 0$ (after [66]).*

(7.128) becomes

$$\frac{\partial^2 v}{\partial \xi_1^2} + \frac{\partial^2 v}{\partial \eta_1^2} = 0, \tag{7.131}$$

of which a solution with radial symmetry is

$$v = \frac{-m \ln (\xi_1^2 + \eta_1^2)}{K} = -m \ln \left(\frac{\xi^2}{K_1'} + \frac{\eta^2}{K_2'} \right), \tag{7.132}$$

where m is a constant. The flux is, by (7.129),

$$f_\xi = \frac{2m \left(K_2' \xi + A\eta \right)}{K_2' \left(\frac{\xi^2}{K_1'} + \frac{\eta^2}{K_2'} \right)}, \qquad f_\eta = \frac{2m \left(A\xi - K_1' \eta \right)}{K_1' \left(\frac{\xi^2}{K_1'} + \frac{\eta^2}{K_2'} \right)}. \tag{7.133}$$

The total quantity of heat Q crossing a circle of radius a about the origin (per unit thickness of the plate) is

$$Q = \int_0^{2\pi} (f_\xi \cos \theta + f_\eta \sin \theta) \, a \, d\theta, \tag{7.134}$$

and, using (7.133) with $\xi = a \cos \theta$, $\eta = a \sin \theta$, this gives

$$Q = 4\pi m \left(K_1' K_2' \right)^{\frac{1}{2}}, \tag{7.135}$$

which is independent of both a and A. With the value (7.135) of m, (7.132) gives the steady temperature due to supply of heat at the origin at the rate Q.

The isothermals are the family of ellipses

$$\frac{\xi^2}{K_1'} + \frac{\eta^2}{K_2'} = \text{const.} \tag{7.136}$$

The direction of the flux vector is given by

$$\frac{f_\eta}{f_\xi} = \frac{K_2' \left(K_1' \eta - A\xi \right)}{K_1' \left(K_2' \xi + A\eta \right)}. \tag{7.137}$$

If $A = 0$, this direction is radially from the origin (and not normal to the equipotentials) so that the equipotentials and lines of heat flow (curves whose direction at each point is in the direction of the flux vector) are as shown in Figure 7.2 (left).

If $A \neq 0$, the differential equation of the lines of heat flow is

$$\frac{d\eta}{d\xi} = \frac{K_2' \left(K_1'\eta - A\xi \right)}{K_1' \left(K_2'\xi + A\eta \right)}. \tag{7.138}$$

The solution of this is

$$\left(K_1' K_2' \right)^{\frac{1}{2}} \tan^{-1}\left(\eta \left(K_1' \right)^{\frac{1}{2}} / \xi \left(K_2' \right)^{\frac{1}{2}} \right) + \frac{1}{2} A \ln \left(K_1'\eta^2 + K_2'\xi^2 \right) = \text{const.} \tag{7.139}$$

The curves (7.139) are the family of spirals shown in Figure 7.2 (right): if $K_1' = K_2'$, or in the ξ_1, η_1 plane, these spirals are equiangular. Thus, if the so-called 'rotatory' term A is not zero, the directions of heat flow from a point source in an infinite plate are as shown in Figure 7.2 (right). It follows that if a slit is cut in a radial direction in the plate, heat will not be able to flow in these spirals so that there should be a difference in temperature between the two sides of the slit. In [66] it is stated that A is small (less than one thousandth of K_1 or K_2).

In [66] also the variation of thermal conductivity and the flux vector in anisotropic solids are studied. This is not repeated here.

7.3   Thermal boundary layers

In Section 5.3.2.4 the basic ideas of the Prandtl boundary layer theory have been presented. These considerations have referred only to the velocity field. However, one of the important problems of the theory of boundary layers is the coupling of mechanical and thermal effects. Then, the so-called thermal boundary layers appear. They emerge in many practical applications such as phase transformations (melting, evaporation, solidification, condensation, etc.) but also problems of heat transfer between civil engineering constructions and environment, air conditioning systems, etc. contain field equations whose boundary conditions concern transition regions in which thermal boundary layers appear. On pages 60-63 they have been roughly mentioned. The Boussinesq approximation has been introduced, i.e., the assumption that for stationary flows changes of the mass density are sufficiently small to be neglected in the mass balance.

In this section we will investigate the effect of variable physical properties. These properties are the density ρ, the viscosity μ, the isobaric specific heat capacity c_p and the thermal conductivity λ. In the most general case, the properties can depend on the temperature T and the pressure p. A consequence of the dependence of the density and the viscosity on the temperature is a coupling of the velocity field and the temperature field. Our considerations rely on the book by H. Schlichting [338].

If we use a coordinate system in the gravitational field which follows the contour of the body (see the sketch) with the local angle of inclination to the horizontal at position with

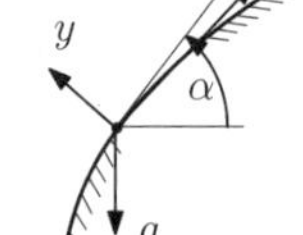

$$g_x = -g \sin\alpha, \qquad g_y = -g \cos\alpha, \tag{7.140}$$

then the equations of motion for steady, plane boundary-layer flows in dimensional form read

$$\frac{\partial\left(\rho u\right)}{\partial x} + \frac{\partial\left(\rho v\right)}{\partial y} \approx 0,$$

$$\rho\left(u\frac{\partial u}{\partial x} + v\frac{\partial u}{\partial y}\right) = -\rho g\sin\alpha - \frac{\partial p}{\partial x} + \frac{\partial}{\partial y}\left(\mu\frac{\partial u}{\partial y}\right), \tag{7.141}$$

$$\rho c_p\left(u\frac{\partial T}{\partial x} + v\frac{\partial T}{\partial y}\right) = \frac{\partial}{\partial y}\left(\lambda\frac{\partial T}{\partial y}\right) + \beta T u\frac{\partial p}{\partial x} + \mu\left(\frac{\partial u}{\partial y}\right)^2.$$

Here, λ is the thermal conductivity and β is the coefficient of thermal expansion. For constant physical properties, the first two equations of (7.141) reduce to (5.90) (of course, without the time derivative). Compare Equations (7.141) also to Equations (5.97) where the Boussinesq approximation has been used.

We consider boundary layers with moderate wall heat transfer and without gravitational effects. The latter means that the Froude number $Fr = \frac{V}{\sqrt{gL}}$ tends to infinity. Deviations of the temperature from T_∞, the temperature of the outer flow, as a consequence of the heat transfer at the wall occur only in the boundary layer. These temperature differences should remain small but still be enough that changes in the physical properties occur. We assume that the physical properties depend only on the temperature.

We repeat here the considerations presented in [338] on how the temperature dependence of the properties is taken into account in the boundary-layer calculation. As an example, the density function $\rho\left(T\right)$ is examined. This function is expanded at the position $T = T_\infty$ in a Taylor series

$$\rho\left(T\right) = \rho_\infty + \left(\frac{d\rho}{dT}\right)_\infty \left(T - T_\infty\right) + \dots. \tag{7.142}$$

With the dimensionless temperature

$$\vartheta = \frac{T - T_\infty}{\Delta T}, \tag{7.143}$$

where ΔT is a reference temperature difference, we have for standard cases

$$\begin{aligned} T_w = \text{const.} : &\quad \Delta T = T_w - T_\infty, \\ q_w = \text{const.} : &\quad \Delta T = q_w L/\lambda_\infty. \end{aligned} \tag{7.144}$$

For the density we then find

$$\rho\left(T\right) = \rho_\infty\left(1 + K_\rho\vartheta\frac{\Delta T}{T_\infty} + \dots\right), \tag{7.145}$$

with the dimensionless physical property

$$K_\rho = \left(\frac{d\rho}{dT}\frac{T}{\rho}\right)_\infty. \tag{7.146}$$

In order to find the boundary-layer solution, a regular perturbation calculation is carried out with the small parameter $\varepsilon = \Delta T/T_\infty$. In analogy to Equation (7.145), we obtain the other physical properties as

$$\begin{aligned} \mu\left(T\right) &= \mu_\infty\left(1 + K_\mu\vartheta\varepsilon + \dots\right), \\ c_p\left(T\right) &= c_{p\infty}\left(1 + K_c\vartheta\varepsilon + \dots\right), \\ \lambda\left(T\right) &= \lambda_\infty\left(1 + K_\lambda\vartheta\varepsilon + \dots\right). \end{aligned} \tag{7.147}$$

Table 7.1: *Physical properties and temperature dependence of air and water.*

Fluid	air			water		
T [K]	293	473	773	273	293	343
T [°C]	20	200	500	0	20	70
$\rho \left[\frac{\text{kg}}{\text{m}^3}\right]$	1.188	0.736	0.450	999.8	998.2	977.8
$\mu \left[\frac{10^{-6}\text{kg}}{\text{ms}}\right]$	18.185	25.850	35.800	1791.5	1001.6	403.9
$\nu \left[\frac{10^{-6}\text{m}^2}{\text{s}}\right]$	15.307	35.122	79.556	1.792	1.004	0.413
$\lambda \left[\frac{10^{-3}\text{W}}{\text{mK}}\right]$	25.721	38.660	56.346	561.1	598.5	663.1
$c_p \left[\frac{\text{kJ}}{\text{kg K}}\right]$	1.014	1.048	1.096	4.219	4.185	4.188
Pr	0.717	0.702	0.696	13.47	7.00	2.55
K_ρ	-1.000	-1.000	-1.000	0.018	-0.061	-0.200
K_μ	0.775	0.696	0.633	-9.264	-7.239	-4.758
K_λ	0.891	0.809	0.726	0.924	0.872	0.404
K_c	0.068	0.076	0.108	-0.226	-0.050	0.052
$\tilde{K}_\rho$	1	1	1	$5 \cdot 10^{-5}$	$5 \cdot 10^{-5}$	$5 \cdot 10^{-5}$
$\tilde{K}_\mu$	$6 \cdot 10^{-4}$	$3 \cdot 10^{-4}$	$1 \cdot 10^{-4}$	$-1 \cdot 10^{-4}$	$-5 \cdot 10^{-5}$	$6 \cdot 10^{-5}$
$\tilde{K}_\lambda$	$2 \cdot 10^{-3}$	$9 \cdot 10^{-4}$	$4 \cdot 10^{-4}$	$1 \cdot 10^{-4}$	$8 \cdot 10^{-5}$	$8 \cdot 10^{-5}$
$\tilde{K}_c$	$2 \cdot 10^{-3}$	$5 \cdot 10^{-4}$	$2 \cdot 10^{-4}$	$-1 \cdot 10^{-4}$	$-7 \cdot 10^{-5}$	$-5 \cdot 10^{-5}$

The values of $K_\rho = -\beta T$, K_μ, K_c and K_λ are given for some substances in Table 7.1 which is an excerpt of Table 3.1 of [338].

Inserting the trial solutions

$$\begin{aligned}
u(x,y) &= u_0(x,y) + \varepsilon \left[K_\rho u_{1\rho}(x,y) + K_\mu u_{1\mu}(x,y)\right], \\
v(x,y) &= v_0(x,y) + \varepsilon \left[K_\rho v_{1\rho}(x,y) + K_\mu v_{1\mu}(x,y)\right], \\
p(x) &= p_0(x) + \varepsilon \left[K_\rho p_{1\rho}(x) + K_\mu p_{1\mu}(x)\right], \\
\vartheta(x,y) &= \vartheta_0(x,y) + \varepsilon \left[K_\rho \vartheta_{1\rho}(x,y) + K_\mu \vartheta_{1\mu}(x,y) \right. \\
&\quad \left. + K_c \vartheta_{1c}(x,y) + K_\lambda \vartheta_{1\lambda}(x,y)\right],
\end{aligned} \tag{7.148}$$

into the boundary-layer equations (7.141), sorting powers of ε and ignoring terms proportional to ε^2, two systems of equations are obtained. The first is the system of equations for the boundary layer with constant physical properties. The second is a set of equations which describes the first approximation to the effect of the temperature dependence of the physical properties. It arises from the terms proportional to ε. In contrast to the first set, this is linear. Thus, its complete solution can be additively formed from four partial solutions, each proportional to the quantities K_ρ, K_μ, K_c and K_λ, respectively.

From the complete solution, finally an expression for the skin-friction coefficient c_f is obtained

$$c_f \sqrt{Re} = F_0(x) + \frac{\Delta T}{T_\infty} \left[K_\rho F_\rho(x, Pr, Ec) + K_\mu F_\mu(x, Pr, Ec)\right], \tag{7.149}$$

where the functions $F_\rho(x)$ and $F_\mu(x)$ also depend on the type of thermal boundary condition at the wall. The Prandtl number Pr already appeared earlier in this book. It was defined as the quotient of the kinematic viscosity and the thermal diffusivity. The Eckert number $Ec := V^2/(c_p T)$ characterizes the effect of the dissipation.

Corresponding formulae are also valid for the wall heat transfer, where terms proportional to K_c and K_λ also appear. The equation for the skin-friction coefficient has the advantage that the effects of the temperature dependence for the four relevant physical properties are separate and therefore can be determined independently of each other.

If the physical properties also depend on the pressure, the perturbation calculation can be correspondingly extended. In practice, mostly only the pressure dependence of the density is of importance. The Taylor series (7.142) is then extended as follows

$$
\rho(T,p) = \rho_\infty + \left(\frac{d\rho}{dT}\right)_\infty (T - T_\infty) + \left(\frac{d\rho}{dp}\right)_\infty (p - p_\infty) + \dots
$$

$$
= \rho_\infty + \left(1 + K_\rho \frac{T - T_\infty}{T_\infty} + \tilde{K}_\rho \frac{p - p_\infty}{p_\infty} + \dots\right),
$$
(7.150)

where p_∞ is the pressure at the reference point. The $\tilde{K}$ parameters for water and air can also be found in Table 7.1.

In dimensionless notation Equation (7.150) has the following form

$$
\frac{\rho(T,p)}{\rho_\infty} = 1 + K_\rho \frac{T - T_\infty}{T_\infty} + \gamma Ma_\infty^2 \frac{p - p_\infty}{\rho_\infty V^2}.
$$
(7.151)

Thus, the effects of the pressure dependence of the density are proportional to the square of the Mach number $Ma_\infty^2 = V^2/c_\infty^2$, where c is the speed of sound.

In practice, two methods are frequently used which were initially developed empirically. With their help results obtained under the assumption of constant physical properties can be corrected with respect to the effect of variable properties. These are the property ratio method and the reference temperature method. Both of them are discussed in details in [338].

Next, compressible boundary layers without gravitational effects are examined. Consider a plane body in a flow of velocity V. Computing the inviscid flow yields the distributions $u_e(x)$ and $T_e(x)$ at the outer edge of the boundary layer. The index e stands for external or edge. Boundary-layer theory now aims to calculate the boundary-layer equations (7.141) (with $g = 0$) for given thermal boundary conditions at the wall. In general either the temperature distribution $T_w(x)$ or the distribution of the heat flux $q_w(x)$ at the wall may be given. Special cases are the standard cases $T_w = \text{const.}$ or $q_w = \text{const.}$ We are interested in the distribution of the skin-friction coefficient, the adiabatic wall temperature, the Nusselt number $Nu := q_w L/(\lambda(T_w - T_\infty))$ or the wall temperature.

For the constitutive relations it is assumed that 1) we consider ideal gases, 2) constant specific heat capacities, 3) a constant Prandtl number, and 4) the viscosity $\mu(T)$ depends only on the temperature. For the viscosity $\mu(T)$ three representations are common: 1) a linear law, 2) a power law, or 3) the so-called Sutherland formula which contains an additional constant dependent on the type of the gas.

If we use these assumptions, we find a particularly simple equation for the specific total enthalpy (index t)

$$
h_t = c_p T_t = h + \frac{1}{2}u^2 = c_p T + \frac{1}{2}u^2.
$$
(7.152)

Here, T_t is the total temperature and $v^2/2$ is neglected in comparison to $u^2/2$, as is permitted inside the boundary layer.

If we multiply the second equation of (7.141) by u and add the resulting equation for the kinetic energy to the third equation of (7.141), we obtain

$$\rho\left(u\frac{\partial h_t}{\partial x} + v\frac{\partial h_t}{\partial y}\right) = \frac{\partial}{\partial y}\left(\frac{\mu}{Pr}\frac{\partial h_t}{\partial y}\right) + \frac{\partial}{\partial y}\left[\left(1 - \frac{1}{Pr}\right)\mu u\frac{\partial u}{\partial y}\right]. \tag{7.153}$$

It can be seen immediately from this equation that $h_t = $ const. for inviscid outer flow ($\mu = 0$). Therefore, at the outer edge of the boundary layer we have

$$c_p T_e + \frac{1}{2}u_e^2 = h_{te} = c_p T_0, \tag{7.154}$$

with T_0 as the total temperature or stagnation temperature of the outer flow.

Equation (7.153) is greatly simplified for the case Pr = 1. Then two simple solutions, the so-called Busemann-Crocco solutions, can be immediately written down [60], [100]:

1. *Adiabatic wall (Pr = 1).*

   The solution of Equation (7.153) is $h_t = h_{te} = $ const. Differentiating (7.152), and using $u_w = 0$ we find

   $$\left(\frac{\partial h_t}{\partial y}\right)_w = c_p\left(\frac{\partial T}{\partial y}\right)_w - \frac{c_p}{\lambda_w}q_w, \tag{7.155}$$

   and therefore $h_t = $ const. also satisfies the condition $q_w = 0$ for the adiabatic wall. In this case Equation (7.152) implies that the temperature $T(u)$ is a quadratic function of the velocity. We have

   $$\frac{T_0 - T(u)}{T_0} = \frac{u^2}{2c_p T_0}. \tag{7.156}$$

   Because of the no-slip condition ($u_w = 0$), the adiabatic wall temperature is equal to the total temperature T_0.

2. *Plate flow (Pr = 1).*

   In this case there is a linear relation between h_t and u of the form

   $$\frac{h_t - h_{te}}{h_{tw} - h_{te}} = 1 - \frac{u}{U_\infty}, \tag{7.157}$$

   because Equations (7.153) for h_t and (7.141)$_2$ for u have the same structure. Thus, the dependence of the temperature $T(u)$ on the velocity is again a second order polynomial

   $$\frac{T_0 - T(u)}{T_0} = \frac{u^2}{2c_p T_0} + \frac{T_0 - T_w}{T_0}\left(1 - \frac{u}{U_\infty}\right), \tag{7.158}$$

   with the wall temperature assumed constant at T_w. For $T_w = T_0$ this formula is again reduced to Equation (7.156) for the adiabatic case.

   If the Mach number of the free stream (temperature T_∞ in the free stream)

   $$Ma_\infty = \frac{U_\infty}{c_\infty} = \frac{U_\infty}{\sqrt{c_p(\gamma - 1)T_\infty}}, \tag{7.159}$$

is introduced, Equation (7.158) can also be written as

$$\frac{T - T_\infty}{T_\infty} = \frac{\gamma - 1}{2} Ma_\infty^2 \left[1 - \left(\frac{u}{U_\infty} \right)^2 \right] + \frac{T_w - T_{ad}}{T_\infty} \frac{u}{U_\infty}, \qquad (7.160)$$

or

$$\frac{T - T_w}{T_\infty} = \frac{\gamma - 1}{2} Ma_\infty^2 \frac{u}{U_\infty} \left(1 - \frac{u}{U_\infty} \right) + \frac{T_\infty - T_w}{T_\infty} \frac{u}{U_\infty}, \qquad (7.161)$$

where, for the adiabatic wall temperature we have

$$T_{ad} = T_0 = T_\infty \left(1 + \frac{\gamma - 1}{2} Ma_\infty^2 \right). \qquad (7.162)$$

If we form the derivative at the wall from Equation (7.160), we obtain

$$q_w = -\lambda_w \left(\frac{\partial T}{\partial y} \right)_w = \frac{(T_w - T_{ad}) \lambda_w \tau_w}{U_\infty \mu_w}, \qquad (7.163)$$

or in dimensionless form

$$Nu = -\lambda_w \left(\frac{\partial T}{\partial y} \right)_w = \frac{q_w L}{\lambda_\infty (T_w - T_{ad})} = \frac{c_f}{2} Re, \qquad (7.164)$$

where $Re = \rho U_\infty L / \mu_\infty$.

This simple relation between the Nusselt number Nu and the skin-friction coefficient c_f is called the Reynolds analogy. It is only true for plate flows at $Pr = 1$, but for arbitrary Mach numbers.

From Equation (7.164) it follows that

$$\begin{aligned}
q_w > 0, \quad & T_w > T_{ad} : \quad && \text{heating: heat from wall to fluid,} \\[4pt]
q_w < 0, \quad & T_w < T_{ad} : \quad && \text{cooling: heat from fluid to wall.}
\end{aligned} \qquad (7.165)$$

7.4 Composite beams with embedded shape memory alloy

In his contribution [210] to the book [12] M. S. Kuczma showed that shape memory alloys (SMAs) constitute a class of functional, smart materials which have found many technological applications and offer innovative solutions in the design of adaptive structures (see, for example, [132], [212], [306], [358]). SMAs are materials which may undergo a temperature- or stress-induced martensitic phase transformation resulting in the shape memory effect and pseudoelastic (superelastic) behavior. A typical example of SMAs is Nickel-Titanium; it exhibits the just mentioned behavior, i.e., is capable of recovering large strains during mechanical loading-unloading cycles conducted at a constant temperature. This unique material response is attributed to a martensitic phase transformation, which is a first-order reversible transformation from a high temperature phase with greater symmetry, called austenite, to a low temperature phase with lower symmetry, called martensite [288]. The martensitic phase transformation can be induced by stresses, changes of temperature, or a magnetic field.

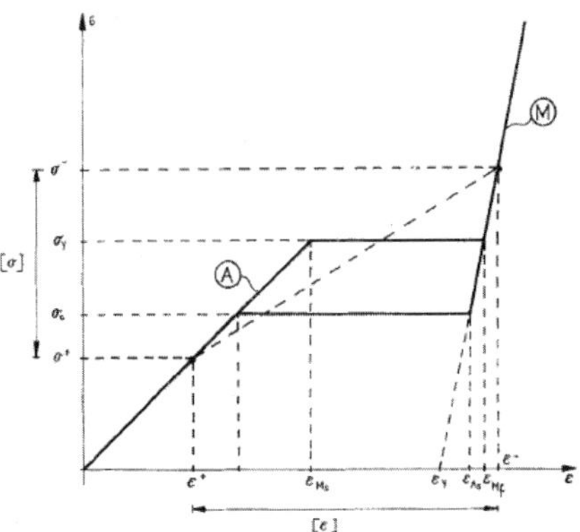

Figure 7.3: *Simplified stress-strain relation for a pseudoelastic body at relatively low temperature (from [430]).*

Many processes in various fields exhibit hysteresis loops. Several of them are treated in the three volume set [42]-[44]. A historical survey on the phenomenon of hysteresis, especially in the field of soil moisture hysteresis, is given in [16]. The first investigations on this subject were related to magnetism and ferroelectric bodies. In [264] it is pointed out that the pseudoelastic behavior means that bodies behave much like plastic bodies at low temperatures, while at high temperatures they are elastic. In particular, such bodies may sustain a residual deformation after a large load has been applied, but upon heating the original configuration is restored. Such bodies have also been called ferroelastic because there is a certain similarity of their load-deformation curves with the field-magnetization curves of a ferromagnet. In Figure 7.3 a simplified stress-strain diagram of a pseudoelastic body is reproduced from [430]. We recognize a virginal elastic curve at small loads and deformations, which corresponds to the austenitic phase of a shape memory alloy, and a plastic yield once the yield stress σ_y is exceeded. The yielding along the horizontal parts of the diagram reflects the possibility of the phase transformation. In contrast to the behavior of plastic bodies, there exists a second elastic branch which occurs at large deformations and which enables the body to support stresses beyond σ_y. In the case of pseudoelasticity shown in Figure 7.3, the deformed material will recover its initial neutral state upon unloading.

In this section, a special problem concerning SMAs is examined. We reproduce from [210] the formulation and numerical solution of a quasi-static, isothermal bending problem of composite beams reinforced with SMA wires. As a special case, this formulation covers the case of a monolithic beam made of shape memory material. The bending problem of SMA beams has been investigated also, e.g. in [302] or [299]. For various other aspects and models of SMAs see e.g. [27], [46], [203], [209], [211], [226], [265], [325], [432] or references therein.

7.4.1 Constitutive relations for a shape memory alloy

In this subsection constitutive relations for shape memory alloys in the range of pseudoelasticity are presented. The model accounts for the characteristic hysteresis loops in both

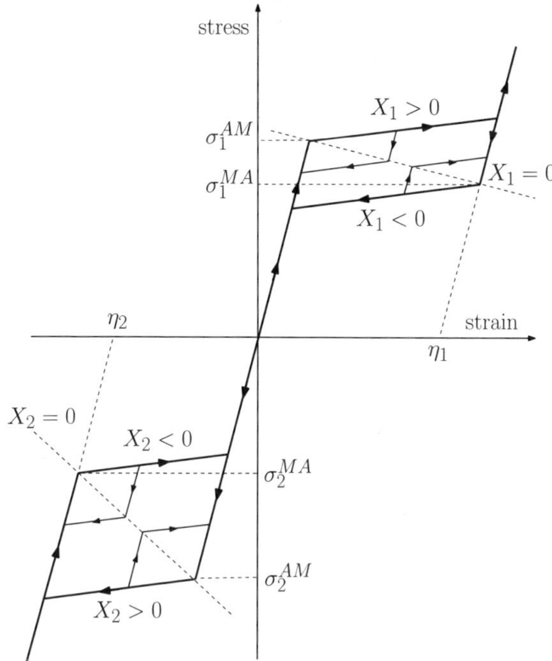

Figure 7.4: *Hysteretic stress-strain diagram of pseudoelasticity (after [210]).*

tension and compression states as illustrated in a typical one-dimensional stress-strain diagram in Figure 7.4. It should be noted that a shape memory alloy is itself a kind of composite in which the concentrations (volume fractions) of its constituents are not known in advance but evolve as a result of the deformation process. In the one-dimensional case of the bending problem considered here, it is assumed that the A/M-mixture is composed of austenite A and two 'averaged' variants of martensite M_1 and M_2. The variant M_1 corresponds to the phase transformation strain η_1 measured in a tensile test, while the variant M_2 corresponds to the phase transformation strain η_2 measured in a compression test, see Figure 7.4. The eigenstrain of austenite η_3 is assumed to be zero. c_1 and c_2 denote the volume fractions of martensite M_1 and M_2, respectively, and $c_3 = 1 - c_1 - c_2$ is the volume fraction of austenite. By definition, the volume fractions satisfy the condition $0 \leq c_i \leq 1$ for $i = 1, 2, 3$. In the studied case of two variants of martensite, at a given material point only one variant can be active, i.e., $c_1 \cdot c_2 = 0$.

We define the averaged specific free energy of the A/M-mixture by

$$\widetilde{W} = \sum_{i=1}^{3} c_i \, W_i + W_{\mathrm{mix}} + I_{[0,1]}(c_1 + c_2), \qquad (7.166)$$

where W_i $(i = 1, 2, 3)$ is the free energy of the particular phase, W_{mix} denotes a mixing energy and $I_{[0,1]}(\cdot)$ is the indicator function of the interval $[0, 1]$ which is a formal imposition of the constraint on the volume fractions c_1 and c_2.

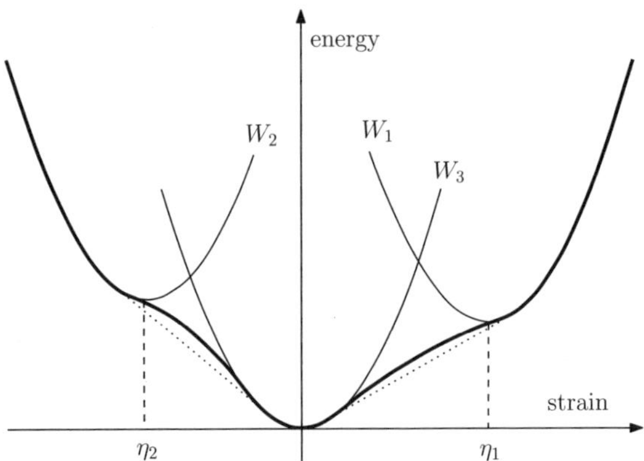

Figure 7.5: *Quasiconvexified three-well energy function (bold line). The dashed line corresponds to convexification (after [210]).*

When adopting the quadratic energy functions for each phase ($i = 1, 2, 3$),

$$W_i = \frac{1}{2} E \left(\varepsilon_i^e\right)^2 + \varpi_i(\theta), \qquad W_{\text{mix}} = \sum_{i=1}^{2} \frac{1}{2} B_i \left(1 - c_i\right)c_i, \tag{7.167}$$

in which $\varepsilon_i^e = \varepsilon - c_i\eta_i$, is the elastic part of the total strain ε and $\varpi_i(\theta)$ is the energy in a stress-free state, we finally obtain the following expression for the quasiconvexified free energy of the A/M-mixture, see Figure 7.5,

$$\widetilde{W}(\varepsilon, \mathbf{c}) = \frac{1}{2} E \left(\varepsilon - \sum_{i=1}^{2} c_i\eta_i\right)^2 + \sum_{i=1}^{3} c_i\varpi_i + \sum_{i=1}^{2} \frac{1}{2} B_i(1 - c_i)c_i + \partial I_{[0,1]}(c_1 + c_2). \tag{7.168}$$

In (7.167) and (7.168), E denotes the Young modulus of austenite and martensite, B_i, η_i ($i = 1, 2$) are material parameters which can be calculated from the stress-strain diagram of pseudoelasticity shown in Figure 7.4; for details we refer to [211] or [265]. The temperature, which is assumed to be constant, is denoted by θ and $\partial I_{[0,1]}(\cdot)$ is a subdifferential of the indicator function.

In order to account for dissipative effects in the martensitic phase transformation process, expressed by the hysteresis loops in Figure 7.4, we make use of the rate of dissipation inequality

$$D = \sigma\dot{\varepsilon} - \dot{\widetilde{W}} - s\dot{\theta} \geq 0. \tag{7.169}$$

In view of the equations for the stress

$$\sigma \equiv \frac{\partial \widetilde{W}}{\partial \varepsilon} = E(\varepsilon - c_1\eta_1 - c_2\eta_2), \tag{7.170}$$

and the entropy

$$s \equiv -\frac{\partial \widetilde{W}}{\partial \theta}, \tag{7.171}$$

Equation (7.169) can be reduced to

$$D = X_1\dot{c}_1 + X_2\dot{c}_2 \geq 0. \tag{7.172}$$

In Equation (7.172) X_i is a thermodynamically conjugate variable to the volume fraction c_i

$$X_i \equiv -\frac{\partial \widetilde{W}}{\partial c_i}, \quad i = 1, 2, \tag{7.173}$$

which is the driving force of the reversible phase transformation $A \to M_i$,

$$X_i(\varepsilon, c_i) = E(\varepsilon - c_i \eta_i)\eta_i - (\varpi_i - \varpi_3) - 0.5\, B_i(1 - 2c_i) - R_i, \quad R_i \in \partial I_{[0,1]}(c_i). \tag{7.174}$$

The driving forces X_i are positive for the forward transformation from austenite to martensite $A \to M_i$ ($i = 1, 2$), negative for the reverse transformation $M_i \to A$, and equal to zero at the characteristic diagonals shown in Figure 7.4.

In reference to the hysteretic behavior, the evolutions of the volume fractions c_i, $i = 1, 2$, are supposed to be governed by the following phase transformation rule (PTR)

$$\begin{aligned}
\text{if } X_i(\varepsilon, c_i) &= \kappa_{3 \to i}(c_i) && \text{then } \dot{c}_i \geq 0, \\
\text{if } X_i(\varepsilon, c_i) &= \kappa_{i \to 3}(c_i) && \text{then } \dot{c}_i \leq 0, \\
\text{if } \kappa_{i \to 3}(c_i) < X_i(\varepsilon, c_i) &< \kappa_{3 \to i}(c_i) && \text{then } \dot{c}_i = 0,
\end{aligned} \tag{7.175}$$

where $\kappa_{3 \to i} \geq 0$, and $\kappa_{i \to 3} \leq 0$ are threshold functions, which are presumed in the form $\kappa_{3 \to i}(c_i) = L_i\, c_i$, $\kappa_{i \to 3}(c_i) = L_i(c_i - 1)$, with L_i being material parameters. The forward or reverse phase transformation commences and continues only then when the driving force reaches and remains equal to the current value of the threshold function. In the interior of the hysteresis loops, the material response is elastic.

7.4.2 Bending of a SMA composite beam

The quasi-static bending problem for a composite beam enforced with SMA wires or strips is considered. Starting point is the classical beam theory, complemented with a layerwise approach in which the cross-section of the beam is divided into a number of layers, N_l. Monolithic shape memory alloy beams are a particular case of the layerwise model.

The deflection of the beam axis is denoted by $w_0 = w_0(x, t)$, where x is the abscissa, $0 \leq x \leq l_b$, and t is a time-like parameter. Making use of the Bernoulli kinematical hypothesis, we can express the normal strain ε as

$$\varepsilon = \varepsilon(x, z, t) = \varepsilon_0(t) - \frac{\partial^2 w_0}{\partial x^2}(z - z_0(t)) = \varepsilon_0 - w_0'' z, \tag{7.176}$$

where z is the through-the-thickness coordinate, ε_0 is a normal strain at the axis located at $z = z_0$. For simplicity, we suppose further that the properties of the cross-section of the beam are symmetrical with respect to the axis at $z_0 = 0$ and presume $\varepsilon_0 = 0$.

The location of layer k in the cross-section is determined by the z-coordinates z_k, z_{k+1} and the height $h_k = z_{k+1} - z_k$. For the matrix and the fibers, denoted by indices m and f, respectively, $E_m^{(k)}, v_m^{(k)}$ and $E_f^{(k)}, v_f^{(k)}$ indicate Young's modulus and volume fraction of layer k. The effective modulus of the composite material in the kth layer is defined by $E_{mf}^{(k)} = v_m^{(k)} E_m^{(k)} + v_f^{(k)} E_f^{(k)}$, and $b = b(z)$ is the width of the beam cross-section. By $\mathcal{I}_f$ we designate the set of indices of the beam layers which are reinforced with SMA fibers.

The normal stress in the matrix is defined by

$$\sigma(x, z, t) = E_m^{(k)} \varepsilon(x, z, t), \quad \text{with } z \in (z_k, z_{k+1}), \tag{7.177}$$

and that in the SMA fibers by

$$\sigma(x, z, t, \mathbf{c}^{(k)}) = E_f^{(k)} \left(\varepsilon(x, z, t) - \eta_1^{(k)} c_1^{(k)}(x, z, t) - \eta_2^{(k)} c_2^{(k)}(x, z, t) \right), \quad z \in (z_k, z_{k+1}).$$
$$(7.178)$$

Combining (7.177) and (7.178), it is supposed that the normal stress in the kth layer with SMA fibers is given by

$$\sigma(x, z, t, \mathbf{c}^{(k)}) = E_{mf}^{(k)} \varepsilon(x, z, t) - v_f^{(k)} E_f^{(k)} \left(\eta_1^{(k)} c_1^{(k)}(x, z, t) + \eta_2^{(k)} c_2^{(k)}(x, z, t) \right), \qquad (7.179)$$

with $z \in (z_k, z_{k+1})$. It should be noted that the values of the volume fractions of martensites M_1 and M_2 in the fibers of the kth layer, $c_1^{(k)}$ and $c_2^{(k)}$, are not known in advance but are additional unknowns of the problem. For all N_f layers with SMA fibers we denote $\mathbf{c} \equiv (\mathbf{c}^{(1)}, \mathbf{c}^{(2)}, \ldots, \mathbf{c}^{(N_f)})$, with $\mathbf{c}^{(k)} \equiv (c_1^{(k)}, c_2^{(k)})$. In the one-dimensional bending problem under consideration, one of $c_1^{(k)}$ and $c_2^{(k)}$ equals zero, i.e., $c_1^{(k)}(x, z, t) \, c_2^{(k)}(x, z, t) = 0$ at all material points (x, z) and all time levels t.

The equilibrium condition of the beam subjected to the load $f(x, t)$ can be written as the variational equation (equation of virtual work)

$$a(w_0, \mathbf{c}; v) = F(t, v) \qquad \forall v \in H_0^2(0, l_b), \qquad (7.180)$$

where the linear and bilinear forms are defined by

$$F(t, v) = \int_0^{l_b} f(x, t) v \, dx, \qquad (7.181)$$

$$
\begin{aligned}
a(w_0, \mathbf{c}; v) \;=\; & \int_0^{l_b} v'' \left(\sum_{k=1}^{N_l} \int_{z_k}^{z_{k+1}} E_{mf}^{(k)} z^2 b \, dz \right) w_0'' \, dx \\
& + \int_0^{l_b} v'' \left(\sum_{k \in \mathcal{I}_f} \int_{z_k}^{z_{k+1}} v_f^{(k)} E_f^{(k)} \eta_1^{(k)} c_1^{(k)} \right) z \, b \, dz \, dx \\
& + \int_0^{l_b} v'' \left(\sum_{k \in \mathcal{I}_f} \int_{z_k}^{z_{k+1}} v_f^{(k)} E_f^{(k)} \eta_2^{(k)} c_2^{(k)} \right) z \, b \, dz \, dx.
\end{aligned}
\qquad (7.182)
$$

The weak form of the phase transformation rule (7.175) of SMA wires in the kth layer is built. Therefore, the phase transformation functions for forward and reverse martensitic phase transformations are defined as

$$
\begin{aligned}
\Phi_{i,+}^{(k)} \;=\; & v_f^{(k)} \int_{z_k}^{z_{k+1}} \left(\kappa_{i,+}^{(k)}(c_i^{(k)}) - X_i(w_0, c_i^{(k)}) \right) b \, dz \geq 0, \\
\Phi_{i,-}^{(k)} \;=\; & v_f^{(k)} \int_{z_k}^{z_{k+1}} \left(X_i(w_0, c_i^{(k)}) - \kappa_{i,-}^{(k)}(c_i^{(k)}) \right) b \, dz \geq 0,
\end{aligned}
\qquad (7.183)
$$

with

$$
\begin{aligned}
\kappa_{i,+}^{(k)}(c_i^{(k)}) \;\equiv\; & \max \{ L_i^{(k)} c_i^{(k)}, 0 \}, \\
\kappa_{i,-}^{(k)}(c_i^{(k)}) \;\equiv\; & \min \{ L_i^{(k)} (c_i^{(k)} - 1), 0 \}.
\end{aligned}
\qquad (7.184)
$$

The forces X_i are expressed in terms of w_0 via (7.174) and (7.176).

We use the following notation

$$\mathbf{T}^{(k)} = \mathrm{col}\,(\Phi_{1,+}^{(k)}, \Phi_{2,+}^{(k)}, \Phi_{1,-}^{(k)}, \Phi_{2,-}^{(k)}), \tag{7.185}$$

$$\mathbf{u}^{(k)} = \mathrm{col}\,(\dot{c}_{1,+}^{(k)}, \dot{c}_{2,+}^{(k)}, \dot{c}_{1,-}^{(k)}, \dot{c}_{2,-}^{(k)}), \tag{7.186}$$

where

$$\dot{c}_{i,+}^{(k)} \equiv \max\,\{\dot{c}_i^{(k)}, 0\}, \qquad \dot{c}_{i,-}^{(k)} \equiv \min\,\{\dot{c}_i^{(k)}, 0\}, \tag{7.187}$$

are positive and negative rates of $c_i^{(k)}$.

Now we can impose the conditions (7.175) in the form of the variational inequality for each kth layer with SMA fibers,

$$\big\langle \mathbf{T}^{(k)}(w_0, \mathbf{c}^{(k)}; \mathbf{u}^{(k)}), \, \mathbf{s} - \mathbf{u}^{(k)} \big\rangle \geq 0 \quad \forall \mathbf{s} \in C, \qquad k \in \mathcal{I}_f, \tag{7.188}$$

where

$$\langle \mathbf{s}, \, \mathbf{v} \rangle = \int_0^{l_b} \mathbf{s}(x) \cdot \mathbf{v}(x)\, dx, \tag{7.189}$$

and C is a positive cone in $(L_2(0, l_b))^4$.

Equation (7.180) and inequalities (7.188) completely describe the evolution of the beam in bending, automatically accounting for the progress of the phase transformation process under study. After the finite element approximation of (7.180) and (7.188), the resulting system of matrix equations and inequalities for finite increments can be solved as a series of linear complementarity problems. This is not shown here, instead we refer to the original work [210] or to [209].

7.4.3 Numerical examples

In order to illustrate the response of Kuczma's shape memory alloy model we reproduce two of his numerical examples presented in [210]. The results pertain a monolithic SMA beam and an epoxy resin beam reinforced with two SMA strips symmetrically disposed across the cross-section of the beam. For both beams a rectangular cross-section of dimensions 2.5×10 mm is assumed. The monolithic beam is divided into 20 layers of equal thickness, whereas the composite beam into five layers whose location is defined by the z_k coordinates: -5.0, -4.75, $-3,75$, 3.75, 4.75, 5.0 mm. Layers 2 and 4 are SMA strips of thickness 1 mm, i.e., $\mathcal{I}_f = \{2, 4\}$. The material data corresponds to the CuZuAl shape memory alloy (cf. [211], [265]): $E_f^{(k)} = 12.30$ GPa, $B_1^{(k)} = B_2^{(k)} = 1.30$ MPa, $L_1^{(k)} = L_2^{(k)} = 1.01 B^{(k)} = 1.313$ MPa, $\eta_1^{(k)} = -\eta_2^{(k)} = 0.0686$, $(\varpi_i^{(k)} - \varpi_3) = 3.756$ MJ/m^3, $i = 1, 2$, and for the epoxy resin: $E_m^{(k)} = 5.00$ GPa.

The first example is a one-span beam of length $l_b = 150$ mm, fixed at both ends and subjected to half-span variable loading ($e = l_b/2$) as shown in Figure 7.6 (left). The history of the concentrated force $P = P(t)$ is displayed in the right panel of this figure.

Figure 7.7 (left) illustrates the distribution of volume fractions of martensites M_1 and M_2 along the beam and through its thickness. As expected the martensitic phase transformation takes place in regions of greatest stress, i.e., at the supports and the application point of the force. These regions of highly localized deformation form the so-called superelastic (pseudoplastic) hinges which allow for considerable rotations and, resulting from these, deflections of the beam, see the right panel of Figure 7.7. In contrast to the

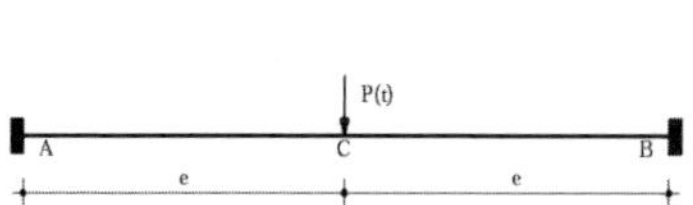

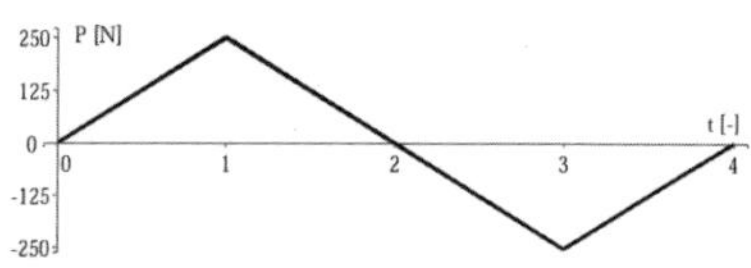

Figure 7.6: *Left: Clamped beam loaded with concentrated force $P = P(t)$ applied at its mid span point C ($e = l_b/2 = 0.075$ m), right: history of the concentrated force $P = P(t)$, from: [210].*

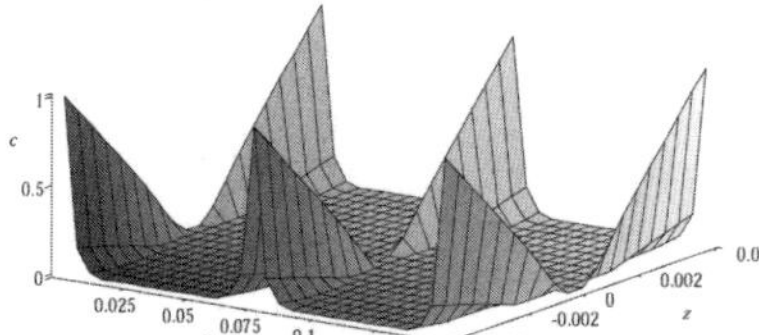

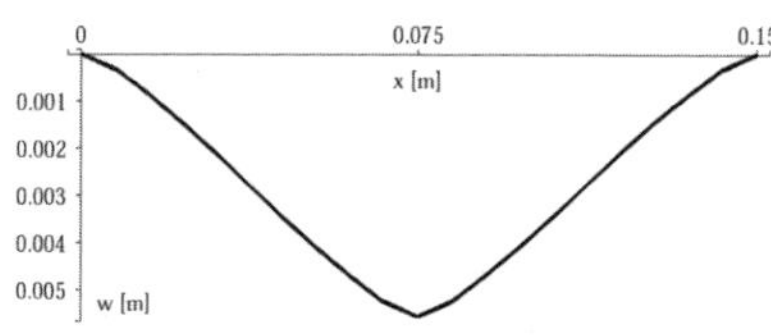

Figure 7.7: *Left: Volume fraction distribution of two variants of martensites, $c \in \{c_1, c_2\}$, in the beam of Figure 7.6 (left) as a function of abscissa $x \in [0, 0.15]$ and through-the-thickness coordinate $z \in [-0.005, 0.005]$ at time $t = 1$ in Figure 7.6 (right); right: displacement $w(x, t)$ of the beam in Figure 7.6 (left) by half-span point load $P = 250$ N at $t = 1$ in Figure 7.6 (right); from [210].*

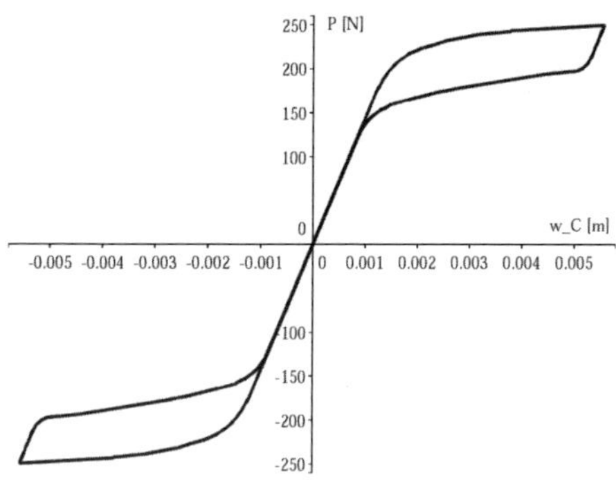

Figure 7.8: *Hysteresis loops in the displacement w_C at the center of the beam shown in Figure 7.6 (left) induced by variable half-span loading $P = P(t)$ with history shown in Figure 7.6 (right); from [210].*

usual plastic hinges, the superelastic hinges are reversible, provided that the stresses in the material will not exceed some critical strength.

The second example illustrates the influence of SMA strips on a hysteretic response of a composite beam. The geometrical data and loading conditions of the composite beam are the same as for the first example. The difference is that instead of a monolithic SMA beam in the second example the beam is made of an epoxy resin and two SMA strips located in cross-section at $z \in (-4.75, -3.75) \cup (3.75, 4.75)$. Figure 7.8 shows the obtained hysteresis loops in the deflection of center point, w_C, and the concentrated load $P(t)$ defined in Figure 7.6 (right).

The presented theoretical model and numerical calculations by M. Kuczma [210] are capable of modeling the hysteretic behavior of SMA beams. The stress-induced martensitic phase transformation in a SMA beam results in the occurrence of reversible superelastic hinges allowing for large reversible deflections of the SMA beam.

Chapter 8

Extended thermodynamics of Jou–Casas-Vázquez–Lebon

Almost simultaneously with the development of extended thermodynamics, initiated by Ingo Müller, I-Shih Liu and Tommaso Ruggeri whose outline we presented in Part I, a new branch of non-equilibrium thermodynamics also called extended thermodynamics started to develop with works of David Jou, José Casas-Vázquez and Georgy Lebon. In both approaches the main idea was the extension of the number of fields but the microscopic motivation and, consequently, the structure of the macroscopic models was different. Ingo Müller and coworkers were mainly motivated by Grad's method of the kinetic theory in which the construction of an approximate solution of Boltzmann's equation was proposed in the form of an expansion with respect to polynomials. This hierarchic microscopic model could be easily replaced by a similar macroscopic model as it was done, for instance, in the first paper of this version of extended thermodynamics of I-Shih Liu and Ingo Müller [235]. It has been shown later that in the case of ideal gases both approaches practically coincide (compare Part I). On the other hand David Jou, José Casas-Vázquez and Georgy Lebon were primarily motivated by the non-equilibrium statistical mechanics and, in particular, by the so-called Fluctuation-Dissipation Theorem. In Section 8.3 we will comment on the main differences between this version of extended thermodynamics and that of Müller/Liu/Ruggeri, as well as their common points.

We repeat below roughly the ideas of the microscopic theory presented by David Jou, José Casas-Vázquez and Georgy Lebon in their book [191]. Some ideas are reproduced and the original notation is kept.

David Jou
1953-

José Casas-Vázquez
1938-

Georgy Lebon
1936-

8.1　Summary of Extended Irreversible Thermodynamics (EIT)

Extended Irreversible Thermodynamics (EIT) goes beyond the classical formulation of thermodynamics in enlarging the space of basic independent variables through the introduction of non-equilibrium variables such as dissipative fluxes. For these additional variables evolution equations have to be found. Whereas the evolution equations for the classical variables are given by the usual balance laws, no general criteria exist concerning the evolution equations of dissipative fluxes, with the exception of the restrictions imposed on them by the second law of thermodynamics. The fluxes, in general, are fast variables which decay to their local-equilibrium values after a short relaxation time. A description of the system in terms of these fast variables is searched for. Phenomena at frequencies comparable to the inverse of the relaxation times of the fluxes are of interest.

The basis to obtain the evolution equations for the fluxes is to generalize classical theories. Thus, the existence of a generalized entropy which depends on the dissipative fluxes and on classical variables is assumed. Restrictions are then imposed by the laws of thermodynamics. Jou, Casas-Vázquez and Lebon [191] propose a physical interpretation of the different contributions to the generalized entropy. From this expression generalized equations of state are derived. They are of interest in the description of non-equilibrium steady states.

As an example, the simple case of a one-component isotropic fluid is considered. In such a system, the heat flux, the bulk viscous pressure, and the viscous pressure tensor are taken as supplementary independent variables, on the same footing as the classical ones. We reproduce here the introduction of the generalized Gibbs equation, of the generalized entropy flux and of the entropy production presented in [191].

8.1.1　The generalized Gibbs equation

As in classical thermodynamics the entropy of a system plays a central role. It is assumed that it depends locally not only on the classical variables, the internal energy u and the specific volume v, but also on the dissipative fluxes $\mathbf{q}$, p^v, and $\overset{0}{\mathbf{P}}{}^v$. The generalized entropy $s(u, v, \mathbf{q}, p^v, \overset{0}{\mathbf{P}}{}^v)$ is assumed to have the following properties:

1. it is an additive quantity;

2. it is a convex function of the whole set of variables, which means a function that lies everywhere below its family of tangent lines; and

3. its rate of production is locally positive.

The hypothesis of a generalized macroscopic entropy depending on the above mentioned dissipative fluxes was advanced by S. Machlup and L. Onsager [240] in an indirect way. During the 1960s a more direct formulation applied to fluids was developed by R. E. Nettleton [272] and by I. Müller [260]. Since then much research has been done in this field, for an overview see e.g. [354, 262].

In differential form, the entropy is written as

$$ds = \left(\frac{\partial s}{\partial u}\right) du + \left(\frac{\partial s}{\partial v}\right) dv + \left(\frac{\partial s}{\partial \mathbf{q}}\right) \cdot d\mathbf{q} + \left(\frac{\partial s}{\partial p^v}\right) dp^v + \left(\frac{\partial s}{\partial \overset{0}{\mathbf{P}}{}^v}\right) : d\overset{0}{\mathbf{P}}{}^v. \tag{8.1}$$

In analogy with the classical theory, a non-equilibrium absolute temperature θ and a non-equilibrium thermodynamic pressure π are introduced

$$
\theta^{-1}(u, v, \mathbf{q}, p^v, \overset{0}{\mathbf{P}^v}) = \left(\frac{\partial s}{\partial u}\right)_{v, \mathbf{q}, p^v, \overset{0}{\mathbf{P}^v}},
$$
$$
\theta^{-1}\pi(u, v, \mathbf{q}, p^v, \overset{0}{\mathbf{P}^v}) = \left(\frac{\partial s}{\partial v}\right)_{u, \mathbf{q}, p^v, \overset{0}{\mathbf{P}^v}}.
$$
$$(8.2)$$

If it is expressed in terms of its natural extensive variables, the entropy is a thermodynamic potential. Therefore, rather than the fluxes themselves, one should use as independent variables $v\mathbf{q}$, vp^v, and $v\overset{0}{\mathbf{P}^v}$. These variables are extensive in the following sense: for two identical systems of volume V_1 and V_2, respectively, crossed by the same heat flux $\mathbf{q}$, the variable $V\mathbf{q}$ is additive, i.e., $V_{tot}\mathbf{q} = V_1\mathbf{q} + V_2\mathbf{q}$ but not $\mathbf{q}$ itself. The modification in the choice of variables only affects the definition of the non-equilibrium pressure, which should be defined as the derivative of the entropy with respect to the volume at constant $v\mathbf{q}$ rather than at constant $\mathbf{q}$.

The quantities θ and π can be expanded around their local-equilibrium counterparts T and p according to

$$
\theta^{-1} = T^{-1}(u, v) + \mathcal{O}\left(\mathbf{q} \cdot \mathbf{q}, p^v p^v, \overset{0}{\mathbf{P}^v} : \overset{0}{\mathbf{P}^v}\right),
$$
$$
\theta^{-1}\pi = T^{-1}p(u, v) + \mathcal{O}\left(\mathbf{q} \cdot \mathbf{q}, p^v p^v, \overset{0}{\mathbf{P}^v} : \overset{0}{\mathbf{P}^v}\right),
$$
$$(8.3)$$

where $\mathcal{O}$ refers to corrective terms which are of second order in the fluxes and whose explicit form can be found in [191]. They are usually negligible and will therefore be omitted in the following. The remaining partial derivatives in (8.1) are

$$
\frac{\partial s}{\partial \mathbf{q}} = -T^{-1}v\boldsymbol{\alpha}_1(u, v, \mathbf{q}, p^v, \overset{0}{\mathbf{P}^v}),
$$
$$
\frac{\partial s}{\partial p^v} = -T^{-1}v\alpha_0(u, v, \mathbf{q}, p^v, \overset{0}{\mathbf{P}^v}),
$$
$$
\frac{\partial s}{\partial \overset{0}{\mathbf{P}^v}} = -T^{-1}v\overset{0}{\boldsymbol{\alpha}_2}(u, v, \mathbf{q}, p^v, \overset{0}{\mathbf{P}^v}).
$$
$$(8.4)$$

The minus sign and the factor $T^{-1}v$ are introduced for later convenience. In (8.4), $\boldsymbol{\alpha}_1$, α_0 and $\overset{0}{\boldsymbol{\alpha}_2}$ are vector, scalar, and tensor fields, respectively. For isotropic systems, T, p, and α_0 are functions of u, v, and the algebraic invariants of $\mathbf{q}$ and $\overset{0}{\mathbf{P}^v}$, namely

$$
I_1 = \operatorname{tr}\overset{0}{\mathbf{P}^v}, \qquad I_2 = \operatorname{tr}(\overset{0}{\mathbf{P}^v})^2, \qquad I_3 = \operatorname{tr}(\overset{0}{\mathbf{P}^v})^3,
$$
$$
I_4 = \mathbf{q} \cdot \mathbf{q}, \qquad I_5 = \mathbf{q} \cdot \overset{0}{\mathbf{P}^v} \cdot \mathbf{q}, \qquad I_6 = \mathbf{q} \cdot (\overset{0}{\mathbf{P}^v})^2 \mathbf{q}.
$$
$$(8.5)$$

For simplicity it is assumed that the coefficients $\boldsymbol{\alpha}_1$, α_0 and $\overset{0}{\boldsymbol{\alpha}_2}$ are linear in the fluxes, so that

$$
\boldsymbol{\alpha}_1 = \alpha_{10}\mathbf{q}, \qquad \alpha_0 = \alpha_{00}p^v, \qquad \overset{0}{\boldsymbol{\alpha}_2} = \alpha_{21}\overset{0}{\mathbf{P}^v},
$$
$$(8.6)$$

where α_{10}, α_{00} and α_{21} are scalar functions of u and v. They have to be specified in terms of physical parameters. This is not done here, instead we refer to the literature in this field, e.g. [191]. With the preceding expressions, the generalized Gibbs equation becomes

$$ds = T^{-1}du + T^{-1}pdv - T^{-1}v\alpha_{00}p^v dp^v - T^{-1}v\alpha_{10}\mathbf{q}\cdot d\mathbf{q} - T^{-1}v\alpha_{21}\overset{0}{\mathbf{P}^v} : d\overset{0}{\mathbf{P}^v}. \tag{8.7}$$

From (8.7) and the balance equations of energy and mass

$$\rho\dot{u} = -\nabla\cdot\mathbf{q} - \mathbf{P}^T : \nabla\mathbf{v}, \qquad \rho\dot{v} = \nabla\cdot\mathbf{v}, \tag{8.8}$$

where u the specific internal energy, $\mathbf{v}$ the velocity field, $\mathbf{q}$ the heat flux vector, $\mathbf{P}$ the pressure tensor and $v = 1/\rho$ the specific volume, one obtains for $\dot{s}$, the material time derivative of s,

$$\begin{aligned}
\rho\dot{s} = &-T^{-1}\nabla\cdot\mathbf{q} - T^{-1}p^v\nabla\cdot\mathbf{v} - T^{-1}\overset{0}{\mathbf{P}^v} : \overset{0}{\mathbf{V}} - T^{-1}\alpha_{00}p^v\dot{p}^v \\
&-T^{-1}\alpha_{10}\mathbf{q}\cdot\mathbf{q} - T^{-1}\alpha_{21}\overset{0}{\mathbf{P}^v} : (\overset{0}{\mathbf{P}^v})^{\cdot}.
\end{aligned} \tag{8.9}$$

This can be written in the general form of a balance equation:

$$\rho\dot{s} + \nabla\cdot\mathbf{J}^s = \sigma^s, \tag{8.10}$$

in which the expressions for the entropy flux $\mathbf{J}^s$ and the entropy production σ^s have to be specified.

8.1.2　The generalized entropy flux and entropy production

For isotropic systems, the most general vector depending on the variables u, v, $\mathbf{q}$, $\overset{0}{\mathbf{P}^v}$ and p^v is, up to second order terms in the fluxes,

$$\mathbf{J}^s = \beta\mathbf{q} + \beta' p^v\mathbf{q} + \beta''\overset{0}{\mathbf{P}^v}\cdot\mathbf{q}, \tag{8.11}$$

where the coefficients β, β' and β'' are functions of u and v. The coefficient β must be identified as T^{-1} to recover the results of the classical theory of irreversible processes; accordingly

$$\mathbf{J}^s = T^{-1}\mathbf{q} + \beta' p^v\mathbf{q} + \beta''\overset{0}{\mathbf{P}^v}\cdot\mathbf{q}. \tag{8.12}$$

The entropy production is derived from the general form (8.9) of the entropy balance equation. Together with (8.9) and (8.12), Equation (8.10) leads to the following expression for the entropy production

$$\begin{aligned}
\sigma^s = &\mathbf{q}\cdot\left(\nabla T^{-1} + \beta''\nabla\cdot\overset{0}{\mathbf{P}^v} + \beta'\nabla p^v - T^{-1}\alpha_{10}\dot{\mathbf{q}}\right) \\
&+ p^v[-T^{-1}\nabla\cdot v - T^{-1}\alpha_{00}\dot{p}^v\nabla\cdot(\beta'\mathbf{q})] \\
&+ \overset{0}{\mathbf{P}^v} : \left\{-T^{-1}\overset{0}{\mathbf{V}} - T^{-1}\alpha_{21}(\overset{0}{\mathbf{P}^v})^{\cdot} + [\nabla(\beta''\mathbf{q})^s]\right\}.
\end{aligned} \tag{8.13}$$

It has the structure of a bilinear form

$$\sigma^s = \mathbf{q}\cdot\mathbf{X}_1 + p^v X_0 + \overset{0}{\mathbf{P}^v} : \overset{0}{\mathbf{X}}_2, \tag{8.14}$$

consisting of a sum of products of the fluxes $\mathbf{q}$, p^v and $\overset{0}{\mathbf{P}^v}$ and their conjugate generalized forces $\mathbf{X}_1$, X_0 and $\overset{0}{\mathbf{X}}_2$. Equation (8.13) is similar to the expressions obtained in the

Classical Irreversible Thermodynamics (CIT), mainly developed by the Belgian-Dutch school headed by Ilya Prigogine, but it contains additional terms depending on the time and space derivatives of the fluxes.

In [191] it is noted that, defining the proper form of the forces $\mathbf{X}_1$, $\overset{o}{\mathbf{X}}_2$ and X_0 there exists a class of transformations of the time derivatives of $\mathbf{q}$ and $\overset{o}{\mathbf{P}^v}$ which leaves the entropy production invariant. The authors mention the following example of such a transformation

$$D_a\mathbf{q} = \dot{\mathbf{q}} + a\mathbf{W} \cdot \mathbf{q},$$
$$D_b\overset{o}{\mathbf{P}^v} = (\overset{o}{\mathbf{P}^v})^{\cdot} + b(\mathbf{W} \cdot \overset{o}{\mathbf{P}^v} - \overset{o}{\mathbf{P}^v}\cdot\mathbf{W}),$$

(8.15)

where $\mathbf{W}$ is an antisymmetric tensor, for instance, the antisymmetric part of the velocity gradient. It is easy to verify that

$$\mathbf{q} \cdot D_a\mathbf{q} = \mathbf{q} \cdot \dot{\mathbf{q}}, \qquad \overset{o}{\mathbf{P}^v} : D_b\overset{o}{\mathbf{P}^v} = \overset{o}{\mathbf{P}^v} : (\overset{o}{\mathbf{P}^v})^{\cdot}.$$

(8.16)

This means that the expression of the entropy production remains unchanged when general derivatives of the form (8.15) are used instead of the material time derivatives of the fluxes.

Jou et al. [191] point out that thermodynamics cannot give any information about the coefficients a and b in Equations (8.15), since they do not appear explicitly either in the entropy production or in the Gibbs equation. However, they state that they can be determined by other means: (1) from general invariance requirements, such as the frame-indifference principle, which leads to $a = b = 1$, and in this case the derivatives defined by (8.15) coincide with the corotational time derivative; (2) from a microscopic description of matter, e.g. the kinetic theory of gases; a comparison with kinetic theory yields $a = b = -1$; (3) from experiments on rotating systems but such experiments turn out to be very difficult in view of the smallness of the terms involved; (4) from the requirement that the reversible part of the evolution equations must have a Hamiltonian form [191].

In order to obtain evolution equations for the fluxes compatible with the positiveness of σ^s, the forces $\mathbf{X}_1$, X_0 and $\overset{o}{\mathbf{X}}_2$ are expressed as functions of the fluxes. Linear relations between fluxes and forces are assumed. Then, up to the linear approximation in the fluxes,

$$\mathbf{X}_1 = \mu_1\mathbf{q}, \qquad X_0 = \mu_0 p^v, \qquad \overset{o}{\mathbf{X}}_2 = \mu_2\overset{o}{\mathbf{P}^v},$$

(8.17)

where the coefficients μ_i may depend on u and v but not on the fluxes. With (8.17) inserted into (8.14), one is led to

$$\sigma^s = \mu_1\mathbf{q} \cdot \mathbf{q} + \mu_0 p^v p^v + \mu_2\overset{o}{\mathbf{P}^v} : \overset{o}{\mathbf{P}^v}.$$

(8.18)

The requirement that σ^s must be positive leads to the restrictions

$$\mu_1 \geq 0, \qquad \mu_0 \geq 0, \qquad \mu_2 \geq 0.$$

(8.19)

In [191] it is discussed that inclusion of nonlinear terms raises some important conceptual questions concerning the interpretation of the second law. For instance, a nonlinear force implies a positive entropy production and this, in turn, provides a useful limitation on the domain of validity of the constitutive equations and on their possible forms.

8.1.3   Evolution equations of the fluxes

Identifying the forces as the conjugate terms of the fluxes in (8.13) and substituting these expressions in (8.17), one obtains in the linear approximation (products like $\mathbf{q} \cdot \nabla u$ and $\mathbf{q} \cdot \nabla v$ are omitted) the following set of evolution equations

$$\nabla T^{-1} - T^{-1}\alpha_{10}D_a\mathbf{q} = \mu_1\mathbf{q} - \beta''\nabla \cdot \overset{0}{\mathbf{P}}{}^v - \beta'\nabla p^v, \tag{8.20}$$

$$-T^{-1}\nabla \cdot \mathbf{v} - T^{-1}\alpha_{00}\dot{p}^v = \mu_0 p^v - \beta'\nabla \cdot \mathbf{q}, \tag{8.21}$$

$$-T^{-1}\overset{0}{\mathbf{V}} - T^{-1}\alpha_{21}D_b\overset{0}{\mathbf{P}}{}^v = \mu_2\overset{0}{\mathbf{P}}{}^v - \beta''(\nabla\mathbf{q})^s, \tag{8.22}$$

where, according to the result (8.16), the material time derivative has been replaced by more general objective derivatives.

The main features issued from the above thermodynamic formalism, according to [191] are:

1. The positiveness of the coefficients μ_1, μ_0 and μ_2.

2. The equality of the cross terms relating $\mathbf{q}$ with $\nabla \cdot \overset{0}{\mathbf{P}}{}^v$ and $\overset{0}{\mathbf{P}}{}^v$ with $(\nabla\mathbf{q})^s$ on the one side, $\mathbf{q}$ with ∇p^v and p^v with $\nabla \cdot \mathbf{q}$ on the other. The equality of these coefficients, confirmed by the kinetic theory, belongs to a class of higher-order Onsager relations [284, 285].

3. The equality of the coefficients β' and β'' appearing in the second-order terms of the entropy flux and the coefficients of the cross terms in the evolution equations (8.20)-(8.22). This result is confirmed by kinetic theory (see Chapter 7 of Part I).

In (8.20)-(8.22) several coefficients are introduced which have to be physically specified. Assume first a situation characterized by stationary and homogeneous fluxes (i.e., their time and space derivatives may be neglected). Equations (8.20)-(8.22) then reduce to

$$\nabla T^{-1} = \mu_1\mathbf{q}, \qquad -T^{-1}\nabla \cdot \mathbf{v} = \mu_0 p^v, \qquad -T^{-1}\overset{0}{\mathbf{V}} = \mu_2\overset{0}{\mathbf{P}}{}^v. \tag{8.23}$$

Comparison with the Fourier and Newton-Stokes laws

$$\mathbf{q} = -\lambda\nabla T, \qquad p^v = -\zeta\nabla \cdot \mathbf{v}, \qquad \overset{0}{\mathbf{P}}{}^v = -2\eta\overset{0}{\mathbf{V}}, \tag{8.24}$$

leads to the identifications

$$\mu_1 = (\lambda T^2)^{-1}, \qquad \mu_0 = (\zeta T)^{-1}, \qquad \mu_2 = (2\eta T)^{-1}, \tag{8.25}$$

with λ, ζ and η the thermal conductivity, bulk viscosity and shear viscosity, respectively.[1]

If non-stationary (but homogeneous) fluxes are assumed and, in addition, the time derivatives are identified with the material time derivatives, (8.20)-(8.22) reduce to

$$\nabla T^{-1} - T^{-1}\alpha_{10}\dot{\mathbf{q}} = (\lambda T^2)^{-1}\mathbf{q}, \tag{8.26}$$

---

[1]See page 48 for a footnote on the denotation of different viscosities. In this chapter the original notation of [191] is used. It coincides with the notation used in Chapter 5, only that the bulk viscosity ζ is called the volume viscosity and that the shear viscosity η is called the dynamic shear viscosity.

$$-T^{-1}\nabla \cdot \mathbf{v} - T^{-1}\alpha_{00}\dot{p}^v = (\zeta T)^{-1}p^v, \tag{8.27}$$

$$-T^{-1}\overset{0}{\mathbf{V}} - T^{-1}\alpha_{21}(\overset{0}{\mathbf{P}^\nu})\dot{} = (2\eta T)^{-1}\overset{0}{\mathbf{P}^v}. \tag{8.28}$$

These equations can be identified with the so-called Maxwell-Cattaneo laws (see [247], [68])

$$\tau_1 \dot{\mathbf{q}} + \mathbf{q} = -\lambda\nabla T, \tag{8.29}$$

$$\tau_0 \dot{p}^v + p^v = -\zeta\nabla \cdot \mathbf{v}, \tag{8.30}$$

$$\tau_2(\overset{0}{\mathbf{P}^v})\dot{} + \overset{0}{\mathbf{P}^v} = -2\eta\overset{0}{\mathbf{V}}, \tag{8.31}$$

where τ_1, τ_0 and τ_2 are the relaxation times of the respective fluxes. One is then led to the identifications

$$\alpha_{10} = \tau_1(\lambda T)^{-1}, \qquad \alpha_{00} = \tau_0\zeta^{-1}, \qquad \alpha_{21} = \tau_2(2\eta)^{-1}. \tag{8.32}$$

In terms of λ, ζ and η and the relaxation times τ_1, τ_0 and τ_2 the evolution equations (8.20)-(8.22) take the following form, where the expressions of the time derivatives (8.15) have been used

$$\tau_1 D_a\mathbf{q} = -(\mathbf{q} + \lambda\nabla T) + \beta''\lambda T^2\nabla \cdot \overset{0}{\mathbf{P}^v} + \beta'\lambda T^2\nabla p^v, \tag{8.33}$$

$$\tau_0 \dot{p}^v = -(p^v + \zeta\nabla \cdot \mathbf{v}) + \beta'\zeta T\nabla \cdot \mathbf{q}, \tag{8.34}$$

$$\tau_2 D_h\overset{0}{\mathbf{P}^v} = -(\overset{0}{\mathbf{P}^v} + 2\eta\overset{0}{\mathbf{V}}) + 2\beta''\eta T(\overset{0}{\nabla}\mathbf{q})^s. \tag{8.35}$$

Some values of the coefficients appearing in (8.33)-(8.35) are reported in [272].

8.1.4 Non-equilibrium equations of state and convexity requirements

In the preceding subsections the second-order contributions of the fluxes in the expressions of the temperature and the pressure have been neglected. Now, having identified the parameters α_{10}, α_{00} and α_{21} in physical terms, such contributions can be evaluated.

Insertion of the identifications (8.32) into (8.7) yields

$$ds = \theta^{-1}du + \theta^{-1}\pi dv - \frac{v\tau_1}{\lambda T^2}\mathbf{q} \cdot d\mathbf{q} - \frac{v\tau_0}{\zeta T}p^v dp^v - \frac{v\tau_2}{2\eta T}\overset{0}{\mathbf{P}^v} : d\overset{0}{\mathbf{P}^v}, \tag{8.36}$$

where the generalized absolute temperature θ and the thermodynamic pressure π have been used instead of their respective local-equilibrium approximations T and p. Integrability of (8.36) means that

$$\frac{\partial\theta^{-1}}{\partial\mathbf{q}} = -\left(\frac{\partial(v\tau_1/\lambda T^2)}{\partial u}\right)\mathbf{q}, \qquad \frac{\partial(\theta^{-1}\pi)}{\partial\mathbf{q}} = -\left(\frac{\partial(v\tau_1/\lambda T^2)}{\partial v}\right)\mathbf{q},$$

$$\frac{\partial\theta^{-1}}{\partial\overset{0}{\mathbf{P}^v}} = -\left(\frac{\partial(v\tau_2/2\eta T)}{\partial u}\right)\overset{0}{\mathbf{P}^v}, \qquad \frac{\partial(\theta^{-1}\pi)}{\partial\overset{0}{\mathbf{P}^v}} = -\left(\frac{\partial(v\tau_2/2\eta T)}{\partial v}\right)\overset{0}{\mathbf{P}^v}, \tag{8.37}$$

$$\frac{\partial\theta^{-1}}{\partial p^v} = -\left(\frac{\partial(v\tau_0/\zeta T)}{\partial u}\right)p^v, \qquad \frac{\partial(\theta^{-1}\pi)}{\partial p^v} = -\left(\frac{\partial(v\tau_0/\zeta T)}{\partial v}\right)p^v.$$

As a consequence of these two sets of expressions and keeping in mind that for vanishing values of the fluxes one must recover the local-equilibrium values of T and p, one obtains

$$\theta^{-1} = T^{-1} - \frac{1}{2}\left[\frac{\partial(v\tau_1/\lambda T^2)}{\partial u}\mathbf{q}\cdot\mathbf{q} + \frac{\partial(v\tau_0/\zeta T)}{\partial u}(p^v)^2 + \frac{\partial(v\tau_2/2\eta T)}{\partial u}\overset{0}{\mathbf{P}^v}:\overset{0}{\mathbf{P}^v}\right],$$

$$\tag{8.38}$$

$$\theta^{-1}\pi = T^{-1}p - \frac{1}{2}\left[\frac{\partial(v\tau_1/\lambda T^2)}{\partial v}\mathbf{q}\cdot\mathbf{q} + \frac{\partial(v\tau_0/\zeta T)}{\partial v}(p^v)^2 + \frac{\partial(v\tau_2/2\eta T)}{\partial v}\overset{0}{\mathbf{P}^v}:\overset{0}{\mathbf{P}^v}\right].$$

These expressions can be considered as non-equilibrium equations of state for the temperature and pressure. In [191] some consequences of introducing a generalized absolute temperature, pressure and chemical potential are discussed (see also [67]).

Now, the attention is turned to the requirements stemming from the convexity of the entropy. Such a property implies that the matrix of its second derivatives with respect to the whole set of variables must be negative definite. It amounts to demanding the stability of the (local) equilibrium state. When the dissipative fluxes are zero, one recovers the classical conditions for stability,

$$c_v = T\left(\frac{\partial s}{\partial T}\right)_v > 0, \qquad \kappa_T = -\frac{1}{v}\left(\frac{\partial v}{\partial p}\right)_T > 0, \tag{8.39}$$

with c_v the specific heat at constant volume and κ_T the isothermal compressibility. When the fluxes are taken into account, the convexity of the entropy implies that

$$\frac{v\tau_1}{\lambda T^2} > 0, \qquad \frac{v\tau_0}{\zeta T} > 0, \qquad \frac{v\tau_2}{2\eta T} > 0. \tag{8.40}$$

As the local-equilibrium temperature T is positive and since the second law requires that the dissipative coefficients λ, ζ and η are all positive, it turns out from (8.40) that the relaxation times τ_1, τ_0 and τ_2 must be positive. Note that otherwise causality would be violated.

In [191] it is shown for an ideal monatomic gas that the convexity of the entropy introduces supplementary restrictions. It is demonstrated that the matrix of the second differential of the entropy s is negative only for values of the fluxes lower than some critical values. Simultaneously, the domain of applicability of EIT is indicated: It seems to be well suited for wavelengths greater than the mean free path. The authors remark that it would not be realistic to conceive a thermodynamic description for length-scales shorter than the mean free path, or time scales shorter than the mean collision time; at these length-scales and time-scales the behavior of the particles is individual and ballistic rather than collective because of the lack of a sufficient number of collisions. Therefore, the result that the generalized entropy remains convex for lengths and time intervals at least of the order of the mean free path and the collision time, respectively, makes plausible the hypotheses underlying EIT. Further, the authors state that the positive character of the relaxation times implies the symmetric hyperbolicity of the set of evolution equations. The relation between symmetric hyperbolicity and the existence of a convex function whose production has a definite sign, that is, the existence of an entropy function, has been examined from a general mathematical point of view by several authors (e.g. I. Müller and T. Ruggeri [262, 263], K. O. Friedrichs and P. D. Lax [130], T. Ruggeri [327], T. Ruggeri and A. Strumia [332]). Such a connection between hyperbolicity and the existence of a convex entropy is of great interest in non-equilibrium thermodynamics, since it indicates that the condition of hyperbolicity implies the existence of an entropy function. In [191] the reciprocal point of view has been adopted: starting from the hypothesis of the existence of a generalized entropy, the authors have arrived at hyperbolic equations.

8.2 Microscopic foundations

8.2.1 Kinetic theory of gases

In Chapter 3 of [191] David Jou, José Casas-Vázquez and Georgy Lebon aimed to provide a microscopic interpretation of EIT by means of the kinetic theory of gases. Hypotheses and the main results of EIT should be justified, i.e., (a) the choice of the dissipative fluxes as independent variables, (b) the form of the generalized Gibbs equation

$$s(u, v, \mathbf{q}, p^v, \overset{0}{\mathbf{P}}{}^v) = s_{eq}(u, v) - \frac{v\tau_1}{2\lambda T^2}\mathbf{q} \cdot \mathbf{q} - \frac{v\tau_0}{2\zeta T}(p^v)^2 - \frac{v\tau_2}{4\eta T}\overset{0}{\mathbf{P}}{}^v : \overset{0}{\mathbf{P}}{}^v, \tag{8.41}$$

where $s_{eq}(u, v)$ is the local-equilibrium entropy (Equation (8.41) is obtained starting from the differential form of the generalized entropy (8.36), integration from the local-equilibrium value and use of (8.38)) and (c) the form (8.12) of the entropy flux is motivated. Moreover, the evolution equations for the fluxes (8.33)-(8.35) as well as the relations between the transport coefficients in these equations are justified. The main focus lies on the linear terms of the evolution equations, nonlinear terms are only occasionally looked at since thermodynamics, in its present state, is not able to put explicit restrictions on the nonlinear terms.

The main ideas of David Jou, José Casas-Vázquez and Georgy Lebon presented in [191] are reproduced here.

8.2.1.1 Basic concepts

Ideal or highly diluted monatomic gases are considered. The basis for the analysis is the distribution function $f(\mathbf{r}, \mathbf{c}, z)$, which accounts for the number of particles between $\mathbf{r}$ and $\mathbf{r} + d\mathbf{r}$ with velocity between $\mathbf{c}$ and $\mathbf{c} + d\mathbf{c}$ at time t. The evolution of $f(\mathbf{r}, \mathbf{c}, t)$ is described by the well-known Boltzmann equation, which takes into account the effects of binary collisions between particles, and neglects collisions involving more than two particles, a quite plausible hypothesis in dilute gases [72, 139, 153, 451].

The Boltzmann equation has the form

$$\frac{\partial f}{\partial t} + \mathbf{c} \cdot \frac{\partial f}{\partial \mathbf{r}} + \frac{\mathbf{F}}{m} \cdot \frac{\partial f}{\partial \mathbf{c}} = \int d\tilde{\mathbf{c}} \int d\Omega |\mathbf{c} - \tilde{\mathbf{c}}|\sigma(\mathbf{c} - \tilde{\mathbf{c}}, \theta)[f'\tilde{f}' - f\tilde{f}]. \tag{8.42}$$

Here, f, $\tilde{f}$, f' and $\tilde{f}'$ stand for $f(\mathbf{r}, \mathbf{c}, t)$, $f(\mathbf{r}, \tilde{\mathbf{c}}, t)$, $f(\mathbf{r}, \mathbf{c}', t)$ and $f(\mathbf{r}, \tilde{\mathbf{c}}', t)$ respectively; m is the mass of the particles and $\mathbf{F}$ the external force acting on the particles; $\sigma(\mathbf{c} - \tilde{\mathbf{c}}, \theta)$ is the differential cross-section of the collisions between the particles, one of them with initial velocity $\mathbf{c}$ and the other with initial velocity $\tilde{\mathbf{c}}$, to give as final velocities after collision $\mathbf{c}'$ and $\tilde{\mathbf{c}}'$; θ is the angle between $\mathbf{c}$ and $\mathbf{c}'$; $d\Omega$ is the differential solid angle around θ.

Only a single species of molecules is considered. If different types of molecules are included, a separate distribution function for each type must be introduced, and the collision term couples them. Furthermore, spinless molecules are dealt with and internal degrees of freedom are neglected.

The Boltzmann equation is a nonlinear, integro-differential equation and difficult to solve. However, several general consequences can be drawn from it even without solving it explicitly. The reason is a useful symmetry property of the collision term. (8.42) is written as

$$\frac{\partial f}{\partial t} + \mathbf{c} \cdot \frac{\partial f}{\partial \mathbf{r}} + \frac{\mathbf{F}}{m} \cdot \frac{\partial f}{\partial \mathbf{c}} = J(f), \tag{8.43}$$

with $J(f)$ the collision term. Let $\psi(\mathbf{c})$ be an arbitrary function of the molecular velocity $\mathbf{c}$. Then, the following relation is satisfied

$$\int \psi(\mathbf{c})J(f)d\mathbf{c} = \frac{1}{4}\int [\psi(\mathbf{c}) + \psi(\tilde{\mathbf{c}}) - \psi(\mathbf{c}') - \psi(\tilde{\mathbf{c}}')]J(f)d\mathbf{c}. \tag{8.44}$$

This equality is a consequence of the next three relations. First, it can immediately be seen that

$$\int \psi(\mathbf{c})J(f)d\mathbf{c} = \int d\mathbf{c} \int d\tilde{\mathbf{c}} \int d\Omega |\mathbf{c} - \tilde{\mathbf{c}}|\sigma[f'\tilde{f}' - f\tilde{f}]\psi(\mathbf{c}) = \int \psi(\tilde{\mathbf{c}})J(f)d\mathbf{c}. \tag{8.45}$$

The equality is obvious, for in (8.45) $\mathbf{c}$ and $\tilde{\mathbf{c}}$ are dummy quantities, since they are integrated over all their possible values; hence an exchange between $\mathbf{c}$ and $\tilde{\mathbf{c}}$ is irrelevant.

A second equality is obtained after replacing $\mathbf{c}$ and $\tilde{\mathbf{c}}$ by $\mathbf{c}'$ and $\tilde{\mathbf{c}}'$; it yields

$$\int \psi(\mathbf{c}')J(f')d\mathbf{c}' = \int d\mathbf{c}' \int d\tilde{\mathbf{c}}' \int d\Omega'|\mathbf{c}' - \tilde{\mathbf{c}}'|\sigma'[f\tilde{f} - f'\tilde{f}']\psi(\mathbf{c}')$$

$$= -\int d\mathbf{c} \int d\tilde{\mathbf{c}} \int d\Omega |\mathbf{c} - \tilde{\mathbf{c}}|\sigma[f'\tilde{f}' - f\tilde{f}]\psi(\mathbf{c}) = -\int \psi(\tilde{\mathbf{c}})J(f)d\mathbf{c}. \tag{8.46}$$

This equality results clearly from the properties $d\mathbf{c}d\tilde{\mathbf{c}} = d\mathbf{c}'d\tilde{\mathbf{c}}'$, $\mathbf{c}-\tilde{\mathbf{c}} = \mathbf{c}'-\tilde{\mathbf{c}}'$ and $\sigma = \sigma(\mathbf{c}-\tilde{\mathbf{c}}, \theta) = \sigma(\mathbf{c}'-\tilde{\mathbf{c}}', \theta') = \sigma'$. The first two of these equalities are valid for elastic binary collisions, while the third is satisfied when the intermolecular potential is invariant under spatial rotations and reflections and under time reversal.

A third equality follows from the exchange of the dummy quantities $\mathbf{c}'$ and $\tilde{\mathbf{c}}'$:

$$\int \psi(\mathbf{c}')J(f')d\mathbf{c}' = \int \psi(\tilde{\mathbf{c}}')J(\tilde{f}')d\tilde{\mathbf{c}}'. \tag{8.47}$$

Thus, combining (8.45) to (8.47) the key relation (8.44) is obtained.

Balance equations   The first important consequence of (8.44) is the possibility of deriving the hydrodynamic balance equations for mass, momentum, and energy from Boltzmann's equation. Note that the mass density ρ, the mean velocity $\mathbf{v}$, and the internal energy u per unit mass are defined in terms of the distribution function as

$$\rho(\mathbf{r}, t) = \int mf(\mathbf{r}, \mathbf{c}, t)d\mathbf{c},$$

$$\rho(\mathbf{r}, t)\mathbf{v}(\mathbf{r}, t) = \int m\mathbf{c}f(\mathbf{r}, \mathbf{c}, t)d\mathbf{c}, \tag{8.48}$$

$$\rho(\mathbf{r}, t)u(\mathbf{r}, t) = \int \frac{1}{2}m(\mathbf{c} - \mathbf{v}) \cdot (\mathbf{c} - \mathbf{v})f(\mathbf{r}, \mathbf{c}, t)d\mathbf{c}.$$

Evolution equations for these quantities are obtained from the evolution equation for f. In doing this, one must consider the influence of the collision term. Note, however, that in view of (8.44) one may write

$$\int m J(f) d\mathbf{c} = \frac{1}{4} \int [m + m - m - m] J(f) d\mathbf{c} = 0,$$

$$\int m\mathbf{c} J(f) d\mathbf{c} = \frac{1}{4} [m\mathbf{c} + m\tilde{\mathbf{c}} - m\mathbf{c}' - m\tilde{\mathbf{c}}'] J(f) d\mathbf{c} = 0, \tag{8.49}$$

$$\int m c^2 J(f) d\mathbf{c} = \frac{1}{4} \int [m c^2 + m\tilde{c}^2 - m c'^2 - m\tilde{c}'^2] J(f) d\mathbf{c} = 0.$$

The vanishing of these integrals follows from the property that mass, momentum and kinetic energy are collisional-invariant, i.e., they do not change in elastic binary collisions.

From the definitions (8.48) and Relations (8.49), the balance equations for mass, momentum and energy can be derived from the Boltzmann equation. Multiplying each term of (8.42) by m, $m\mathbf{c}$ and $\frac{1}{2}mc^2$, respectively, and integrating over $\mathbf{c}$ yields

$$\frac{\partial \rho}{\partial t} + \nabla \cdot (\rho \mathbf{v}) = 0, \tag{8.50}$$

$$\frac{\partial (\rho \mathbf{v})}{\partial t} = \nabla \cdot (\rho \mathbf{v}\mathbf{v} + \mathbf{P}) = 0, \tag{8.51}$$

$$\frac{\partial}{\partial t}\left(\rho u + \frac{1}{2}\rho v^2\right) + \nabla \cdot \left[\rho(u + \frac{1}{2}v^2 + \mathbf{P} \cdot \mathbf{v} + \mathbf{q})\right] = 0, \tag{8.52}$$

with $\mathbf{P}$ and $\mathbf{q}$ defined as

$$\mathbf{P} = \int m\mathbf{C}\mathbf{C} f d\mathbf{c}, \qquad \mathbf{q} = \int \frac{1}{2}m C^2 \mathbf{C} f d\mathbf{c}, \tag{8.53}$$

where $\mathbf{C} = \mathbf{c} - \mathbf{v}$ is the relative velocity of the molecules with respect to the mean motion of the gas.

Equations (8.50) to (8.52) turn out to be the well-known balance equations of hydrodynamics, provided that one identifies $\mathbf{P}$ and $\mathbf{q}$ defined by (8.53) as the pressure tensor and the heat flux vector, respectively.

Since at equilibrium f is an isotropic function of $\mathbf{C}$, the pressure tensor reduces to $\mathbf{P} = p\mathbf{U}$ with $\mathbf{U}$ the identity tensor and $p = \frac{1}{3}\int mC^2 f d\mathbf{c}$ the equilibrium pressure. It is found from the last definition of (8.48) of the internal energy that $p = \frac{2}{3}\rho u$. The macroscopic thermal equation of state for ideal gases leads then to the definition of the absolute equilibrium temperature: $p = \frac{2}{3}\rho u = nk_B T$, with n the number of particles per unit volume and k_B the Boltzmann constant.

The H-Theorem and the second law   Another important result which can be derived from Boltzmann's equation is the so-called H-Theorem which allows to obtain a microscopic expression for the entropy and accordingly for the second law. η is defined as follows

$$\rho(\mathbf{r}, t)\eta(\mathbf{r}, t) = \int f(\mathbf{r}, \mathbf{c}, t) \ln f(\mathbf{r}, \mathbf{c}, t) d\mathbf{c}. \tag{8.54}$$

The evolution equation for this quantity may be derived by multiplying term by term the Boltzmann equation (8.42) by $\ln f$ and integrating over $\mathbf{c}$. In this way, one obtains

$$\frac{\partial (\rho \eta)}{\partial t} + \nabla \cdot (\mathbf{J}^\eta + \rho \eta \mathbf{v}) = \sigma^\eta, \tag{8.55}$$

with the flux $\mathbf{J}^\eta = \int \mathbf{C} f \ln f \, d\mathbf{c}$ and the production term $\sigma^\eta = \int J(f) \ln f \, d\mathbf{c}$, which can be written, by virtue of (8.44) as

$$
\begin{aligned}
\sigma^\eta &= \frac{1}{4} \int d\mathbf{c} \int d\tilde{\mathbf{c}} \int d\Omega |\mathbf{c} - \tilde{\mathbf{c}}| \sigma [f'\tilde{f}' - f\tilde{f}][\ln f + \ln \tilde{f} - \ln f' - \ln \tilde{f}'] \\
&= \frac{1}{4} \int d\mathbf{c} \int d\tilde{\mathbf{c}} \int d\Omega |\mathbf{c} - \tilde{\mathbf{c}}| \sigma [f'\tilde{f}' - f\tilde{f}] \ln[f\tilde{f}/(f'\tilde{f}')].
\end{aligned}
\tag{8.56}
$$

Inspecting the product of the quantities $[f'\tilde{f}' - f\tilde{f}]$ and $\ln[f\tilde{f}/(f'\tilde{f}')]$ it can be seen immediately that $\sigma^\eta \leq 0$. This equality holds only for $f'\tilde{f}' = f\tilde{f}$, i.e., for $J(f) = 0$, which corresponds to equilibrium situations. Note that the equilibrium distribution function is obtained by realizing that, since $f_{eq}\tilde{f}_{eq} = f'_{eq}\tilde{f}'_{eq}$, then $\ln f_{eq}$ is a collisional invariant and can therefore be expressed as a linear combination of m, $m\mathbf{c}$ and $\frac{1}{2}mc^2$. The constants appearing in such a combination can be determined in terms of p, $\mathbf{v}$ and T. The result is the well-known Maxwell-Boltzmann distribution function

$$
f_{eq} = n \left(\frac{m}{2\pi k_B T} \right)^{3/2} \exp\left[-\frac{mC^2}{2k_B T} \right].
\tag{8.57}
$$

The H-Theorem, stating the negative character of σ^η, allows to express the entropy in terms of the distribution function f. It suggests that the entropy s per unit mass may be defined as $\rho s = -A\rho\eta + B$, where A and B are constants. Their values can be determined by comparing the equilibrium value of s with the value of η when the equilibrium distribution function (8.57) is substituted in the definition (8.54). It turns out that $A = k_B$, the Boltzmann constant. The value of B is not so important because generally only entropy changes are relevant. The above considerations suggest the following expressions for the entropy and the entropy flux, respectively

$$
\rho s = -k_B \int f \ln f \, d\mathbf{c}, \qquad \mathbf{J}^s = -k_B \int \mathbf{C} f \ln f \, d\mathbf{c}.
\tag{8.58}
$$

In the preceding, microscopic expressions for the entropy and the entropy flux in non-equilibrium situations in terms of the non-equilibrium distribution function have been derived. This distribution function could be expanded so that it depends on a small parameter, for instance the ratio of the relaxation time to the macroscopic time, the ratio of the mean free path to a characteristic length of the macroscopic inhomogeneities, the higher-order moments of the velocity distribution function, etc. but this is not shown here (see [191] for details).

Two important models, namely the Chapman-Enskog and Grad models, have been proposed to solve the Boltzmann equation in non-equilibrium situations. In the Chapman-Enskog approach [72], f is expressed in terms of the first five moments n, $\mathbf{v}$, and T and their gradients. In Grad's model [139, 153], f is developed in terms of its moments with respect to the molecular velocity. Note that, in view of definitions (8.53), $\mathbf{P}$ and $\mathbf{q}$ are directly related to the moments of the velocity distribution function (the scalar viscous pressure p^v vanishes in an ideal gas). Therefore, the mean values of $\mathbf{q}$ and of $\overset{0}{\mathbf{P}}{}^v$ are considered in Grad's theory as independent variables, so that Grad's theory is closer to the macroscopic developments of extended irreversible thermodynamics than Chapman-Enskog's. It can thus be asserted that both extended irreversible thermodynamics and Grad's 13-moment method make use of the same independent variables. In [191] Grad's method is presented in details.

8.2.2 Fluctuation theory

In the chapter on fluctuation theory in [191] the focus lies on the generalized Gibbs equation, which is the central thermodynamic result of the version of extended thermodynamics presented by the authors. They aim to show that the generalized entropy plays an essential role in the description of fluctuations around equilibrium states not only of conserved variables but also of dissipative fluxes. The study of fluctuations provides not only a link between entropy and probability but it also allows to determine non-classical entropy coefficients in terms of the equilibrium distribution function. Investigations of fluctuations around equilibrium led to the Onsager reciprocity relations of non-equilibrium thermodynamics.

8.2.2.1 Einstein's formula. Second moments of equilibrium fluctuations

The consequences of a replacement of the classical entropy by a generalized one on the relation between entropy and probability is studied. While some authors construct non-equilibrium ensembles in the Gibbsian sense, in [191] a way is followed which starts from the Einstein relation for the probability of fluctuations. For isolated systems, the probability of fluctuations of thermodynamic variables with respect to equilibrium reference values is related to the change in entropy ΔS associated with the fluctuation

$$\Delta S = S - S(\text{equilibrium}), \tag{8.59}$$

where S is the total entropy of a system of mass M and volume V. The relation between the probability and the entropy change was proposed by Einstein by inverting the well-known Boltzmann-Planck formula for the entropy

$$S = k_B \ln W, \tag{8.60}$$

with the probability of the macro-state W and the Boltzmann constant k_B. It follows that

$$W \approx \exp(\Delta S/k_B). \tag{8.61}$$

If the fluctuations are small enough, the quantity ΔS may be expanded

$$\Delta S = (\delta S)_{eq} + \frac{1}{2}(\delta^2 S)_{eq}, \tag{8.62}$$

where δS and $\delta^2 S$ are the first and second differentials of the entropy. Since for isolated systems the entropy possesses its maximum at equilibrium it follows that $(\delta S)_{eq} = 0$ and $(\delta^2 S)_{eq} \leq 0$. Substitution into Equation (8.61) leads to

$$W \approx \exp\left[\frac{1}{2}\delta^2 S/k_B\right]. \tag{8.63}$$

This is the Einstein formula for the probability of fluctuations. This relation is valid for moments up to second order but it fails for higher order moments [62].

If non-isolated systems are considered, the surroundings must be incorporated and A. Einstein [110] suggested, expressing the probability of fluctuations as

$$W \approx \exp(-\Delta\mathcal{A}/k_B T_0), \tag{8.64}$$

where the new quantity $\mathcal{A}$, the availability, is defined as

$$\mathcal{A} = U - T_0 S + p_0 V, \tag{8.65}$$

in which T_0 and p_0 are the constant temperature and pressure of the surrounding. $\mathcal{A}$ is not a property of the system alone but of the system in a given environment. If the system and the surroundings are in thermal and mechanical equilibrium, $T = T_0$ and $p = p_0$, and the maximum amount of work which can be extracted from the system during exchanges with the surroundings, $\Delta\mathcal{A}$ can be described by

$$\Delta\mathcal{A} = \Delta U - T\Delta S + p\Delta V = \Delta G, \tag{8.66}$$

where G is the Gibbs function, so that (8.64) takes the form

$$W \approx \exp(-\Delta G/k_B T). \tag{8.67}$$

Analogously, for a system with fixed V and T, $\Delta\mathcal{A} = \Delta F$, with F the Helmholtz free energy, and the probability of fluctuations is

$$W \approx \exp(-\Delta F/k_B T). \tag{8.68}$$

The latter two expressions are valid for large and small fluctuations. Restricting the study to small fluctuations, $\Delta\mathcal{A}$ can be expanded in a series, $\Delta\mathcal{A} = \delta\mathcal{A}_{eq} + \frac{1}{2}(\delta^2\mathcal{A})_{eq} + ...$, where the first-order term is zero at equilibrium because G or F at fixed T and p or fixed V and T, respectively, are at a minimum. Thus,

$$W \approx \exp\left[-\frac{1}{2}\delta^2 G/(k_B T)\right], \qquad W \approx \exp\left[-\frac{1}{2}\delta^2 F/(k_B T)\right]. \tag{8.69}$$

Recalling that $F = U - TS$ and studying the fluctuations of U at constant T (with U the independent variable and $S = S(U)$) follows $\delta^2 F = -T\delta^2 S$. Insertion into (8.69) leads to (8.63) but $\delta^2 S$ is computed at constant T and V. Analogously, $\delta^2 G = -T\delta^2 S$ leads to (8.63) but with $\delta^2 S$ computed at constant T and p. This shows clearly that (8.63) does not apply only to isolated systems but to many other situations, keeping in mind, that according to external constraints, different quantities have to be fixed.

Now, the generalized entropy of extended irreversible thermodynamics is used to obtain information about fluctuations around equilibrium states [189], [192]. The second differential of the generalized entropy may be derived directly from the Gibbs equation (8.41), and when this expression is inserted into (8.63) one obtains

$$W(\delta u, \delta v, \delta p^v, \delta\mathbf{q}, \delta\overset{0}{\mathbf{P}}{}^v) \approx \exp\left\{\frac{M}{2k_B}\left[(T^{-1})_u(\delta u)^2 + 2(T^{-1})_v\delta u\delta v + (T^{-1}p)_v(\delta v)^2\right.\right.$$

$$\left.\left. -\frac{v\tau_0}{\zeta T}\delta p^v\delta p^v - \frac{v\tau_1}{\lambda T^2}\delta\mathbf{q}\cdot\delta\mathbf{q} - \frac{v\tau_2}{2\eta T}\delta\overset{0}{\mathbf{P}}{}^v : \delta\overset{0}{\mathbf{P}}{}^v\right]\right\}, \tag{8.70}$$

where the subscripts u and v stand for partial derivatives with respect to u and v, respectively.

To obtain the second moments of the fluctuations from (8.70) it is important to know that for a multivariant Gaussian distribution function $W \approx \exp\left[-\frac{1}{2}E_{ij}\delta x_i\delta x_j\right]$ the second moments read $\langle\delta x_i\delta x_j\rangle = (E^{-1})_{ij}$. The matrix corresponding to the second derivatives of the entropy with respect to u and v is $E_{uv} = -\frac{M}{k_B} \begin{pmatrix} (T^{-1})_u & (T^{-1})_v \\ (T^{-1})_v & (T^{-1}p)_v \end{pmatrix}$. From (8.70) it becomes obvious that the fluctuations of the classical variables u and v are uncoupled with the fluctuations of the fluxes. Taking into account the just mentioned relations, the second moments of the internal fluctuations of u and v may be obtained from (8.70) as

$$\begin{aligned}
\langle \delta u \delta u \rangle &= k_B c_p T^2 - 2k_B T^2 pv\alpha + k_B T p^2 v \kappa_T, \\
\langle \delta u \delta v \rangle &= k_B T^2 v\alpha - k_B T pv\kappa_B, \\
\langle \delta v \delta v \rangle &= k_B T v \kappa_T,
\end{aligned} \tag{8.71}$$

where c_p is the specific heat at constant pressure, α the coefficient of thermal expansion, and κ_T the isothermal compressibility. The second moments of the fluctuations of the fluxes read

$$\begin{aligned}
\langle \delta q_i \delta q_j \rangle &= k_B \lambda T^2 (\tau_1 V)^{-1} \delta_{ij}, \\
\langle \delta p^v \delta p^v \rangle &= k_B \zeta T (\tau_0 V)^{-1}, \\
\langle \delta P_{ij}^v \delta P_{ij}^v \rangle &= k_B \eta T (\tau_2 V)^{-1} \Delta_{ijkl},
\end{aligned} \tag{8.72}$$

with the Kronecker symbol δ_{ij} and $\Delta_{ijkl} = \left(\delta_{ik}\delta_{jl} + \delta_{il}\delta_{jk} - \frac{2}{3}\delta_{ij}\delta_{kl} \right)$. Expressions (8.72) relate the dissipative coefficients λ, ζ and η to the fluctuations of the fluxes with respect to equilibrium. These relations may be interpreted in two different ways: either by stating that the dissipative coefficients determine the strength of the fluctuations or saying that the fluctuations determine the dissipative coefficients. Further information in this context can be found in [191].

8.2.2.2 Ideal gases

As was done for the version of extended thermodynamics of Müller, Ruggeri and Liu, investigated in Part I of this book, also for EIT the example of ideal gases is studied. The coefficients appearing in the non-classical part of the generalized entropy are determined by using the expressions for the second moments derived in the previous subsection and the equilibrium distribution function for ideal gases.

In [191] the microscopic expressions for $\mathbf{q}$ and $\mathbf{P}^v$ are given as

$$\mathbf{q} = \int \frac{1}{2} mC^2 \mathbf{C} f \, dc, \qquad P^v = \int \left(m\mathbf{C}\mathbf{C} - \frac{1}{3} pU \right) f \, dc, \tag{8.73}$$

where f is the distribution function, $\mathbf{c}$ the velocity of a molecule, $\mathbf{v}$ the mean velocity, $\mathbf{C} = \mathbf{c} - \mathbf{v}$ the molecular velocity relative to the mean motion, m the mass of a molecule, U the identity tensor and p the thermodynamic pressure.

The fluctuations of $\mathbf{q}$ near equilibrium are due to the fluctuations δf of the distribution function and the fluctuations $\delta \mathbf{v}$ of the mean velocity. Up to first-order terms in the fluctuations one has

$$\delta\mathbf{q} = \int \frac{1}{2} mC^2 \mathbf{C}\delta f \, dc - \left[\int \frac{1}{2} mC^2 f_{eq} dc \right] \delta\mathbf{v} - \left[\int m\mathbf{C}\mathbf{C} f_{eq} dc \right] \cdot \delta\mathbf{v}. \tag{8.74}$$

For an ideal gas the second and third integrals on the right-hand side give, respectively, $\rho u \delta \mathbf{v}$ and $p\delta \mathbf{v}$, and hence

$$\delta\mathbf{q} = \int \frac{1}{2} mC^2 \mathbf{C}\delta f \, dc - \rho h\delta\mathbf{v}, \tag{8.75}$$

where $h = u + (p/\rho)$ is the enthalpy per unit mass, which for a classical ideal gas takes the form $h = \frac{5}{2}k_B T/m$.

Since $\delta \mathbf{v}$ is, by definition, taken from

$$\rho \delta \nu = \int m \mathbf{C} \delta f \, d\mathbf{c}, \tag{8.76}$$

Equation (8.75) for the fluctuation of the heat flux can be written as

$$\delta q = \int \left(\frac{1}{2}mC^2 - \frac{5}{2}k_B T \right) \mathbf{C} \delta f \, d\mathbf{c}. \tag{8.77}$$

Using this expression, usually called the subtracted heat flux, to compute the second moments of the fluctuations, it is found that

$$\langle \delta \mathbf{q} \delta \mathbf{q} \rangle = \int d\mathbf{c} \int d\mathbf{c}' \left(\frac{1}{2}mC^2 - \frac{5}{2}k_B T \right) \mathbf{C} \left(\frac{1}{2}mC'^2 - \frac{5}{2}k_B T \right) \mathbf{C}' \left\langle \delta f(\mathbf{C}) \delta f(\mathbf{C}') \right\rangle . \tag{8.78}$$

For the fluctuations of the non-diagonal components of the viscous pressure tensor, an analogous derivation yields

$$\delta \mathbf{P}^v = \int m \mathbf{CC} \delta f \, d\mathbf{c}, \tag{8.79}$$

from which

$$\left\langle \delta P_{ij}^v \delta P_{kl}^v \right\rangle = \int d\mathbf{c} \int d\mathbf{c}' m C_i C_j m C_k' C_l' \left\langle \delta f(\mathbf{C}) \delta f(\mathbf{C}') \right\rangle . \tag{8.80}$$

The expression for the second moments of the fluctuations of the distribution function may be obtained [218] from the Einstein formula (8.63) when Boltzmann's formula for the entropy,

$$S = -k_B V \int f \ln f \, d\mathbf{c}, \tag{8.81}$$

where V is the volume of the system, is used. According to (8.63) and (8.81),

$$\langle \delta f(\mathbf{r}, \mathbf{c}, t) \delta f(\mathbf{r}', \mathbf{c}', t) \rangle = \frac{1}{V} f_{eq}(\mathbf{r}, \mathbf{c}) \delta(\mathbf{r} - \mathbf{r}') \delta(\mathbf{c} - \mathbf{c}'). \tag{8.82}$$

As a consequence, (8.78) and (8.80) reduce to

$$\langle \delta \mathbf{q} \delta \mathbf{q} \rangle = \frac{1}{V} \int d\mathbf{c} \left(\frac{1}{2}mC^2 - \frac{5}{2}k_B T \right)^2 \mathbf{CC} f_{eq},$$

$$\left\langle \delta P_{ij}^v \delta P_{kl}^v \right\rangle = \frac{1}{V} \int d\mathbf{c} \, m^2 C_i C_j C_k C_l f_{eq}. \tag{8.83}$$

Inserting the Maxwell-Boltzmann distribution function

$$f_{eq} = n \left(\frac{m}{2\pi k_B T} \right)^{3/2} \exp \left[-\frac{mC^2}{2k_B T} \right], \tag{8.84}$$

into both expressions of (8.83) and substituting the corresponding results in the first and third expression of (8.72), one obtains for the ratios τ_1/λ and τ_2/η

$$\frac{\tau_1 v}{\lambda T^2} = \frac{2}{3}(p^2 T)^{-1}, \qquad \frac{\tau_2 v}{2\eta T} = \frac{1}{2} v(pT)^{-1}. \tag{8.85}$$

The present theory may also be applied to degenerate ideal gases. In this case, the entropy reads [218]

$$S = -k_B V \int \left[f \ln \frac{f}{y} \mp y \left(1 \pm \frac{f}{y} \right) \ln \left(1 \pm \frac{f}{y} \right) \right] d\mathbf{c}, \qquad (8.86)$$

where the upper sign applies for Bose-Einstein statistics and the lower one for Fermi-Dirac statistics; $y = (2s + 1)/h^3$ for particles with spin s and mass m different from zero, and $y = 2s/h^3$ for particles with vanishing mass. Here, h is the Planck constant. Combining the Einstein formula (8.63) with (8.86), one obtains for the second moments of the fluctuations of the distribution function

$$\langle \delta f(\mathbf{r}, \mathbf{c}, t) \delta f(\mathbf{r}', \mathbf{c}', t) \rangle = -\frac{1}{V} \left(\frac{\partial f}{\partial \alpha} \right)_{eq} \delta(\mathbf{r} - \mathbf{r}') \delta(\mathbf{c} - \mathbf{c}'), \qquad (8.87)$$

where the equilibrium distribution function is given by

$$f_{eq} = \frac{y}{\exp(\alpha + x^2) \mp 1}, \qquad (8.88)$$

with $\mathbf{x} = [m/(2k_B T)]^{1/2} \mathbf{C}$ and $\alpha = -\mu/k_B T$, μ being the chemical potential. The expressions for $\mathbf{q}$ and $\mathbf{P}^v$ are identical with those of the preceding case, but now, of course, the expressions for the enthalpy h and pressure p are those corresponding to degenerate gases.

The result is presented in terms of the functions $I_n(\alpha)$ defined as

$$I_n^{\pm}(\alpha) = \int_0^\infty \frac{x^n}{\exp(\alpha + x^n) \pm 1} \, dx, \qquad (8.89)$$

which satisfy the relation $dI_n/d\alpha = -(1/2)(n - 1)I_{n-2}(\alpha)$. Concerning the quantities $(\tau_1 v/\lambda T^2)$ and $(\tau_2 v/2\eta T)$, it is found that

$$\frac{\tau_1 v}{\lambda T^2} = \left\{ \frac{5}{3} \frac{k_B T^{7/2}}{mA} \left[\frac{7}{5} \frac{k_B T}{m} I_6(\alpha) - \frac{5}{3} \frac{k_B T^{5/2}}{m\rho A} I_4^2(\alpha) \right] \right\}^{-1}, \qquad (8.90)$$

and

$$\frac{\tau_2 v}{2\eta T} = \frac{3mA}{4k_B T^{7/2} I_4(\alpha)}, \qquad (8.91)$$

with $A^{-1} = 4\pi 2^{1/2} y R^{3/2} m$ and R the gas constant. These results coincide with those derived by Liu and Müller [235] (compare also Part I) by an altogether different method. In the limit of high α, (8.90), (8.91) reduce to (8.85).

The results of this subsection are in agreement with those of the kinetic theory of ideal gases presented in Subsection 8.2.1. The practical interest of the theory of fluctuations is outlined by these results, which show that it is possible to calculate the extra terms of entropy from equilibrium statistical mechanics without going necessarily through non-equilibrium kinetic theory. For dilute real gases interaction effects between particles have to be taken into account (see [188]).

8.3 Final comments

The version of extended thermodynamics presented in this chapter is conceptually analogous in the main fundamental aspects to the Müller/Liu/Ruggeri version (see Part I), namely: (a) both of them use fluxes as additional independent variables (not only heat flux, viscous pressure, diffusion flux and electric current but also higher-order fluxes when necessary); (b) both of them use a generalized entropy and a generalized entropy flux, deeply related to the relaxational terms and the non-local terms in generalized transport equations; (c) both of them impose the requirements of the positiveness of the generalized entropy production as the statement for the second law of thermodynamics.

The main differences are, on one side, that Jou/Casas-Vázquez/Lebon extend the classical version of non-equilibrium thermodynamics, whereas Müller/Liu/Ruggeri extend rational thermodynamics, but the results of both extensions coincide in many situations. Both theories complement each other in some aspects. In particular, the Jou/Casas-Vázquez/Lebon version presented here complements the Müller/Liu/Ruggeri in three main aspects:

1. it has been applied not only to ideal gases, photons, phonons and electrons, but also to a wide diversity of other systems, as real gases, liquids, semiconductor devices, polymer solutions, colliding atomic nuclei [190], laminar and turbulent superfluids [193], and nanosystems [78, 348];

2. it looks for the microscopic grounds not only in the kinetic theory of gases, but also in fluctuation theory, Fluctuation-Dissipation Theorems, maximum-entropy methods, and non-equilibrium projection operators [239], thus opening the theory to a wider diversity of microscopic approaches;

3. it has explored in detail the non-equilibrium equations of state characterizing the thermodynamics of non-equilibrium steady states, both on conceptual exploration of the meaning of temperature, pressure and chemical potential in non-equilibrium steady states [67, 221], as well as on its practical implications concerning, for instance, the shift of phase diagrams of flowing systems or of systems under a heat flux [190].

This wider generality in applications and foundations is achieved at the price of some loss of the mathematical elegance and predictive power with respect to the numerical values of the coefficients that is achieved in the Müller/Liu/Ruggeri version. Thus, both theories complement each other in the applications and microscopic foundations, and coincide in the use of the fluxes as independent variables, with an extended form of entropy, entropy flux and entropy production. The generalized transport equations (or constitutive equations) derived from them are especially useful in nanosystems, in fast perturbations (atomic nuclei collisions, for instance), and in systems with long relaxation times (as superfluids and polymers).

Chapter 9

Dislocations

9.1   Introduction

One of the difficult questions of materials science some 100 years ago was the elucidation of the mechanism of plastic deformation of crystalline bodies. In the 30th of the XX century it was shown by A. H. Cottrell [90], E. Orowan [286], [287], M. Polanyi [296], G. I. Taylor [377], that dislocations, line defects in crystalline bodies, are the source of plastic deformation. Since this discovery a new branch of plasticity has been developed – crystal plasticity. It began with works of E. Schmid [340], W. Boas [341], G. I. Taylor [377] and yielded important results in the field of evolution of plastic anisotropy, textures, cold rolling and forge techniques of metals.

In this chapter, we present some properties of discrete dislocations as well as a continuum model of these defects in crystalline materials. However, the theory of discrete dislocations found also an application in other fields of continua such as in modeling of rupture of tectonic plates yielding earthquakes. This application shall be briefly presented at the end of this chapter.

As we demonstrate further, the dislocation is a line defect characterized by the so-called Burgers vector which, in turn, is defined in a crystal by the so-called Burgers contour. We present these notions on a few simplified cartoons and then we introduce their description stemming from the linear theory of elasticity.

It should be mentioned that apart from dislocations which are of the main concern in this chapter various other defects in materials play an important role in practical applications. One of them is the so-called extra matter. This notion embraces Frenkel defects, i.e., crystalline vacancies and additional interstitial atoms, and Schottky defects which are pairs of vacancies and additional atoms. These defects cannot be described by Burgers vectors as they are not line defects. Another important class of defects are microcracks whose creation and growth are the subject of damage mechanics. We skip both these classes in this brief presentation.

The simplest form of the dislocation line is defined in an infinite crystal. We show schematically such an ideal cubic crystal in Figure 9.1. In the left panel of this figure we present the defect which is created by removing a half-plane of atoms in the lower part of the crystal. As a result the upper half-space and the lower half-space possess a misfit. In order to correct it, in the vicinity of the horizontal cut the lattice constants (distance between the circles indicated in the figure) must be different. Then the contour created by the connection of neighboring atoms around the line where the half-plane terminates cannot be closed if we make the same number of steps in two perpendicular directions on the plane. The vector connecting the end-points of such a contour is called the Burgers

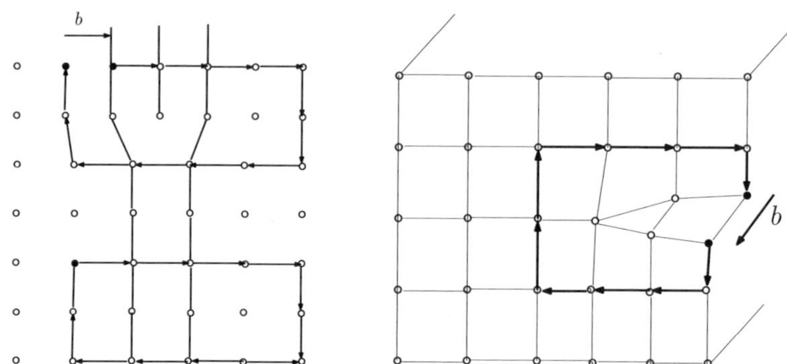

Figure 9.1: *Burgers contours of edge dislocation (left panel) and screw dislocation (right panel).*

vector. It is denoted by **b** and in the example of the left panel it is horizontal. Obviously, a contour which does not enclose the dislocation line, as shown in the lower part of this drawing, is closed. In the case of this dislocation the Burgers vector is perpendicular to the dislocation line and such a dislocation is called edge dislocation. As we see further, the Burgers vector defines a discontinuity of the displacement of a macroscopic continuum theory.

In the right panel we show a schematic of the so-called screw dislocation. The dislocation line is perpendicular to the shown cut and the Burgers vector is parallel to this line.

In Figure 9.2 we show the combination of many edge dislocations. Another schematic of the screw dislocation is shown in the middle panel of Figure 9.3.

The above described construction is demonstrated, again, in the Föll cartoons[1] of Figure 9.3. The right panel shows the combination of edge and screw dislocations in which the vector **b** describing the misfit is neither perpendicular to the dislocation line (edge dislocation) nor parallel to this line (screw dislocation). It forms rather an angle of 60 degrees with this line. Obviously, depending on the combination, this angle may be arbitrary.

Dislocation lines carry both an accumulated energy and self-equilibrated stresses in the reference state of the body. They try to minimize this energy by minimizing the length. This means that in the infinite ideal crystal they form straight lines. However, in real finite crystals this is not possible. Consequently, they must either terminate on the boundaries or, which is mostly the case in reality, they must form closed circuits. This is also the reason for their motion. Under loading – shearing in the plane of misfit, these lines change the curvature and, in the attempt to minimize the length, they are shifted along this plane. The closed circuit of dislocation is almost in all its points neither edge nor screw dislocation. For instance, for a constant Burgers vector and a circular circuit at two points the dislocation is of edge type, at two other opposite points it is of screw type and everywhere else it is none of those.

We proceed to the mathematical description of dislocations.

The most general approach has been proposed by B. Bilby, R. Bullough and E. Smith [47] (see also some related references in the book of M. Zorawski [458]). As the construc-

---

[1]after Helmut Föll; *Defects in crystals*, http://www.tf.uni-kiel.de/matwis/amat/def_en/index.html

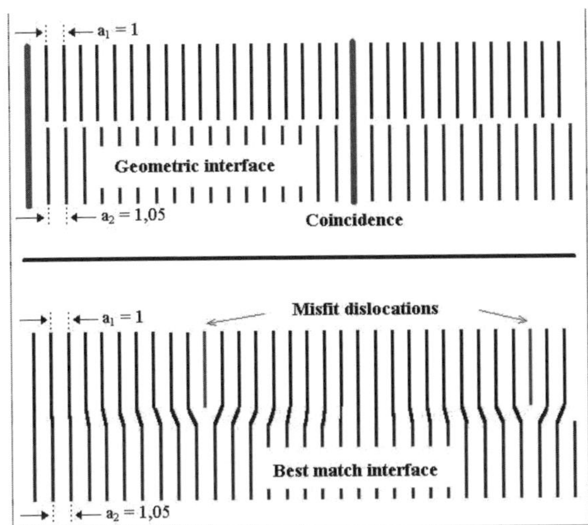

Figure 9.2: *Schematic picture of a two-dimensional misfit along the horizontal line yielding edge dislocation line every 20 steps (http://www.tf.uni-kiel.de/matwis/amat/semi_en/kap_5/illustr/misfit_dislocations.gif).*

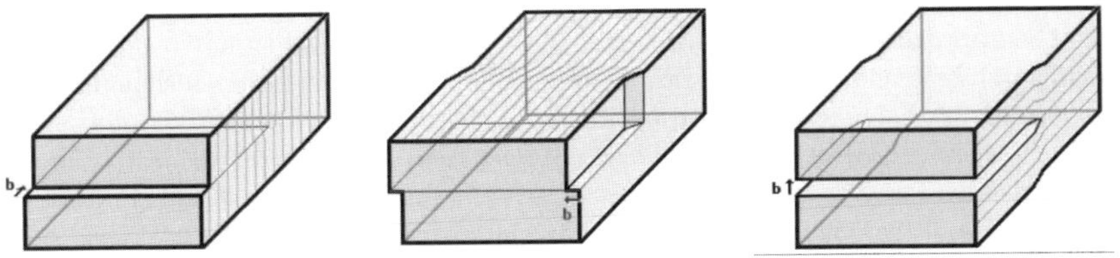

Figure 9.3: *Three characteristic types of dislocations: edge (left panel), screw (middle panel), and "sixty degree" (right panel).*

tion of the Burgers contour indicates, the geometry of a body with dislocation cannot be described by the projection into the Euclidean space. Two solutions of this problem have been proposed. The first one was proposed by R. A. Toupin (see the reprint of six memoirs from 1968 on the theory of dislocations which contains works of W. Noll, R. A. Toupin and C.-C. Wang [274]). He made the assumption that the mapping of the body with dislocations into the Euclidean space of motion does not yield a classical deformation gradient by the differentiation of this mapping (i.e., the faithful local configuration in the terminology of R. A. Toupin) but the so-called unfaithful local configurations which describe the parallel transport of vectors between tangent spaces. The parallel transport in the manifold of the body is defined approximately in the following way. If we choose a point $\mathbf{X} = X^K \mathbf{g}_K$ and a neighboring point ${}^*\mathbf{X} = \left(X^K + dX^K\right) \mathbf{g}_K$, then the parallel shift of the vector $\mathbf{V}$ from the point $\mathbf{X}$ to the point ${}^*\mathbf{X}$ is defined by the following relation

$$ {}^*V^K = V^K - \Gamma^K_{LM} V^L dX^M, \tag{9.1} $$

where Γ^K_{LM} is the so-called connection. In the case of metric spaces it is symmetric: $\Gamma^K_{LM} = \Gamma^K_{ML}$ and given by the differentiation of the metric tensor (compare Appendix A.2).

In the case of the parallel transport of the vector along the closed contour such a connection yields the vector identical with the initial one. Otherwise, if the connection is not symmetric, it contains as well an antisymmetric part $\Gamma^K_{[LM]}$ called the tensor of torsion and the parallel transport along the closed contour produces a vector different than at the beginning of the shift.

Such a description can also be introduced in Euclidean spaces by means of aholonomic vectors (i.e., vectors which are not basis vectors of any coordinate system) related to the crystallographic directions [458]. Such aholonomic basis vectors produce nonmetric connections which means that these connections possess an antisymmetric part – a torsion.

An alternative approach to nonlinear modeling of dislocations by means of a generalization of the symmetry group of a solid for which field equations are derived by means of the variation of the Lagrangian was proposed by A. Kadič and D. G. B. Edelen [194]. They propose also some rather elementary extensions to account for the irreversibility.

Another approach is suggested by the attempt of A. Einstein in the model combining electromagnetism with gravity. In the case of dislocations one can introduce a non-Euclidean space with teleparallelism in which the crystallographic lattice would be ideal (without dislocations) and the body would be stress-free in the natural configuration. The latter is not the case for the approach of Toupin and others in which Euclidean spaces are used. The approach with the use of spaces with teleparallelism has not been developed beyond purely geometrical considerations (compare [458]). Therefore we show here only some features of the approach based on Euclidean spaces.

Instead of the tensor of torsion it is customary to use the tensor of dislocation density β introduced by J. F. Nye [280]. Within the frame of a linear model described in Cartesian coordinates these two quantities are related in the following way

$$ \beta^{ij} = -\epsilon^{ikl} \Gamma^j_{[kl]}. \tag{9.2} $$

We proceed to the formal definition of the dislocation density appropriate for the application within the frame of linear elasticity. There are two possibilities. One of them was proposed by C. Somigliana in 1914 [360] and it is based on the notion of the dislocation line. Another one was introduced by V. Volterra in 1907 [412] and it is using the notion of a singular surface on which the displacement vector $\mathbf{u}$ is discontinuous. Both definitions yield similar models but they are not equivalent. A detailed discussion of the problem can be found in the article of Z. Mossakowska: *Self-equilibrated stresses and dislocations* published, unfortunately, in Polish [359]. An introduction to the subject is also available in the book of T. Mura [267].

9.2 Continuum with dislocations

The closed curve D is said to be the dislocation line (dislocation loop) in a continuum if a line integral along any sufficiently small closed circuit B circumventing once the curve D possesses the property

$$\oint_B d\mathbf{u} = \mathbf{b}, \tag{9.3}$$

for an arbitrary loading of the continuum. Obviously, $\mathbf{u}$ is the displacement vector and $\mathbf{b} \neq \mathbf{0}$ is called the Burgers vector of the dislocation line D. According to the Stokes Theorem,[2] for the differentiable displacement field $\mathbf{u}(\mathbf{x}, t)$ we can write the integral in the form

$$\oint_B \frac{\partial u_i}{\partial x_j} dx_j = \int_{S_B} \epsilon_{kij} \frac{\partial^2 u_i}{\partial x_i \partial x_j} n_k dS = 0, \tag{9.4}$$

where S_B is a surface spanned on the curve B. Hence, the definition (9.3) is nontrivial only for displacement fields which are discontinuous on the surface S_B. This relates the Somigliana definition of the dislocation to the surface definition introduced by Volterra.

For energetic reasons $\mathbf{b}$ is usually the shortest translation vector of the lattice; e.g. $|\mathbf{b}| = a/2 < 110 >$ for the fcc lattice.[3] The assumption that the Burgers contour B is sufficiently small means that it encloses only one dislocation line D and that it is intersecting only once a surface span by the line D. This is schematically shown in Figure 9.4. The sign of the Burgers vector $\mathbf{b}$ is defined by the right screw rule shown also in Figure 9.4.

One of the surfaces related to the dislocation line is the cylinder for which the curve D is the directrix and whose generatrix[4] are straight lines in the direction of the Burgers vector $\mathbf{b}$. This is the so-called gliding surface along which dislocations move most frequently because the resistance of the crystal to such a motion is in this surface the smallest (conservative motion). Another possibility which requires much more energy

---

[2]G. Stokes proved the theorem that for all fields $\mathbf{v}$ differentiable on an oriented surface S with the normal vector $\mathbf{n}$ the following relation holds

$$\int_S (\text{rot } \mathbf{v}) \cdot \mathbf{n} dS = \oint_{\partial S} \mathbf{v} \cdot d\mathbf{x}, \quad \text{i.e.,} \quad \int_S \epsilon_{ijk} \frac{\partial v_k}{\partial x_j} n_i dS = \oint_{\partial S} v_i dx_i,$$

where ∂S is the boundary curve of the surface S.

[3]The face-centered cubic unit cell (abbreviated either by fcc or cF) is a cube with an atom at each corner of the unit cell and an atom situated in the middle of each face of the unit cell.

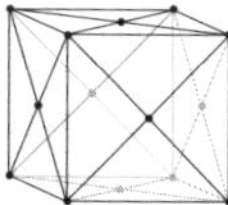

[4]Surfaces may be machined by tools which, in general, provide two kinds of general motion. The line generated by the primary motion (cutting motion) is called the generatrix, while the line representing the secondary motion (feed motion) is called the directrix (for details see [215]).

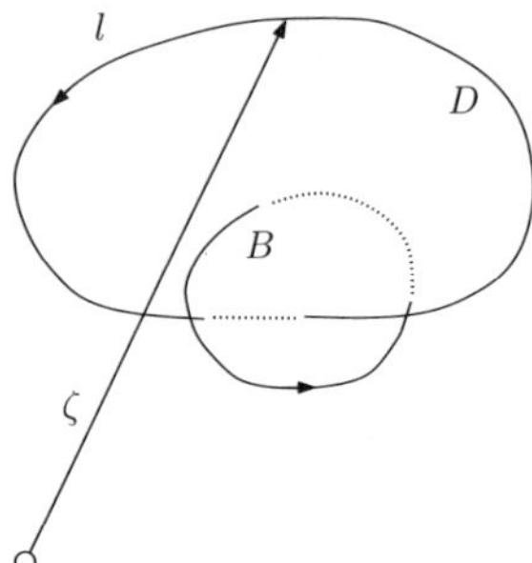

Figure 9.4: *Dislocation loop D, Burgers contour B, and the sign convention.*

is the so-called climbing of dislocations (nonconservative motion). The latter requires some atomic diffusion processes. For details we refer to numerous books on the subject (e.g. J. Weertman, J. R. Weertman [420], D. Hull [167], J. D. Eshelby [119], A. M. Kosevich [205].

We proceed to describe the geometry of the dislocation line. It is described by the position vector

$$\mathbf{x} = \zeta\left(l, t\right), \tag{9.5}$$

where l is a parameter along the line. Then the tangent vector and the velocity of dislocation are defined by the relations

$$d\zeta_k = \frac{\partial \zeta_k}{\partial l} dl \quad \Rightarrow \quad t_k\left(\mathbf{x}, t\right) = \oint_D \delta\left(\mathbf{x} - \zeta\left(l, t\right)\right) d\zeta_k, \tag{9.6}$$

$$\dot{\zeta}_k = \frac{\partial \zeta_k}{\partial t},$$

where the vector field $\mathbf{t}\left(\mathbf{x}, t\right)$ is given in the whole continuum but it is different from zero only on dislocation lines.

In order to describe continuous fields in the presence of dislocations we introduce the tensor of distortion $\boldsymbol{\beta}$ which smears out the Burgers condition (9.3). This tensor coincides with the gradient of the displacement in simply connected domains $\mathcal{P}$ which do not contain dislocation loops: $\mathcal{P} \cap D = \varnothing$. Hence, we define

$$du_j = \beta_{ij} dx_i, \quad \text{with} \quad \beta_{ij}\left(\mathbf{x}, t\right) = \frac{\partial u_j}{\partial x_i} \quad \text{for} \quad \mathbf{x} \in \mathcal{P}. \tag{9.7}$$

Then the condition (9.3) has the form

$$b_j = \oint_B \beta_{ij} dx_i = \int_{S_B} \epsilon_{kli} \frac{\partial \beta_{ij}}{\partial x_l} n_k dS, \tag{9.8}$$

where the second relation follows from the Stokes Theorem for an arbitrary surface S_B spanned on the curve B.

This relation allows to smear out the field of distortion. Namely, for the dislocation loop D intersecting a surface S_B in the point $\mathbf{x}$, as shown in Figure 9.5, we use the

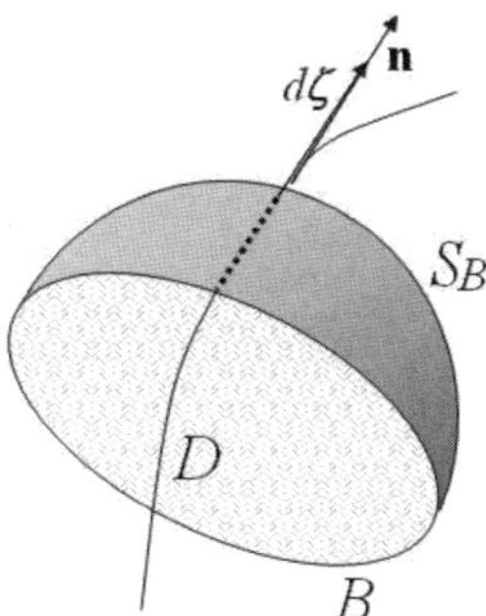

Figure 9.5: *Orientations of Burgers surface S_B and dislocation loop D.*

following identity

$$\int\limits_{S_B} \left(\oint\limits_D \delta\left(\mathbf{x} - \zeta\right) d\zeta_k \right) n_k dS = \begin{cases} 1 & \text{for the same orientation of } \mathbf{n} \text{ and } d\zeta, \\ -1 & \text{for the opposite orientation of } \mathbf{n} \text{ and } d\zeta, \\ 0 & \text{for } D \text{ not intersecting } S_B. \end{cases} \tag{9.9}$$

Then (9.8) can be written in the form

$$b_j \int\limits_{S_B} \left(\oint\limits_D \delta\left(\mathbf{x} - \zeta\right) d\zeta_k \right) n_k dS = \int\limits_{S_D} \epsilon_{kli} \frac{\partial \beta_{ij}}{\partial x_l} n_k dS. \tag{9.10}$$

This relation must hold for an arbitrary surface S_B spanned on the Burgers contour B. Consequently,

$$\epsilon_{kli} \frac{\partial \beta_{ij}}{\partial x_l} = b_j t_k, \tag{9.11}$$

where we have used the definition (9.6) of the vector $\mathbf{t}$ tangent to the dislocation loop D. This is the differential form of the Burgers relation (9.3). As the vector $\mathbf{t}$ is different from zero only on dislocation lines the above relation states that the distortion $\boldsymbol{\beta}$ has a vector potential beyond the dislocation line (rot $\boldsymbol{\beta} = \mathbf{0}$), i.e., $\boldsymbol{\beta} = \operatorname{grad} \mathbf{u}$ as we have already mentioned before.

The tensor

$$\boldsymbol{\alpha} = \mathbf{b} \otimes \mathbf{t} \quad \text{i.e.} \quad \alpha_{ij} = b_i t_j, \tag{9.12}$$

is called the tensor of dislocation density. Obviously, it satisfies the relation

$$\epsilon_{ilk} \frac{\partial \beta_{kj}}{\partial x_l} - \alpha_{ij} = 0. \tag{9.13}$$

The above considerations allow one to write the full set of equations which determine the distortion $\boldsymbol{\beta}$ and other fields of a linear elastic continuum caused by a given dislocation line. In the static case, the problem is quite simple. We use the equilibrium condition

$$\frac{\partial T_{ij}}{\partial x_j} = 0, \tag{9.14}$$

which follows from the momentum balance equation and Hooke's law

$$T_{ij} = c_{ijkl}\beta_{kl}, \qquad c_{ijkl} = \lambda\delta_{ij}\delta_{kl} + \mu\left(\delta_{ik}\delta_{jl} + \delta_{il}\delta_{jk}\right). \tag{9.15}$$

We have to incorporate Relation (9.13) which describes the loading by the dislocation. To do so we differentiate the equilibrium condition and subsequently we substitute (9.15)

$$c_{ijkl}\frac{\partial^2\beta_{kl}}{\partial x_j\partial x_p} = 0.$$

Now we multiply (9.13) by ϵ_{iql}, and use the identity

$$\epsilon_{kpl}\epsilon_{kqi}\frac{\partial\beta_{ij}}{\partial x_q} = \left(\delta_{pq}\delta_{il} - \delta_{qi}\delta_{qj}\right)\frac{\partial\beta_{ij}}{\partial x_q} = \frac{\partial\beta_{lj}}{\partial x_p} - \frac{\partial\beta_{pj}}{\partial x_l} = \epsilon_{kpl}\alpha_{kj}, \tag{9.16}$$

which follows from the contracted epsilon identity: $\epsilon_{ijk}\epsilon_{imn} = \delta_{jm}\delta_{kn} - \delta_{jn}\delta_{km}$. Finally, we have

$$c_{ijkl}\frac{\partial^2\beta_{pl}}{\partial x_i\partial x_k} = c_{ijkl}\epsilon_{qpk}\frac{\partial\alpha_{ql}}{\partial x_i}. \tag{9.17}$$

Together with the condition

$$e_{kl} = \frac{1}{2}\left(\beta_{kl} + \beta_{lk}\right), \tag{9.18}$$

Equation (9.17) fully describes the problem. Once we know β_{ij} we can find the stresses from Hooke's law (9.15). As this field of stresses follows only from the presence of the dislocation without any external load we say that it is self-equilibrated.

The solutions of this equilibrium problem are very important because they determine the stress concentration in the vicinity of the line defect. They can be found by means of the Green function of linear elasticity (see Appendix C). We quote here only the result of W. G. Burgers for the displacement $\mathbf{u}$ in the case of a dislocation loop D with constant Burgers vector $\mathbf{b}$. It has the form

$$u_k = -\frac{1}{4\pi}b_k\oint_D\frac{\epsilon_{ijl}r_jk_l}{r\left(r - r_ik_i\right)}d\zeta_i + \frac{1}{4\pi}\epsilon_{kij}b_i\oint_D\frac{d\zeta_j}{r} + \frac{1}{4\pi}\frac{\lambda+\mu}{\lambda+2\mu}\epsilon_{ijl}b_j\frac{\partial}{\partial x_k}\oint_D\frac{r_l}{r}d\zeta_i,$$

$$r_k = x_k - \zeta_k\left(l\right), \qquad r = \sqrt{r_kr_k}, \tag{9.19}$$

where $\mathbf{k}$ is the unit vector perpendicular to the plane of the loop D.

Only in exceptional cases one can perform analytically the integration in the above relation. It can be done for straight line dislocations. In such a case, one obtains, for instance, the following components of the stresses

1. screw dislocation given by $\mathbf{l} = (0,0,1)$ where $\mathbf{l}$ points in the direction of the dislocation line D, and $\mathbf{b} = (0,0,b)$

$$\sigma_x = \sigma_y = \sigma_z = \tau_{xy} = 0, \tag{9.20}$$

$$\tau_{xz} = \frac{\mu b}{2\pi}\frac{y}{x^2 + y^2}, \qquad \tau_{yz} = -\frac{\mu b}{2\pi}\frac{x}{x^2 + y^2},$$

2. edge dislocation given by $\mathbf{l} = (0, 0, 1)$, $\mathbf{b} = (b, 0, 0)$

$$\tau_{xz} = \tau_{yz} = 0,$$

$$\sigma_x = \frac{b}{2\pi} \frac{2\mu(\lambda+\mu)}{\lambda+2\mu} \frac{y(3x^2+y^2)}{(x^2+y^2)^2},$$

$$\sigma_y = -\frac{b}{2\pi} \frac{2\mu(\lambda+\mu)}{\lambda+2\mu} \frac{y(x^2-y^2)}{(x^2+y^2)^2}, \tag{9.21}$$

$$\sigma_{xz} = \frac{b}{2\pi} \frac{2\mu\lambda}{\lambda+2\mu} \frac{y}{x^2+y^2},$$

$$\sigma_{xy} = -\frac{b}{2\pi} \frac{2\mu(\lambda+\mu)}{\lambda+2\mu} \frac{x(x^2-y^2)}{(x^2+y^2)^2}.$$

Solutions for dislocations in some anisotropic media can be found in the explicit form as well.

We do not present details of the dynamic theory of dislocations. Some aspects of this theory can be found in the earlier quoted work of Z. Mossakowska [359] as well as in the article of H. Zorski [459]. The Burgers condition holds true also in this general case but one has to cope with the problem of elimination of the acceleration term in the equation of motion. One can derive an additional equation for the evolution of the distortion β

$$\frac{\partial \beta_{kl}}{\partial t} - \frac{\partial v_l}{\partial x_k} = J_{kl}, \tag{9.22}$$

where J_{kl} is the dislocation flux given by the relation

$$J_{kl} = b_l \epsilon_{kij} \oint_{D(t)} \dot{\zeta}_i \delta\left(\mathbf{x} - \zeta\left(l, t\right)\right) d\zeta_j. \tag{9.23}$$

Some universal solutions are known also in this case but we shall not quote them in this book.

9.3 On plasticity of metals

As we have mentioned at the beginning of this chapter, the vehement research on continua with dislocations was connected with the discovery that plastic deformations of metals are related to the redistribution and production of dislocations. At low temperatures, i.e., temperatures below approximately 70% of the melting point temperature, dislocations are moving on characteristic crystallographic planes on which they require the least energy for the motion. During this motion they get stacked on the boundaries of the grains of polycrystals and on other obstacles. One of them may be a point at which more than one dislocation appear simultaneously and their Burgers vectors annihilate each other, i.e., $\sum_\alpha \mathbf{b}^{(\alpha)} = 0$, where α numbers the dislocations in this point. Such knots play an important role in the production of dislocations. Namely, the shear stress which acts in the slip plane of motion of a dislocation is bending a line of dislocation pinned to two such obstacles. The loop is trying to minimize the energy which leads to overhanging shown in Figure 9.6. When the two sides meet, they annihilate because their Burgers vectors are identical but of opposite sign. The loop becomes free to move and the rest of the virginal

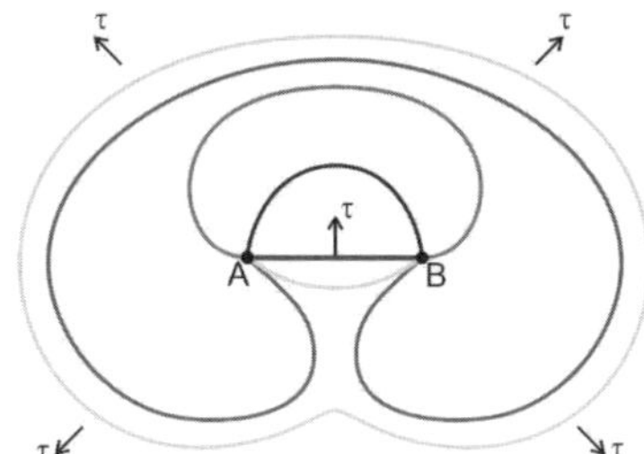

Figure 9.6: *Frank-Read source of dislocation (http://en.wikipedia.org/wiki/Frank-Read_Source#mediaviewer/File:Frank-Read_Source.png).*

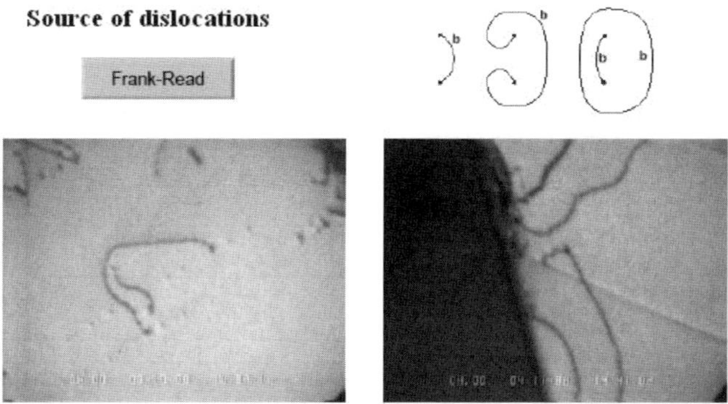

Figure 9.7: *Electron microscope picture of the Frank-Read source. Black traces are dislocations on the surface of the sample.*

line of the dislocation begins the process anew. This is the so-called Frank-Read source of dislocations.

This and similar mechanisms yield the production of dislocations which is an irreversible process related to the increment of plastic strains. During plastic deformations the number density of dislocations may grow from some 10^{10} to $10^{20}/\text{cm}^2$.

Modeling of such processes is based on certain microscopic observations transferred to the level of the continuum. The fundamental role plays here the so-called Orowan equation which relates the rate of shearing to the Burgers vector, the speed of the dislocations and the dislocation density. Together with the evolution equation for this density in which the intensities of the sources are incorporated one obtains a semistructural plasticity model, the so-called crystal plasticity, by use of which many problems of the mechanics of metals could be solved successfully. The introduction to crystal plasticity can be found in Part I. Many details are discussed in the monograph of R. W. K. Honeycombe [165] as well as the book of U. F. Kocks, A. Argon and M. Ashby [202].

At high temperatures the process becomes more complicated because the defects pinned to the grain boundaries begin to move as well. In this range the theory of dislocations as presented above cannot be applied anymore.

9.4 Dislocations in geophysics

The origin of various models of dislocations goes back to the defective structure of crystalline bodies such as metals. However, we have seen that these models describe line defects in continua independently of a particular crystalline lattice. Only some indications concerning the Burgers vector bear on crystallography. Therefore one can apply such models in those cases in which a description of a discontinuous displacement field is needed. This is indeed the case in modeling of earthquakes. Most likely it was A. E. H. Love in 1945 who proposed to apply the Volterra dislocation model [411] in the description of earthquakes. The problem of seismic sources was discussed by A. V. Vvedenskaya [413], J. A. Steketee [364] and others. Some details can be found in the book of A. Udias [398]. The book of Aki and Richards [5] contains a modern presentation of the subject.

The mechanism of earthquake rupture may be more complicated than this which can be described by the Volterra dislocation. It is related to crack formation and it is coupled to complicated tectonic processes which we do not discuss in this book. Readers interested in these problems are referred to an article of J. Rice [310]. We leave out the discussion of the structure of forces acting in the fault – according to Aki, Richards [5] the so-called double couple theory seems to be prevailing, and limit the attention to modeling a slip in the fault and its action on the vicinity.

A fault surface S_D with the boundary $\partial S_B = D$ lies in a linear isotropic medium and it is assumed to be perpendicular to the x_3-axis. A slip is presumed to take place in the direction of a unit vector $\mathbf{l} = l_1\mathbf{e}_1 + l_2\mathbf{e}_2$. Then the displacement vector possesses a discontinuity

$$\Delta\mathbf{u} = \mathbf{u}^+ - \mathbf{u}^- = \mathbf{b} = |\Delta\mathbf{u}|\,\mathbf{l}, \tag{9.24}$$

where $\mathbf{b}$ is the Burgers vector for the dislocation line D. Hence, the Volterra dislocation allows one to take over all results of the dislocation theory in the description of such a fault defect. Strictly speaking, the definition of the Volterra dislocation requires as well that the derivatives of $\mathbf{u}$ on S_D are continuous which means, of course, also the continuity of the stress. The field of displacement created by the dislocation yields a system of self-equilibrated stresses and, for this reason, an accompanying distortion may be considered as a field of initial deformations e_{ij}^0 given by the relation

$$e_{ij}^0 = -b_j \int\limits_{S_D} \delta\left(\mathbf{x} - \xi\right) n_i dS. \tag{9.25}$$

This relation yields immediately the notion of the moment tensor density m_{ij} given by Hooke's law for the initial deformation

$$m_{ij} = c_{ijkl}e_{kl}^0, \tag{9.26}$$

and, consequently, a definition of the forces appearing in the equation for the real displacement $\mathbf{u}$

$$X_k^0 = -\frac{\partial m_{kl}}{\partial x_l}, \quad \rho\frac{\partial^2 u_i}{\partial t^2} = c_{ijkl}\frac{\partial^2 u_k}{\partial x_l \partial x_j} + X_i^0. \tag{9.27}$$

Obviously, the tensor of material parameters for isotropic materials c_{ijkl} has the form $(9.15)_2$. Now, the dynamic Green function $((\text{C.47}) -$ see Appendix C.3) yields the solutions for the displacement. For instance, for the source which is the Heaviside function $\mathbf{b}H\left(t\right)$

we obtain the displacement in the far field approximation in the following form

$$u_i^L = \frac{M_0}{4\pi\rho c_L^3 r}\left(n_k l_l + n_l l_k\right)\frac{x_i x_k x_l}{r^3}\delta\left(t - \frac{r}{c_L}\right), \quad n_k \mathbf{e}_k = \mathbf{e}_3, \quad \mathbf{l} = l_1\mathbf{e}_1 + l_2\mathbf{e}_2,$$

$$u_i^S = \frac{M_0}{4\pi\rho c_T^3 r}\left(n_k l_l + n_l l_k\right)\left(\delta_{ik} - \frac{x_i x_k}{r^2}\right)\frac{x_l}{r}\delta\left(t - \frac{r}{c_T}\right), \tag{9.28}$$

where $M_0 = \eta b\,(\text{area}\,S_D)$ is the seismic moment, η denotes the rigidity modulus. These are two arrivals, longitudinal and transversal, in a point with the distance r from the source.

The model indicated above, specifies various notions appearing in connection with earthquakes, such as the seismic moment and its decomposition into various forces acting on the plane of the defect including the above-mentioned double couples. However, in many respects it seems to be too simplified. For instance, it does not contain any criteria for the rupture. We shall not elaborate this subject any further.

Chapter 10

Acoustic waves

10.1   Preliminaries

In Part I we have discussed extensively the problem of the thermodynamical construction of models which yield hyperbolic field equations. In particular, in Section 8.1 we have indicated that this feature of models of extended thermodynamics leads to well-posedness of initial value Cauchy problems. This property has a particular bearing in mathematics as well as in thermodynamics. In the latter it indicates some possibilities of the construction of evolution equations for quantities describing the deviation from the thermodynamical equilibrium without the necessity of additional boundary conditions with unclear physical meaning (e.g. compare the works of I-Shih Liu [232, 233, 236] on the problem of boundary conditions in extended thermodynamics).

The construction of dynamical solutions of the field equations of continuum thermodynamics is only very seldom possible by analytical methods. Various approximations including numerical methods must be used. However, certain properties of such solutions can be found without full knowledge of their form. Apart from the mathematical properties such as functional spaces, to which these solutions must belong, one can often find the time dependent range of the disturbance, a decay of the solution in time, properties of refraction, reflection and transmission of waves through interfaces, etc. In particular, hyperbolic field equations have the nice physical property that they describe the propagation of disturbances with a finite speed. Hence, they yield in a natural way the notion of waves in the system. The time dependent domain of disturbed initial fields is limited by the surface which we call the wave front. This surface is a locus of certain singularities of the fields and their derivatives which characterize a particular class of waves. Both the shape of the wave front as well as its speed can often be found without constructing the full dynamical solution of the field equations.

The above remark does not mean that we would not use in many practical applications models which are not hyperbolic. For instance, the classical theory of heat conduction or the classical theory of diffusion are based on parabolic equations which do not yield the notion of the wave front. Obviously, for pragmatic reasons these models are used in many practical applications and lead to results which agree in excellent way with observations. Values of the temperature field or the field of concentration in particular points and instances of time coincide in the majority of cases with experimental data within an acceptable range of accuracy.

A prominent example of a model which does not lead to the creation of acoustic wave fronts is the Navier-Stokes theory in which the stress tensor is given by Relation (5.44), i.e.,

$$\mathbf{T} = -p\left(\rho, T\right)\mathbf{1} + \lambda\left(\rho, T\right)\left(\mathrm{tr}\,\mathbf{D}\right)\mathbf{1} + 2\eta\left(\rho, T\right)\mathbf{D}. \tag{10.1}$$

Acoustic waves are characterized by the wave front on which the velocity is continuous but the acceleration is not. The latter yields a discontinuity of the velocity gradient (e.g. compare (3.57) in Part I). Then Hadamard's Lemma (Formula (2.88), Part I) leads to the following jump conditions on the wave front (we return further to the discussion of this issue)

$$[\![2\mathbf{D}]\!] = -c\left(\mathbf{a}\otimes\mathbf{n} + \mathbf{n}\otimes\mathbf{a}\right), \quad [\![\mathrm{tr}\,\mathbf{D}]\!] = -c\,\mathbf{a}\cdot\mathbf{n}, \tag{10.2}$$

where c is the speed of propagation of the wave front, $\mathbf{n}$ is the unit normal vector to the surface of the front and $\mathbf{a}$ denotes the so-called amplitude of the discontinuity. Hence, the discontinuity of the stress tensor is as follows

$$[\![\mathbf{T}]\!] = -c\left\{\lambda\mathbf{a}\cdot\mathbf{n}\mathbf{1} + \eta\left(\mathbf{a}\otimes\mathbf{n} + \mathbf{n}\otimes\mathbf{a}\right)\right\}. \tag{10.3}$$

Simultaneously, the continuity of the velocity yields the continuity of the traction (Poisson's condition) on the wave front: $[\![\mathbf{T}]\!]\,\mathbf{n} = 0$ (compare Table 4 of Part I). Hence, for $c \neq 0$ we obtain the following conditions

$$\left(\lambda + 2\eta\right)\mathbf{a}\cdot\mathbf{n} = 0, \quad \left(\lambda + \eta\right)\left(\mathbf{a}\cdot\mathbf{n}\right)^{2} + \eta a^{2} = 0, \quad a^{2} = \mathbf{a}\cdot\mathbf{a}. \tag{10.4}$$

It follows from the second law of thermodynamics that $\eta \geq 0$, $3\lambda + 2\eta \geq 0$. Consequently, we obtain $\mathbf{a} \equiv 0$, i.e., the result of Pierre Duhem: acoustic waves are impossible in a linear viscous fluid.

In this chapter we present a few important examples of the waves defined within hyperbolic models. The purpose of these examples is twofold. Namely, we want to introduce the main notions appearing in the description of waves and we want to indicate possibilities of dynamical verification of thermodynamical models by measuring quantities characterizing waves.

Before we proceed to discuss in some details those types of waves which we encounter most frequently in practical applications in materials sciences let us point out the two fundamental very distinct classes of waves following from the properties of wave fronts. The first class contains disturbances which yield the so-called strong singularities or shock waves on the surface of the front. The most characteristic feature of those waves is the discontinuity of the velocity on the front. The second class concerns weak singularities or acoustic (sound) waves and these waves are characterized by the discontinuity of the acceleration or even higher time derivatives of the velocity. We skip in this book the discussion of shock waves. However, in order to point out certain characteristic features of those waves we present here the most elementary physical example of such a wave – a one-dimensional shock in the ideal gas. Many details concerning the properties of those waves can be found in the classical monograph on the subject by G. B. Whitham [427] (in particular Chapter 6).

Let us recall the field equations for an ideal gas for the one-dimensional case. We choose the fields of the mass density ρ, the velocity $\mathbf{v} = v\mathbf{e}_1$ and the entropy η. Then the local field equations beyond the singular surface follow from the conservation laws of mass, momentum and energy (5.7), where the stress tensor reduces to the pressure

$\mathbf{T} = -p\mathbf{1}$. The one-dimensional set of equations has the following form in regular points

$$\frac{\partial \rho}{\partial t} + v\frac{\partial \rho}{\partial x} + \rho\frac{\partial v}{\partial x} = 0,$$

$$\rho\left(\frac{\partial v}{\partial t} + v\frac{\partial v}{\partial x}\right) + \frac{\partial p}{\partial x} = 0, \tag{10.5}$$

$$\rho\left(\frac{1}{2}\rho v^2 + \rho\varepsilon\right) + \frac{\partial}{\partial x}\left[\left(\frac{1}{2}\rho v^2 + \rho\varepsilon\right) + pv\right] = 0.$$

For adiabatic processes the energy conservation can be easily replaced by the entropy conservation. We have to use the Gibbs relation $Td\eta = d\varepsilon + pd\left(\dfrac{1}{\rho}\right)$. Then, instead of $(10.5)_3$ we can write

$$\frac{\partial \eta}{\partial t} + v\frac{\partial \eta}{\partial x} = 0. \tag{10.6}$$

On the singular surface of the shock wave the above local equations must be replaced by jump conditions (compare Table 4 of Part I)

$$[\![\rho\,(c - v)]\!] = 0,$$

$$[\![\rho\,(c - v)\,v - p]\!] = 0, \tag{10.7}$$

$$\left[\!\left[\rho\,(c - v)\left(\frac{1}{2}\rho v^2 + \rho\varepsilon\right)v\right] - pv\right]\!\! = 0.$$

These are the so-called Rankine-Hugoniot conditions for the case of an ideal gas. It is easy to check that they must be nonlinear if the shock wave is to be a moving singularity.

It should be stressed that we cannot replace the energy jump condition $(10.7)_3$ by the entropy condition as we did in regular points because the Gibbs equation does not hold on the surface of the shock wave. Further we calculate the change of entropy across the shock.

In the case of a given state ahead of the wave, say ρ_1, v_1, p_1, the Rankine-Hugoniot conditions form a set of three equations for four unknowns ρ_2, v_2, p_2, c. We can choose one of the unknown quantities as a control variable. We call it the shock parameter or the intensity of the shock wave. Let us choose as the intensity the difference of the pressure

$$z = \frac{p_2 - p_1}{p_1}. \tag{10.8}$$

To be more specific we consider a polytropic gas for which the constitutive relations have the following form

$$\varepsilon = \frac{1}{\gamma - 1}\frac{p}{\rho}, \quad h = \varepsilon + \frac{p}{\rho} = \frac{\gamma}{\gamma - 1}\frac{p}{\rho}, \quad c_{ad}^2 = \gamma\frac{p}{\rho}, \quad \gamma = \frac{c_p}{c_v}, \tag{10.9}$$

where the ratio of the specific heats γ is known as the adiabatic index. c_{ad} is the speed of sound under adiabatic conditions.

Bearing the above relations in mind, we can write the quantities characterizing the shock wave in the vicinity of the front in the following form

$$M = \frac{c - c_{ad1}}{c_{ad1}} = \sqrt{1 + \frac{\gamma + 1}{2\gamma}z}, \qquad \frac{v_2 - v_1}{c_{ad1}} = \frac{z}{\gamma\sqrt{1 + \frac{\gamma + 1}{2\gamma}z}},$$

$$\frac{\rho_2}{\rho_1} = \frac{1 + \frac{\gamma + 1}{2\gamma}z}{1 + \frac{\gamma - 1}{2\gamma}z}, \qquad \frac{c_{ad2}}{c_{ad1}} = \sqrt{\frac{(1 + z)\left(1 + \frac{\gamma - 1}{2\gamma}z\right)}{1 + \frac{\gamma + 1}{2\gamma}z}}, \tag{10.10}$$

where M is the Mach number and the index 2 denotes quantities behind the shock. These relations form the foundation for the construction of the solution in the region disturbed by the shock wave. We do not go into any details of such an analysis. However, one more property of the case under consideration should be presented. As we mentioned, we are now in the position to calculate the change of the entropy on the wave front. For the polytropic gas the entropy is given by the relation

$$\eta = c_v \ln\left(\frac{p}{\rho^\gamma}\right). \tag{10.11}$$

Hence, bearing $(10.10)_3$ in mind, we obtain immediately

$$\frac{\eta_2 - \eta_1}{c_v} = \ln \frac{(1 + z)\left(1 + \frac{\gamma - 1}{2\gamma}z\right)^\gamma}{\left(1 + \frac{\gamma + 1}{2\gamma}z\right)^\gamma} > 0. \tag{10.12}$$

The inequality corresponds, obviously, to the second law of thermodynamics on the front of the shock wave. It is fulfilled for $z > 1$, i.e., in the case of a polytropic gas the shock must be compressive: $p_2 > p_1$.

It should be mentioned that the propagation of shock waves in substances more complicated than ideal gases yields a much more complicated structure of shocks. For instance, it is rather self-evident that in solids there must exist expansion shocks. Various cases were discussed in details by Peter Chen (e.g. his Appendix 4A to the Truesdell's book [392]). A practical application of such models is, for example, the description of Kolsky-Hopkinson bars used in the testing of materials at a high strain rate (e.g. [273]).

We do not go into any further details of the theory of shock waves in this book. All further wave considerations concern weak discontinuity waves, i.e., waves whose fronts are carrying at the most a discontinuity of the acceleration.

10.2　Propagation of acoustic waves in nonlinear materials with memory

The thermodynamical theory of acoustic waves in materials with memory has been constructed in the '60s in a series of papers of B. D. Coleman, M. E. Gurtin and I. R. Herrera [81]-[86]. The first results were one-dimensional but then the three-dimensional theory was developed. We present here only a hint on the structure of this theory based on the presentation of C. A. Truesdell [392]. It is based on a general approach of Coleman's thermodynamics to materials with memory. For the so-called thermokinetic process $\{\mathbf{F}^t, T^t, \gamma^t\}$, $\gamma = \mathrm{Grad}\,T/T$, where the index t denotes the history: $\mathbf{F}^t = \mathbf{F}\,(t - s)$, $0 \leq s < \infty$, etc., Coleman's thermodynamics based on the Clausius-Duhem inequality (Formula (5.74) of Part I) yields the following results. The Helmholtz free energy ψ is a functional on the history of the deformation gradient and the temperature. It is the thermodynamical potential for the stresses, entropy and internal energy, i.e., these three quantities are given by Fréchet derivatives[1] (uniquely defined linear operators) of the free energy

$$\psi\,(t) = \mathcal{F}\,(\mathbf{F}^t, T^t)\,, \quad \mathbf{P}\,(t) = \mathcal{D}_{\mathbf{F}}\mathcal{F}\,(\mathbf{F}^t, T^t)\,, \quad \eta = -\mathcal{D}_T\mathcal{F}\,(\mathbf{F}^t, T^t)\,, \quad \varepsilon = \psi + T\eta,$$
$$\text{i.e.} \quad \dot{\psi}\,(t) = \mathrm{tr}\left(\left(\mathcal{D}_{\mathbf{F}}\mathcal{F}\right)\dot{\mathbf{F}}^T\right) + \left(\mathcal{D}_T\mathcal{F}\right)\dot{T}.$$
$$(10.13)$$

From the exploitation of the second law of thermodynamics there remains the dissipation inequality which is immaterial for the analysis of weak discontinuity waves.

We consider the propagation of the weak discontinuity (acoustic) wave in the above specified material. This wave is characterized by the wave front, i.e. a singular surface encompassing the disturbed domain in the reference configuration on which the velocity, the deformation gradient and the temperature are continuous

$$[\![\mathbf{v}]\!] = 0, \quad [\![\mathbf{F}]\!] = 0, \quad [\![T]\!] = 0, \tag{10.14}$$

but the acceleration is discontinuous. Then, according to Maxwell's Theorem (Formula (2.93) and Section 3.3.4 in Part I), we have

$$\left[\!\!\left[\frac{\partial \mathbf{v}}{\partial t}\right]\!\!\right] = U^2\mathbf{A}, \quad [\![\mathrm{Grad}\,\mathbf{F}]\!] = \mathbf{A}\otimes\mathbf{N}\otimes\mathbf{N}, \quad \left[\!\!\left[\dot{\mathbf{F}}\right]\!\!\right] = -U\mathbf{A}\otimes\mathbf{N},$$
$$(10.15)$$

$$[\![\mathrm{Grad}\,T]\!] = \Theta\mathbf{N}, \quad \left[\!\!\left[\dot{T}\right]\!\!\right] = -U\Theta,$$

where U is the speed of propagation, $\mathbf{N}$ the unit normal vector to the wave front, $\mathbf{A}$ is the acceleration amplitude of the discontinuity and Θ denotes the thermal amplitude.

Simultaneously, if we evaluate the limits of the momentum and energy conservation laws on the wave front we arrive at the following relations

$$\rho\left[\!\!\left[\frac{\partial \mathbf{v}}{\partial t}\right]\!\!\right] = [\![\mathrm{Div}\,\mathbf{P}]\!]\,, \quad [\![\mathrm{Div}\,\mathbf{P}]\!] = -\frac{1}{U}\left[\!\!\left[\dot{\mathbf{P}}\right]\!\!\right]\mathbf{N},$$
$$(10.16)$$

$$\rho T\,[\![\dot{\eta}]\!] = [\![\mathrm{Div}\,\mathbf{Q}]\!]\,, \quad [\![\mathrm{Div}\,\mathbf{Q}]\!] = -\frac{1}{U}\left[\!\!\left[\dot{\mathbf{Q}}\right]\!\!\right]\mathbf{N},$$

where $\mathbf{Q}$ denotes the heat flux vector. Eliminating the divergencies between these equations yields

$$\left[\!\!\left[\dot{\mathbf{P}}\right]\!\!\right]\mathbf{N} + \rho U\left[\!\!\left[\frac{\partial \mathbf{v}}{\partial t}\right]\!\!\right] = 0,$$
$$(10.17)$$

$$\left[\!\!\left[\dot{\mathbf{Q}}\right]\!\!\right]\mathbf{N} + \rho T U\,[\![\dot{\eta}]\!] = 0.$$

---

[1]The Fréchet derivative is a derivative defined on Banach spaces. It is commonly used to generalize the derivative of a real-valued function of a single real variable to the case of a vector-valued function of multiple real variables, and to define the functional derivative used widely in the calculus of variations.

Consequently, we have to find only relations between time derivatives. These follow from the constitutive relations (10.13)

$$\frac{1}{\rho}\left[\!\left[\dot{\mathbf{P}}\right]\!\right] = \mathcal{D}_{\mathbf{F},\mathbf{F}}\mathcal{F}\left(\left[\!\left[\dot{\mathbf{F}}\right]\!\right]\right) + \mathcal{D}_{\mathbf{F},T}\mathcal{F}\left(\left[\!\left[\dot{T}\right]\!\right]\right),$$

$$\left[\!\left[\dot{\eta}\right]\!\right] = -\operatorname{tr}\left(\mathcal{D}_{\mathbf{F},T}\mathcal{F}\left(\left[\!\left[\dot{\mathbf{F}}\right]\!\right]\right)\right) - \mathcal{D}_{T,T}\mathcal{F}\left(\left[\!\left[\dot{T}\right]\!\right]\right). \tag{10.18}$$

Now, accounting for the definitions of the amplitudes (10.15) we can easily reduce the above set of equations to the following conditions

$$\left(\widetilde{\mathbf{A}}\left(\mathbf{N}\right) - U^2 \mathbf{1}\right)\mathbf{A} + \Theta\mathbf{d}\left(\mathbf{N}\right) = 0,$$

$$\left[\!\left[\dot{\eta}\right]\!\right] = U\left(\mathbf{d}\left(\mathbf{N}\right)\cdot\mathbf{A} - C\Theta\right), \tag{10.19}$$

where

$$\widetilde{\mathbf{A}}\mathbf{A} = \mathcal{D}_{\mathbf{F},\mathbf{F}}\mathcal{F}\left(\left[\!\left[\mathbf{A}\otimes\mathbf{N}\right]\!\right]\right)\mathbf{N}, \quad \mathbf{d}\left(\mathbf{N}\right) = \mathcal{D}_{\mathbf{F},T}\mathcal{F}\left(\mathbf{N}\right), \quad C = -\mathcal{D}_{T,T}\mathcal{F}. \tag{10.20}$$

The above relations determine the propagation conditions of two classes of waves. First, for homothermal waves we obtain the classical propagation condition

$$\Theta = 0 \quad\Rightarrow\quad \left(\widetilde{\mathbf{A}}\left(\mathbf{N}\right) - U^2\mathbf{1}\right)\mathbf{A} = 0. \tag{10.21}$$

In the second case, $\left[\!\left[\dot{\eta}\right]\!\right] = 0$, the waves are called homocaloric. If $C \neq 0$ we can eliminate Θ from the set and obtain again the classical propagation condition with the modified acoustic tensor

$$\hat{\mathbf{A}}\left(\mathbf{N}\right) = \widetilde{\mathbf{A}}\left(\mathbf{N}\right) + \frac{1}{C}\mathbf{d}\left(\mathbf{N}\right)\otimes\mathbf{d}\left(\mathbf{N}\right). \tag{10.22}$$

Generalized Fresnel-Hadamard-Duhem Theorem:

The Helmholtz free energy functional $\mathcal{F}$ determines for a given history of the thermokinetic process unique homothermal and homocaloric acoustic tensors $\hat{\mathbf{A}}\left(\mathbf{N}\right), \widetilde{\mathbf{A}}\left(\mathbf{N}\right)$. Either of these tensors determines the other through (10.22). The proper directions of these tensors are the homothermal and homocaloric acoustic axes; for each of these, there is at least one orthogonal triad. The corresponding proper numbers are the homothermal and homocaloric acoustic numbers; these are always real.

Further in this chapter we consider particular cases of the above presented model.

10.3   Bulk waves in nonlinear elasticity

10.3.1   Modicum of the wave front description

As we have already discussed in the previous section the hyperbolic set of field equations yields the definition of the disturbed domain at the instant of time t which is encompassed by the boundaries of the system and by a moving nonmaterial surface which we call the wave front. For the purpose of the description of waves in systems undergoing large deformations it is convenient to define this domain in the reference configuration as we did in Section 10.2. Obviously, the propagation of disturbances is observed in the space of current configurations but, due to the continuity and invertibility of the function of motion $\mathbf{f}(\mathbf{X}, t)$, we can project the current disturbed domains on the reference configuration. We denote such a projection of the wave front by $\mathcal{S} = \mathcal{S}(t)$. Some details of the geometry and kinematics of such a surface can be found in Sections 2.6 and 3.3 of Part I. This surface may be described either in terms of two parameters Ξ^1, Ξ^2 (surface coordinates) and the time t

$$\mathbf{X} = \mathbf{S}\left(\Xi^1, \Xi^2, t\right), \tag{10.23}$$

or by the time at which the surface $\mathcal{S}$ is crossing a chosen material point $\mathbf{X}$

$$t = \Psi(\mathbf{X}). \tag{10.24}$$

Clearly, this relation follows from the three parametric relations (10.23) ($\mathbf{S} = S^K \mathbf{e}_K$) if we eliminate the surface parametrization Ξ^1, Ξ^2. The vectors

$$\left\{\frac{\partial \mathbf{S}}{\partial \Xi^A}\right\}, \quad A = 1, 2, \tag{10.25}$$

are tangent to the front $\mathcal{S}$. Simultaneously, for any chosen instant of time t Relation (10.24) indicates $d\Psi = 0$, i.e.,

$$\operatorname{Grad} \Psi \cdot d\mathbf{S} = 0. \tag{10.26}$$

Hence, since $d\mathbf{S}$ are tangent to $\mathcal{S}$ the vector

$$\mathbf{N} = \frac{\operatorname{Grad} \Psi}{|\operatorname{Grad} \Psi|}, \tag{10.27}$$

is the unit vector orthogonal to $\mathcal{S}$.

Substitution of (10.23) in (10.24) and subsequent differentiation with respect to time yields the identity

$$\frac{\partial \mathbf{S}}{\partial t} \cdot \operatorname{Grad} \Psi = 1. \tag{10.28}$$

The derivative $\partial \mathbf{S}/\partial t$ is the local velocity of the surface $\mathcal{S}$. Hence, the above relation means that within the above description of kinematics of the front one can specify only the normal component of the speed of the front with respect to the reference configuration

$$U = \frac{\partial \mathbf{S}}{\partial t} \cdot \mathbf{N} = \frac{1}{|\operatorname{Grad} \Psi|} \quad \Rightarrow \quad \operatorname{Grad} \Psi = \frac{\mathbf{N}}{U}. \tag{10.29}$$

As discussed in Part I (Formula (2.88)) sufficiently smooth functions satisfy the so-called Hadamard conditions on singular surfaces. These conditions replace the classical chain rule of differentiation which holds if the singular surface is absent. They yield the important Maxwell Theorem (Formula (2.93), Part I) which we use in this section.

Let us first consider a wave front on which the deformation gradient $\mathbf{F}$ and, consequently, the velocity $\mathbf{v}$ are continuous. Such waves, as we have already mentioned, are called acceleration waves. For those waves Maxwell's Theorem yields the following conditions on the wave front

$$[\![\operatorname{Grad} \mathbf{F}]\!] = \boldsymbol{A}\mathbf{N}, \quad \left[\!\!\left[\frac{\partial \mathbf{F}}{\partial t}\right]\!\!\right] = -U\boldsymbol{A},$$

$$\boldsymbol{A} = A^{kK} \mathbf{e}_k \otimes \mathbf{e}_K = \mathbf{A} \otimes \mathbf{N} = A^k N^K \mathbf{e}_k \otimes \mathbf{e}_K \tag{10.30}$$

$$[\![\operatorname{Grad} \mathbf{v}]\!] = \mathbf{B} \otimes \mathbf{N}, \quad \left[\!\!\left[\frac{\partial \mathbf{v}}{\partial t}\right]\!\!\right] = -U\mathbf{B}, \quad \mathbf{B} = B^k \mathbf{e}_k.$$

Relations $(10.30)_2$ follow from the symmetry of the gradient $\operatorname{Grad} F$.

The following relation follows immediately by taking the limit on both sides of the surface $\mathcal{S}$

$$\frac{\partial \mathbf{F}}{\partial t} = \operatorname{Grad} v \quad \Rightarrow \quad \mathbf{B} = -U\mathbf{A}. \tag{10.31}$$

Collecting the above results we have for the front of the acceleration wave (compare (10.15))

$$[\![\operatorname{Grad} \mathbf{F}]\!] = \mathbf{A} \otimes \mathbf{N} \otimes \mathbf{N}, \quad \left[\!\left[\frac{\partial \mathbf{F}}{\partial t}\right]\!\right] = -U\mathbf{A} \otimes \mathbf{N}, \quad \left[\!\left[\frac{\partial \mathbf{v}}{\partial t}\right]\!\right] = U^2\mathbf{A}. \tag{10.32}$$

The vector $\mathbf{A}$ which determines the jump of the acceleration on the front of the acceleration wave is called the amplitude of the acceleration wave.

Now we use the above relations in the description of the jump of the momentum balance equation. Bearing the constitutive relation for the nonlinear elastic material (5.29), Part I, in mind, we have for isothermal processes

$$\rho \left[\!\left[\frac{\partial \mathbf{v}}{\partial t}\right]\!\right] = [\![\operatorname{div} \mathbf{P}]\!], \quad \mathbf{P} = \rho\frac{\partial \psi}{\partial \mathbf{F}}, \tag{10.33}$$

where ρ is the mass density in the reference configuration, $\mathbf{P}$ denotes the first Piola-Kirchhoff stress tensor, ψ is the Helmholtz free energy. For isothermal processes it depends only on the deformation gradient $\mathbf{F}$. Consequently, the above condition can be written in the form

$$\left(Q_{ij} - \rho U^2 \delta_{ij}\right) A^j = 0, \quad Q_{ij} = \frac{\partial P_i^{.K}}{\partial F_L^j} N_K N_L = A_{ij}^{K\,L} N_K N_L,$$

$$A_{ij}^{K\,L} = \rho\frac{\partial^2 \psi}{\partial F_{.K}^i \partial F_{.L}^j} N_K N_L, \tag{10.34}$$

which follows from the continuity condition

$$\left[\!\left[\frac{\partial P_i^{.K}}{\partial F_L^j} N_K N_L\right]\!\right] = \left[\!\left[\rho\frac{\partial^2 \psi}{\partial F_{.K}^i \partial F_{.L}^j}\right]\!\right] = 0. \tag{10.35}$$

The symmetric tensor $\mathbf{Q} = Q^{ij}\mathbf{e}_i \otimes \mathbf{e}_j$ is called the acoustic tensor for the propagation in the direction $\mathbf{N}$. The symmetry of the acoustic tensor implies the existence of three mutually orthogonal amplitudes $\mathbf{A}$. They correspond to three positive eigenvalues ρU^2. They are real due to the convexity of the Helmholtz free energy function. A similar result follows if we consider an isentropic acoustic wave $\eta = \text{const.}$ rather than the isothermal wave.

Equation (10.34) is called the Fresnel-Hadamard propagation condition and it determines possible directions and speeds of propagation. The amplitude does not follow from this condition and it must be determined by an appropriate transport equation which we present later. The propagation condition (10.34) can also be written in the following form

$$\left(A_{ij}^{K\,L}\frac{\partial \Psi}{\partial X^K}\frac{\partial \Psi}{\partial X^L} - \rho\delta_{ij}\right) A^j = 0, \tag{10.36}$$

where we have used Relation (10.27). This is a nonlinear first order differential equation which together with initial conditions determines the wave front $t = \Psi\left(\mathbf{X}\right)$. The above

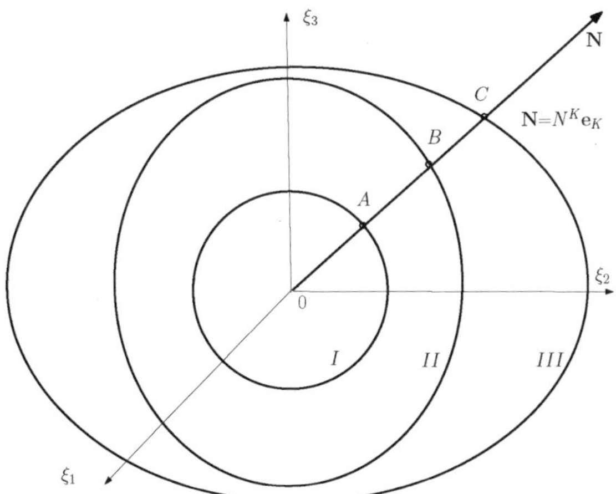

Figure 10.1: *Slowness surfaces following from the propagation condition* (10.36).

equation yields six solutions $\pm\overset{1}{\Psi}, \pm\overset{2}{\Psi}, \pm\overset{3}{\Psi}$. If we denote $\xi_K = \partial\Psi/\partial X^K$ then in the $\{\xi\}$-space the solutions of (10.36) correspond to the three sheets schematically shown in Figure 10.1. Due to the relation $\xi_K = N_K/U$ (compare (10.29)) the length of the radius vector $\xi = \xi^K e_K$ is equal to $1/U$. For a chosen $\mathbf{N}$, the three solutions $\overset{1}{U}, \overset{2}{U}, \overset{3}{U}$ correspond to the three endpoints A, B, C of this vector.

The above construction yields a method of analysis of the incident wave encountering an interface. One can easily construct reflected and transmitted waves in a way similar to the linear optics. We skip here further details and refer the interested reader to the article of Z. Wesolowski [422].

10.3.2 An approximate solution in the vicinity of the front

The construction of the wave front solves the first part of the problem of propagation. The second part is concerned with the evolution of the field in the vicinity of the front, and, consequently, the evolution of the amplitude of the wave. We demonstrate the strategy of solving this problem on the example of the displacement $\mathbf{u}(\mathbf{X}, t)$ yielding small deformations which is superimposed on an arbitrary static deformation, i.e., we construct the equations of motion for the following function

$$\widetilde{\mathbf{f}}(\mathbf{X}, t) = \mathbf{f}(\mathbf{X}) + \mathbf{u}(\mathbf{X}, t). \tag{10.37}$$

In the fully nonlinear case such a superposition requires the differentiation of the displacement $\mathbf{u}$ with respect to the so-called convective coordinates, i.e., with respect to the coordinates X^K in the coordinate system whose basis vectors are specified by the static deformation created by the motion $\mathbf{f}(\mathbf{X})$ (e.g. compare [143]). In the case of small deformations produced by the displacement $\mathbf{u}(\mathbf{X}, t)$ the conditions on the wave front reduce to partial derivatives with respect to X^K and we do not need to go into details of the calculations in convective coordinates. We obtain the following propagation condition

$$\left(A_{ij}^{KL}\frac{\partial\Psi}{\partial X^K}\frac{\partial\Psi}{\partial X^L} - \rho\delta_{ij}\right)A^j = 0, \tag{10.38}$$

and, hence, this condition is formally identical with this for the full nonlinear motion. The difference lies in the matrix A_{ij}^{KL} which in the present case depends on the initial static deformation gradient $\operatorname{Grad}\mathbf{f}$ and is time-independent. The solution of this eigenvalue problem is assumed to be known because the matrix A_{ij}^{KL} is known for a given initial deformation $\mathbf{f}(\mathbf{X})$. Consequently, we may choose the eigenvector $\mathbf{A}$ to be a unit vector and the quantity which determines the amplitude of the wave reduces to a scalar multiplier of this eigenvector. This multiplier determines the magnitude of the amplitude of the small disturbance.

We follow here the method proposed by Z. Wesolowski (e.g. [422, 423]) and seek the solution in the vicinity of the front by an expansion in a series with respect to the distance from the wave front. To this aim we introduce the variable ς which describes the distance from the front

$$\varsigma\left(X^K,t\right) = \Psi\left(X^K\right) - t. \tag{10.39}$$

The solution is sought in the form of the following series

$$u^i = \sum_{\nu=0}^{\infty} Z_{\nu+2}\left(\varsigma\right) g_\nu^i\left(X^K,t\right), \quad \mathbf{u} = u^i\mathbf{e}_i,$$

$$\tag{10.40}$$

$$Z_\nu = \frac{1}{\nu!}\left[\frac{1}{2}\left(\varsigma + |\varsigma|\right)\right]^\nu.$$

Hence, the displacement $\mathbf{u}$ is identically zero for $\varsigma < 0$, i.e., ahead of the front and it is specified by the series of vector functions $g_\nu^i, \nu = 0,\ldots,\infty$, behind the front which we have to find.

It is easy to check that the functions Z_ν satisfy the following recursive relations

$$\frac{dZ_\nu}{d\varsigma} = Z_{\nu-1}, \quad Z_0 = \frac{dZ_1}{d\varsigma} = H\left(\varsigma\right), \tag{10.41}$$

where $H\left(\varsigma\right)$ is the Heaviside function. Bearing these relations in mind we see that the following conditions are satisfied on the wave front

$$\left[\!\left[\frac{\partial u^i}{\partial t}\right]\!\right] = 0, \quad \left[\!\left[\frac{\partial u^i}{\partial X^K}\right]\!\right] = 0,$$

$$\tag{10.42}$$

$$\left[\!\left[\frac{\partial^2 u^i}{\partial X^K \partial X^K}\right]\!\right] = -Z_0\frac{\partial\Psi}{\partial X^K}\frac{\partial\Psi}{\partial X^L}g_0^i, \quad \left[\!\left[\frac{\partial^2 u^i}{\partial t^2}\right]\!\right] = -Z_0 g_0^i,$$

where (10.29) has been used. Hence, the amplitude of the discontinuity is given by the relation $-\dfrac{g_0^i}{U^2}$ and this vector is parallel to the eigenvector $\mathbf{A} = A^i\mathbf{e}_i$.

Substitution of the assumption (10.40) in the equations of motion yields a condition in the form of an infinite series in functions Z_ν which should be zero. We assume additionally that each coefficient of this series is zero. This sufficient condition was proved to yield the solution of the problem [422]. The first two conditions have the form

$$\left(A_{ij}^{KL} \frac{\partial \Psi}{\partial X^K} \frac{\partial \Psi}{\partial X^L} - \rho \delta_{ij} \right) g_0^j = 0,$$

$$\left(A_{ij}^{KL} \frac{\partial \Psi}{\partial X^K} \frac{\partial \Psi}{\partial X^L} - \rho \delta_{ij} \right) g_1^j + \left(A_{ij}^{KL} \frac{\partial \Psi}{\partial X^K} \frac{\partial g_0^j}{\partial X^L} + A_{ij}^{KL} \frac{\partial \Psi}{\partial X^L} \frac{\partial g_0^j}{\partial X^K} + 2\rho \delta_{ij} \frac{\partial g_0^j}{\partial t} \right)$$

$$+ \left(A_{ij}^{KL} \frac{\partial^2 \Psi}{\partial X^K \partial X^L} + \frac{\partial A_{ij}^{KL}}{\partial X^K} \frac{\partial \Psi}{\partial X^L} \right) g_0^j = 0. \tag{10.43}$$

Comparison with (10.38) yields

$$g_0^i = \kappa_0 A^i, \tag{10.44}$$

where the scalar multiplier κ_0, the above mentioned magnitude of the amplitude, must be found. If we take the scalar product of the second equation (10.43) with the vector $\mathbf{A} = A^i \mathbf{e}_i$, the first contribution will vanish identically and we obtain

$$(\rho \delta_{ij} A^i A^j) \frac{\partial \kappa_0}{\partial t} + \left(A_{ij}^{KL} \frac{\partial \varsigma}{\partial X^L} A^i A^j \right) \frac{\partial \kappa_0}{\partial X^K}$$

$$+ \left(A_{ij}^{KL} \frac{\partial \varsigma}{\partial X^L} \frac{\partial A^j}{\partial X^K} + \rho \delta_{ij} \frac{\partial A^j}{\partial t} + A_{ij}^{KL} \frac{\partial^2 \varsigma}{\partial X^K \partial X^L} A^j + \frac{\partial A_{ij}^{KL}}{\partial X^K} \frac{\partial \varsigma}{\partial X^L} A^j \right) A^i \kappa_0 = 0, \tag{10.45}$$

where the symmetry $A_{ij}^{KL} = A_{ij}^{LK}$ has been used. Obviously, Equation (10.45) is a first order partial differential equation for κ_0 with given coefficients. Hence, we can easily solve this equation by the method of characteristics (e.g. [92]; see also Appendix C in [437]). The characteristic curve $\mathcal{C}$ in the space of variables $\{X^K, t\}$ is given by the following differential equations specifying its tangent vector

$$\frac{dX^K}{ds} = A_{ij}^{KL} \frac{\partial \varsigma}{\partial X^L} A^i A^j, \tag{10.46}$$

$$\frac{dt}{ds} = \rho \delta_{ij} A^i A^j,$$

where s is a parameter along the curve $\mathcal{C}$ and the initial condition

$$t(s_0) = \Psi \left(X^K (s_0) \right), \tag{10.47}$$

must be fulfilled. Then Equation (10.45) can be written in the form

$$\frac{d\kappa_0}{ds} + P(s) \kappa_0 = 0, \tag{10.48}$$

where

$$P(s) = \left(A_{ij}^{KL} \frac{\partial \varsigma}{\partial X^L} \frac{\partial A^j}{\partial X^K} + \rho \delta_{ij} \frac{\partial A^j}{\partial t} + A_{ij}^{KL} \frac{\partial^2 \varsigma}{\partial X^K \partial X^L} A^j + \frac{\partial A_{ij}^{KL}}{\partial X^K} \frac{\partial \varsigma}{\partial X^L} A^j \right) A^i. \tag{10.49}$$

Let us mention that the wave front (i.e., $\varsigma = 0$) possesses the orthogonal vector in the four-dimensional space-time $\left\{ X^K, t \right\}$ given by the components

$$\left(\frac{\partial \varsigma}{\partial X^1}, \frac{\partial \varsigma}{\partial X^2}, \frac{\partial \varsigma}{\partial X^3}, \frac{\partial \varsigma}{\partial t} \right) = \left(\frac{\partial \Psi}{\partial X^1}, \frac{\partial \Psi}{\partial X^2}, \frac{\partial \Psi}{\partial X^3}, -1 \right).$$

Hence, the scalar product of this vector with the vector tangent to the characteristic curve $\mathcal{C}$ given by Relations (10.46) is zero. As the initial condition (10.47) requires that one point of $\mathcal{C}$ belongs to the wave front $\mathcal{S}$ this means that the whole curve $\mathcal{C}$ lies on the wave front $\mathcal{S}$.

Obviously, Equation (10.48) can be immediately solved and we obtain the following relation for the magnitude κ_0

$$\kappa_0 = C_0 \exp \left(- \int_0^s P(s)\, ds \right), \tag{10.50}$$

where C_0 is a constant. According to (10.42) the amplitude of the wave is given by

$$-\frac{\kappa_0}{U^2} A^i = -\frac{C_0}{U^2} A^i \exp \left(- \int_0^s P(s)\, ds \right). \tag{10.51}$$

It propagates along the projection of the curve $\mathcal{C}$ on the space of X^K-variables. This projection is called the acoustic ray in analogy to optical rays. The acoustic rays are usually not perpendicular to the wave fronts.

In a similar manner we can calculate further coefficients of the series (10.40). We skip the details of this procedure here and refer to the original work of Z. Wesolowski [422].

Let us mention that the particular form (10.40) of the functions Z_ν was immaterial for the above considerations. We have made use only of the recursive relations (10.41). Consequently, we can also introduce, for instance, the functions

$$T_\nu(\varsigma) = \frac{1}{(i\omega)^\nu} e^{i\omega\varsigma}, \tag{10.52}$$

where ω is an arbitrary real parameter. Obviously, these functions satisfy the recursive relations as well and, hence, we can write another solution for the displacement $\mathbf{u} = u^i \mathbf{e}_i$ in the following form

$$u^i = \sum_{\nu=0}^\infty \frac{1}{(i\omega)^\nu} g_\nu^i e^{i\omega\varsigma} \equiv e^{i\omega(\Psi - t)} \sum_{\nu=0}^\infty \frac{1}{(i\omega)^\nu} g_\nu^i. \tag{10.53}$$

This relation forms the basis of the so-called spectral representation of linear waves which is closely related to the Fourier analysis and the representation of waves by the superposition of monochromatic waves. We return to this point further in this chapter.

10.4   Water waves and surface waves in linear solids

10.4.1   Introduction

The construction of the solution of dynamical problems of continua requires, apart from the field equations, discussed in many places of this book, the formulation of two types of conditions. One of them describes the source of the waves, i.e., a disturbance placed on a certain subdomain of the body and acting in a given way in time and, the second concerns the construction of boundary conditions for finite domains. The first condition will not be discussed in any extent in this book. There are many references in which particular classes of sources of waves and corresponding solutions are discussed. For instance, for seismic waves one should mention the classical books of K. Aki and P. G. Richards [5] or A. Ben-Menahem and S. J. Singh [38]. Certain features of this problem are indicated in the construction of the dynamical Green function for linear elasticity which we were using, for instance, in Chapter 9 for the description of dislocations (see also Appendix C.3).

The construction of boundary conditions splits into two problems. One of them concerns conditions on the wave front, i.e., on a non-material surface separating the disturbed and undisturbed regions. These conditions were already discussed on the basis of the classical Hadamard conditions. The second problem concerns conditions on material surfaces, i.e., either on a true boundary of the body or on an interface separating two different materials. The conditions on such surfaces do not differ from those conditions which we formulate for static or steady-state problems and they are related to dynamical compatibility conditions on singular surfaces. Incident waves on material interfaces yield two important effects caused by the boundary conditions. One is the transition of the wave from one side of the interface to the other and this is related to the creation of new modes of propagation. The other one is the reflection which may lead to new modes of the propagation as well. In some cases these modes may combine and produce a new type of wave which propagates along the interface and decays very fast with the distance from the surface. Such surface waves have a very important practical bearing because they carry the energy within a narrow layer near the interface rather than in the bulk. This means that they preserve large amplitudes on much larger distances than usual bulk waves. For this reason they are the most dangerous part of the dynamics of earthquakes and, simultaneously, they may be easier measured in nondestructive measuring techniques. Further in this chapter we present a rather extensive review of surface waves. This review is based on the article of K. Wilmanski published in the CISM book [444].

10.4.2 Water waves in an ideal incompressible fluid model

There exists a vast literature on this subject (e.g. [38], [91], [112], [427]). In contrast to all other waves considered in this chapter which are produced by an unspecified source far away from the space-time point $(\mathbf{x}, \mathbf{t})$ of analysis, water waves result from the action of gravity. In most cases the modeling of water waves must be nonlinear. Tidal waves, in particular tsunami, possess properties characteristic for shock waves which we have mentioned at the beginning of this chapter. However, many linear models of shallow water waves, waves in basins of bounded extent, standing waves etc. have a great practical bearing as well. The classical monograph of J. V. Wehausen and E. V. Laitone [421] is still a very good reference on these subjects. Many modern contributions, particularly, on tsunami in different regions of the Earth, appear but a comprehensive presentation of this issue is still missing.

We present here the linear version of the theory of water waves because they are well known to all who observed the motion of water on the beach [385], and, simultaneously, they possess all properties characteristic for surface waves.

We consider an ideal (inviscid) incompressible fluid which means that its mass density ρ is constant. We assume that the motion is irrotational, i.e., the velocity $\mathbf{v}(\mathbf{x}, t)$ possesses

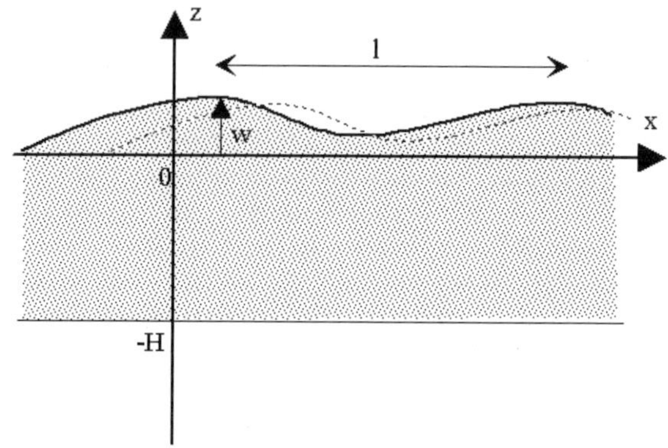

Figure 10.2: *Water profile at the instant of time t (from [214]).*

a potential ϕ. Then the mass conservation yields

$$\Delta^2 \phi = 0, \quad \mathbf{v} = \operatorname{grad} \phi, \quad \rho = \text{const.} \tag{10.54}$$

Simultaneously, the momentum balance equation has the form

$$\rho \left(\frac{\partial \mathbf{v}}{\partial t} + \mathbf{v} \cdot \operatorname{grad} \mathbf{v} \right) = -\operatorname{grad} p + \rho \mathbf{b}, \tag{10.55}$$

where p is the pressure and $\mathbf{b}$ the mass force. Integration of this equation with respect to the spacial variable $\mathbf{x}$ yields the Bernoulli equation

$$\frac{\partial \phi}{\partial t} + \frac{1}{2} v^2 - \varphi + \frac{p}{\rho} = C\left(t \right), \quad \mathbf{b} = \operatorname{grad} \varphi, \tag{10.56}$$

where $C\left(t \right)$ is an arbitrary function of time and we have used the identity $\mathbf{v} \cdot \operatorname{grad} \mathbf{v} = \frac{1}{2} \operatorname{grad} v^2$, $v^2 \equiv \mathbf{v} \cdot \mathbf{v}$, which holds for irrotational velocity fields.

We investigate the motion of the fluid shown schematically in Figure 10.2.

For the two-dimensional motion assumed in this problem the upper boundary surface is described by the relations

$$f_s\left(\mathbf{x}, t \right) = z - w\left(x, t \right) = 0$$
$$\Rightarrow \quad \mathbf{n} = \frac{\operatorname{grad} f_s}{|\operatorname{grad} f_s|} \approx \mathbf{e}_z, \quad \mathbf{n} \cdot \mathbf{v}_s = -\frac{\frac{\partial f_s}{\partial t}}{|\operatorname{grad} f_s|} \approx \frac{\partial w}{\partial t} \mathbf{e}_z, \tag{10.57}$$

where $\mathbf{n}$ is the unit outward normal vector of the surface, $\mathbf{v}_s$ is the velocity of this surface, and $\mathbf{e}_z$ the unit vector in the direction of the z-axis. The approximation in the above relations means that we linearize the problem with respect to the elevation function w and its derivatives.

Let p_a denote the atmospheric pressure acting on the fluid. Then, under the assumption that we can neglect the surface tension, the continuity of the mass flux on the surface (the surface is material with respect to the fluid) and the Bernoulli equation (10.56) yield

$$(\mathbf{v} - \mathbf{v}_s) \cdot \mathbf{n} = 0, \qquad p_a = -\rho g w - \rho \left(\frac{\partial \phi}{\partial t} \right)_{z=w} + \rho C\left(t \right), \tag{10.58}$$

where the nonlinear contribution $\frac{1}{2}v^2$ has been neglected.[2]

Instead of the potential ϕ, we use the potential $\phi' = \phi + \frac{p_a}{\rho}t - \int C(t)\,dt$ because $\mathbf{v} = \operatorname{grad}\phi \equiv \operatorname{grad}\phi'$. Then, omitting the prime,

$$w = -\frac{1}{g}\left.\frac{\partial\phi}{\partial t}\right|_{z=0}, \tag{10.59}$$

where we place the boundary at $z = 0$ instead of $z = w$ which yields an error of the same order of magnitude as in other approximations of this problem.

Simultaneously, the first condition (10.58) and Relation $(10.57)_3$ yield

$$\frac{\partial w}{\partial t} = \left.\frac{\partial\phi}{\partial z}\right|_{z=0}. \tag{10.60}$$

Combination of (10.59) and (10.60) yields the following kinematic boundary condition for ϕ

$$\frac{\partial^2\phi}{\partial t^2} + g\frac{\partial\phi}{\partial z} = 0 \quad\text{at}\quad z = 0. \tag{10.61}$$

The second boundary condition which must be fulfilled by the solutions of Equation (10.54) is formulated at the bottom

$$\mathbf{v}\cdot\mathbf{e}_z|_{z=-H} = \left.\frac{\partial\phi}{\partial z}\right|_{z=-H} = 0. \tag{10.62}$$

We can now summarize the above mentioned in the modicum of linear water waves (see box on the next page).

Now we are in the position to make an ansatz for solutions of the problem. We seek it in the form of a wave progressive in the x-direction

$$\phi(x,z,t) = \left(Ae^{kz} + Be^{-kz}\right)\cos\left(kx - \omega t\right). \tag{10.63}$$

Then, according to (10.59), the elevation w satisfies the one-dimensional wave equation.

Boundary conditions (10.61) and (10.62) yield

$$-\omega^2\left(A + B\right) + gk\left(A - B\right) = 0,$$

$$Ae^{-kH} - Be^{kH} = 0. \tag{10.64}$$

Consequently, from the determinant of this homogeneous set we obtain the following dispersion relation

$$\omega^2 = gk\tanh\left(kH\right). \tag{10.65}$$

---

[2]Approximations which we make in this derivation are based on the comparison with the wavelength $l\ \left(= \dfrac{2\pi}{k},\ k - \text{wave number}\right)$, wave period $T\ \left(= \dfrac{2\pi}{\omega},\ \omega - \text{frequency}\right)$, and the amplitude a. Namely

$$|\mathbf{v}| \sim \frac{a}{T}, \quad \left|\frac{\partial\mathbf{v}}{\partial t}\right| \sim \frac{a}{T^2}, \quad |\mathbf{v}\cdot\operatorname{grad}\mathbf{v}| \sim \frac{a^2}{T^2 l} \quad\Rightarrow$$

v^2 can be neglected if $a \ll l$. Simultaneously

$$|w| \sim a, \quad \left|\frac{\partial w}{\partial x}\right| \sim \frac{a}{l} \ll 1.$$

> *Modicum of linear water waves*
> (conditions satisfied by the velocity potential ϕ)
>
> $$\Delta^2\phi = 0 \quad \text{in the fluid,}$$
>
> $$\frac{\partial\phi}{\partial z} = 0 \quad \text{on the bottom } z = -H,$$
>
> $$\frac{\partial^2\phi}{\partial t^2} + g\frac{\partial\phi}{\partial z} = 0 \quad \text{on the free surface } z = 0,$$
>
> $$w = -\frac{1}{g}\frac{\partial\phi}{\partial t} \quad \text{on the free surface } z = 0,$$
>
> $$\frac{\partial w}{\partial t} = \frac{\partial\phi}{\partial z} \quad \text{on the free surface } z = 0.$$
>
> Among the last three conditions two are independent.

Simultaneously, the potential can be written in the form

$$\phi = \phi_{\max}\frac{\cosh k\,(H+z)}{\cosh kH}\cos\left(kx - \omega t\right), \tag{10.66}$$

where $\phi_{\max}$ is a constant of integration.

According to (10.59) we obtain for the elevation

$$w = -w_{\max}\sin\left[k\left(x - c_{ph}t\right)\right], \quad c_{ph} := \frac{\omega}{k}, \quad w_{\max} := \frac{\omega}{g}\phi_{\max}. \tag{10.67}$$

Hence, the elevation changes in the x-direction as it were a wave moving with the phase velocity

$$c_{ph} = \frac{\omega}{k} = \sqrt{\frac{gl}{2\pi}\tanh\left(2\pi\frac{H}{l}\right)}, \quad l := \frac{2\pi}{k}, \tag{10.68}$$

where l is the wavelength. The phase velocity of this wave depends on the frequency ω (or on the wave number k) and, therefore, the wave is called dispersive. This property is characteristic for all surface waves which propagate in systems with a characteristic length scale (e.g. the depth of the layer, the characteristic length of heterogeneous materials whose properties depend on the location in space, etc.).

In order to find the orbits of the material points we use the following relation for displacements

$$\frac{\partial u_x}{\partial t} = v_x = \frac{\partial\phi}{\partial x} \quad\Rightarrow\quad u_x = -\frac{k}{\omega}\phi_{\max}\frac{\cosh k\,(H+z)}{\cosh kH}\cos\left(kx - \omega t\right),$$

$$\tag{10.69}$$

$$\frac{\partial u_z}{\partial t} = v_z = \frac{\partial\phi}{\partial z} \quad\Rightarrow\quad u_z = -\frac{k}{\omega}\phi_{\max}\frac{\sinh k\,(H+z)}{\cosh kH}\sin\left(kx - \omega t\right).$$

Elimination of time yields

$$\frac{u_x^2}{\alpha_x^2} + \frac{u_z^2}{\alpha_z^2} = 1, \tag{10.70}$$

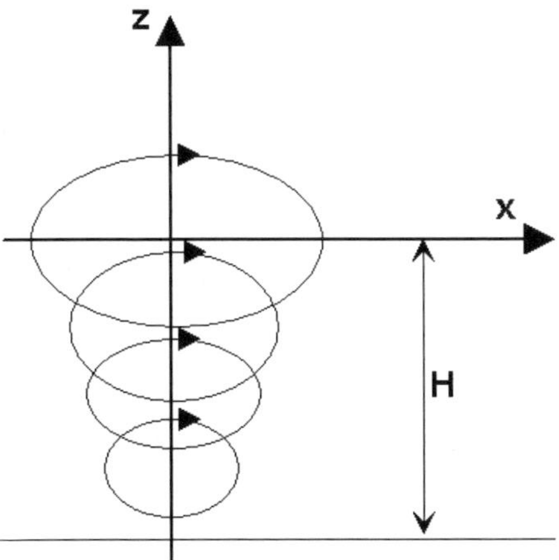

Figure 10.3: *Orbits of particles given by (10.70). Particles travel in clockwise (prograde) direction (from [214]).*

where

$$\alpha_x := \frac{k}{\omega}\phi_{\max}\frac{\cosh k\,(H+z)}{\cosh kH}, \qquad \alpha_z := \frac{k}{\omega}\phi_{\max}\frac{\sinh k\,(H+z)}{\cosh kH}. \tag{10.71}$$

Consequently, the orbit of each particle is an ellipse with semiaxes α_x, α_z. The largest ellipse appears at $z=0$, and at the bottom $z=-H$ it degenerates into a straight line. The orbits are schematically shown in Figure 10.3.

In the short-wave limit (deep water): $kH \to \infty$, we have $\tanh kH \approx 1$, i.e., $\omega^2 \approx gk$. Consequently, the phase velocity is given by the relation

$$c_{ph} = \frac{\omega}{k} = \sqrt{\frac{g}{k}} = \frac{g}{\omega}. \tag{10.72}$$

Simultaneously, the velocity potential becomes

$$\phi \approx \phi_{\max}e^{kz}\cos\left(kx - \omega t\right). \tag{10.73}$$

Hence, the motion of the fluid is negligible at the depth of about a wavelength $l = \frac{2\pi}{k}$. For this reason, these waves are called surface waves.

The above dispersive wave gives rise to a structure of propagation which has a very important bearing. The arrival of such waves to receivers is observed in the form of wave packages rather than in the form of single monochromatic waves or wave fronts. In order to illustrate this property on our simple example of deep water waves we consider the wave consisting of a narrow band of frequencies near the middle frequency ω_0 rather than a single frequency considered above. The solution (10.73) must now be replaced by the Fourier integral which accounts for all frequencies entering the band

$$\phi\left(x,t\right) = \frac{1}{2\pi}\int_{-\infty}^{\infty}\phi_{\max}\left(\omega\right)e^{kz}\cos\left(kx - \omega t\right)d\omega \tag{10.74}$$

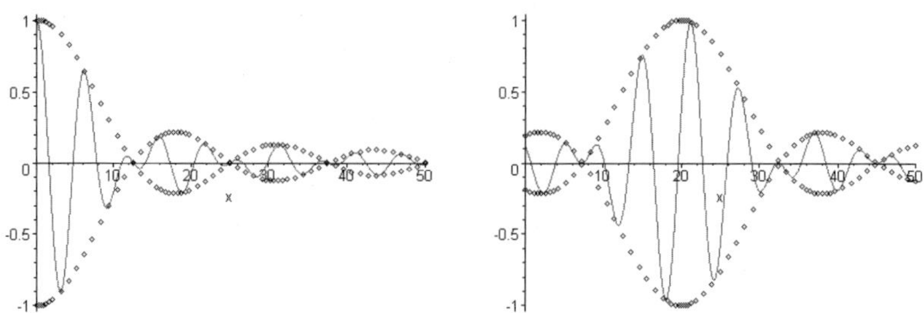

Figure 10.4: *Narrow band wave: configurations at two different instances of time (arbitrary units); (from [214]).*

$$\approx \frac{1}{2\pi} \phi_{\max}\left(\omega_0\right) e^{k_0 z} \int\limits_{-\infty}^{\infty} \cos\left(kx - \omega t\right) d\omega,$$

where

$$k_0 = \frac{\omega_0^2}{g}. \tag{10.75}$$

Integration yields for small $\left|\frac{\Delta\omega}{\omega_0}\right|$

$$\phi = \frac{\Delta\omega}{2\pi} \phi_{\max}\left(\omega_0\right) e^{k_0 z} M\left(\omega_0, \Delta\omega\right) \cos\left[k_0\left(x - c_{ph}t\right)\right], \tag{10.76}$$

where

$$M\left(\omega_0, \Delta\omega\right) := \frac{\sin\left[\frac{\Delta\omega}{c_g}\left(x - c_g t\right)\right]}{\frac{\Delta\omega}{c_g}\left(x - c_g t\right)}, \tag{10.77}$$

and we have the relations

$$k - k_0 = \frac{1}{g}\left(\omega^2 - \omega_0^2\right) \approx \frac{1}{c_g}\left(\omega - \omega_0\right),$$

$$\tag{10.78}$$

$$c_{ph} = \frac{g}{\omega_0}, \quad c_g := \left.\frac{d\omega}{dk}\right|_{\omega=\omega_0} = \frac{1}{2}c_{ph} = \frac{g}{2\omega_0}.$$

The quantity M is called the modulator, and as shown on the example in Figure 10.4, it has an extremum at $x - c_g t = 0$. The modulator is an envelope of the band of waves and propagates with the group velocity $c_g = \dfrac{d\omega}{dk}$. The carrier which in our example is described by the cosine function in (10.76) describes the motion at the frequency ω_0 with the phase velocity c_{ph}.

In Figure 10.4 we show the wave in two instances of time. Clearly, due to different velocities, the shape of the full wave plotted as the solid line moved differently from the envelope of the modulator indicated by dotted curves.

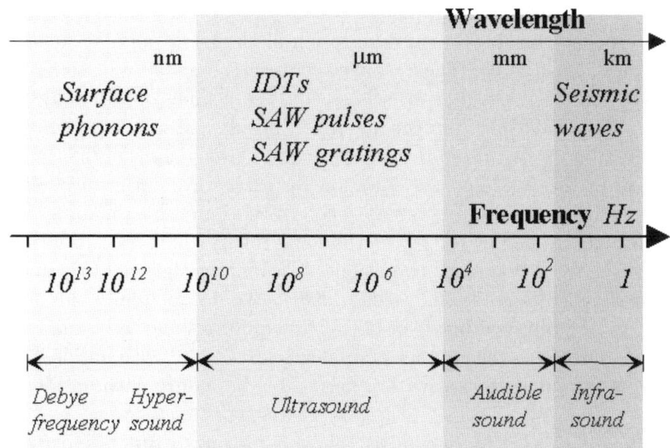

Figure 10.5: *Ranges of frequencies for surface waves (from [214]).*

10.4.3 Surface waves in linear elastic solids

Surface and guided waves are created on free surfaces and interfaces through the superposition of incident, transmitted and reflected bulk waves. The reason for the existence of these various modes is the difference in the material properties of bodies in contact which yields the so-called mode conversion. Surface waves exist in an extremely wide range of frequencies over some 10 orders of magnitude. This is schematically shown in Figure 10.5.

Acoustic surface waves (SAWs) were discovered in 1885 by Rayleigh (John William Strutt) [303] and they were the main subject of studies for some decades due to their appearance in form of seismic waves. The development of interdigital transducers (IDTs) and focused short laser pulses around 1960 extended their study and applications from the low frequency range to the ultrasound. Recent research work goes beyond this frequency range even to the hypersound region corresponding to quantized lattice vibrations (surface phonons).

This section is devoted to the presentation of models of basic types of surface waves in homogeneous elastic materials. We concentrate on models appropriate for the frequency regime of ultra-, audible- and infrasound. The reader interested in applications of surface waves to testing electronic materials should consult the review article of R. M. White [426]. Modeling, numerical evaluation and experimental verification of surface waves in heterogeneous materials, both elastic and viscoelastic, do not enter at all this book in spite of the fact that they are particularly important in nondestructive testing of soils (e.g. [128], [213], [321]). One should mention an area of application of surface waves in geotechnics which shall not be quoted further either in spite of its rash development in the last few years. It is related to the detection of buried objects, particularly land mines (e.g. [344], [455], [456]).

Theoretical and numerical details concerning the propagation of surface waves in two-component porous materials are presented in the article of B. Albers [10] or in the book [13] and we skip this subject entirely in the present book. We leave out as well the problem of nonlinear surface waves. This field of research develops in the recent years very vehemently (e.g. compare [244], [289]).

This section aims at the theoretical description of the most popular surface waves, in particular Rayleigh, Love and Stoneley waves. We introduce as well some terminology used in this theory which is peculiar for seismology, geotechnics, and nondestructive testing in materials science. We leave out the important problem of sources of waves. However, an extensive presentation of this subject can be found elsewhere ([5], [38], [334]).

The derivation of the fundamental equation for each wave – the dispersion relation, follows from the field equations and boundary conditions for each case anew. In the literature, this is sometimes unified by the ray method and the so-called transition and reflection conditions, following from boundary conditions, which altogether yield the method of the constructive interference (e.g. [5], [57], [349], [398]). We do not apply this method here. Instead, we discuss the physical inside of both field equations and of boundary conditions for such different systems as the contact of a linear elastic solid with an ideal fluid versus, say, the contact of porous materials with a permeable boundary with an ideal fluid.

10.4.3.1   Bulk waves and Rayleigh waves on plane boundaries of linear elastic homogeneous materials

The surface wave described in Section 10.4.2 is not typical for solids. As we see in the rest of this chapter, models of surface waves appear primarily as a combination of bulk waves which, in turn, follow as solutions of hyperbolic field equations.

We illustrate this statement by the classical example of the Rayleigh wave [303] (for example, see as well: [1], [204], [409], [410]).

Let us consider a linear elastic material described by the following equations for the unknown fields of velocity $\mathbf{v}\,(\mathbf{x}, t)$ and deformation $\mathbf{e}\,(\mathbf{x}, t)$

- momentum balance

$$\rho \frac{\partial \mathbf{v}}{\partial t} = \operatorname{div} \mathbf{T}, \tag{10.79}$$

- Hooke's law (constitutive relation for the Cauchy stress tensor $\mathbf{T}$)

$$\mathbf{T} = \lambda \operatorname{tr} \mathbf{e} \mathbf{1} + 2\mu \mathbf{e}, \tag{10.80}$$

  where $\mathbf{e}$ is the Almansi-Hamel tensor of small deformations (i.e., $\|\mathbf{e}\| \ll 1$)[3], λ and μ are the Lamé moduli,

- kinematic compatibility condition

$$\frac{\partial \mathbf{e}}{\partial t} = \operatorname{sym} \operatorname{grad} \mathbf{v}. \tag{10.81}$$

---

[3]As usual, the norm of the tensor $\mathbf{e}$ is defined by its eigenvalues which are identical with the principal stretches $\lambda_{(i)}$

$$\det \left(\mathbf{e} - \lambda_{(i)} \mathbf{1}\right) = 0, \quad i = 1, 2, 3,$$

$$\|\mathbf{e}\| := \max \left\{ \left|\lambda_{(1)}\right|, \left|\lambda_{(2)}\right|, \left|\lambda_{(3)}\right| \right\}.$$

Let us first seek bulk waves, described by the above equations in an infinite medium. It is sufficient to exploit the case of monochromatic waves

$$\mathbf{v} = \mathbf{V} e^{i(\mathbf{k}\cdot\mathbf{x}-\omega t)} \equiv \mathbf{V}\exp\left[ik\left(\frac{1}{k}\mathbf{k}\cdot\mathbf{x} - c_{ph}t\right)\right], \quad k := \sqrt{\mathbf{k}\cdot\mathbf{k}},$$

$$(10.82)$$

$$\mathbf{e} = \mathbf{E} e^{i(\mathbf{k}\cdot\mathbf{x}-\omega t)} \equiv \mathbf{E}\exp\left[ik\left(\frac{1}{k}\mathbf{k}\cdot\mathbf{x} - c_{ph}t\right)\right], \quad c_{ph} := \frac{\omega}{k},$$

where ω is a given frequency, $\mathbf{k}$ the wave vector, $k = \sqrt{\mathbf{k}\cdot\mathbf{k}}$ is the wave number, and $\mathbf{V}$ and $\mathbf{E}$ are constant amplitudes. Obviously, the unit vector $\mathbf{n} = \mathbf{k}/k$ points in the propagation direction and, as it is a constant vector, the wave is plane.

Substitution in (10.81) yields the following relation between the amplitudes

$$-\omega\mathbf{E} = \frac{1}{2}\left(\mathbf{k}\otimes\mathbf{V} + \mathbf{V}\otimes\mathbf{k}\right). \tag{10.83}$$

Hence, the momentum balance (10.79) leads to the following equation

$$-\omega\rho\mathbf{V} = \lambda\operatorname{tr}\mathbf{E}\mathbf{k} + 2\mu\mathbf{E}\mathbf{k} = -\lambda\frac{1}{\omega}\mathbf{V}\cdot\mathbf{k}\mathbf{k} - \mu\frac{1}{\omega}\left(\mathbf{V}\cdot\mathbf{k}\mathbf{k} + k^2\mathbf{V}\right). \tag{10.84}$$

We split this equation into the components parallel to $\mathbf{k}$ and perpendicular to $\mathbf{k}$

$$\left(\rho\omega^2 - (\lambda+2\mu)k^2\right)\mathbf{V}\cdot\mathbf{k} = 0, \tag{10.85}$$

$$\left(\rho\omega^2 - \mu k^2\right)\left(\mathbf{V} - \frac{1}{k^2}\mathbf{V}\cdot\mathbf{k}\mathbf{k}\right) = 0. \tag{10.86}$$

Obviously, we obtain two solutions

$$1)\ \mathbf{V} = \frac{1}{k^2}\mathbf{V}\cdot\mathbf{k}\mathbf{k} \quad\Rightarrow\quad \left(\frac{\omega}{k}\right)^2 = \frac{\lambda+2\mu}{\rho}, \tag{10.87}$$

$$2)\ \mathbf{V}\cdot\mathbf{k} = 0 \quad\Rightarrow\quad \left(\frac{\omega}{k}\right)^2 = \frac{\mu}{\rho}. \tag{10.88}$$

The first solution describes longitudinal waves or P-waves (P for primary; the amplitude is parallel to the direction of propagation: $\mathbf{V}\|\mathbf{k}$) whose phase velocity is

$$c_L = \sqrt{\frac{\lambda+2\mu}{\rho}}, \tag{10.89}$$

and this is not dependent on the frequency. Consequently, the wave is non-dispersive – each monochromatic wave propagates with the same velocity.

The second solution describes transversal waves or S-waves (S for secondary or shear; the amplitude is perpendicular to the direction of propagation: $\mathbf{V}\cdot\mathbf{k} = 0$) whose phase velocity is

$$c_T = \sqrt{\frac{\mu}{\rho}}. \tag{10.90}$$

Consequently, this wave is also non-dispersive.

The system of equations of linear elasticity is hyperbolic provided $\mu > 0, \lambda + 2\mu > 0$. Then, both velocities of the above described bulk waves are real.

Let us mention in passing that the separation of the eigenvalue problem presented above can be done in many other ways. Depending on a particular problem, they appear in the literature on classical elasticity as well as in other problems of mechanics. Let us mention two of them.

If we introduce the field of displacement $\mathbf{u}(\mathbf{x}, t)$ the problem is described by Equations (10.79) and (10.80) due to the following relations

$$\mathbf{v} = \frac{\partial \mathbf{u}}{\partial t}, \quad \mathbf{e} = \frac{1}{2}\left(\operatorname{grad}\mathbf{u} + (\operatorname{grad}\mathbf{u})^T\right), \tag{10.91}$$

and Relation (10.81) is identically satisfied. Simultaneously, the vector $\mathbf{u}$ can be represented as a sum of a potential part $\mathbf{u}_L$ and a solenoidal part $\mathbf{u}_T$ which, as can be shown by a straightforward calculation, satisfy the following relations

$$\mathbf{u} = \mathbf{u}_L + \mathbf{u}_T, \quad \operatorname{curl}\mathbf{u}_L = 0, \quad \operatorname{div}\mathbf{u}_T = 0,$$

$$\frac{\partial^2\mathbf{u}_L}{\partial t^2} = c_L^2\Delta^2\mathbf{u}_L, \quad \frac{\partial^2\mathbf{u}_T}{\partial t^2} = c_T^2\Delta^2\mathbf{u}_T. \tag{10.92}$$

Hence, we obtain two wave equations whose solutions have the form of the two waves discussed before.

One can use as well the following identity satisfied by any differentiable vector field (the so-called Helmholtz Decomposition Theorem)

$$\mathbf{u} = \operatorname{grad}\varphi + \operatorname{curl}\psi, \tag{10.93}$$

where φ and ψ are the so-called scalar and vector potentials. Again, one can easily show that they satisfy the following equations

$$\frac{\partial^2\varphi}{\partial t^2} = c_L^2\Delta^2\varphi, \quad \frac{\partial^2\psi}{\partial t^2} = c_T^2\Delta^2\psi. \tag{10.94}$$

We obtain again the same result.

We proceed to construct a solution for a semiinfinite linear elastic medium, i.e., a medium with a boundary. The presence of a boundary leads to important wave effects. First of all, as we have already mentioned, there are bulk wave reflection and transmission phenomena. We shall not discuss them in any details in this book. An interested reader should consult, for instance, [5], [349], [398].

In order to analyze the problem we need boundary conditions on the plane boundary. We choose Cartesian coordinates with the z-axis perpendicular to the boundary. The boundary is defined by $z = 0$. We consider the case of the boundary free of loading. Hence,

$$\mathbf{Tn}\big|_{z=0} = 0, \quad \mathbf{n} = -\mathbf{e}_z, \quad \mathbf{u}\big|_{z\to\infty} = 0, \tag{10.95}$$

where $\mathbf{e}_z$ is the unit basis vector of the z-axis. This means that the z-axis is oriented into the medium.

We seek the solution by splitting the displacement $\mathbf{u}$ into the potential and solenoidal parts, $\mathbf{u}_L, \mathbf{u}_T$. Then we make the following ansatz

$$\mathbf{u}_L = A_L e^{-\gamma z}e^{i(kx-\omega t)}\mathbf{e}_x + B_L e^{-\gamma z}e^{i(kx-\omega t)}\mathbf{e}_z,$$

$$\mathbf{u}_T = A_T e^{-\beta z}e^{i(kx-\omega t)}\mathbf{e}_x + B_T e^{-\beta z}e^{i(kx-\omega t)}\mathbf{e}_z, \tag{10.96}$$

where A_L, A_T, B_L and B_T are constant amplitudes. Thus, we anticipate a progressive wave solution in the x-direction and the decay of the solution in the z-direction provided both γ and β are positive. If this should not be the case, a solution in the form of a surface wave would not exist. Obviously, the form of the solution (10.96) indicates that particles move in the xz-plane. This is the characteristic feature of Rayleigh waves.

Substitution of the above solution in the wave equations $(10.92)_{4,5}$ yields the compatibility conditions

$$\frac{\gamma^2}{k^2} = 1 - \frac{c_R^2}{c_L^2}, \quad \frac{\beta^2}{k^2} = 1 - \frac{c_R^2}{c_T^2}, \quad c_R := \frac{\omega}{k}. \tag{10.97}$$

Now conditions $(10.92)_{2,3}$ characterizing the potential and solenoidal parts of the displacement lead to the following form of the solution

$$B_L = i\frac{\gamma}{k} A_L \quad \Rightarrow \quad \mathbf{u}_L = \left(\mathbf{e}_x + i\frac{\gamma}{k}\mathbf{e}_z\right) A_L e^{-\gamma z} e^{i(kx-\omega t)},$$

$$\tag{10.98}$$

$$B_T = i\frac{k}{\beta} A_T \quad \Rightarrow \quad \mathbf{u}_T = \left(\mathbf{e}_x + i\frac{k}{\beta}\mathbf{e}_z\right) A_T e^{-\beta z} e^{i(kx-\omega t)}.$$

In order to find the amplitudes we apply the boundary conditions (10.95). Obviously, the last condition, the so-called Sommerfeld condition, is satisfied identically provided both γ and β are positive. The remaining relations yield

$$\left(c_L^2 - 2c_T^2\right) \frac{\partial \mathbf{u} \cdot \mathbf{e}_x}{\partial x} + c_L^2 \frac{\partial \mathbf{u} \cdot \mathbf{e}_z}{\partial z}\bigg|_{z=0} = 0,$$

$$\tag{10.99}$$

$$\frac{\partial \mathbf{u} \cdot \mathbf{e}_x}{\partial z} + \frac{\partial \mathbf{u} \cdot \mathbf{e}_z}{\partial x}\bigg|_{z=0} = 0.$$

Substitution of (10.98) gives rise to the following homogeneous set for the unknown constants

$$\left(2 - \frac{c_R^2}{c_T^2}\right) A_L + 2A_T = 0,$$

$$\tag{10.100}$$

$$2\frac{\gamma\beta}{k^2} A_L + \left(2 - \frac{c_R^2}{c_T^2}\right) A_T = 0.$$

Hence, if we choose the positive signs for β and γ in the roots following from the compatibility relations (10.97), the determinant leads to the following Rayleigh dispersion relation

$$\mathcal{P}_R := \left(2 - \frac{c_R^2}{c_T^2}\right)^2 - 4\sqrt{1 - \frac{c_R^2}{c_T^2}}\sqrt{1 - \frac{c_R^2}{c_L^2}} = 0. \tag{10.101}$$

Clearly, the solutions c_R of this equation are independent of the frequency ω. In other words, Rayleigh waves in a semiinfinite medium are nondispersive.

We show (see also e.g. [38]) that this equation possesses one real solution $c_R < c_T$. Namely, it can be easily written in the form

$$f(y, n) := y^3 - 8y^2 + 8(3 - 2n)y - 16(1 - n) = 0, \quad y := \frac{c_R^2}{c_T^2}, \quad n := \frac{c_T^2}{c_L^2}. \tag{10.102}$$

This function is concave in the interval $(0, 1)$ for $n \in (0, 0.5)$, furthermore $f(0, n) = -16(1 - n) < 0$ and $f(1, 0) = 1$. Consequently, there exists one root in the interval determined by the condition $c_R < c_T$. However, it may possess two other real roots bigger than 1 for $c_T/c_L \equiv \sqrt{\mu/(\lambda + 2\mu)}$ bigger than approximately 0.57. It is instructive to calculate this coefficient in terms of the Poisson ratio ν

$$\nu = \frac{\lambda}{2(\lambda + \mu)} \equiv \frac{1}{2}\frac{1 - 2\dfrac{\mu}{\lambda + 2\mu}}{1 - \dfrac{\mu}{\lambda + 2\mu}} \quad \Rightarrow \quad n \equiv \frac{\mu}{\lambda + 2\mu} = \frac{1}{2}\frac{1 - 2\nu}{1 - \nu}, \tag{10.103}$$

i.e.,

$$0 < \nu \le \frac{1}{2} \quad \Rightarrow \quad \frac{1}{2} > \frac{\mu}{\lambda + 2\mu} \ge 0. \tag{10.104}$$

Consequently, the value $\sqrt{\mu/(\lambda + 2\mu)} = \sqrt{0.5} \approx 0.707$ corresponds to the minimum value of ν equal to zero. For growing ν the fraction $\mu/(\lambda + 2\mu)$ decays. Hence, the range $0.570 < \sqrt{\mu/(\lambda + 2\mu)} < 0.707$ is physically not empty.

Clearly, there exist real solutions smaller than one for all values of ν, and two real solutions bigger than one (i.e., the corresponding velocity is bigger than the shear velocity) for $\nu \gtrsim 0.26309$. It is easy to check that these real solutions of (10.102) are not solutions of the dispersion relation (10.101) and they follow due to taking the square of this equation. However, they do become solutions of the dispersion relation if we substitute for β and γ values following from (10.97) with opposite signs: $\mathrm{sign}\,\beta = -\,\mathrm{sign}\,\gamma$. It is claimed (e.g. [1]: "... these roots are extraneous; they arise from the rationalization process of squaring.") that such solutions do not possess any physical meaning.

We show further that these solutions correspond to the so-called leaky (evanescent) waves intermediate between surface waves and bulk waves which lose their energy to both Rayleigh wave and two usual body waves (e.g. [271]). They were observed experimentally and explain the existence of certain critical values of the angle of incidence [345]. Some details on the properties of leaky waves can also be found in the books of Brekhovskikh and Godin [56], Viktorov [410], and possible practical applications in nondestructive testing in [26].[4]

The solution, which is slower than the shear velocity, is shown in Figure 10.6, where the fraction c_R/c_T is plotted as a function of the fraction of bulk velocities c_T/c_L. Clearly,

---

[4]There is an analytical method to check how many solutions a dispersion relation may possess. It is based on the *argument principle* of complex analysis. Namely, an arbitrary function $f(z)$ possesses the root z_0 if $f(z_0) = 0$. An analytic function possesses the pole z_0 of order n if the function $(z - z_0)^n f(z)$ is holomorphic at z_0 (i.e., this function possesses the derivative in a neighborhood of z_0). According to the argument principle, if $f(z)$ is meromorphic in a region R enclosed by a contour Γ, then

$$\frac{1}{2\pi i}\int_{\Upsilon} \frac{dw}{w} = N - P, \quad w := f(z), \quad \Upsilon := f(\Gamma),$$

and N is the number of complex roots of $f(z)$, and P the number of poles in Γ. A meromorphic function $f(z)$ is defined as a single valued function analytic in all but possibly a discrete subset of its domain, and at those singularities it must go to infinity like a polynomial.

Achenbach [1] defines the function

$$R(s) := \left(2s^2 - s_T^2\right)^2 + \left(s_L^2 - s^2\right)^{\frac{1}{2}}\left(s_T^2 - s^2\right)^{\frac{1}{2}} = 0,$$

$$s_L = \frac{1}{c_L}, \quad s_T = \frac{1}{c_T}, \quad s = \frac{1}{c},$$

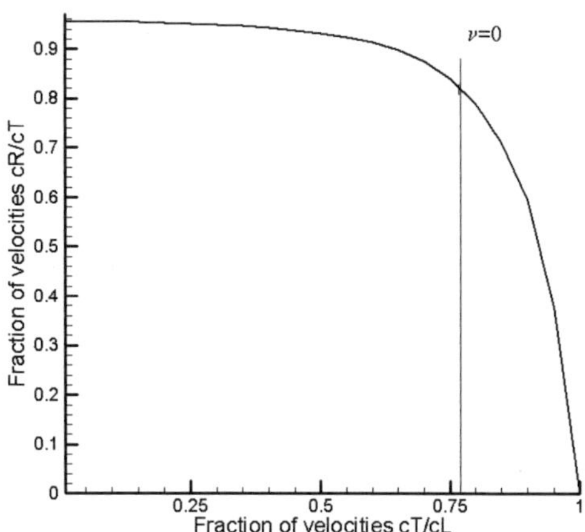

Figure 10.6: *Dimensionless velocity of the Rayleigh wave c_R/c_T as a function of the fraction of shear and longitudinal velocities c_T/c_L (from [214]).*

the value of the Rayleigh velocity lies very near to the velocity of shear waves for most values of the material parameters.

Let us consider the orbits of the particles of Rayleigh waves. According to $(10.92)_1$, and (10.98) the displacement is given by the relations

$$u_x = \left(A_L e^{-\gamma z} + A_T e^{-\beta z}\right) e^{i(kx-\omega t)},$$

$$(10.105)$$

$$u_z = i\left(\frac{\gamma}{k} A_L e^{-\gamma z} + \frac{k}{\beta} A_T e^{-\beta z}\right) e^{i(kx-\omega t)}.$$

We choose the motion in which the x-component propagates as $\cos(kx-\omega t)$. Then, taking the real parts of the above relations and eliminating time, we obtain

$$\frac{(\operatorname{Re} u_x)^2}{\alpha_x^2} + \frac{(\operatorname{Re} u_z)^2}{\alpha_z^2} = 1,$$

$$(10.106)$$

where s is the slowness, and proves that the relation

$$\frac{1}{2\pi i} \int_C \frac{dv}{v} = N - P, \quad v = R(s),$$

where C is a sufficiently large contour around the origin, yields a single solution for c^2 and it is real and positive.

The above method is particularly useful for complicated dispersion relations and its results considerably simplify the numerical effort.

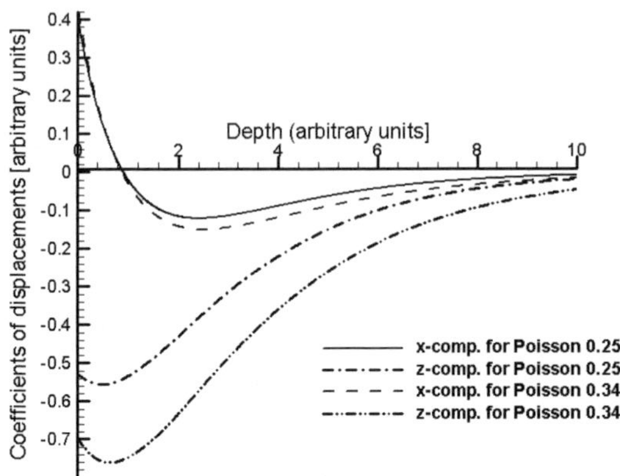

Figure 10.7: *Coefficients of the components of the displacement (semiaxes of the displacement ellipse) as functions of the depth z for two values of Poisson's ratio $\nu = 0.25$ and 0.34 (i.e., $c_T/c_L = 0.5773$ and 0.4924); (from [214]).*

where

$$\alpha_x := A_L \mathrm{e}^{-\gamma z} + A_T \mathrm{e}^{-\beta z},$$

$$\alpha_z := \frac{\gamma}{k} A_L \mathrm{e}^{-\gamma z} + \frac{k}{\beta} A_T \mathrm{e}^{-\beta z}, \tag{10.107}$$

and

$$\operatorname{Re} u_x = \alpha_x \cos\left(kx - \omega t\right), \quad \operatorname{Re} u_z = -\alpha_z \sin\left(kx - \omega t\right). \tag{10.108}$$

Hence, as in the case of water waves (compare (10.69)), the orbit of each particle is an ellipse with semiaxes $|\alpha_x|$, $|\alpha_z|$. However, the motion is anticlockwise (retrograde). The semiaxes decay exponentially with the depth (see Figure 10.7).

Modicum of Rayleigh waves on a plain boundary

$$\mathcal{P}_R := \left(2 - \frac{c_R^2}{c_T^2}\right)^2 - 4\sqrt{1 - \frac{c_R^2}{c_T^2}}\sqrt{1 - \frac{c_R^2}{c_L^2}} = 0 \qquad - \quad \text{Rayleigh dispersion relation,}$$

where $c_R := \dfrac{\omega}{k}$.

$$\operatorname{Re} u_x = \alpha_x \cos\left(kx - \omega t\right), \quad \operatorname{Re} u_z = -\alpha_z \sin\left(kx - \omega t\right) \qquad - \text{displacement,}$$

where

$$\alpha_x := A_L \mathrm{e}^{-\gamma z} + A_T \mathrm{e}^{-\beta z}, \quad \alpha_z := \frac{\gamma}{k} A_L \mathrm{e}^{-\gamma z} + \frac{k}{\beta} A_T \mathrm{e}^{-\beta z}.$$

Obviously, the vector of the amplitude of the Rayleigh wave lies on the xz-plane. This plane is defined by the propagation direction x (called in seismology the P direction, for primary), and the direction of decay z (called in seismology SV-direction: for shear vertical). It is called in seismology the plane of incidence or the sagittal plane in materials sciences. In the general case, the progressive wave solution in the x-direction may also possess an amplitude in the direction perpendicular to the plane of incidence – the so-called SH-direction (for shear horizontal). This part of the solution would have the form

$$u_y\left(x, z, t\right) = C\mathrm{e}^{-\delta z}\mathrm{e}^{i(kx-\omega t)}. \tag{10.109}$$

Such waves are called Love waves. It is easy to check that the boundary condition (10.95) yields immediately $C = 0$. Consequently, Love waves do not exist in homogeneous semiinfinite media. As we see in the next section they do exist in layers on semiinfinite media, i.e., they are guided waves rather than surface waves.

10.4.3.2 Rayleigh waves on cylindrical boundaries

The problem of surface waves on cylindrical surfaces is quite common in geotechnics and it appears, for instance, in the analysis of waves in boreholes. We present a simple example of a solid cylinder in order to see the influence of curvature on the propagation of surface waves.

Let us investigate the problem of propagation of surface waves in an infinite cylinder of the radius R, i.e., waves which satisfy the following conditions:

1. the surface of the cylinder is free of stresses,

2. waves propagate in the circumferential direction and decay in the radial direction.

This problem has been solved in 1958 by I. A. Viktorov [407, 408, 410] and it demonstrates the influence of the curvature of the surface on the propagation of surface waves.

It is convenient to use the potentials φ and ψ in this case. As the problem is two-dimensional the vector potential has only one component ψ_z in the direction of the axis of the cylinder. These potentials satisfy Equations (10.94) which have the following form in polar coordinates

$$\frac{\partial^2 \varphi}{\partial t^2} = c_L^2 \left\{ \frac{1}{r}\frac{\partial}{\partial r}\left(r\frac{\partial \varphi}{\partial r}\right) + \frac{1}{r^2}\frac{\partial^2 \varphi}{\partial \theta^2} \right\},$$

$$\tag{10.110}$$

$$\frac{\partial^2 \psi_z}{\partial t^2} = c_T^2 \left\{ \frac{1}{r}\frac{\partial}{\partial r}\left(r\frac{\partial \psi_z}{\partial r}\right) + \frac{1}{r^2}\frac{\partial^2 \psi_z}{\partial \theta^2} \right\},$$

where r and θ are the radius and the angle coordinate, respectively.

It is convenient to consider the extension of the problem to the infinite interval for the angle θ: $-\infty < \theta < \infty$. Then, the axis $r = 0$ is the branch cut of infinite order. This extension allows to consider the propagation of surface waves on an infinite plane of repetitions of the circumferential surface of the cylinder. In practical applications it may correspond, for instance, to spiral surface waves whose amplitude is weakly dependent on the z-variable.

In polar coordinates the physical radial and circumferential components of the displacement are given by the relations

$$u_r = \frac{\partial \varphi}{\partial r} + \frac{1}{r}\frac{\partial \psi_z}{\partial \theta}, \qquad u_\theta = \frac{1}{r}\frac{\partial \varphi}{\partial \theta} - \frac{\partial \psi_z}{\partial r}. \tag{10.111}$$

We seek solutions in the form

$$\varphi = \Phi\left(r\right)e^{i(p\theta-\omega t)}, \quad \psi_z = \Psi\left(r\right)e^{i(p\theta-\omega t)}, \tag{10.112}$$

where p plays the role of the angular wave number. It is related to the wavelength l by the relation $p = 2\pi R/l$. The usual wave number is then given by the relation $k = 2\pi/l \equiv p/R$. Due to the extension of the domain for angle θ, the variable p can be considered as continuous. Certainly, the limit $p \to 0$ corresponds to very long waves ($l \gg R$) and the limit $p \to \infty$ is the limit of short waves.

Substitution of the above ansatz in Equations (10.110) yields the following Bessel equations (see Appendix D)

$$\frac{d^2\Phi}{d\xi_L^2} + \frac{1}{\xi_L}\frac{d\Phi}{d\xi_L} + \left(1 - \frac{p^2}{\xi_L^2}\right)\Phi = 0, \quad \xi_L := r\frac{\omega}{c_L} \equiv \frac{r}{R}p\frac{c}{c_L},$$

$$\frac{d^2\Psi}{d\xi_T^2} + \frac{1}{\xi_T}\frac{d\Psi}{d\xi_T} + \left(1 - \frac{p^2}{\xi_T^2}\right)\Psi = 0, \quad \xi_T := r\frac{\omega}{c_T} \equiv \frac{r}{R}p\frac{c}{c_T}, \quad c := \frac{\omega R}{p}. \tag{10.113}$$

The solution of this set which remains finite in the middle points of the cylinder has the following form

$$\Phi = AJ_p\left(\xi_L\right), \quad \Psi = BJ_p\left(\xi_T\right), \tag{10.114}$$

where A and B are constants and J_p is the Bessel function of order p.

The constants appearing in the solution should be determined from the boundary condition on the free surface $r = R$. These boundary conditions – the boundary is free of stresses – have the following form in cylindrical coordinates

$$(c_L^2 - 2c_T^2)\left(\frac{\partial^2\varphi}{\partial r^2} + \frac{1}{r}\frac{\partial\varphi}{\partial r} + \frac{1}{r^2}\frac{\partial^2\varphi}{\partial\theta^2}\right)$$

$$+2c_T^2\left(\frac{\partial^2\varphi}{\partial r^2} + \frac{1}{r}\frac{\partial^2\psi_z}{\partial r\partial\theta} - \frac{1}{r^2}\frac{\partial\psi_z}{\partial\theta}\right) = 0, \quad \text{for} \quad r = R, \tag{10.115}$$

$$\frac{2}{r}\frac{\partial^2\varphi}{\partial r\partial\theta} - \frac{1}{r^2}\frac{\partial\varphi}{\partial\theta} + \frac{1}{r^2}\frac{\partial^2\psi_z}{\partial\theta^2} - \frac{\partial^2\psi_z}{\partial r^2} + \frac{1}{r}\frac{\partial\psi_z}{\partial r} = 0,$$

i.e., the radial stress and the shear stress are zero at the circumferential surface $r = R$.

Substitution of (10.114) in (10.112) and, subsequently, in (10.115) yields a homogeneous set of equations for the constants A, B

$$b_{11}A + b_{12}iB = 0, \quad b_{21}A + b_{22}iB = 0, \tag{10.116}$$

where[5]

$$b_{11} := -2pJ_p^L - \frac{c}{c_T}\frac{c_L}{c_T}p\left(2p + 1 - 2\frac{c_T^2}{c_L^2}\right)J_{p+1}^L + \frac{c^2}{c_T^2}p^2J_{p+2}^L,$$

---

[5]We have used here the following identity for Bessel functions

$$\frac{dJ_p\left(x\right)}{dx} = \frac{p}{x}J_p\left(x\right) - J_{p+1}\left(x\right).$$

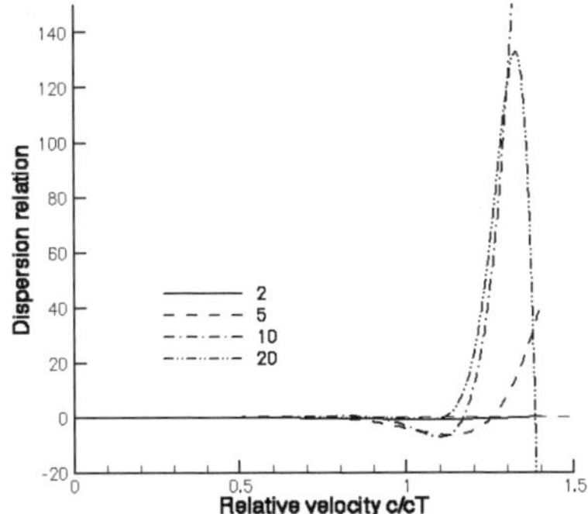
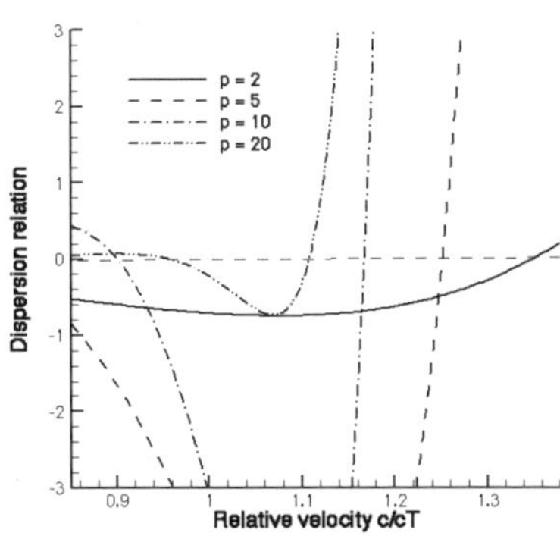

Figure 10.8: *Zeros of the dispersion relation* (10.118) *for four different values of the wave number p: 2, 5, 10, 20. The right panel is the magnification of the part of the left panel (from [214]).*

$$b_{12} := 2 \left[(p^2 - 1) J_p^T - \frac{c}{c_T} p^2 J_{p+1}^T \right], \quad b_{21} := p(2p-1) J_p^L - 2\frac{c}{c_T}\frac{c_T}{c_L} p^2 J_{p+1}^L,$$

$$(10.117)$$

$$b_{22} := 2p(p-1) J_p^T - \frac{c}{c_T} p(2p-1) J_{p+1}^T + \frac{c^2}{c_T^2} J_{p+2}^T,$$

$$J_p^L := J_p\left(p\frac{c}{c_L} \right), \quad J_p^T := J_p\left(p\frac{c}{c_T} \right),$$

and similarly for the other Bessel functions.

Consequently, for the existence of nontrivial solutions the determinant of the set (10.116) must be zero and we obtain the following dispersion relation for this problem

$$\mathcal{P}_{RC}(c, p) := b_{11}b_{22} - b_{12}b_{21} = 0. \tag{10.118}$$

This relation is plotted in Figure 10.8 for the data $c_T/c_L = 0.4924$ which corresponds to Poisson's ratio $\nu = 0.34$. This value has been chosen to coincide with the value chosen by Viktorov [410] in his numerical example. Not visible in Figure 10.7 are infinitely many zero points for relative velocities bigger than $c/c_T > c_L/c_T = 2.0310$.

Rayleigh waves on cylinders are waves whose velocity c is bigger than c_T and they correspond to the first zero points above the point $\frac{c}{c_T} = 1$. This is demonstrated in the right panel of Figure 10.8 which is a magnification of the diagram in the left panel. These solutions of the dispersion relation exist for $p > 1$ (i.e., $l < 2\pi R$) and decay to the Rayleigh velocity on the plane boundary for $p \to \infty$. This is a purely geometrical effect. The velocity of surface waves can be bigger than the velocity of shear waves because the path of the surface wave is longer than the path of the shear wave in the case of curved boundaries. For instance, for points of the boundary lying on the same diameter, the shear wave covers the distance $2R$ and the surface wave the distance πR. Consequently, the arrival time of the shear wave is shorter than the arrival time of the surface wave and, in this way, we fulfill the condition for constructive interference.

This effect was demonstrated by Viktorov and he claims that the influence of the Poisson ratio (i.e., relative bulk velocity $\frac{c_T}{c_L}$) is rather small.

Numerous zero points in this problem indicate that, in addition to Rayleigh waves, some additional waves may exist due to interactions of bulk waves with the surface. We shall discuss further some aspects of this property of dispersion relations.

The problem of propagation of surface waves on cylindrical surfaces has an important bearing in nondestructive testing of shafts. These circumferential waves belong then to the class of guided waves (e.g. [1], [230]). The dispersion relation for surface waves in a hollow cylinder with an inner shaft was obtained numerically by C. Valle, J. Qu and L. J. Jacobs [400]. They compare the first five modes of propagation of these waves with the corresponding modes for the solid and hollow cylinders and with the Rayleigh wave on the plane boundary. It comes out that the behavior of the waves in a layered cylinder coincides with this of the solid cylinder in the low frequency regime and it differs substantially in the range of high frequencies. In this range, the first mode tends asymptotically to the Rayleigh mode on the plane surface, while the second mode approaches in this limit a Rayleigh mode appearing on sliding interfaces. Further references can be found in the original work [400].

10.4.4 Waves in a layer of an ideal fluid and Love waves on plane boundaries

10.4.4.1 Layer of an ideal compressible fluid on a semiinfinite rigid body

Heterogeneity of the system in the form of layers has great influence on the modes of surface waves which may appear on the interfaces. In order to appreciate this influence on the propagation of surface waves, we investigate first a simple example of a layer of an ideal fluid $-\infty < x < \infty$, $0 \leq z \leq H$. The upper surface $z = H$ (z-axis is oriented upward in this case) is free of loading and the lower surface $z = 0$ is in contact with a rigid body. The problem is described by the equations of mass and momentum conservation

$$\frac{\partial \rho}{\partial t} + \rho_0 \operatorname{div} \mathbf{v} = 0, \quad \rho_0 \frac{\partial \mathbf{v}}{\partial t} = -\operatorname{grad} p, \quad p = p_0 + \kappa (\rho - \rho_0), \qquad (10.119)$$

where ρ_0 and p_0 are reference constant values of the mass density and pressure, respectively, and κ denotes a constant compressibility coefficient of the fluid.

Simple manipulations lead to the following wave equation for the pressure p

$$\frac{\partial^2 p}{\partial t^2} = \kappa \nabla^2 p, \quad (x, z) \in (-\infty, \infty) \times (0, H). \qquad (10.120)$$

The solution of this equation must satisfy the following boundary conditions

$$p(x, z = H, t) = 0, \quad v_z(x, z = 0, t) = 0. \qquad (10.121)$$

As before, we seek the solution in the form of a monochromatic wave of the frequency ω

$$p = \left(A e^{irkz} + B e^{-irkz}\right) e^{i(kx - \omega t)}. \qquad (10.122)$$

Then, due to the momentum balance, the second boundary condition can be replaced by the following one

$$\frac{\partial p}{\partial z}(x, z = 0, t) = 0. \qquad (10.123)$$

Substitution of (10.122) in Equation (10.120) yields the compatibility relation

$$r^2 = \frac{c_{ph}^2}{c^2} - 1, \quad c_{ph} := \frac{\omega}{k}, \quad c := \sqrt{\kappa}. \tag{10.124}$$

Simultaneously, the evaluation of the boundary conditions with the ansatz (10.122) yields the set of homogeneous algebraic relations for the constants A and B

$$Ae^{irkH} + Be^{-irkH} = 0,$$

$$A - B = 0. \tag{10.125}$$

Consequently, the determinant of this set must be equal to zero and we obtain the dispersion relation

$$\cos(rkH) = 0. \tag{10.126}$$

In order to obtain nontrivial solutions we have to require that r is real. This means, however, that the dispersion relation given in terms of a periodic function yields infinitely many solutions. Each solution is called a mode of propagation. This is the characteristic feature of heterogeneous systems.

Simultaneously, it follows from (10.126) that the phase velocities c_{ph} are bigger than the velocity of propagation c appearing in the wave equation for the pressure (10.120)$(\kappa \equiv c^2)$, and that they go to infinity as the frequency approaches certain critical values. If we require that waves of the form (10.122) do exist, then this seems to violate the basic property of the hyperbolic problem. This result follows from the assumption that the foundation of the fluid is a rigid body in which all disturbances propagate with an infinite velocity. As we see further, a modification of the boundary condition (10.121)$_2$ for the case of contact with an elastic body, which we make for the so-called Love waves, eliminates this paradox.

In details, the solution of Equation (10.126) yields immediately the following relation between the phase velocity and the frequency

$$c_{ph} = \frac{c}{\sqrt{1 - \frac{\omega_{cr}^2}{\omega^2}}}, \quad \omega_{cr} := \left(n + \frac{1}{2}\right)\pi\frac{c}{H}, \quad n = 1, 2, \ldots \,. \tag{10.127}$$

This relation is illustrated in Figure 10.9.

10.4.4.2   Love waves on plane boundaries

The paradox of infinite phase velocities does not appear anymore in the case of surface waves which propagate in an elastic layer over an elastic half-space. Transversal waves in such a system have been described in 1911 by Love [237]. We proceed to present briefly these results. They form the simplest illustration of the problem of surface waves in heterogeneous solids.

We consider the propagation of a wave whose amplitude has only an $\mathbf{e}_y$-component $u_y \equiv \mathbf{u} \cdot \mathbf{e}_y$ (perpendicular to the (x, z)-plane; hence, it corresponds to the SH amplitude for waves with P-SV incident plane in seismological terminology). The body consists of a layer of thickness H in the z-direction in which the mass density is ρ' and the velocity of shear waves is c'_T. This layer is connected to the elastic half-space $z \leq 0$ whose mass

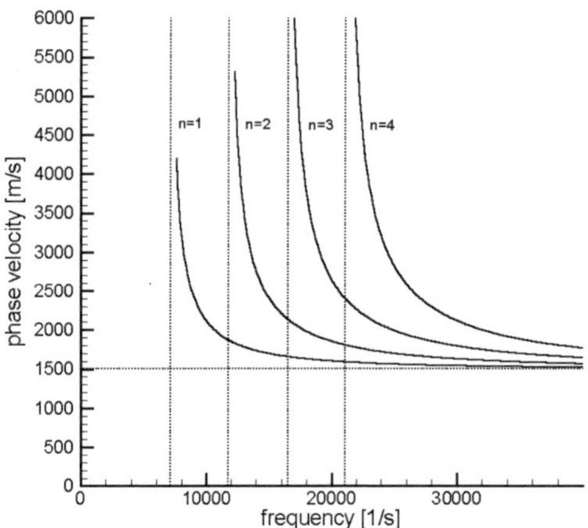

Figure 10.9: *Phase velocity for a layer of an ideal fluid on a rigid body. Data: $c = 1500$ m/s, $H = 1$ m. Modes: $n = 1, 2, 3, 4$; (from [214]).*

density is ρ, and the velocity of shear waves is c_T. We seek the solution of the wave equations

$$\frac{\partial^2 u'_y}{\partial t^2} = c'^2_T \nabla^2 u'_y, \quad 0 < z < H,$$

$$\frac{\partial^2 u_y}{\partial t^2} = c^2_T \nabla^2 u_y, \quad z < 0,$$

(10.128)

in the form

$$u'_y = \left(A' e^{iks'z} + B' e^{-iks'z} \right) e^{i(kx-\omega t)}$$

$$\equiv 2 \left(\operatorname{Re} A' \cos ks'z - \operatorname{Im} A' \sin ks'z \right) e^{i(kx-\omega t)}, \tag{10.129}$$

$$u_y = B e^{ksz} e^{i(kx-\omega t)},$$

i.e., as a monochromatic wave which propagates in the direction of the x-axis with the frequency ω, wave number k in this direction, and with the phase velocity $c := \frac{\omega}{k}$. The wave should decay in the z-direction, i.e., s must be positive. We check now if the ansatz (10.129) can fulfill Equations (10.128), and the following boundary conditions

1) shear stress on the plane $z = H$ is equal to zero, i.e.,

$$\frac{\partial u'_y}{\partial z} (x, z = H, t) = 0, \tag{10.130}$$

2) shear stress and the displacement must be continuous on the interface $z = 0$

$$\rho' c_T'^2 \frac{\partial u_y'}{\partial z}\left(x, z = 0, t\right) = \rho c_T^2 \frac{\partial u_y}{\partial z}\left(x, z = 0, t\right),$$

$$u_y'\left(x, z = 0, t\right) = u_y\left(x, z = 0, t\right). \tag{10.131}$$

Substitution of the ansatz (10.129) in Equations (10.128) yields

$$s'^2 = \frac{c^2}{c_T'^2} - 1, \quad s^2 = 1 - \frac{c^2}{c_T^2}, \quad c \equiv \frac{\omega}{k}. \tag{10.132}$$

Boundary condition (10.130) leads immediately to the following relation for the displacement in the layer

$$u_y' = 2 \operatorname{Re} A' \frac{\cos\left(ks'\left(H - z\right)\right)}{\cos\left(ks'H\right)} e^{i(kx - \omega t)}. \tag{10.133}$$

Then the boundary conditions (10.131) yield a homogeneous set of two algebraic relations for the constants $\operatorname{Re} A'$ and B. Consequently, its determinant must be zero and this condition yields the Love dispersion relation

$$\omega = \frac{c}{Hs'}\left[\arctan\left(\frac{\rho c_T^2 s}{\rho' c_T'^2 s'}\right) + n\pi\right], \quad n = 1, 2, 3, \ldots, \tag{10.134}$$

where both s and s' must be real, i.e.,

$$c_T' \leq c \leq c_T. \tag{10.135}$$

This is the condition for the existence of Love waves. Hence, Love waves can propagate solely in layers which are softer than the foundation. In addition there exist infinitely many modes of propagation whose existence is limited from below by corresponding critical frequencies. All these modes are dispersive because the phase velocities depend on the frequency given by the inverse relation to (10.134).

Modicum of Love waves

$$\omega = \frac{c}{Hs'}\left[\arctan\left(\frac{\rho c_T^2 s}{\rho' c_T'^2 s'}\right) + n\pi\right], \quad n = 1, 2, 3, \ldots, \quad - \text{ Love dispersion relation,}$$

where

$$s^2 = 1 - \frac{c^2}{c_T^2}, \quad s'^2 = \frac{c^2}{c_T'^2} - 1, \quad c_T' \leq c \leq c_T.$$

$$u_y' = u_{y0}' \frac{\cos\left(ks'\left(H - z\right)\right)}{\cos\left(ks'H\right)} e^{i(kx - \omega t)}, \quad - \text{ displacement,}$$

with

$$u_{y0}' = 2 \operatorname{Re} A'.$$

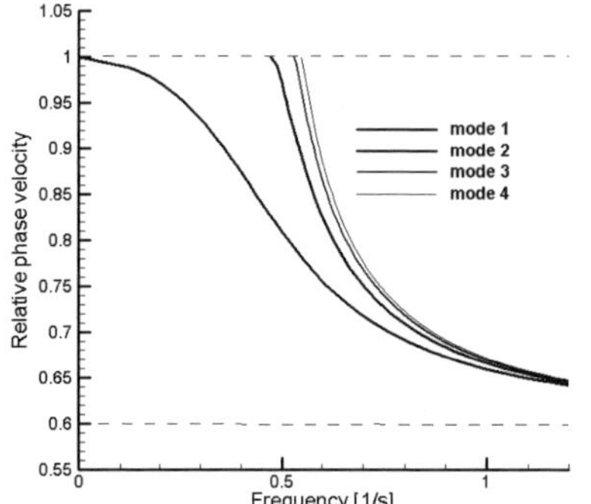
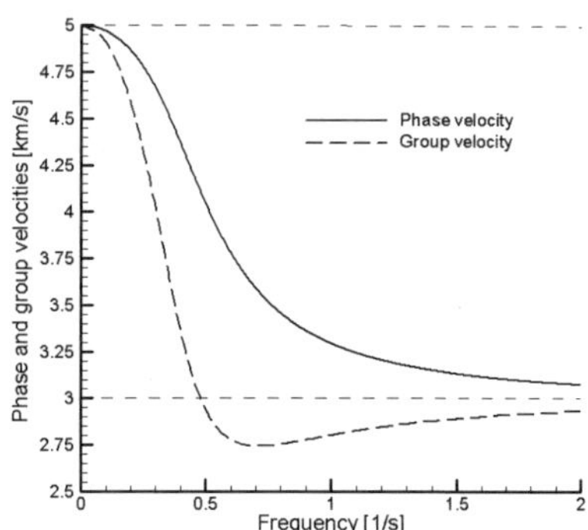

Figure 10.10: *Left: relative phase velocities of Love waves for four modes ($n = 0, 1, 2, 3$); right: phase and group velocity of the first mode of Love wave (from [214]).*

In Figure 10.10 we show an example of the solution of Relation (10.134) for the following data:

$$c_T = 5\,\frac{\mathrm{km}}{\mathrm{s}}, \quad c'_T = 3\,\frac{\mathrm{km}}{\mathrm{s}}, \quad \frac{\rho'}{\rho} = 0.875, \quad H = 10\,\mathrm{km}. \tag{10.136}$$

In the left panel we plot the relative velocities c_{ph}/c_T as functions of the frequency ω for four values of n.

The data were chosen to be identical with the data used in Box 7.2 by Aki and Richards [5]. For these data the phase and group velocities of the first mode are shown in the right panel of Figure 10.10.

The group velocity as a function of the phase velocity was calculated by means of the following relation

$$c_g = \frac{\dfrac{d\omega}{dc_{ph}}c_{ph}^2}{\dfrac{d\omega}{dc_{ph}}c_{ph} - \omega}, \tag{10.137}$$

which follows immediately from the definitions. Then we can directly use Relation (10.134) and invert variables $c_{ph} \to \omega$ in a graphical program. In this way, we avoid a numerical differentiation.

10.4.5   Rayleigh waves in a layer of elastic material

Now we investigate a problem similar to the propagation of Love waves, i.e., the propagation in a semiinfinite elastic body with a layer of thickness H and different material properties but we assume the amplitude of the wave to lie in the P-SV-plane rather than in the SH-direction. This is the same assumption as in the case of Rayleigh waves on the plane boundary of a semiinfinite medium.

The material properties are as follows. The mass density of the semiinfinite medium is ρ, the velocities of the longitudinal and transversal waves c_L, c_T, and in the layer they are ρ', c_L', c_T'. We use the potentials for the description of the displacements, i.e.,

$$u_x = \frac{\partial\varphi}{\partial x} + \frac{\partial\psi_y}{\partial z}, \quad u_z = \frac{\partial\varphi}{\partial z} - \frac{\partial\psi_y}{\partial x} \quad \text{for} \quad z \leq 0,$$

$$u_x' = \frac{\partial\varphi'}{\partial x} + \frac{\partial\psi_y'}{\partial z}, \quad u_z' = \frac{\partial\varphi'}{\partial z} - \frac{\partial\psi_y'}{\partial x} \quad \text{for} \quad 0 \leq z \leq H. \tag{10.138}$$

Anticipating the existence of a surface wave in the semiinfinite medium, we make the following ansatz for the solution

$$\varphi = A e^{krz} e^{i(kx-\omega t)}, \quad \psi_y = B e^{ksz} e^{i(kx-\omega t)},$$

$$\varphi' = \left(A_1' \sin\left(kr'z \right) + A_2' \cos\left(kr'z \right) \right) e^{i(kx-\omega t)}, \tag{10.139}$$

$$\psi_y' = \left(B_1' \sin\left(ks'z \right) + B_2' \cos\left(ks'z \right) \right) e^{i(kx-\omega t)}.$$

These potentials fulfill wave equations of the form (10.94) with the appropriate material parameters. This requirement yields

$$r^2 = 1 - \frac{c^2}{c_L^2}, \quad s^2 = 1 - \frac{c^2}{c_T^2}, \quad c := \frac{\omega}{k},$$

$$r'^2 = \frac{c^2}{c_L'^2} - 1, \quad s'^2 = \frac{c^2}{c_T'^2} - 1. \tag{10.140}$$

Solutions of the problem should fulfill the following boundary conditions

1. The stress vector on the free surface $z = H$ must be zero

$$\rho' c_L'^2 \left(\frac{\partial^2\varphi'}{\partial x^2} + \frac{\partial^2\varphi'}{\partial z^2} \right) - \rho' c_T'^2 \left(\frac{\partial^2\varphi'}{\partial x^2} + \frac{\partial^2\psi_y'}{\partial x \partial z} \right) = 0,$$

$$\rho' c_T'^2 \left(2\frac{\partial^2\varphi'}{\partial x \partial z} + \frac{\partial^2\psi_y'}{\partial z^2} - \frac{\partial^2\psi_y'}{\partial x^2} \right) = 0. \tag{10.141}$$

2. The displacement and the stress vector must be continuous on the interface $z = 0$

$$\frac{\partial\varphi}{\partial x} + \frac{\partial\psi_y}{\partial z} = \frac{\partial\varphi'}{\partial x} + \frac{\partial\psi_y'}{\partial z}, \quad \frac{\partial\varphi}{\partial z} - \frac{\partial\psi_y}{\partial x} = \frac{\partial\varphi'}{\partial z} - \frac{\partial\psi_y'}{\partial x}, \tag{10.142}$$

$$\rho c_L^2 \left(\frac{\partial^2\varphi}{\partial x^2} + \frac{\partial^2\varphi}{\partial z^2} \right) - \rho c_T^2 \left(\frac{\partial^2\varphi}{\partial x^2} + \frac{\partial^2\psi_y}{\partial x \partial z} \right)$$

$$= \rho' c_L'^2 \left(\frac{\partial^2\varphi'}{\partial x^2} + \frac{\partial^2\varphi'}{\partial z^2} \right) - \rho' c_T'^2 \left(\frac{\partial^2\varphi'}{\partial x^2} + \frac{\partial^2\psi_y'}{\partial x \partial z} \right), \tag{10.143}$$

$$\rho c_T^2 \left(2\frac{\partial^2\varphi}{\partial x \partial z} + \frac{\partial^2\psi_y}{\partial z^2} - \frac{\partial^2\psi_y}{\partial x^2} \right) = \rho' c_T'^2 \left(2\frac{\partial^2\varphi'}{\partial x \partial z} + \frac{\partial^2\psi_y'}{\partial z^2} - \frac{\partial^2\psi_y'}{\partial x^2} \right). \tag{10.144}$$

3. The solution vanishes for $z \to -\infty$.

These conditions form a homogeneous set of algebraic equations for the constants $iA, B, iA_1', B_1', iA_2'$ and B_2'. The matrix of coefficients has the following form

$$\mathbf{D}_L \tag{10.145}$$

$$
:= \begin{bmatrix}
0 & 1 & s' & 0 & -1 & -s \\
r' & 0 & 0 & -1 & -r & 1 \\
0 & \frac{\rho'}{\rho}\left(\frac{c^2}{c_T^2} - 2\frac{c_T'^2}{c_T^2}\right) & -2\frac{\rho'}{\rho}\frac{c_T'^2}{c_T^2}s' & 0 & 2 - \frac{c^2}{c_T^2} & 2s \\
2\frac{\rho'}{\rho}\frac{c_T'^2}{c_T^2}r' & 0 & 0 & 2 - \frac{c^2}{c_T'^2} & -2r & -\left(2 - \frac{c^2}{c_T^2}\right) \\
\left(2 - \frac{c^2}{c_T'^2}\right)S_{r'} & \left(2 - \frac{c^2}{c_T'^2}\right)C_{r'} & s'C_{s'} & -s'S_{s'} & 0 & 0 \\
2r'C_{r'} & -2r'S_{r'} & \left(2 - \frac{c^2}{c_T'^2}\right)S_{s'} & \left(2 - \frac{c^2}{c_T'^2}\right)C_{s'} & 0 & 0
\end{bmatrix},
$$

where for typographical reasons we have introduced the notation

$$S_{r'} = \sin kr'H, \quad C_{r'} = \cos kr'H, \quad S_{s'} = \sin ks'H, \quad C_{s'} = \cos ks'H. \tag{10.146}$$

Consequently, the determinant of this matrix yields the dispersion relation

$$\mathcal{P}_{RL} := \det \mathbf{D}_L = 0. \tag{10.147}$$

We skip here a complicated analysis. However, two properties of the problem are immediately visible in the above relations. First of all, if there exists a solution of the above dispersion relation then, similarly to Love waves, it is dispersive and there exist infinitely many modes due to the contributions of trigonometric functions to Equation (10.147). Secondly, Relations (10.140) and the boundary condition in infinity yield the following necessary condition for the existence of solutions in the form of surface waves

$$c_L \geq c_T \geq c \geq c_L' \geq c_T'. \tag{10.148}$$

Further, we discuss in some details the problem of existence of solutions only for the much simpler problem of Stoneley waves.

10.4.6   Stoneley waves

10.4.6.1   Interface of two semiinfinite elastic solids

Waves on an interface between two elastic solids were considered for the first time by Robert Stoneley in 1924 [368]. The interface investigated in this work is a classical ideal plane boundary between two linearly elastic isotropic solids and the boundary conditions follow from the jump conditions on the ideal singular surface. This means that all surface effects like surface tension, boundary layers, intermediate layers of a different material, etc. are neglected. In contrast to Love waves, the amplitudes of the wave should decay with the distance from the interface in both directions: for the coordinate z going to $+\infty$ and to $-\infty$ (see Figure 10.11). Formally, one can construct the solution of the problem by taking the infinite limit of depth H of the layer in the solution for Rayleigh waves in a semiinfinite medium. However, following the work of Stoneley, we formulate the problem anew.

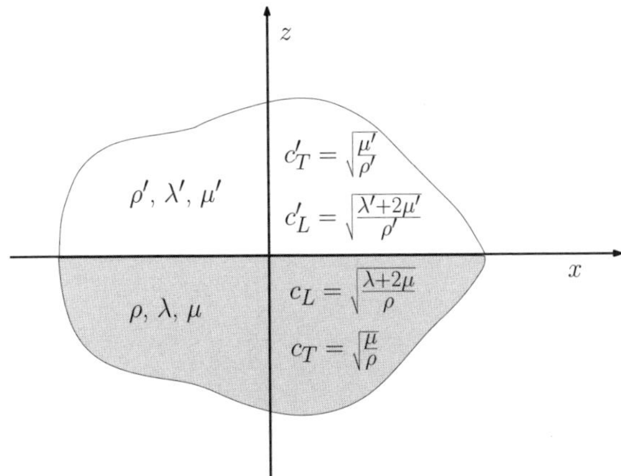

Figure 10.11: *Conditions for the Stoneley wave on the interface between two elastic solids (after [214]).*

We consider two semiinfinite elastic media whose properties are described, as before, by the mass density ρ', the velocities of propagation of the bulk waves: c'_L, c'_T in the upper part $(z > 0)$, and, respectively, ρ, c_L, c_T in the lower part $(z < 0)$. The bulk waves satisfy Equations (10.94) for the potentials with appropriate constants in each part of the system. For technical reasons, it is convenient to choose the potentials $\varphi, \psi \equiv \psi_y, \varphi', \psi' \equiv \psi'_y$ for the components u_x, u_z, u'_x, u'_z of the displacement and the components of displacement u_y, u'_y as unknown. For the two-dimensional problem, the latter would be given only by the difference of the derivatives of the two components of the vector potentials: $\psi_x(x, z), \psi_z(x, z)$ and, correspondingly, by $\psi'_x(x, z), \psi'_z(x, z)$. Certainly, this can be replaced by combined functions u_y, u'_y.

Consequently, we make the following ansatz for the solution

$$\varphi = Ae^{krz}e^{ik(x-ct)} \quad \text{for} \quad z < 0, \quad \varphi' = A'e^{-kr'z}e^{ik(x-ct)} \quad \text{for} \quad z > 0,$$

$$\psi = Be^{ksz}e^{ik(x-ct)} \quad \text{for} \quad z < 0, \quad \psi' = B'e^{-ks'z}e^{ik(x-ct)} \quad \text{for} \quad z > 0, \qquad (10.149)$$

$$u_y = Ce^{kqz}e^{ik(x-ct)} \quad \text{for} \quad z < 0, \quad u'_y = C'e^{-kqs'z}e^{ik(x-ct)} \quad \text{for} \quad z > 0.$$

Hence, the wave is supposed to propagate in the x-direction, and the problem is assumed to be two-dimensional.

Substitution in Equations (10.94) yields the following compatibility conditions for the exponents of the potentials

$$r^2 = 1 - \frac{c^2}{c_L^2}, \quad s^2 = 1 - \frac{c^2}{c_T^2},$$

$$r'^2 = 1 - \frac{c^2}{c_L'^2}, \quad s'^2 = 1 - \frac{c^2}{c_T'^2}. \tag{10.150}$$

As we see in a moment, the relations for q and q' are immaterial.

The boundary conditions follow from the continuity of the displacements on the interface

$$[\![\mathbf{u}]\!] \equiv \mathbf{u}' - \mathbf{u} = 0 \quad \text{for} \quad z = 0, \tag{10.151}$$

and from the jump condition for the stress vector on the plane boundary $z = 0$ as well as from the Sommerfeld conditions for $|z| \to \infty$

$$[\![\mathbf{T}]\!]\,\mathbf{n} \equiv (\mathbf{T}' - \mathbf{T})\,\mathbf{n} = 0, \quad \text{for} \quad z = 0, \tag{10.152}$$

$$\lim_{z \to \infty} \mathbf{u}' = 0, \quad \lim_{z \to -\infty} \mathbf{u} = 0,$$

where $\mathbf{n} = \mathbf{e}_z$ is the unit vector normal to the interface.

Hence, for real k, Stoneley waves may exist solely if r, s, q, r', s', q' are nonnegative. The boundary condition for the shear stresses in y-direction yields immediately

$$C = C' = 0, \tag{10.153}$$

i.e., a Stoneley wave with the amplitude in SH-direction (horizontal polarization) does not exist. The amplitude must lie in the vertical P-SV incident plane.

Bearing Relations (10.93) in mind, we obtain for $z = 0$

$$(iA' - s'B') - (iA + sB) = 0,$$

$$(-r'iA' + B') - (riA + B) = 0, \tag{10.154}$$

$$[-\rho'c_L'^2 (1 - r'^2) + 2\rho'c_T'^2]\,(iA') - 2\rho'c_T'^2 s'B'$$

$$= [-\rho c_L^2 (1 - r^2) + 2\rho c_T^2]\,(iA') + 2\rho c_T^2 sB, \tag{10.155}$$

$$\rho'c_T'^2\,[-2iA'r' + B'(1 + s'^2)] = \rho c_T^2\,[2iAr + B(1 + s^2)].$$

As usual, this set of homogeneous algebraic equations yields the dispersion relation. It can be written in the following compact form

$$\mathbf{D}_{St}\mathbf{X} = 0, \quad \mathbf{X} := [iA', B', iA, B]^T, \tag{10.156}$$

where the matrix $\mathbf{D}_{St}$ is defined as follows

$$\mathbf{D}_{St} := \begin{bmatrix} -1 & \sqrt{1 - \dfrac{c^2}{c_T'^2}} & 1 & \sqrt{1 - \dfrac{c^2}{c_T^2}} \\[2em] \sqrt{1 - \dfrac{c^2}{c_L'^2}} & -1 & \sqrt{1 - \dfrac{c^2}{c_L^2}} & 1 \\[2em] -\dfrac{\rho'\,c_T'^2}{\rho\,c_T^2}\left(2 - \dfrac{c^2}{c_T^2}\right) & 2\dfrac{\rho'\,c_T'^2}{\rho\,c_T^2}\sqrt{1 - \dfrac{c^2}{c_T'^2}} & 2 - \dfrac{c^2}{c_T^2} & 2\sqrt{1 - \dfrac{c^2}{c_T^2}} \\[2em] 2\dfrac{\rho'\,c_T'^2}{\rho\,c_T^2}\sqrt{1 - \dfrac{c^2}{c_L'^2}} & -\dfrac{\rho'\,c_T'^2}{\rho\,c_T^2}\left[2 - \dfrac{c^2}{c_T'^2}\right] & 2\sqrt{1 - \dfrac{c^2}{c_L^2}} & 2 - \dfrac{c^2}{c_L^2} \end{bmatrix}. \tag{10.157}$$

The determinant of this matrix gives rise to the dispersion relation for Stoneley waves

$$\mathcal{P}_{St} := \det \mathbf{D}_{St} = 0. \tag{10.158}$$

One important property of Stoneley waves follows immediately from the inspection of the matrix $\mathbf{D}_{St}$. Namely, if there exists a real solution $\frac{c}{c_T}$ of Equation (10.158) then it is independent of the frequency. Hence, like Rayleigh waves, Stoneley waves are nondispersive.

The question of the existence of Stoneley waves is not trivial. Already in the original work [368] Stoneley has shown that for certain combinations of material parameters these waves may not exist (see also: [1], [61]).

An extensive study of the existence of this wave was also carried out by Ginzbarg and Strick whose paper [136] contains many graphs of ranges of existence.

10.4.6.2 Interface of a semiinfinite elastic solid and a semiinfinite ideal fluid

The interface between a linear elastic solid and an ideal fluid yields the problem of surface waves which can be obtained from the above problem of Stoneley waves as a particular case. The equations for this problem were formulated by Scholte in 1947 [343] and, for this reason, these waves are sometimes called Scholte waves.

We derive the dispersion relation again from the governing equations indicating some points which are common with the modeling of surface waves for porous materials with permeable boundaries.

We choose the same coordinates as before and assume that the fluid in the range $z > 0$ is ideal, i.e., we have the following linear equations for the fields ρ' and $\mathbf{v}'$ in this region

$$\frac{\partial \rho'}{\partial t} + \rho_0' \operatorname{div} \mathbf{v}' = 0, \quad \rho_0' \frac{\partial \mathbf{v}'}{\partial t} = -\operatorname{grad} p', \quad p' = p_0' + \kappa' \left(\rho' - \rho_0' \right), \tag{10.159}$$

where κ' is the compressibility coefficient. Consequently,

$$\frac{\partial^2 \mathbf{v}'}{\partial t^2} = c_L'^2 \operatorname{grad} \operatorname{div} \mathbf{v}', \quad c_L'^2 = \kappa'. \tag{10.160}$$

It is customary to introduce for technical reasons the displacement vector $\mathbf{u}'$ also for the fluid component

$$\mathbf{v}' = \frac{\partial \mathbf{u}'}{\partial t} \quad \Rightarrow \quad \frac{\partial^2 \mathbf{u}'}{\partial t^2} = c_L'^2 \operatorname{grad} \operatorname{div} \mathbf{u}' + \mathbf{C}\left(\mathbf{x}\right), \tag{10.161}$$

where $\mathbf{C}\left(\mathbf{x}\right)$ is an arbitrary function of the spacial variable. We can redefine the displacement in the following way

$$\mathbf{u}' \to \mathbf{u}' + \mathbf{U}\left(\mathbf{x}\right), \quad \operatorname{grad} \operatorname{div} \mathbf{U} = -\mathbf{C}, \tag{10.162}$$

and eliminate this arbitrary function entirely because we are interested only in the time derivative of $\mathbf{u}'$.

It is easy to see that only the potential part of $\mathbf{u}'$ is of interest. Hence,

$$\mathbf{u}' = \operatorname{grad} \varphi' \quad \Rightarrow \quad \frac{\partial^2 \varphi'}{\partial t^2} = c_L'^2 \nabla^2 \varphi', \tag{10.163}$$

where an arbitrary function of time was incorporated in the potential.

Hence, the solution for monochromatic waves in this case can be written in the form (10.149), (10.150) in which $B' = 0$.

We have to modify also the boundary conditions. Instead of the continuity of motion on the interface between two solids we have now only the condition of continuity of the normal velocity

$$\mathbf{v}' \cdot \mathbf{n} = \frac{\partial \mathbf{u}'}{\partial t} \cdot \mathbf{n} \quad \text{for} \quad z = 0. \tag{10.164}$$

For an ideal fluid the tangential component of the velocity can be arbitrary and, consequently, we have a scalar condition instead of the vector condition (10.151).

The condition for the stresses must be modified as well because now we have the shear stress in the fluid identically zero and the normal component of stresses is reduced to the pressure. The latter has to be written in terms of the displacement potential φ'. Namely, it follows by integration of the mass balance $(10.159)_1$

$$\rho' = \rho'_0 \left(1 - \operatorname{div} \mathbf{u}'\right) = \rho'_0 \left(1 - \nabla^2 \varphi'\right) \quad \Rightarrow \quad p' = p'_0 - \rho'_0 c_L'^2 \nabla^2 \varphi'. \tag{10.165}$$

The boundary conditions for the stresses at the boundary $z = 0$ have now the form

$$\rho'_0 c_L'^2 \nabla^2 \varphi' = \rho c_L^2 \left(\frac{\partial^2 \varphi}{\partial x^2} + \frac{\partial^2 \varphi}{\partial z^2}\right) - 2\rho c_T^2 \left(\frac{\partial^2 \varphi}{\partial x^2} + \frac{\partial^2 \psi}{\partial x \partial z}\right),$$

$$\tag{10.166}$$

$$2\frac{\partial^2 \varphi}{\partial x \partial z} + \frac{\partial^2 \psi}{\partial z^2} - \frac{\partial^2 \psi}{\partial x^2} = 0.$$

Substitution of Relations (10.149) yields again a homogeneous set of equations

$$\mathbf{D}_{Sch}\mathbf{X} = 0, \quad \mathbf{X} := [iA', iA, B]^T, \tag{10.167}$$

where

$$\mathbf{D}_{Sch} := \begin{bmatrix} \sqrt{1 - \dfrac{c^2}{c_L'^2}} & \sqrt{1 - \dfrac{c^2}{c_L^2}} & 1 \\[2ex] \dfrac{\rho'_0}{\rho}\dfrac{c^2}{c_T^2} & 2 - \dfrac{c^2}{c_T^2} & 2\sqrt{1 - \dfrac{c^2}{c_T^2}} \\[2ex] 0 & 2\sqrt{1 - \dfrac{c^2}{c_L^2}} & 2 - \dfrac{c^2}{c_T^2} \end{bmatrix}. \tag{10.168}$$

The determinant of this matrix yields immediately the following Stoneley (Scholte) dispersion relation

Modicum of the Stoneley-Scholte wave

$$\sqrt{1 - \frac{c^2}{c_L'^2}}\,\mathcal{P}_R + \frac{\rho'_0}{\rho}\frac{c^4}{c_T^4}\sqrt{1 - \frac{c^2}{c_L^2}} = 0, \quad \text{Stoneley (Scholte) dispersion relation}$$

$$\mathcal{P}_R := \left(2 - \frac{c^2}{c_T^2}\right)^2 - 4\sqrt{1 - \frac{c^2}{c_L^2}}\sqrt{1 - \frac{c^2}{c_T^2}}.$$

The last relation is identical with the dispersion function (10.101) defining Rayleigh waves.

It is obvious that the velocity c is independent of the frequency ω and, consequently, also this surface wave is nondispersive.

The problem of existence of Stoneley-Scholte waves has been rather intensively investigated (e.g. [369], [370]). It can be shown that a single real solution exists if $c'_L < c_R$, where c_R is the Rayleigh velocity and it is smaller than c'_L, i.e., it is the smallest velocity of propagation appearing in the system.

One of the important properties of the Stoneley-Scholte wave is that the energy is carried by this wave primarily in the fluid. This can be proven by the analysis of the amplitudes which we do not present here.

The propagation of surface waves in the system described above but either with a weak anisotropy of the solid or the fluid under a hydrostatic pressure has been investigated by Norris and Sinha [277]. They have used a perturbation method and found explicit corrections to the velocity of Scholte waves.

10.4.6.3 Interface of a semiinfinite elastic solid and a layer of an ideal fluid

In this case, one can consider waves intermediate between Stoneley and Rayleigh varying with the height H of the layer. We quote here only a finite result which can be derived in the same way as the above presented examples of surface waves. We obtain [410]

$$\sqrt{\frac{c^2}{c'^2_L} - 1}\,\mathcal{P}_R + \frac{\rho'}{\rho}\frac{c^4}{c^4_T}\sqrt{1 - \frac{c^2}{c^2_L}}\,\tan\left(\frac{\omega H}{c}\sqrt{\frac{c^2}{c'^2_L} - 1}\right) = 0, \tag{10.169}$$

where $\mathcal{P}_R$ is again the Rayleigh dispersion function.

Clearly, these waves are dispersive, there exist infinitely many surface modes whose velocity is bigger than the velocity in the fluid.

10.5 A few remarks on leaky waves

It has been seen already in the Rayleigh dispersion relation that for some values of material parameters some solutions $k(\omega)$ may be complex. This means that the factor $e^{i(kx-\omega t)} = e^{-\operatorname{Im}kx}e^{i(\operatorname{Re}kx-\omega t)}$ of the ansatz for the solution yields a decay of the amplitude characteristic for the dissipation. However, the whole system is reversible – there are no losses of energy due to the heat transfer or some other means of dissipation. Consequently, such solutions can be admissible solely under the condition that the energy of this mode of propagation must be transferred on some other modes – bulk or surface waves. If this is indeed the case we say that the wave losing the energy is leaky.

These complex solutions of the Rayleigh dispersion relation have been discussed by C. T. Schroeder and W. R. Scott [345]. They have shown that these solutions may belong to five classes and one of these classes contains the classical real solution of the problem. Other classes contain combinations of surface-like contributions and bulk-like (shear) waves which do not decay to zero in infinity. Consequently, they can exist only in finite parts of the medium and they must transfer energy to other modes of propagation. Such (pseudosurface) waves have been observed experimentally. They appear in the vicinity of wave sources and their existence explains some critical phenomena (e.g. the critical angle of incidence) on the surface.

We present briefly an example of such a wave considered by Viktorov [410]. It corresponds to one of the complex solutions of Equation (10.169). Viktorov found analytically the complex solution of this equation under the assumption of small Poisson's ratio ν. He was able to show that this solution yields attenuation of the wave $\operatorname{Im} k \sim \nu^4$ and the velocity is of the order of the longitudinal wave c_L. The wave consists of longitudinal and transversal parts. The amplitude of the longitudinal part decays exponentially with the depth (i.e., it is indeed a surface wave). The transversal part behaves like a bulk wave propagating from the boundary under the angle $\beta = \frac{\pi}{4}$. The energy of this wave is transferred on the transversal bulk wave.

The existence of leaky waves has an important practical bearing. For instance, A. N. Norris [275] has presented a very neat model explaining the phenomenon of backscattering of ultrasound from a fluid-solid interface. It is shown that the energy radiated back into the fluid, where the incident beam originates, can be observed when the angle of incidence lies near the so-called leaky wave angle. Hence, it is proven that the backscattering is due to a leaky wave reflection zone.

The existence of leaky modes depends on the distribution of complex roots of the dispersion relation. R. A. Phinney [293] has designed a fairly general method (a generalization of the so-called Rosenbaum method) to find these roots by the analysis of a single integral expression. The method refers to properties of such an integral when the path of integration changes the Riemann surface.

10.6   Bulk and surface waves in viscoelastic solids

Propagation of acoustic waves in viscoelastic solids is an intensive research subject since some 20 years. The main motivation for these investigations is the modeling of soils and rocks by means of viscoelastic materials. Clearly, such materials are porous and, therefore their macroscopic behavior is influenced by the motion of the fluid components in the channels. These diffusion processes yield naturally a viscous character of the material modeled by a single component continuum. Applications of such viscoelastic models in geomechanics are obvious. Further in this book we present some properties of porous materials when they are modeled by multicomponent continua. In this section we present briefly the most characteristic features of acoustics for the single component model. We follow here the presentation of Carlo Lai [214]. We limit the attention to isotropic material satisfying the elastic-viscoelastic correspondence principle (Section 5.5.2). Then assuming the harmonic form of the time dependence

$$\mathbf{u}_L = \widetilde{\mathbf{u}}_L e^{i\omega t}, \quad \mathbf{u}_T = \widetilde{\mathbf{u}}_T e^{i\omega t}, \tag{10.170}$$

we can write the following equations for viscoelastic solids, corresponding to Equations (10.92)

$$-\omega^2 \widetilde{\mathbf{u}}_L = V_L^2 \Delta^2 \widetilde{\mathbf{u}}_L, \quad -\omega^2 \widetilde{\mathbf{u}}_T = V_T^2 \Delta^2 \widetilde{\mathbf{u}}_L, \tag{10.171}$$

where V_L and V_T are complex longitudinal and transversal velocities which replace the constant real velocities c_L and c_T of linear elasticity. They are given by the relations (compare (5.242))

$$V_L = \sqrt{\frac{\bar{K}(\omega) + \dfrac{4}{3}\bar{\mu}(\omega)}{\rho}}, \quad V_T = \sqrt{\frac{\bar{\mu}(\omega)}{\rho}}. \tag{10.172}$$

Similarly to linear elasticity, Equations (10.171) show that distortional and volume deformations in linear, viscoelastic materials are uncoupled. A general solution of Equations

(10.171) can be written in the form

$$\mathbf{u}_{L,T}\left(\mathbf{x},t\right) = \mathbf{U}_{L,T}^{1}\exp\left[i\left(\mathbf{k}_{L,T}\cdot\mathbf{x}-\omega t\right)\right] + \mathbf{U}_{L,T}^{2}\exp\left[i\left(\mathbf{k}_{L,T}\cdot\mathbf{x}-\omega t\right)\right],\qquad(10.173)$$

where the wave vectors $\mathbf{k}_{L,T} = \mathrm{Re}\,\mathbf{k}_{L,T} - i\alpha_{L,T}$, $\alpha_{L,T} \equiv -\mathrm{Im}\,\mathbf{k}_{L,T}$ are complex. The real part $\mathrm{Re}\,\mathbf{k}_{L,T}$ determines the direction of propagation and the imaginary part $\alpha_{L,T}$ the direction of attenuation. These two vectors do not need to be parallel [5]. However, if they are, we say that the wave is simple or homogeneous. Non-simple waves may arise as a result of boundary effects (reflection, refraction of monochromatic waves, etc.).

As an illustration let us consider the simple example of one-dimensional wave propagation. The complex wave vector can be written in the form

$$k_{L,T} = \frac{\omega}{V_{L,T}} = \frac{\omega}{c_{L,T}} - i\alpha_{L,T},\qquad(10.174)$$

where $c_{L,T}$ and $\alpha_{L,T}$ are the real-valued physical velocities and attenuation coefficients. The attenuation measures the spatial decay of the amplitude. We can write these quantities in the form

$$c_L\left(\omega\right) = \left[\mathrm{Re}\sqrt{\frac{\rho}{\bar{K}\left(\omega\right)+\dfrac{4}{3}\bar{\mu}\left(\omega\right)}}\,\right]^{-1},\quad \alpha_L = \omega\,\mathrm{Im}\sqrt{\frac{\rho}{\bar{K}\left(\omega\right)+\dfrac{4}{3}\bar{\mu}\left(\omega\right)}},$$
$$\qquad(10.175)$$

$$c_T\left(\omega\right) = \left[\mathrm{Re}\sqrt{\frac{\rho}{\bar{\mu}\left(\omega\right)}}\,\right]^{-1},\quad \alpha_L = \omega\,\mathrm{Im}\sqrt{\frac{\rho}{\bar{\mu}\left(\omega\right)}}.$$

Two material functions are required to specify the constitutive parameters of a linear, isotropic viscoelastic material but they cannot be prescribed arbitrarily. As we have already indicated in Section 5.5.2 (see (5.225)) they are linked by the Kramers-Krönig relation. This relation is justified by the dissipation in viscoelastic processes. It is convenient to make this dependence more explicit. It is customary to measure the dissipation of monochromatic waves by means of the following fraction

$$D_{L,T}\left(\omega\right) = \frac{\Delta W_{L,T}^{dissip}\left(\omega\right)}{8\pi W_{L,T}^{ave}\left(\omega\right)},\qquad(10.176)$$

where $D_{L,T}\left(\omega\right)$ is a dimensionless parameter called the material damping ratio, $\Delta W_{L,T}^{dissip}$ is the energy dissipated (per unit volume) during a cycle of harmonic oscillation and $W_{L,T}^{ave}$ is the average stored energy per one cycle of harmonic oscillation. Then, the phase speed and attenuation (10.175) can be written in the form

$$c_{L,T} = c_{L,T}^{el}\sqrt{\frac{2\left(1+4D_{L,T}^{2}\left(\omega\right)\right)}{1+\sqrt{1+4D_{L,T}^{2}\left(\omega\right)}}},\quad \alpha_{L,T} = \frac{\omega}{c_{L,T}}\frac{\sqrt{1+4D_{L,T}^{2}\left(\omega\right)}-1}{2D_{L,T}\left(\omega\right)}.\qquad(10.177)$$

In many works the material damping ratio is replaced by the so-called quality factor

$$Q_{L,T}\left(\omega\right) = \frac{1}{2D_{L,T}\left(\omega\right)}.\qquad(10.178)$$

We complete this section with a brief remark on surface waves. Even though Rayleigh waves do exist on free surfaces of viscoelastic materials their properties are not described anymore by any simple dispersion relation. The numerical analysis of equations describing such waves has a particular bearing in practical applications in geotechnics and geophysics and details can be found in the above quoted work of Carlo Lai [214].

Much beyond the scope of this book lies the full analysis of surface waves on interfaces of viscoelastic materials. This is important in the description of stratified materials. One should quote here numerous works on this subject published by G. Caviglia and A. Morro (see the work [69] for further references).

Chapter 11

Interactions of ponderable bodies with electromagnetic fields

11.1 Preliminaries

The main purpose of this chapter is to demonstrate a simple example of modeling a thermomechanical system interacting with an electromagnetic field. The models of matter which we have considered up to now in this book play an important role in engineering applications. However, modern technologies yield discoveries of effects and devices whose description requires more sophisticated models than these considering thermomechanical properties alone. Among them interactions of the electromagnetic field with ponderable bodies belong to the practically most important processes in the macrophysics. Simultaneously, many questions of the construction of macroscopic models can be answered only with the help of modern continuum thermodynamics. Such is the problem of the choice of an appropriate definition of the stress tensor in thermomechanical bodies interacting with electromagnetic fields, the structure of the vector of energy transfer, the general form of constitutive relations in which electromagnetic fields contribute to thermomechanical constitutive quantities and many others. We have indicated the importance of these problems in Remark 3 of Chapter 14 of Part I. In the present chapter we extend this subject and discuss some issues of plasmas.

Electromagnetic interactions with ponderable bodies belong to three large classes whose modeling differs essentially:

Firstly, *electromagnetic interactions with solids*. These can be roughly divided into the following groups (see the article of R. A. Grot [146] and the monographs of A. C. Eringen and G. A. Maugin [117, 118]):

1. Dielectrics (i.e., electrical insulators which can be polarized by an applied electric field),

2. Magnetic materials (ferromagnets or permanent magnets, ferrimagnets, paramagnets, diamagnets according to the possible degree of alignment of magnetic ions at a given temperature),

3. Conductors,

4. Semiconductors.

This is not an exhaustive classification as materials change their reaction on the electromagnetic field in dependence on the range of magnitude of such fields as stresses or temperature. Correspondingly, their modeling changes as well. It happens quite often that some effects cannot be modeled without accounting for relativistic or quantum effects (e.g. spin-spin interactions in micromagnetic materials; compare: [243]). Due to the complexity of those models it is frequently the case that some additional simplifications are made. For instance, it is often sufficient to consider static theories in which all phenomena are independent of time, quasi-static theories in which the thermomechanical part is time dependent but the electromagnetic fields are described by time-independent Maxwell equations or dynamical theories in which the formulation of the thermomechanical part is Galilean invariant and the electromagnetic part is described in the nonrelativistic limit of low velocities.

Secondly, *electromagnetic interactions with fluids.* They form either the class of electrolytes (i.e., fluid mixtures of charged and neutral particles; e.g. compare the book of H. Falkenhagen [123]; in contrast to plasmas in which the negative charge is carried by electrons electrolytes consist of positive and negative ions; they may easily exist at low temperatures in contrast to plasmas which require high temperatures for ionization) or the so-called electrorheological fluids (see the book of K. Hutter, A. A. F. van de Ven and A. Ursescu [174]). The latter are mixtures of an electrically neutral fluid with suspended particles which can be polarized by the electric current. When the current is switched off the fluid possesses a small viscosity. Under the influence of the electric current the polarizable particles arrange in columns perpendicular to the flow yielding a very substantial increase in viscosity. This spectacular property is used, for instance, in actuators based on such a "smart" material. However, even for nonviscous heat conducting, dielectric and magnetizable fluids electromagnetic interactions yield important phenomena such as a magnetocaloric effect (i.e., cooling due to an adiabatic demagnetization), or magnetostriction and piezoelectricity due to a density dependence on the magnetization and polarization (see Chapter 9 in the book of I. Müller [261]).

Thirdly, *processes in cold and hot plasmas.* These gaseous mixtures of charged and neutral particles usually require a description by various kinetic theories (e.g. the books of Yu. L. Klimontovich or N. A. Krall and A. W. Travelpiece [200, 207]). However, some practically important properties, in particular a stability analysis of various flows, can also be investigated by a continuum model.

In all models of the above listed processes various nonlinearities appear in the natural way. For instance, such material properties as energy conduction or viscosity are influenced by the intensity of electric and magnetic fields. For example, for isotropic semiconductors the heat conductivity is the sum of two contributions: the electronic heat conductivity K_e and the thermal conductivity K_T, where the first contribution depends on the electric conductivity, the degeneracy of the gas of electrons etc. (e.g. see Wiedemann-Franz relation [238]). This plays an important role in the conversion of temperature differences into the electric voltage (thermoelectric generators using the thermoelectric or Seebeck effect), or in measurements of the temperature by thermocouples.

On the other hand, the theory of miscible mixtures as presented in Chapter 11 of Part I gives rise to a possibility of generalization to model mixtures of charged particles which, in turn, yields the magnetohydrodynamics (MHD) of plasmas. This important model allows to investigate, for instance, various instabilities of hot plasma [31, 63] which arrest the development of technologies of controlled thermofusion.

Many piezoelectric devices applied in non-destructive testing, transducers in ink-jet printers and many more render as well systems in which the interactions of a ponderable body and an electromagnetic field are of primary importance.

In this chapter we present a few facets of the coupled field theory (e.g. [246]) which is based on the classical thermomechanics extended on electromagnetic interactions. The latter are described by a non-relativistic version of Maxwell's equations. In particular, we demonstrate how the mixture theory of miscible mixtures presented in Chapter 11 of Part I can be extended to yield the MHD-theory of plasmas.

11.2 Primer of Maxwell theory of electromagnetism

11.2.1 Governing equations

In this section we present a sketch of the theory of electromagnetism developed by James Clerk Maxwell. Details can be found in many classical books on this subject (e.g. [183, 290]). The fundamental idea of this theory is based on the assumption that interactions of charged particles are carried by a field induced by these particles rather than directly by Coulomb forces as assumed in the earlier model, created by Coulomb in 1785. From this idea follows that charged particles are the source of electromagnetic fields. We consider the description of an electromagnetic field which acts on a continuous medium in its current configuration. We use the Eulerian description and the motion of this continuum is described by the velocity field $\mathbf{v}(\mathbf{x}, t)$.

One of the fundamental questions was the description of charged particles in the form suitable for a continuum. This problem was solved by Paul Dirac in 1930 in his book on quantum mechanics. Dirac has introduced the so-called delta-function, a distribution which allows to write concentrated charges in a form appropriate for the field description In general the distribution of charges in the space is described by a differential form $\mathfrak{R} = (\varrho_{123} dx^1 \wedge dx^2 \wedge dx^3)$,[1] the so-called spacial density of charges specifying the total charge Q in the volume $\mathcal{P}_t$ by the integral

$$Q = \int_{\mathcal{P}_t} \mathfrak{R}, \tag{11.1}$$

which, to possess proper transformation properties must be the 0-current. This means that the density of charges can be replaced by the scalar quantity $q = \varrho_{123}/\sqrt{g}$, where g is the determinant of the metric tensor of the chosen coordinates

$$Q = \int_{\mathcal{P}_t} q\, dv. \tag{11.2}$$

In the case of a charge concentrated in a single point, say $\mathbf{x} = 0$, it has the form

$$q = Q_0 \delta(\mathbf{x}), \tag{11.3}$$

where Q_0 is a constant and $\delta(\mathbf{x})$ is the above mentioned Dirac delta-function. It has the property to be zero everywhere except of $\mathbf{x} = 0$ and $\int \delta(\mathbf{x})\, dv = 1$ where the domain of

---

[1] For the mathematical definition, properties and physical applications of such forms see, for instance, the book of H. Flanders [126].

integration is arbitrary but it must contain the point $\mathbf{x} = 0$. Generalizations on charges distributed on lines or surfaces are immediate.

We use here SI units.[2] Hence, the density of charges has the unit 1 Coulomb per cubic meter 1 $[\mathrm{C/m^3}] = 1\ [\mathrm{A{\cdot}s/m^3}]$ where one Ampere (1 [A]) is defined as 6.28×10^{18} electrons passing by the point of the measurement in one second.

The electromagnetic field can be described by a different choice of objects (see the comparison of different choices in Section 4.1 of the book [174]). In the modern formulation of electrodynamics (e.g. the book of R. S. Ingarden and A. Jamiołkowski [182]) it is described by the following four differential forms

1. the electromotive intensity (the effective electric field) $\mathfrak{E}=\mathcal{E}_i dx^i$ [3] which describes the electric force acting on a unit charge,

2. the electric induction $\mathfrak{D}=\frac{1}{2}\mathcal{D}_{ij}dx^i \wedge dx^j$ (introduced by M. Faraday and J. Henry, based on the discovery of H. C. Oersted in 1820) which describes the number of lines of the electric field crossing the surface spanned by $dx^i \wedge dx^j$ (it is the surface perpendicular to the vector $\epsilon_{kij}dx^i dx^j$, i.e., to the vector product of the vectors $dx^i \mathbf{e}_i$ and $dx^j \mathbf{e}_j$). It is more often described by the vector of electric induction ($\mathbf{D}$, a dual form to $\mathfrak{D}$) defined by the relation

$$D_i = \frac{1}{2}\sqrt{g}\epsilon_{ijk}D^{jk}, \quad D_i = \mathbf{D} \cdot \mathbf{e}_i, \tag{11.4}$$

where ϵ_{ijk} is the permutation symbol and $\sqrt{g}$ denotes the square root of the determinant of the metric tensor,

3. the effective magnetic field or the magnetomotive intensity $\mathfrak{H} = \mathcal{H}_i dx^i$,

4. the magnetic flux density or the magnetic induction $\mathfrak{B}=\frac{1}{2}\mathcal{B}_{ij}dx^i \wedge dx^j$. Again, it may be described by the dual form to $\mathfrak{B}$, denoted by $\mathbf{B}$, which is called the vector of magnetic induction

$$B_i = \frac{1}{2}\sqrt{g}\epsilon_{ijk}B^{jk}, \quad B_i = \mathbf{B} \cdot \mathbf{e}_i. \tag{11.5}$$

The integrals of the above forms (see Section 5.7 in [126]) possess direct physical interpretations which we proceed to present.

The integral along the curve $\mathcal{C}_t$

$$U = \int_{\mathcal{C}_t} \mathfrak{E} = \int_{\mathcal{C}_t} \mathcal{E}_i dx^i = \int_{\mathcal{C}_t} \mathcal{E} \cdot \mathbf{h} ds, \quad \mathbf{h} = h^i \mathbf{e}_i = \frac{\partial x^i}{\partial s}\mathbf{e}_i, \quad \mathcal{E}_i = \mathcal{E} \cdot \mathbf{e}_i, \tag{11.6}$$

describes the work done by displacing the unit charge along this curve in the electric field of strength $\mathfrak{E}$. If the curve is closed, the integral

---

[2] This differs considerably from the so-called Gaussian system in which the charge unit, the statcoulomb, can be written in terms of mechanical units: 1 [statC] $= 1\ [\mathrm{g^{0.5}{\cdot}\ cm^{1.5}{\cdot}\ s^{-1}}]$. The speed of light c appears directly in Maxwell's equations in Gaussian units while in the SI units it enters the Maxwell equations only through the product of the electric permittivity and the magnetic permeability $\mu_0\varepsilon_0 = c^{-2}$ which we discuss further in this chapter.

[3] For the purpose of this section we use arbitrary frames of reference; Cartesian frames would require unnecessarily complicated forms for surface and line elements.

$$\oint_{\mathcal{C}_t} \mathfrak{E} = \oint_{\mathcal{C}_t} \mathcal{E} \cdot \mathbf{h} dl_t, \tag{11.7}$$

is the electromotive force acting on the linear conductor along the curve $\mathcal{C}_t$. Its unit is $1\,[\mathrm{V}] = 1\,[\mathrm{m}^2\cdot\mathrm{kg}/\mathrm{A}\cdot\mathrm{s}^2]$ and the unit of the strength of the effective electric field follows to be $[\mathcal{E}] = 1\,[\mathrm{V/m}]$.

For an arbitrary surface $\mathcal{S}_t$ the integral of the electric induction $\boldsymbol{\Psi} = \int_{\mathcal{S}_t} \mathfrak{D}$ specifies a number of lines of the field $\mathfrak{D}$ crossing this surface. By means of the vector of electric induction the vector $\boldsymbol{\Psi}$ can be written in the form

$$\boldsymbol{\Psi} = \int_{S_t} \mathbf{D} \cdot \mathbf{n} dS_t, \tag{11.8}$$

and its unit is 1 Coulomb $[\mathrm{C}]$ while the unit of the electric induction is $1\,[\mathrm{C/m}^2] = 1\,[\mathrm{A}\cdot\mathrm{s/m}^2]$.

The above quantities characterize the electric field.

The magnetic field strength, or simply the magnetic field, $\mathfrak{H}$ defines the so-called magnetomotoric force which characterizes the magnetic force acting on a closed curve (the loop) $\mathcal{C}_t$

$$\oint_{\mathcal{C}_t} \mathfrak{H} = \oint_{\mathcal{C}_t} \mathcal{H} \cdot \mathbf{h} dl_t, \quad \mathcal{H}_i = \mathcal{H} \cdot \mathbf{e}_i, \tag{11.9}$$

whose unit is 1 Ampere $[\mathrm{A}]$.

On the other hand, the integral over an arbitrary surface $\mathcal{S}_t$: $\Phi = \int_{\mathcal{S}_t} \mathfrak{B}$ defines the current of the field of magnetic induction $\mathfrak{B}$ crossing this surface. Making use of the vector of magnetic induction (11.5) we have for this current

$$\Phi = \int_{S_t} \mathbf{B} \cdot \mathbf{n} dS_t, \tag{11.10}$$

and its unit is 1 Weber $[\mathrm{Wb}]$.[4] We should recall that the presented further balance laws for electromagnetic fields acting on ponderable bodies involve velocities of material particles of those bodies. Therefore it is sometimes convenient to replace the above effective electric and magnetic fields by the so-called Minkowskian electric and magnetic fields (see: the work of H. Minkowski from the year 1908 [253])

$$\begin{aligned} E_i &= \mathcal{E}_i - \epsilon_{ijk} v^j B^k, \quad \text{i.e.} \quad \mathbf{E} = \mathcal{E} - \mathbf{v} \times \mathbf{B}, \\ H_i &= \mathcal{H}_i + \epsilon_{ijk} v^j D^k \quad \text{i.e.} \quad \mathbf{H} = \mathcal{H} + \mathbf{v} \times \mathbf{D}, \quad \mathbf{v} = v^i \mathbf{e}_i, \end{aligned} \tag{11.11}$$

where $\mathbf{v}$ is the vector of velocity.

We proceed to present the set of equations for the determination of the above four fields. It should be mentioned that the electromagnetic field in the vacuum requires only

---

[4] "The *Weber*, 1 $[\mathrm{Wb}] = 1\,[(\mathrm{kg}\cdot\mathrm{m}^2)/(\mathrm{s}^2\cdot\mathrm{A})]$, is the magnetic flux which, linking a circuit of one turn, would produce in it an electromotive force of 1 Volt if it were reduced to zero at a uniform rate in 1 second". For comparison, the magnetic flux density of the earth is approximately $5 \cdot 10^{-5}$ T, $1\mathrm{T} = 1\mathrm{Wb/m}^2$. "A particle carrying a charge of 1 Coulomb and passing through a magnetic field of 1 Tesla at a speed of 1 meter per second perpendicular to said field experiences a force of 1 Newton" (CIPM, 1946: Resolution 2 / Definitions of Electrical Units. International Committee for Weights and Measures (CIPM) Resolutions. International Bureau of Weights and Measures (BIPM). 1946).

Table 11.1: *Fields of the Maxwell model.*

electromotive intensity = effective electric field	$\mathfrak{E} = \mathcal{E}_i dx^i$	$[\mathcal{E}] = 1\ [\text{V/m}]$
electric induction = dielectric displacement	$\mathfrak{D} = \frac{1}{2} D_{ij} dx^i \wedge dx^j,$ $D_i = \frac{1}{2}\sqrt{g}\epsilon_{ijk} D^{jk}$	$[\mathbf{D}] = 1\ [\text{C/m}^2] = 1\ [\text{A·s/m}^2]$
magnetomotive intensity = effective magnetic field	$\mathfrak{H} = \mathcal{H}_i dx^i$	$[\mathcal{H}] = 1\ [\text{A/m}]$
magnetic induction = magnetic flux density	$\mathfrak{B} = \frac{1}{2} B_{ij} dx^i \wedge dx^j$ $B_i = \frac{1}{2}\sqrt{g}\epsilon_{ijk} B^{jk}$	$[\mathbf{B}] = 1\ [\text{T}]$ $= 1\ [\text{Wb}]/[\text{m}^2]$

two fields for its description. Four fields are needed when the electromagnetic field is interacting with the ponderable body. We return to this point further in this chapter.

The first equation describes the conservation law for the magnetic induction and it is motivated by a simple experiment. Namely, the magnetic flux density $\mathbf{B} = B^i \mathbf{e}_i$ is assumed to satisfy the following Faraday law of electromagnetic induction

$$\frac{d}{dt} \int_{\mathcal{S}_t} \mathbf{B}\cdot\mathbf{n} dS_t + \int_{\partial\mathcal{S}_t} \mathcal{E} \cdot \mathbf{h} dl_t = 0, \tag{11.12}$$

where $\mathcal{S}_t$ is an arbitrary material smooth surface with the boundary line $\partial\mathcal{S}_t$ and $\mathbf{n} = n^i \mathbf{e}_i$, $\mathbf{h} = h^i \mathbf{e}_i$ denote the unit normal vector to the surface and the unit tangent vector to the boundary line, respectively (compare Relations (3.32) and (4.44-49) of Part I). The physical justification of this law is demonstrated in Figure 11.1 illustrating the so-called Lenz experiment. According to the above law the varying magnetic flux interacting with a conductor generates the electrical current and this induces a counter magnetic flux that opposes the magnetic flux generating the current.

Equation (11.12) can be written in local form. In a particular case when we apply this equation to a closed surface of the subbody $\mathcal{P}_t$, i.e., $\partial\mathcal{P}_t = \mathcal{S}_t$ we have (the magnetic Gauss law)

$$\frac{d}{dt} \int_{\partial P_t} \mathbf{B}\cdot\mathbf{n} dS_t = 0 \quad \Longrightarrow \quad \int_{\partial P_t} \mathbf{B}\cdot\mathbf{n} dS_t = 0, \tag{11.13}$$

in which the integration constant was set equal zero. Then, in regular points and on singular surfaces, respectively, we obtain the Gauss-Faraday law

$$\operatorname{div}\mathbf{B} = 0 \quad \text{and} \quad [\![\mathbf{B} \cdot \mathbf{n}]\!] = 0. \tag{11.14}$$

On the other hand, Equation (11.12) in its full generality can be written in regular points and on singular surfaces in the following form

$$\int_{\mathcal{S}_t} \left(\frac{\partial \mathbf{B}}{\partial t} + \operatorname{rot}(\mathbf{B} \times \mathbf{v} + \mathcal{E}) \right) \cdot \mathbf{n}\, dS_t = 0 \quad \text{and} \quad \mathbf{n} \times [\![\mathbf{B} \times \mathbf{v} + \mathcal{E}]\!] - [\![\mathbf{B}]\!]\, c_\mathcal{S} = 0, \tag{11.15}$$

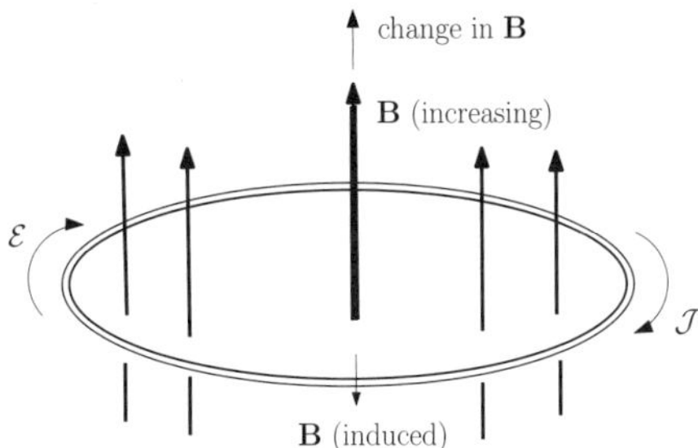

Figure 11.1: *Schematic to Lenz experiment; magnetic field* **B** *normal to the surface of the loop (Faraday's law).*

where[5] $c_{\mathcal{S}}$ is the normal speed of the surface and the operator rot acts, for instance, in the following way

$$\operatorname{rot}\mathcal{E} = \epsilon^{ijk}\frac{\partial\mathcal{E}_k}{\partial x^j}\mathbf{e}_i, \quad \text{i.e.,} \quad \int_{\partial\mathcal{S}_t}\mathcal{E}\cdot\mathbf{h}\,dl_t = \int_{\mathcal{S}_t}\operatorname{rot}\mathcal{E}\cdot\mathbf{n}\,d\mathcal{S}_t, \tag{11.16}$$

in Cartesian coordinates with the basis vector $\mathbf{e}_i$. In order to prove this relation one has to use the Kelvin relation (Formula (A.82) in Appendix A of Part I). The derivation can also be found in the books of P. Chadwick [70] or K. Hutter and K. Jöhnk [173].

It is convenient to use the Minkowskian electric field in Equation (11.15). Then, this equation yields the following local form of the law of conservation of the magnetic flux, the so-called Faraday law,

$$\frac{\partial\mathbf{B}}{\partial t} + \operatorname{rot}\mathbf{E} = \mathbf{0} \quad \text{and} \quad \mathbf{n}\times[\![\mathbf{E}]\!] - [\![\mathbf{B}]\!]\,c_{\mathcal{S}} = 0, \tag{11.17}$$

in the regular points and on the singular surface, respectively.

The second fundamental law of the electromagnetic field is the conservation of charge. It is assumed to have the form

$$\frac{d}{dt}\int_{\mathcal{P}_t}\mathfrak{R} + \oint_{\partial\mathcal{P}_t}\mathcal{J}^i n_i dS_t = 0, \tag{11.18}$$

where $\mathcal{J} = \mathcal{J}^i\mathbf{e}_i$ is the so-called conductive current. Now we are in the position to identify two of the electromagnetic fields of the above table. The dielectric displacement **D** is assumed to be related to the electric charge density $\mathfrak{R}$ in the following manner (the Gauss law)

$$\int_{\mathcal{P}_t}\mathfrak{R} = \oint_{\partial\mathcal{P}_t}D^i n_i dS_t. \tag{11.19}$$

---

[5]The time differentiation of the surface integral in (11.12) yields (see page 64 of the book [261])

$$\frac{d}{dt}\int_{\mathcal{S}_t}\mathbf{B}\cdot\mathbf{n}\,dS_t = \int_{\mathcal{S}_t}\left(\frac{\partial B^i}{\partial t} + v^k\frac{\partial B^i}{\partial x^k} - B^k\frac{\partial v^i}{\partial x^k} + B^i\frac{\partial v^k}{\partial x^k}\right)n_i dS_t$$

$$\equiv \int_{\mathcal{S}_t}\left(\frac{\partial B^i}{\partial t} + v^i\frac{\partial B^k}{\partial x^k} + \epsilon^{ijk}\frac{\partial}{\partial x^k}\left(\epsilon_{jmn}B^m v^n\right)\right)n_i dS_t.$$

Bearing (11.14) in mind we obtain the integrand in the form (11.15).

Hence, the dielectric displacement is the vector potential for the electric charge. Now, for an arbitrary material surface $\mathcal{S}_t$ we can write

$$\frac{d}{dt}\int_{\mathcal{S}_t} D^i n_i dS_t + \int_{\mathcal{S}_t} \mathcal{J}^i n_i dS_t = \int_{\partial\mathcal{S}_t} \mathcal{H}^i h_i dl_t, \tag{11.20}$$

where $\mathcal{H} = \mathcal{H}^i \mathbf{e}_i$ is the effective magnetic field. This is the so-called Ampère law. Obviously, for the closed surface $\mathcal{S}_t = \partial\mathcal{P}_t$ we obtain (11.18) from (11.20).

As before we can transfer Equations (11.18) and (11.19) to their local form. We obtain the following local conservation of charge equation and the Gauss law for the dielectric displacement

$$\frac{\partial q}{\partial t} + \operatorname{div} \mathbf{J} = 0, \quad \mathbf{J} = \mathcal{J} + q\mathbf{v},$$

$$\operatorname{div} \mathbf{D} = q, \quad \text{and} \quad [\![\mathbf{D}]\!] \cdot \mathbf{n} = 0, \tag{11.21}$$

the latter equations in regular points and on singular surfaces, respectively. Simultaneously, Equation (11.20) yields the Ampère-Oersted law

$$\frac{\partial \mathbf{D}}{\partial t} - \operatorname{div} \mathbf{H} + \mathbf{J} = \mathbf{0} \quad \text{and} \quad \mathbf{n} \times [\![\mathbf{H}]\!] + [\![\mathbf{D}]\!] c_{\mathcal{S}} = 0. \tag{11.22}$$

In the last equation the surface charge and the surface flux may appear in addition to the jumps of the fields across the singular surface. This may, for instance, happen on electrodes. We do not go into any details of those surface effects.

Obviously, it is an easy task to check that not all relations presented above are independent. For instance, the conservation of charge (11.19) follows immediately from (11.21). The set of Maxwell equations as quoted in the works on the subject are listed in the juxtaposition below.

Table 11.2: *The Maxwell electromagnetic equations.*

Global form	Local form
$\displaystyle\int_{\partial\mathcal{P}_t} \mathbf{B}\cdot\mathbf{n}dS_t = 0$	$\operatorname{div} \mathbf{B} = 0$
$\displaystyle\int_{\partial\mathcal{P}_t} \left(\frac{\partial \mathbf{B}}{\partial t} + \mathbf{rot\,E}\right)\cdot\mathbf{n}dS_t = 0$	$\dfrac{\partial \mathbf{B}}{\partial t} + \mathbf{rot\,E} = \mathbf{0}$
$\displaystyle\int_{\partial\mathcal{P}_t} D^i n_i dS_t = \int_{\mathcal{P}_t} qdv$	$\operatorname{div} \mathbf{D} = q$
$\displaystyle\int_{\partial\mathcal{P}_t} \left(\frac{\partial \mathbf{D}}{\partial t} - \mathbf{rot\,H} + \mathbf{J}\right)\cdot\mathbf{n}dS_t = 0$	$\dfrac{\partial \mathbf{D}}{\partial t} - \mathbf{rot\,H} + \mathbf{J} = \mathbf{0}$

The set of Maxwell equations is underdetermined. In addition, it is coupled with quantities describing the ponderable body which require additional material relations. Consequently, we have to consider a constitutive problem and its invariance properties. Further, we shall do so for the particular case of plasmas. However, we should note

Table 11.3: *Examples of electric and magnetic susceptibility.*

	electric susceptibility χ	magnetic susceptibility κ
vacuum	0	0
water	87	$-9.035 \cdot 10^{-6}$
diamond	4.5-9	$-2.2 \cdot 10^{-5}$
graphite	9-14	$-1.1 \cdot 10^{-5} - -8.3 \cdot 10^{-4}$ (anisotropic!)
Al	–	$2.2 \cdot 10^{-5}$

already here that this problem of invariance cannot be consistently solved within the classical thermomechanics coupled with the electromagnetic field described by the general Maxwell relations. We illustrate this problem below. Even in the simplest case of universal Maxwell-Lorentz aether relations the question of invariance is present. These relations couple the fields $\mathbf{B}$, $\mathbf{E}$, with $\mathbf{D}$, $\mathbf{H}$. Namely, in an inertial frame,

$$\mathbf{D} = \varepsilon_0 \mathbf{E}, \quad \mathbf{H} = \frac{1}{\mu_0} \mathbf{B}, \tag{11.23}$$

where the dielectric permittivity constant ε_0 and the constant of magnetic permeability μ_0 have the values

$$\varepsilon_0 = \frac{10^7}{4\pi c^2} \left[\text{A·s/V·m}\right], \quad \mu_0 = 4\pi \cdot 10^{-7} \left[\text{V·s/A·m}\right], \quad \frac{1}{\varepsilon_0 \mu_0} = c^2, \tag{11.24}$$

with $c = 0.29978 \cdot 10^9$ m/s – the speed of light in vacuum. Similar relations with parameters ε and μ depending on the material are assumed for various isotropic materials in a chosen inertial system. In spite of formal problems with the invariance they are used to classify electro- and magnetosensitive materials. One usually introduces the dimensionless relative permeabilities ε_w and μ_w

$$\begin{aligned} \varepsilon &= \varepsilon_0 \varepsilon_w = \varepsilon_0 \left(1 + \chi\right), \\ \mu &= \mu_0 \mu_w = \mu_0 \left(1 + \kappa\right), \end{aligned} \tag{11.25}$$

where χ is the electric susceptibility and κ the magnetic susceptibility. Then,

$$\mathbf{D} = \varepsilon_0 \varepsilon_w \mathbf{E} = \varepsilon_0 \mathbf{E} + \mathbf{P}, \qquad \mathbf{P} = \varepsilon_0 \chi \mathbf{E}. \tag{11.26}$$

The vector $\mathbf{P}$ is called the polarization vector. It is a characteristic field for dielectric materials which are electrically non-conductive but polarizable, i.e., they create their own electric field due to the action of the external electromagnetic field.

Practically all materials have the magnetic susceptibility almost equal to zero. For $\kappa > 0$ they are called paramagnetic materials and for $\kappa < 0$ they are called diamagnetic materials. We have

$$\mathbf{B} = \mu_0 \mu_w \mathbf{H} = \mu_0 \mathbf{H} + \mathbf{M}, \quad \mathbf{M} = \mu_0 \kappa \mathbf{H}. \tag{11.27}$$

The vector $\mathbf{M}$ is called the magnetization.

In addition to paramagnetic and diamagnetic materials there exists a third group of ferromagnetic materials for which the fields $\mathbf{B}$ and $\mathbf{H}$ cannot be assumed to be proportional to each other. This is caused by the creation of the so-called Weiss domains of uniformly oriented magnetic momenta.

Now let us turn our attention to one of the main problems of the construction of models of coupled fields which is the invariance of thermomechanical and electromagnetic quantities. The former should be specified by constitutive relations which are frame-indifferent (materially objective; compare Section 5.2 of Part I). This means that these relations should be invariant with respect to the group of Eulerian transformations

$$\mathbf{x}^* = \mathbf{O}^*\left(t^*\right)\mathbf{x} + \mathbf{c}^*\left(t^*\right), \quad t^* = t, \quad \mathbf{O}^{*-1} = \mathbf{O}^{*T}, \tag{11.28}$$

where $\mathbf{O}^*\left(t^*\right)$ is the time-dependent orthogonal matrix of rotations of the frame with star with respect to the frame without star. For instance, in the case of rotation about the x^1-axis of the Cartesian frame with the rotation angle α this matrix has the form

$$\mathbf{O}^* = O_{ij}^*\mathbf{e}_i \otimes \mathbf{e}_j, \qquad \left(O_{ij}^*\right) = \begin{pmatrix} 1 & 0 & 0 \\ 0 & \cos\alpha & \sin\alpha \\ 0 & -\sin\alpha & \cos\alpha \end{pmatrix}. \tag{11.29}$$

Obviously, Relation (11.28) indicates the following transformation rules for velocities and accelerations (compare Relations (3.5) of Part I or (3.29) of this part)

$$\mathbf{v}^* = \mathbf{O}^*\mathbf{v} + \dot{\mathbf{O}}^*\mathbf{x} + \dot{\mathbf{c}}, \quad \mathbf{a}^* = \mathbf{O}^*\mathbf{a} + 2\dot{\mathbf{O}}^*\mathbf{v} + \ddot{\mathbf{O}}^*\mathbf{x} + \ddot{\mathbf{c}}, \quad \dot{\mathbf{O}}^* = \frac{d\mathbf{O}^*}{dt}, \quad \text{etc.} \tag{11.30}$$

The classical balance equations are invariant only with respect to the Galilean subgroup of transformations in which the matrix $\mathbf{O}^*$ and the vector $\mathbf{c}^*$ are constant

$$\mathbf{x}^* = \mathbf{O}^*\mathbf{x} + \mathbf{c}^*, \quad \mathbf{O}^*, \mathbf{c}^* - \text{const.} \tag{11.31}$$

This subgroup defines a class of inertial reference frames. It is well-known (e.g. compare Chapter 3 of Part I or page 27 here) that the Eulerian transformation of, for instance, the equations of motion yields additional forces (Coriolis force, centrifugal force, Euler force and inertial force of relative translation – Chapter 4 of Part I) in the non-inertial description.

The quantities $\mathbf{D}$, $\mathbf{E}$, $\mathbf{H}$ and $\mathbf{B}$ entering Relations (11.23) are not invariant with respect to the above Eulerian group of transformation and, consequently, they are not objective. Hence, the constitutive relations (11.23) and similar relations with different permittivity constants are not frame-indifferent. This is, obviously, a violation of one of the fundamental principles of continuum thermodynamics. The same problem appears in the Ohm law $\mathbf{J} = \sigma\mathbf{E}$ where the constant conductivity σ appears.

In order to see the transformation rules satisfied by the electromagnetic fields we consider a combination of Maxwell's equations which have the structure of balance laws. Namely, the scalar multiplication of Equation (11.17) by $-\mathbf{H}$ and of Equation (11.22) by $\mathbf{E}$, the use of the Maxwell-Lorentz relations and subsequent addition yields

$$\frac{\partial}{\partial t}\left(\frac{1}{2}\left(\mathbf{E}\cdot\mathbf{D}+\mathbf{B}\cdot\mathbf{H}\right)\right) + \text{div}\left(\mathbf{E}\times\mathbf{H}\right) = -\mathbf{E}\cdot\mathbf{J}. \tag{11.32}$$

Similarly, the cross multiplication of Equation (11.17) by $\mathbf{B}$ and the use of Equations (11.14) and (11.21) yield together with the Maxwell-Lorentz relations the following equation

$$\frac{\partial}{\partial t}\left(\mathbf{D}\times\mathbf{B}\right) + \text{div}\left(\frac{1}{2}\left(\mathbf{E}\cdot\mathbf{D}+\mathbf{B}\cdot\mathbf{H}\right)\mathbf{1} - \mathbf{E}\otimes\mathbf{D} - \mathbf{B}\otimes\mathbf{H}\right) = -q\mathbf{E} - \mathbf{J}\times\mathbf{B}. \tag{11.33}$$

The left-hand sides of these equations have the structure of balance laws. Consequently, the right-hand sides can be interpreted as sources: the vector $q\mathbf{E} + \mathbf{J} \times \mathbf{B}$ describes the reaction force of the material body on the action of the electromagnetic field and the scalar $\mathbf{E} \cdot \mathbf{J}$ describes the reaction power of the material body on the action of the electromagnetic field. Using this interpretation in the equations of momentum and energy conservation yields the following form of these laws

$$\frac{\partial}{\partial t}\left(\rho\mathbf{v} + \mathbf{D} \times \mathbf{B}\right) + \mathrm{div}\left(\rho\mathbf{v} \otimes \mathbf{v} - \mathbf{T} + \frac{1}{2}\left(\mathbf{E} \cdot \mathbf{D} + \mathbf{B} \cdot \mathbf{H}\right)\mathbf{1} - \mathbf{E} \otimes \mathbf{D} - \mathbf{B} \otimes \mathbf{H}\right) = 0,$$

$$\frac{\partial}{\partial t}\left(\rho\left(\varepsilon + \frac{1}{2}v^2\right) + \frac{1}{2}\left(\mathbf{E} \cdot \mathbf{D} + \mathbf{B} \cdot \mathbf{H}\right)\right) + \mathrm{div}\left(\rho\left(\varepsilon + \frac{1}{2}v^2\right)\mathbf{v} + \mathbf{q} - \mathbf{T}^T\mathbf{v} + \mathbf{E} \times \mathbf{H}\right) = 0.$$
$$(11.34)$$

Hence, this thermodynamical argument yields the following interpretations:

1. $\mathbf{D} \times \mathbf{B}$ is the contribution of the electromagnetic field to the momentum density,

2. $\mathbf{T}^M = -\frac{1}{2}\left(\mathbf{E} \cdot \mathbf{D} + \mathbf{B} \cdot \mathbf{H}\right)\mathbf{1} + \mathbf{E} \otimes \mathbf{D} + \mathbf{B} \otimes \mathbf{H}$ is the so-called Maxwell stress tensor, the contribution of the electromagnetic field to the stress tensor; let us notice that this contribution is a symmetric tensor in the case of the Maxwell-Lorentz relations. It may not be so if these relations describe anisotropic media. Then, the skew-symmetric part of the stress tensor yields the necessity of couple-stresses (for articles on couple-stresses in elastic materials see e.g. R. A. Toupin [383] or by R. D. Mindlin and H. F. Tiersten [252]),

3. $\frac{1}{2}\left(\mathbf{E} \cdot \mathbf{D} + \mathbf{B} \cdot \mathbf{H}\right)$ is the energy density of the electromagnetic field,

4. $\mathbf{q}^M = \mathbf{E} \times \mathbf{H}$ is the energy flux due to the electromagnetic field, the so-called Poynting vector.

We conclude that the quantity q must be an objective scalar and, for an arbitrary current $\mathbf{J}$, $\mathbf{E} + \frac{1}{q}\mathbf{J} \times \mathbf{B} \sim \mathbf{E} + \mathbf{v} \times \mathbf{B}$ must be an objective vector. If we account for the definition (11.21), this yields the following rules for the electromagnetic quantities

$$\mathbf{E}^* + \mathbf{v}^* \times \mathbf{B}^* = \mathbf{O}^*\left(\mathbf{E} + \mathbf{v} \times \mathbf{B}\right), \qquad \mathcal{J}^* = \mathbf{O}^*\mathcal{J}, \qquad (11.35)$$

which, bearing the transformation rule for the velocity (compare, for instance Relations (3.29)) in mind, yields the following relations

$$\mathbf{E}^* = \mathbf{O}^*\left(\mathbf{E} - \mathbf{O}^{*T}\left(\mathbf{W}^*\left(\mathbf{x}^* - \mathbf{c}^*\right) + \dot{\mathbf{c}}^*\right) \times \mathbf{B}\right),$$
$$\mathbf{B}^* = \left(\det\mathbf{O}^*\right)\mathbf{O}^*\mathbf{B}, \qquad \mathbf{W}^* = \dot{\mathbf{O}}^*\mathbf{O}^{*T}, \qquad (11.36)$$

$$\mathbf{J}^* = \mathbf{O}^*\mathbf{J} + q\left(\mathbf{W}^*\left(\mathbf{x}^* - \mathbf{c}^*\right) + \dot{\mathbf{c}}^*\right). \qquad (11.37)$$

We skip here the discussion of the other transformation rules. Details can be found in the book of I. Müller [261]. However, it is clear that the Maxwell-Lorentz aether relations and their extensions yielding a very important classification of materials sensitive on the action of the electromagnetic field make sense exactly only in static cases when all transformations are also time-independent. We return to this problem in the next section.

We conclude this section with the remark that the above discussed fields can be easily transformed to the Lagrangian description. We do not need it in this book and therefore we shall not quote it here and refer the reader to the book of K. Hutter, A. A. F. van de Ven and A. Ursescu [174].

11.2.2   Lorentz invariance of the Maxwell equations

We return now to the problem of invariance of relations for electromagnetic quantities. As we see further, the Maxwell equations are invariant with respect to the Lorentz transformation group of the special relativity. In order to introduce this property, it is convenient to define the four-dimensional space-time of the special relativity. The spatial Cartesian coordinates in this space will be denoted[6] by $z^k = x^k$, $k = 1, 2, 3$ and $z^0 = t$. The structure of this four-dimensional space is specified by the metric form

$$(ds)^2 = g_{AB} dz^A dz^B = -c^2 (dt)^2 + \delta_{kl} dx^k dx^l, \quad A, B = 0, 1, 2, 3, \tag{11.38}$$

i.e.,

$$(g_{AB}) = \begin{pmatrix} -c^2 & 0 & 0 & 0 \\ 0 & 1 & 0 & 0 \\ 0 & 0 & 1 & 0 \\ 0 & 0 & 0 & 1 \end{pmatrix}, \tag{11.39}$$

where we apply the summation convention to both families of indices k and A. Due to the postulates

1. a particle which moves with a constant velocity in one inertial frame must move with constant velocity in the other inertial frame (the equivalence of inertial reference frames),

2. the spherical wave of light in vacuum moving with the speed c in one frame must move with the speed c in all other frames as well,

the form (11.39) is invariant with respect to the transformation of inertial frames. We define such a transformation by the linear relation with constant coefficients

$$z^{*A} = \alpha^A_{.B} z^B + z^A_0. \tag{11.40}$$

This leads to the transformation of the metric form

$$g_{AB} dz^{*A} dz^{*B} = g_{AB} \alpha^A_{.C} \alpha^B_{.D} dz^C dz^D. \tag{11.41}$$

The above conditions yield then the orthogonality of the transformation matrix

$$g_{AB} \alpha^A_{.C} \alpha^B_{.D} = g_{CD}, \tag{11.42}$$

i.e., $\alpha^T = \alpha^{-1}$. Its spatial part defines the rotation of spatial coordinates. It is convenient to denote this part in the same way as we do for the rotations in the non-relativistic formulation, i.e., $\alpha^{kl} \mathbf{e}_k \otimes \mathbf{e}_l = \mathbf{O}^*$.

Bearing the invariance of the metric form (11.41) in mind we can easily find the general form of the transformation (11.40). Due to the symmetry of the metric g_{AB} only ten out of 16 equations (11.42) are independent. Hence, the solution depends on six parameters. If $\mathbf{V}$ denotes the constant velocity of the frame with star with respect to the frame without star in the Euclidean space of positions (three parameters) and $\mathbf{O}^*$ is the rotation matrix

---

[6] In Part I we have used the variable ct instead of t. It is only a matter of convenience.

in the space (again three parameters) then this transformation has the following form (e.g. [256, 323])

$$\mathbf{x}^* = \mathbf{O}^*\mathbf{x} + \mathbf{O}^*\mathbf{V}\left((\gamma - 1)\left(\mathbf{x} \cdot \mathbf{V}\right)/V^2 - \gamma t\right), \quad t^* = \gamma\left(t - \mathbf{x} \cdot \mathbf{V}/c^2\right),$$

$$\gamma = \frac{1}{\sqrt{1 - V^2/c^2}}, \quad V^2 = \mathbf{V} \cdot \mathbf{V}, \tag{11.43}$$

and we call it the homogeneous Lorentz transformation even though it has been discovered by H. Poincaré.

Hendrik A. Lorentz	*Henri Poincaré*	*Michael Faraday*
1853-1928	*1854-1912*	*1791-1867*

In the case of heterogeneous transformations one has to replace the spatial vector $\mathbf{x}^*$ by the vector $\mathbf{x}^* - \mathbf{c}^*$, where $\mathbf{c}^*$ is an arbitrary constant vector

In the particular case of the Lorentz transformation in which the x^1-axis and x^2-axis remain unchanged and the x^{*3}-axis moves with respect to the x^3-axis with the speed V (hence, there is no rotation), we obtain

$$x^{*1} = x^1, \quad x^{*2} = x^2, \quad x^{*3} = \gamma\left(x^3 - Vt\right), \quad t^* = \gamma\left(t - x^3\frac{V}{c^2}\right), \tag{11.44}$$

which yields the following form of transformation matrices

$$\left(\frac{\partial z^{*A}}{\partial z^B}\right) = \left(\alpha^A_{.B}\right) = \begin{pmatrix} \gamma & 0 & 0 & -\gamma\frac{V}{c^2} \\ 0 & 1 & 0 & 0 \\ 0 & 0 & \gamma & 0 \\ -\gamma V & 0 & 0 & \gamma \end{pmatrix},$$

$$\left(\frac{\partial z^A}{\partial z^{*B}}\right) = \left(\overset{-1}{\alpha}{}^A_{.B}\right) = \begin{pmatrix} \gamma & 0 & 0 & \gamma\frac{V}{c^2} \\ 0 & 1 & 0 & 0 \\ 0 & 0 & \gamma & 0 \\ \gamma V & 0 & 0 & \gamma \end{pmatrix}, \tag{11.45}$$

because, obviously, the inverse transformation corresponds in this particular case to the change of direction of the speed $V \to -V$. This transformation is proper orthogonal because $\det\left(\frac{\partial z^{*A}}{\partial z^B}\right) = 1$. However, in the general case the Lorentz transformation may also describe the inversion and then this determinant is equal to -1.

It can be immediately seen that in the non-relativistic limit ($c \to \infty$, i.e., $\left|\dfrac{V}{c}\right| \to 0$, $\gamma \to 1$) we obtain from (11.44) the Galilean transformation.

The space-time formulation of the Maxwell model allows one to look at the transformation properties from a different viewpoint. Let us first introduce the following electromagnetic field tensor

$$
(\varphi_{AB}) = \begin{pmatrix} 0 & -E_1 & -E_2 & -E_3 \\ E_1 & 0 & B_3 & -B_2 \\ E_2 & -B_3 & 0 & B_1 \\ E_3 & B_2 & -B_1 & 0 \end{pmatrix}. \tag{11.46}
$$

It is easy to check that for the transformation defined by (11.40) this object transforms according to the rule

$$
\varphi^{*}_{AB} = \overset{-1}{\alpha}{}^{C}_{.A} \overset{-1}{\alpha}{}^{D}_{.B} \varphi_{CD}, \tag{11.47}
$$

i.e., it fulfills the transformation rule of covariant components of the tensor.

On the other hand, the charge-current vector defined by the relation

$$
\left(\sigma^{A}\right) = \left(q, J_1, J_2, J_3\right), \tag{11.48}
$$

transforms in the following way

$$
\sigma^{*A} = \left(\det\left(\overset{-1}{\alpha}{}^{D}_{.C}\right)\right) \alpha^{A}_{.B}\sigma^{B}, \tag{11.49}
$$

which means that the charge-current vector is axial. Due to the orthogonality of the matrix α the coefficient in this relation is ± 1.

Now, the second group of functions appearing in the Maxwell equations forms the world tensor of the charge-current potential

$$
\left(\eta^{AB}\right) = \begin{pmatrix} 0 & D_1 & D_2 & D_3 \\ -D_1 & 0 & H_3 & -H_2 \\ -D_2 & -H_3 & 0 & H_1 \\ -D_3 & H_2 & -H_1 & 0 \end{pmatrix}. \tag{11.50}
$$

It transforms according to the rule

$$
\eta^{*AB} = \left(\det\left(\overset{-1}{\alpha}{}^{D}_{.C}\right)\right) \alpha^{A}_{.C}\alpha^{B}_{.D}\eta^{CD}. \tag{11.51}
$$

Hence, it is a tensor with the weight 1, similarly to the charge-current vector.

Bearing the above definitions in mind we can write the set of Maxwell equations in the following compact form

$$
\epsilon^{ABCD}\frac{\partial \varphi_{CD}}{\partial z^{B}} = 0, \qquad \frac{\partial \eta^{AB}}{\partial z^{B}} = \sigma^{A}, \tag{11.52}
$$

where

$$
\epsilon^{ABCD} = \begin{cases} -1 & \text{if } A, B, C, D \text{ is an even permutation of } 0123, \\ 1 & \text{if } A, B, C, D \text{ is an odd permutation of } 0123, \\ 0 & \text{else}, \end{cases} \tag{11.53}
$$

is the four-dimensional permutation symbol. Equations (11.52) would be invariant under the transformation described above if the relation between the electromagnetic field tensor and the charge-current potential

$$\eta^{AB} = \eta^{AB}\left(\varphi_{CD}\right),\tag{11.54}$$

were invariant. Such are the Maxwell-Lorentz aether relations which in the present notation have the form

$$\eta^{AB} = \frac{1}{\mu_0}g^{AC}g^{BD}\varphi_{CD}, \quad \left(g^{AB}\right) = \begin{pmatrix} -\frac{1}{c^2} & 0 & 0 & 0 \\ 0 & 1 & 0 & 0 \\ 0 & 0 & 1 & 0 \\ 0 & 0 & 0 & 1 \end{pmatrix},\tag{11.55}$$

where the inverse metric g^{AB} is invariant according to the definition of the Lorentz transformation.

Constitutive equations (11.54) for arbitrary materials are not so simple anymore and their invariance with respect to the Lorentz group of transformations yields nontrivial representation problems. They will not be considered in this book but it is clear that low velocities of the medium are sufficient for the approximation of Lorentz invariance by the Galilean group for the mechanical part of the model. This is relatively well satisfied for solids in which even the electrons possess rather low velocities. This approximation may not be so good for plasmas as we see further and in some cases it requires a full relativistic formulation. Some examples of such models but without electromagnetic couplings can be found in the book of I. Müller and T. Ruggeri [263] and in the papers quoted there.

11.3 On thermodynamics of coupled fields

The coupling of thermomechanical and electromagnetic fields for a single component solid means that a thermodynamical model for the following fields must be constructed

$$\mathcal{V}_{\text{Lagrange}} = \{\mathbf{f}, T, \mathbf{E}, \mathbf{B}\},\tag{11.56}$$

in the Lagrangian description. In this set, the vector $\mathbf{f}$ is, obviously the function of motion $\mathbf{x} = \mathbf{f}\left(\mathbf{X}, t\right)$, where $\mathbf{x}$ is the current position of the material point $\mathbf{X}$. On the other hand, for fluids we have

$$\mathcal{V}_{\text{Euler}} = \{\rho, \mathbf{v}, T, \mathbf{E}, \mathbf{B}\},\tag{11.57}$$

in the Eulerian description. In this set $\rho = \rho\left(\mathbf{x}, t\right)$ is the mass density at the current position $\mathbf{x}$ and $\mathbf{v} = \mathbf{v}\left(\mathbf{x}, t\right)$ the velocity at this position.

The field equations for those fields follow from the mass, momentum and energy conservation laws coupled with the Maxwell equations. We write them only in Eulerian description because we shall refer to them in this form in the following part of this chapter. We have

$$\frac{\partial\rho}{\partial t} + \text{div}\left(\rho\mathbf{v}\right) = 0,$$

$$\frac{\partial\left(\rho\mathbf{v}\right)}{\partial t} + \text{div}\left(\rho\mathbf{v}\otimes\mathbf{v} - \mathbf{T}\right) = q\mathbf{E} + \mathbf{J}\times\mathbf{B} + \rho\mathbf{b},$$

$$\frac{\partial \left(\rho \varepsilon\right)}{\partial t} + \operatorname{div}\left(\rho \mathbf{v}\varepsilon + \mathbf{q}\right) = \operatorname{tr}\left(\mathbf{T}\operatorname{grad}\mathbf{v}\right) + \mathbf{J}\cdot\mathbf{B},$$

$$\frac{\partial \mathbf{B}}{\partial t} + \operatorname{rot}\mathbf{E} = 0, \quad \operatorname{div}\mathbf{B} = 0, \tag{11.58}$$

$$\frac{\partial \mathbf{D}}{\partial t} - \operatorname{rot}\mathbf{H} + \mathbf{J} = 0, \quad \operatorname{div}\mathbf{D} = q.$$

The first three equations are, obviously, the balance equations of mass, momentum and internal energy (compare Table 4 in Part I). The remaining equations are Maxwell's equations (Table 11.2). Certainly, we can write the momentum and energy balance equations in the divergent form if we eliminate the Lorentz force $q\mathbf{E} + \mathbf{J}\times\mathbf{B}$ in the momentum balance, the source $\mathbf{J}\cdot\mathbf{B}$ in the energy balance and, simultaneously, add to the stress tensor $\mathbf{T}$ the Maxwell contribution $-\frac{1}{2}\left(\mathbf{E}\cdot\mathbf{D} + \mathbf{B}\cdot\mathbf{H}\right)\mathbf{1} + \mathbf{E}\otimes\mathbf{D} + \mathbf{B}\otimes\mathbf{H}$ and add to the heat flux the Poynting vector $\mathbf{E}\times\mathbf{H}$. We have to correct then the notions of momentum and internal energy as explained above.

Obviously, we have to deal with a closure problem. Constitutive relations for the following quantities must be postulated

$$\mathcal{F} = \{\mathbf{T}, \mathbf{P}, \mathbf{M}, \varepsilon, \mathbf{q}\}, \tag{11.59}$$

in addition to the Maxwell-Lorentz aether relations. Here $\mathbf{P}$ and $\mathbf{M}$ denote, as introduced by (11.26) and (11.27), the polarization and magnetization vectors, respectively. These relations have to be objective and they have to satisfy the second law of thermodynamics. We have indicated the problems with objectivity in the previous section. For the non-relativistic model the difficulties with the Maxwell equations can be ignored. This is done in the majority of the models in which interactions of electromagnetic fields and solids are considered. We proceed to formulate the thermodynamical restriction. As extensively discussed in Part I, the thermodynamical restriction consists of two conditions – the entropy inequality and the condition on the ideal wall. The first one means that the inequality

$$\frac{\partial \left(\rho \eta\right)}{\partial t} + \operatorname{div}\left(\rho \eta \mathbf{v} + \mathbf{h}\right) \geq 0, \tag{11.60}$$

should hold for all solutions of the field equations while the entropy density η and the entropy flux $\mathbf{h}$ are constitutive functions. This condition is evaluated by I. Müller [261] for neutral, nonconducting fluids for which the set of constitutive variables is $\{\rho, \mathbf{v}, T, \operatorname{grad} T, \mathbf{E}, \mathbf{B}\}$. K. Hutter [170, 171, 174] derived the model for viscous nonconducting heat fluids for which the constitutive variables are $\{\rho, \mathbf{v}, T, \mathbf{D}_v, \mathbf{E}, \mathbf{B}\}$. The rate of strain tensor $\mathbf{D}_v = \frac{1}{2}\left(\mathbf{L} + \mathbf{L}^T\right)$, $\mathbf{L} = \operatorname{grad}\mathbf{v}$ is defined by Relation (3.7) of Part I or (3.15) here, and the index "v" is here introduced to distinguish this quantity from the dielectric displacement. This model has been used to describe the behavior of electrorheological fluids mentioned in the introduction to this chapter. Models less formal from the thermodynamical point of view were considered for dielectric solids since the pioneering works of R. A. Toupin [382, 384] and for piezoelectric solids by R. D. Mindlin [251]. Thermodynamical foundations for various models of solids and fluids are also discussed in the monographs of A. C. Eringen and G. A. Maugin [117, 118]. Finally, many applications of linearized models, particularly in the theory of vibrations of plates, propagation of bulk and surface waves can be found in the books of K. Hutter, A. A. F. van de Ven and A. Ursescu as well as W. Nowacki [174, 279]. In this chapter we are primarily interested in applications of the thermodynamical theory of mixtures to the description of interactions with the electromagnetic fields. For this reason, we limit the attention to the magnetohydrodynamics applied to plasmas and we discuss this subject in some details in the forthcoming sections.

11.4 Magnetohydrodynamics of plasmas

11.4.1 Phenomenology

It is claimed that almost 99% of the visible matter in the Universe appears as plasma. Plasma consists of ionized and neutral particles, in contrast to electrolytes, it does not possess a fluid carrier and it satisfies the very important principle of being electrically quasi-neutral. This means that in relatively small domains the sum of positive and negative charges yields the zero resultant charge. We shall discuss this point in some details further in this section.

The form of appearance and processes characteristic for different domains of parameters of plasma are very different. In Figure 11.2 we show roughly different forms of plasma in the domains characterized by the energy density or, equivalently, the temperature $(1 \text{ [eV]} = 11\ 600 \text{ [K]}^7)$ and the number of electrons in 1 m^3. Each domain requires a different model, some of them may be macroscopic, some other require kinetic description and for many one has to account for quantum effects. The range of interest for the thermodynamical modeling is denoted by MHD: magnetohydrodynamics. In spite of the smallness of this domain in Figure 11.2 this model is frequently applied to other sorts of plasma and yields very good results indeed.

It is claimed in the books of J. P. Goedbloed [137, 138] that the two large application domains of plasma physics, the research of laboratory plasmas (nuclear fusion[8]) and of astrophysical plasmas (sun, magnetospheres, stellar coronas, pulsars, accretion disks, etc.), may be described from the single point of view of magnetohydrodynamics (MHD). This yields effective methods and insights for the interpretation of plasma phenomena on all scales, from the laboratory to the Universe. It can be seen in Figure 11.2 that these domains of applications lie rather far from the MHD-domain.

In the subsequent sections we present a multi-component model of plasma and a simplification of this model for a two-component fully ionized plasma. Then the MHD-model

---

[7]1 [eV] – 1 electronvolt is a unit of energy equal to 1.60210^{-19} [J] (joule). It is the amount of energy gained by a single electron moved across an electric potential difference of one volt.

[8]"Research into controlled fusion aims to demonstrate that it is a valid option for generating power in the long term future in an environmentally, politically and economically acceptable way. Controlled fusion is a process in which light nuclei fuse together to form heavier ones: during this process a very large amount of energy is released. For a fusion reactor it is planned to use the two isotopes of hydrogen: deuterium (D) and tritium (T), which fuse together much more readily than any other combination of light nuclei according to the following reaction: $D_2 + T_3$ ß He_4 + n + 17.6 MeV. The end products are helium and neutrons (n). The total energy liberated by fusing one gram of a 50:50% mixture of deuterium and tritium is 94000 kWh, which is 10 million times more than from the same mass of oil. 80% of this energy is carried by the neutrons with an energy of 14 MeV while the remaining 20% is carried by the helium nucleus. Most of this energy eventually becomes heat to be stored or converted by conventional means into electricity. The temperature at which fusion reactions start to become significant are above a few tens of millions of degrees. For the D-T reaction, the optimal temperature is of the order of 70-200 million degrees. At such temperatures the D-T fuel is in the plasma state. Deuterium is very abundant on the earth and can be extracted from water (0.034 g/l). Tritium does not occur naturally, since its half-life is only 12.3 years, but it can be regenerated from lithium using the neutrons produced by the D-T fusion reactions." (Appendix D: *"The basis of controlled fusion"* in [101])

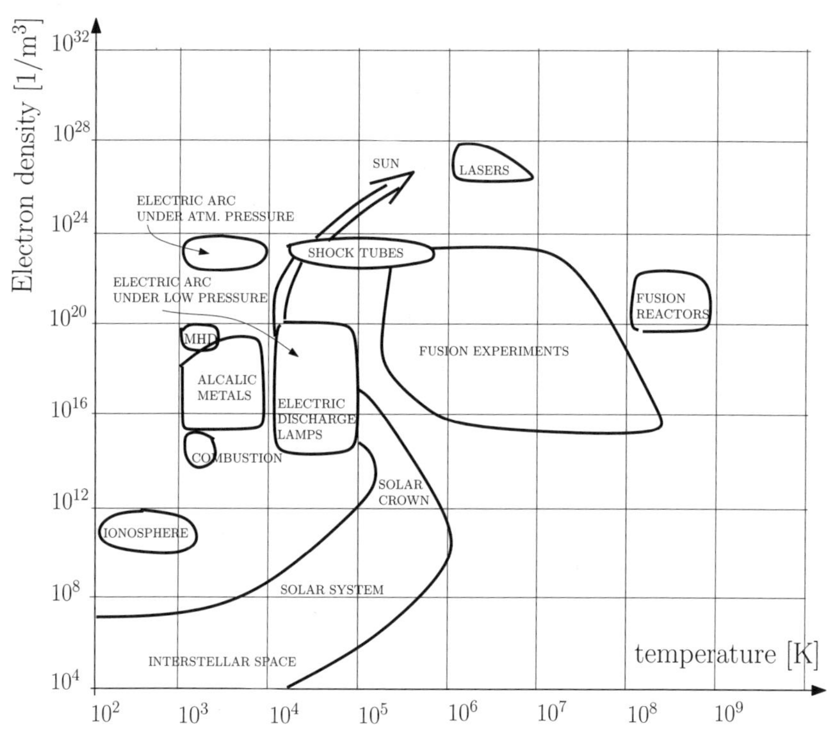

Figure 11.2: *Different domains of plasma appearing in nature.*

as a one-component consequence of the multi-component description is discussed. For this model we present also some special flows important in practical applications.

11.4.2 Characteristic parameters

In order to estimate various contributions to models of plasma we have to introduce some characteristic parameters of the processes of interaction.

1. *Larmor frequency*

We begin with the investigation of a simple example of motion of a charged particle in the electromagnetic field. Namely, we consider the motion of a single particle of the mass m and of the charge Q in the homogeneous electric field $\mathbf{E}$ and the homogeneous magnetic flux $\mathbf{B}$. Let us split the electric field into the part $\mathbf{E}_\parallel$ parallel to $\mathbf{B}$ and the part $\mathbf{E}_\perp$ perpendicular to $\mathbf{B}$, i.e.,

$$\mathbf{E} = \mathbf{E}_\parallel + \mathbf{E}_\perp, \quad \mathbf{E}_\parallel = \frac{\mathbf{E} \cdot \mathbf{B}}{B^2}\mathbf{B}, \quad \mathbf{E}_\parallel \cdot \mathbf{E}_\perp = 0. \tag{11.61}$$

The similar representation is introduced for the velocity $\mathbf{v}$ of the particle

$$\mathbf{v} = \mathbf{v}_\parallel + \mathbf{v}_\perp, \quad \mathbf{v}_\parallel = \frac{\mathbf{v} \cdot \mathbf{B}}{B^2}\mathbf{B}, \quad \mathbf{v}_\parallel \cdot \mathbf{v}_\perp = 0. \tag{11.62}$$

Then the equation of motion of the charge is, of course, given by the momentum conservation law with the external force in the form of the Lorentz force in which the electric current is replaced for this particular case by the quantity $Q\mathbf{v}$, i.e.,

$$m\frac{d\mathbf{v}}{dt} = Q\left(\mathbf{E} + \mathbf{v} \times \mathbf{B}\right). \tag{11.63}$$

According to the above representations this equation can be written in the form of the following set

$$m\frac{d\mathbf{v}_\parallel}{dt} = Q\mathbf{E}_\parallel, \quad m\frac{d\mathbf{v}_\perp}{dt} = Q\left(\mathbf{E}_\perp + \mathbf{v}_\perp \times \mathbf{B}\right). \tag{11.64}$$

Hence,

$$\frac{d}{dt}\left(\frac{1}{2}mv_\parallel^2\right) = Q\mathbf{v}_\parallel \cdot \mathbf{E}_\parallel, \quad \frac{d}{dt}\left(\frac{1}{2}mv_\perp^2\right) = Q\mathbf{v}_\perp \cdot \mathbf{E}_\perp. \tag{11.65}$$

The first equation describes the accelerated motion in the direction of the vector $\mathbf{B}$ (i.e., $\mathbf{E}_\parallel$). In order to investigate the second equation (11.65) we introduce the new system of coordinates, moving with the constant velocity $\mathbf{w}$ with respect to the old one and perpendicular to the vector $\mathbf{B}$. Then (compare (11.30) and (11.35))

$$\mathbf{v}'_\perp = \mathbf{v}_\perp - \mathbf{w}, \quad \mathbf{B}' = \mathbf{B}, \quad \mathbf{E}'_\perp = \mathbf{E}_\perp + \mathbf{w} \times \mathbf{B}. \tag{11.66}$$

Consequently, the transformed equation of motion has the form

$$m\frac{d\mathbf{v}'_\perp}{dt} = Q\left(\mathbf{E}'_\perp + \mathbf{v}'_\perp \times \mathbf{B}\right). \tag{11.67}$$

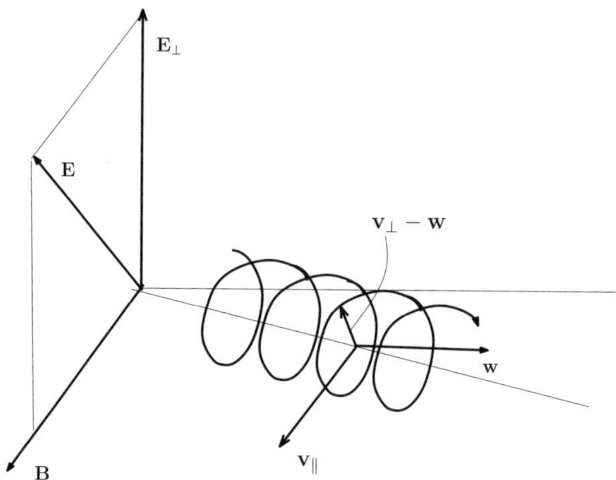

Figure 11.3: *Schematic to the hodograph of a charged particle in the constant electromagnetic field.*

Let us define the relative velocity in such a way that $\mathbf{E}'_\perp = 0$. Then

$$\mathbf{E}'_\perp \times \mathbf{B} = 0 = \mathbf{E} \times \mathbf{B} + \mathbf{B} \times (\mathbf{B} \times \mathbf{w}) = \mathbf{E} \times \mathbf{B} + \underbrace{(\mathbf{w} \cdot \mathbf{B})\mathbf{B}}_{=0} - B^2 \mathbf{w}$$

$$\Rightarrow \mathbf{w} = \frac{\mathbf{E} \times \mathbf{B}}{B^2}. \tag{11.68}$$

The velocity $\mathbf{w}$ is called the drift velocity. Consequently, Equation (11.67) yields

$$m\frac{d\mathbf{v}'_\perp}{dt} = Q\mathbf{v}'_\perp \times \mathbf{B}, \quad \text{i.e.,} \quad \frac{d(\mathbf{v}_\perp - \mathbf{w})}{dt} = \omega_L(\mathbf{v}_\perp - \mathbf{w}) \times \frac{\mathbf{B}}{B}, \quad \omega_L = \frac{QB}{m}. \tag{11.69}$$

This equation describes the precession ($|\mathbf{v}_\perp - \mathbf{w}| = \text{const.}$) around the direction of the magnetic flux $\frac{\mathbf{B}}{B}$ with the so-called Larmor (cyclotron or gyro-) frequency ω_L. The whole motion of the particle is the superposition of the accelerated motion $\mathbf{v}_\parallel$, the drift $\mathbf{w}$ and the precession $\mathbf{v}_\perp - \mathbf{w}$. It is schematically shown in Figure 11.3.

Typical orders of magnitude for tokamak parameters ($B = 3[\text{T}]$) are $\omega_{L\ \text{electron}} = \omega_C = 5.3 \cdot 10^{11}[1/\text{s}]$ (i.e., 84 GHz) for electrons and $\omega_{L\ \text{ion}} = 2.9 \cdot 10^{8}[1/\text{s}]$ (i.e., 46 MHz) for ions. The quantity ω_C is the cyclotron frequency characteristic for all sorts of plasma.

2. *Separation of charges*

As mentioned in the remarks on the phenomenology of plasmas one of the most fundamental properties of those substances is their electric quasi-neutrality. This yields restrictions on the separation of particles. Such a separation is always connected with a very high energy requirement. For instance, in typical MHD-generators the density of particles is approximately $n \approx 10^{19}$ electrons/m^3. The change of this concentration on 1% at the distance of 1 cm requires an electric field of approximately 10^5 V/cm. If the plasma is not subjected to the action of any external field the separation of charges must be connected with the change of the kinetic energy of the particles. Let us say that d is a typical separation of charges when the neutrality is violated. Then the energy density is of the order of magnitude $\sim e^2 n/\varepsilon_0 d^2 \sim kT$, where e is the charge of the electron, k is the Boltzmann constant. Consequently, $d \approx \sqrt{\varepsilon_0 kT/ne^2} = 69\sqrt{T/n}$ [9] which yields for MHD-generators ($n \approx 10^{19}$ [1/m^3], $T = 2500$ [K]) $d \approx 10^{-6}$ [m]. Hence, the quasi-neutrality is typical for normal circumstances.

3. *Plasma frequency*

The quasi-neutrality yields the notion of the time of separation of charges. We present here a simplified consideration based on the set of equations which we justify further in this chapter. Let us impose a small motion of electrons in the x_1-direction on a steady state of plasma

$$n^e = n_0^e + \delta n^e, \quad n^j = n_0^j,$$
$$\mathbf{v}^e = \delta \mathbf{v}^e = \delta v \mathbf{e}_1, \tag{11.70}$$
$$\mathbf{E} = \delta \mathbf{E} = \delta E \mathbf{e}_1,$$

where n^e, n^j are particle number densities of electrons[10] and ions and the index zero corresponds to the initial steady state. These disturbances must satisfy the following equations

- partial mass balance of electrons; divided by the mass of the electron this yields the equation

$$\frac{\partial n^e}{\partial t} + \operatorname{div}(n^e \mathbf{v}^e) = 0 \quad \Rightarrow \quad \frac{\partial \delta n^e}{\partial t} + n_0^e \frac{\partial \delta v}{\partial x} = 0, \tag{11.71}$$

- partial momentum balance

$$\frac{\partial n^e \mathbf{v}^e}{\partial t} + \operatorname{div}\left(n^e \mathbf{v}^e \otimes \mathbf{v}^e - \frac{1}{\mu^e}\mathbf{T}^e\right) = -en^e \frac{1}{\mu^e}\mathbf{E}$$
$$\Rightarrow n_0^e \frac{\partial \delta v}{\partial t} = -en_0^e \frac{1}{\mu^e}\delta E, \tag{11.72}$$

(compare Chapter 11 of Part I) where we assume that the disturbance does not change essentially the partial stresses $\mathbf{T}^e$; here e is the charge of the electron and μ^e is the atomic weight ("molecular mass"; compare Section 6.1 of Part I) of the electron,

---

[9]The constants are as follows (compare (11.24)): $\varepsilon_0 = 10^7/4 \cdot \pi \cdot c^2 = 0.8854923893 \cdot 10^{-11}$ [As/Vm], $k = 1.38044 \cdot 10^{-23}$ [J/K], $e = 1.602 \cdot 10^{-19}$ [C]. Hence: $\sqrt{\varepsilon_0 \cdot k/e^2} = 69$ [1/$\sqrt{\text{m} \cdot \text{K}}$].

[10]In models of plasmas the particle number densities replace the partial mass densities of the mixture theory. Their definitions are not very precise because such quantum particles as electrons possess a well-defined notion of the rest energy but the "molecular" mass is prescribed only on the basis of Einstein's mass-energy equivalence principle. For instance for electrons, their rest energy is 0.511 [MeV] (mega-electronvolt). This corresponds to the rest mass (atomic weight) $\mu^e = 9.109 \cdot 10^{-28}$ [g] and $n^e = \rho^e/\mu^e$, where ρ^e is the partial mass density of electrons. In this sense the common statement that the rest mass of the electron is approximately 1/1837 of the mass of the proton (atomic weight of hydrogen H is $\mu_H = 1.67329 \cdot 10^{-24}$ [g], i.e., $\mu_H/\mu^e = 1836.96$) refers rather to their energies. However, the large difference in the rest mass of electrons and protons has an important bearing in the macroscopic thermodynamical modeling of plasmas because it yields statements concerning the characteristic times of relaxation to the local equilibrium – due to this difference temperatures of electrons and protons remain different for a long time.

- Maxwell (Ampère-Oersted) equation

$$\varepsilon_0 \frac{\partial \mathbf{E}}{\partial t} = -\mathbf{J} \quad \Rightarrow \quad \varepsilon_0 \frac{\partial \delta E}{\partial t} - e n_0^e \delta v = 0. \tag{11.73}$$

The last two equations yield

$$\varepsilon_0 n_0^e \frac{\partial^2 \delta v}{\partial t^2} = -\frac{e n_0^e}{\mu^e} \frac{\partial \delta E}{\partial t} = -\frac{(e n_0^e)^2}{\mu^e} \delta v. \tag{11.74}$$

Hence, the disturbance of the velocity describes, obviously, harmonic vibrations around the steady state, i.e.,

$$\delta v = V(x)\, e^{i\omega_P t}, \quad \text{where} \quad \omega_P = \sqrt{\frac{e^2 n_0^e}{\varepsilon_0 \mu^e}}, \tag{11.75}$$

is the so-called plasma frequency. For instance, for the particle density of electrons $n^e = 10^{18}$ electrons/m^3 it is approximately 10^{10} Hz. This corresponds to the characteristic time $t_{separ} = \frac{1}{\omega_P} = 0.1$ ns. This is the time of separation of charges in the neutral plasma. If the plasma were not neutral there would be a total charge yielding, for instance, additional interactions with magnetic fields. We discuss these problems in some details further in this chapter.

4. *Debye radius*

The quasi-neutrality yields also the notion of the range of interactions in plasmas which is called the Debye sphere. This is the domain in which one can separate electric charges. In order to define this range we need some simple results of the kinetic theory of plasma which we shall not justify in this book (e.g. see [200]). Let Q be a point charge introduced into a homogeneous plasma in which the electrons have the temperature T^e and the ions the temperature T^i. We consider the case in which the plasma is a mixture of these two components and the partial temperatures are defined by means of kinetic averages from the microscopic equations (compare Relation (7.42) of Part I). The electric field $\mathbf{E}$ in the absence of the magnetic flux $\mathbf{B}$ possesses a potential φ, i.e., we have to fulfill the equations (see Table 11.2)

$$\operatorname{div} \mathbf{D} = q, \quad \mathbf{E} = \operatorname{grad} \varphi, \quad \mathbf{D} = \varepsilon_0 \mathbf{E}, \tag{11.76}$$

with

$$q = q^e \tilde{n}^e + q^i \tilde{n}^i + Q \delta(\mathbf{x}), \tag{11.77}$$

and $\tilde{n}^{e,i}$ being the particle number densities related to the equilibrium densities $n^{e,i}$ (without an electric field) by the relations

$$\tilde{n}^{e,i} = n^{e,i} \exp\left(-\frac{q^{e,i}\varphi}{kT^{e,i}}\right). \tag{11.78}$$

These relations, expressing the equilibrium distribution (Gibbs canonical distribution under the potential energy $q^{e,i}\varphi$) follow from simple statistical considerations (e.g. [307], Section 6.4) which we shall not discuss here any further.

The quasi-neutrality condition has for this case the following form

$$q^e n^e + q^i n^i = 0. \tag{11.79}$$

Consequently, the above equations can be reduced to the following equation for the electric potential

$$\varepsilon_0 \operatorname{div} \operatorname{grad} \varphi + Q\delta\left(\mathbf{x}\right) + q^e n^e \exp\left(-\frac{q^e \varphi}{kT^e}\right) + q^i n^i \exp\left(-\frac{q^i \varphi}{kT^i}\right) = 0. \qquad (11.80)$$

Under the condition that the temperatures are high enough for the approximation of the exponential functions by the first two contributions to the Taylor expansion we arrive at the following Poisson-Boltzmann equation for the potential

$$\varepsilon_0 \operatorname{div} \operatorname{grad} \varphi + Q\delta\left(\mathbf{x}\right) - \left(\frac{\left(q^e\right)^2 n^e}{kT^e} + \frac{\left(q^i\right)^2 n^i}{kT^i}\right)\left(q^e\right)^2 n^e \varphi = 0. \qquad (11.81)$$

This equation can be easily solved by the Fourier transform

$$\varphi\left(\mathbf{x}\right) = \int e^{i\mathbf{k}\cdot\mathbf{x}}\bar{\varphi}\left(\mathbf{k}\right) d\mathbf{k} \quad \text{with} \quad \delta\left(\mathbf{x}\right) = \frac{1}{\left(2\pi\right)^3}\int e^{i\mathbf{k}\cdot\mathbf{x}}d\mathbf{k}. \qquad (11.82)$$

It follows

$$\varphi\left(\mathbf{x}\right) = \frac{Q}{\left(2\pi\right)^3}\int \frac{e^{i\mathbf{k}\cdot\mathbf{x}}}{k^2 + 1/r_D^2}d\mathbf{k}, \qquad (11.83)$$

where

$$\frac{1}{r_D^2} = \frac{1}{\varepsilon_0}\left(\frac{\left(q^e\right)^2 n^e}{kT^e} + \frac{\left(q^i\right)^2 n^i}{kT^i}\right). \qquad (11.84)$$

The roots of the denominator in (11.83) are $k = \pm i/r_D$ which yields finally

$$\varphi\left(\mathbf{x}\right) = \frac{Q}{|\mathbf{x}|}e^{-|\mathbf{x}|/r_D}. \qquad (11.85)$$

It is clear that for $|\mathbf{x}| \gtrsim r_D$ the potential becomes almost identical with this of the vacuum: $Q/|\mathbf{x}|$. Then the influence of the plasma background can be neglected. Hence, we obtain the screening effect resulting from the assumption on quasi-neutrality.

In the particular case $n^e = n^i = n$, $T^e = T^i = T$ (i.e., $-q^e = q^i = e$) we have

$$r_D = \sqrt{\frac{\varepsilon_0 2kT}{e^2 n}} = \frac{1}{\omega_P}\sqrt{\frac{2kT}{\mu^e}}. \qquad (11.86)$$

This quantity is called the Debye radius (Debye length). The quantity $v_{\text{therm}} = \sqrt{2kT/\mu^e}$ is called the thermal velocity. In Table 11.4 (excerpted from the Caltech Course: *Applications of Classical Physics* of Kip S. Thorne) some typical values of the Debye length are quoted.

The thermal velocity allows to define the important notion of the cyclotron radius (gyroradius)

$$R = \frac{v_{\parallel}}{\omega_L} \approx \frac{v_{\text{therm}}}{\omega_L} = \frac{\sqrt{2\mu^e kT}}{eB}, \qquad (11.87)$$

where $v_{\parallel}$ is the speed of the accelerated motion (see Figure 11.2). This radius is of the same order of magnitude as the Debye length. Incidentally, the thermal speed of electrons in tokamaks is of the order of magnitude ($B = 3[\text{T}]$) of $5.9 \cdot 10^7[\text{m/s}]$ which

Table 11.4: *Some parameters of plasmas.*

Plasma	Number density [$1/m^3$]	Temperature of electrons [K]	Magnetic flux [T]	Debye length [m]
gas discharge	10^{16}	10^4	-	10^{-4}
tokamak	10^{20}	10^8	10	10^{-4}
ionosphere	10^{12}	10^3	10^{-5}	10^{-3}
solar wind	10^6	10^5	10^{-9}	10
interstellar medium	10^5	10^4	10^{-10}	10
intergalactic medium	1	10^6	-	10^5

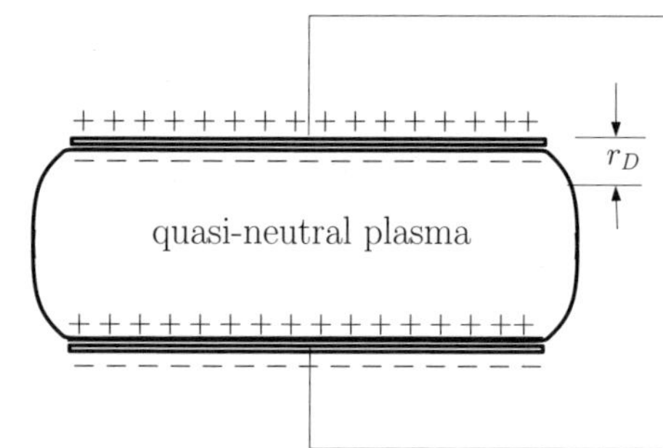

Figure 11.4: *Example of the screening effect caused by the quasi-neutrality of plasma.*

means that $\gamma \approx 0.9613$ and one can apply the non-relativistic version of Maxwell's equations.

The result (11.87) shows that the domain of neutrality of plasmas corresponding to the Representative Elementary Volume of the theory of mixtures (see Chapter 12 of Part I) has a kinematic character (the relation to the characteristic time given by the plasma frequency) and it is temperature dependent.

In Figure 11.4 we show the example of screening. The plasma between the two electrodes forms boundary layers of a thickness approximately of the Debye length r_D remaining neutral in the interior.

11.4.3   MHD-model derived from the theory of miscible mixtures

Let us recall the fundamental equations of the thermodynamics of miscible mixtures which were presented in Chapter 11 of Part I. We denote by $\mu^\alpha, \alpha = 1, ..., A$, the molecular mass (the atomic weight for atoms) of the component α (compare the presentation of this issue in Section 6.1 of Part I). After accounting for the action of the electromagnetic field the revised partial balance equations in their local form and the Maxwell equations are as follows

$$\frac{\partial n^\alpha}{\partial t} + \operatorname{div}\left(n^\alpha \mathbf{v}^\alpha\right) = \frac{\hat{\rho}^\alpha}{\mu^\alpha}, \qquad \sum_{\alpha=1}^{A} \hat{\rho}^\alpha = 0, \quad \sum_{\alpha=1}^{A} \frac{q^\alpha \hat{\rho}^\alpha}{\mu^\alpha} = 0,$$

$$\frac{\partial\left(n^\alpha \mathbf{v}^\alpha\right)}{\partial t} + \operatorname{div}\left(n^\alpha \mathbf{v}^\alpha \otimes \mathbf{v}^\alpha - \frac{1}{\mu^\alpha}\mathbf{T}^\alpha\right)$$
$$= \frac{q^\alpha}{\mu^\alpha} n^\alpha \left(\mathbf{E} + \mathbf{v}^\alpha \times \mathbf{B}\right) + n^\alpha \mathbf{b}^\alpha + \frac{\hat{\mathbf{p}}^\alpha}{\mu^\alpha}, \qquad \sum_{\alpha=1}^{A} \hat{\mathbf{p}}^\alpha = 0,$$

$$\frac{\partial}{\partial t}\left(n^\alpha\left(\varepsilon^\alpha + \frac{1}{2}\mathbf{v}^\alpha \cdot \mathbf{v}^\alpha\right)\right) + \operatorname{div}\left(n^\alpha\left(\varepsilon^\alpha + \frac{1}{2}\mathbf{v}^\alpha \cdot \mathbf{v}^\alpha\right)\mathbf{v}^\alpha + \frac{1}{\mu^\alpha}\mathbf{q}^\alpha - \frac{1}{\mu^\alpha}\mathbf{T}^\alpha\mathbf{v}^\alpha\right)$$
$$= \frac{q^\alpha}{\mu^\alpha} n^\alpha \mathbf{v}^\alpha \cdot \mathbf{E} + n^\alpha \mathbf{b}^\alpha \cdot \mathbf{v}^\alpha + n^\alpha r^\alpha + \frac{\hat{\varepsilon}^\alpha}{\mu^\alpha}, \qquad \sum_{\alpha=1}^{A} \hat{\varepsilon}^\alpha = 0,$$

$$\frac{1}{c^2}\frac{\partial \mathbf{E}}{\partial t} + \mu_0 \mathbf{J} = \operatorname{rot}\mathbf{B}, \qquad \frac{\partial \mathbf{B}}{\partial t} = -\operatorname{rot}\mathbf{E},$$

$$\operatorname{div}\mathbf{B} = 0, \quad \mathbf{J} = \sum_{\alpha=1}^{A} q^\alpha n^\alpha \mathbf{v}^\alpha, \quad \sum_{\alpha=1}^{A} q^\alpha n^\alpha = 0 \quad \Rightarrow \quad \operatorname{div}\mathbf{E} = 0,$$

$$\tag{11.88}$$

where $\hat{\rho}^\alpha, \hat{\mathbf{p}}^\alpha, \hat{\varepsilon}^\alpha$ are mass, momentum and energy sources, $\mathbf{T}^\alpha$ is the thermomechanical partial Cauchy stress tensor, $\hat{\varepsilon}^\alpha$ is the partial internal energy, $\mathbf{q}^\alpha$ the partial heat flux, $\mathbf{b}^\alpha$ the partial body force and r^α the partial energy radiation. We have assumed the Maxwell-Lorentz relations $\mathbf{D} = \varepsilon_0\mathbf{E}$, $\mathbf{B} = \mu_0\mathbf{H}$, $\varepsilon_0\mu_0 = 1/c^2$ to hold.

In comparison with the classical theory of mixtures proposed by C. A. Truesdell [388, 389] the first, thermomechanical, part of this set contains additional terms following from the aforementioned interaction with the electromagnetic field. This is the Lorentz force density in the momentum balance proportional to the charge density of the component α: $(q^\alpha n^\alpha)$ and the Joule heating for to the component α: $(q^\alpha n^\alpha \mathbf{v}^\alpha \cdot \mathbf{E})$ in the energy balance. In addition, for the mass balance it is not only assumed that the mass sources are balanced but also that there is no production of charges. This is the last condition in the first line. The form of the electric current $\mathbf{J}$ is assumed to be very restricted and contains only the diffusive part

$$\mathbf{J} = \sum_{\alpha=1}^{A} q^\alpha n^\alpha \mathbf{v}^\alpha = \sum_{\alpha=1}^{A} q^\alpha n^\alpha \mathbf{u}^\alpha, \quad \mathbf{u}^\alpha = \mathbf{v}^\alpha - \mathbf{v}, \quad \mathbf{v} = \sum_{\alpha=1}^{A} c^\alpha \mathbf{v}^\alpha, \quad c^\alpha = \frac{\mu^\alpha n^\alpha}{\displaystyle\sum_{\beta=1}^{A} \mu^\beta n^\beta}, \tag{11.89}$$

where $\mathbf{u}^\alpha$ is the so-called diffusion velocity, c^α denotes the mass concentration of the component α and the above relation follows from the quasi-neutrality condition. It is equivalent to the assumption that the electric conductivity of each component is assumed to be zero. This is justified in the case of rarefied plasmas which physically consist of concentrated charges moving far apart in vacuum. This assumption yields an additional constraint on the motion. Namely, the conservation of charge (11.21) simplifies due to the quasi-neutrality and this yields

$$\operatorname{div}\mathbf{J} = 0 \quad \Rightarrow \quad \operatorname{div}\left(\sum_{\alpha=1}^{A} q^{\alpha} n^{\alpha}\mathbf{u}^{\alpha}\right) = 0, \tag{11.90}$$

which clearly restricts the diffusion velocities. Again the quasi-neutrality restricts the Gauss equation for the dielectric displacement and yields the last Relation (11.88).

The above set of equations forms the basis for the construction of field equations for the following fields

$$\mathcal{V}_A = \{n^{\alpha}, \mathbf{v}^{\alpha}, T^{\alpha}, \mathbf{E}, \mathbf{B}\}. \tag{11.91}$$

However, we have to solve first, as usual, the closure problem which means that we have to specify the constitutive relations for the following quantities

$$\mathcal{F}_A = \{\hat{\rho}^{\alpha}, \mathbf{T}^{\alpha}, \hat{\mathbf{p}}^{\alpha}, \varepsilon^{\alpha}, \mathbf{q}^{\alpha}, \hat{\varepsilon}^{\alpha}\}. \tag{11.92}$$

In the case that internal energy radiation processes must be included, we need also a constitutive relation for r^{α}.

We present further the details of this construction for a two-component mixture. However, a few remarks concerning the physical interpretation of a general case are appropriate. First of all, the contribution of mass sources describes the production of both partial mass and charges. Such processes in multi-component plasmas may appear indeed as they result from the dissociation, recombination and ionization of ions and neutral particles. These processes appear usually in the range of high temperatures.[11] In many cases of practical interest it can be assumed that the components of plasma behave like ideal gases. Then, the constitutive relations for the partial stresses and the partial energies become simple. We return further to this problem. The momentum sources $\hat{\mathbf{p}}^{\alpha}$ appear primarily due to the diffusion and the presence of temperature gradients. For the two-component model this means among other properties that this source is induced by the electric current. We discuss this issue further. The partial heat flux vectors $\mathbf{q}^{\alpha}$ and the energy sources $\hat{\varepsilon}^{\alpha}$ control the relaxation processes yielding local thermodynamical equilibria with a common temperature for all components. This issue is still controversial in continuum thermodynamics. The second law of thermodynamics requires the continuity of temperatures on the so-called ideal walls (compare Section 11.2 of Part I). This condition is clearly violated in the mixture theory with multiple temperatures. Thus, the partial temperature cannot be controlled on the boundary of the system and this, obviously, yields difficulties with the formulation of boundary value problems for the heat conduction equations. As the states with different partial temperatures are common in plasmas we present the problem of relaxation to the local equilibrium further in this chapter but the general thermodynamics of such systems is still to be developed.

The constitutive relations should, obviously, satisfy the general conditions of continuum thermodynamics

1. the material objectivity (frame indifference),

2. the second law of thermodynamics.

---

[11]The ionization of a neutral hydrogen requires the energy of 13.5 [eV] which corresponds to temperatures $13.5 \cdot 11600 = 156600$ [K]. Usually other substances require even higher temperatures. The ionization may appear at lower temperatures due to the so-called tail in the Maxwell velocity distribution, i.e., high kinetic energies of very few particles. This yields plasmas of a low degree of ionization. They appear, for instance, in processes of laser heating, metal cutting and welding.

Within the frame presented above the first problem is rather simple. The principle of the material objectivity is violated by the Maxwell-Lorentz aether relations but this is not crucial as we have shown before. The relation for the electric current which for the case under discussion is given by the diffusion velocity (see (11.89)) is due to the quasi-neutrality objective. All other conditions are practically trivial and we shall not go here into any details.

The problem of the second law of thermodynamics reduces to the entropy inequality as we indicated in Section 1.3 of this chapter. The condition on ideal walls is still controversial but it seems to be practically not very essential.

Similarly to the usual mixture theory we define the bulk quantities of the model. This concerns, obviously, only thermomechanical quantities because the electromagnetic fields possess already a bulk character. Under the Truesdell condition [388] that the model should fulfill the usual conservation conditions of mass, momentum and energy we obtain (compare Relations (11.9) in Part I)

$$\rho = \sum_{\alpha=1}^{A} \mu^{\alpha} n^{\alpha}, \quad \rho \mathbf{v} = \sum_{\alpha=1}^{A} \mu^{\alpha} n^{\alpha} \mathbf{v}^{\alpha},$$

$$\mathbf{T} = \sum_{\alpha=1}^{A} \left(\mathbf{T}^{\alpha} - \mu^{\alpha} n^{\alpha} \mathbf{u}^{\alpha} \otimes \mathbf{u}^{\alpha} \right), \quad \rho \mathbf{b} = \sum_{\alpha=1}^{A} \mu^{\alpha} n^{\alpha} \mathbf{b}^{\alpha},$$

$$\rho \varepsilon = \sum_{\alpha=1}^{A} \mu^{\alpha} n^{\alpha} \left(\varepsilon^{\alpha} + \tfrac{1}{2} \mathbf{u}^{\alpha} \cdot \mathbf{u}^{\alpha} \right),$$

$$\mathbf{q} = \sum_{\alpha=1}^{A} \left(\mathbf{q}^{\alpha} + \mu^{\alpha} n^{\alpha} \left(\varepsilon^{\alpha} + \tfrac{1}{2} \mathbf{u}^{\alpha} \cdot \mathbf{u}^{\alpha} \right) \mathbf{u}^{\alpha} - \mathbf{T}^{\alpha} \mathbf{u}^{\alpha} \right), \quad \rho r = \sum_{\alpha=1}^{A} \mu^{\alpha} n^{\alpha} \left(r^{\alpha} + \mathbf{b}^{\alpha} \cdot \mathbf{u}^{\alpha} \right),$$

$$(11.93)$$

where the diffusion velocity $\mathbf{u}^{\alpha}$ is defined by Relation (11.89). Obviously, $\mathbf{T}$ denotes the thermomechanical bulk Cauchy stress tensor, ε is the bulk specific internal energy, $\mathbf{q}$ is the bulk heat flux and r the bulk radiation. Then the one-component thermomechanical balance equations have the form (compare (11.58))

$$\frac{\partial \rho}{\partial t} + \operatorname{div} \left(\rho \mathbf{v} \right) = 0,$$

$$\frac{\partial \left(\rho \mathbf{v} \right)}{\partial t} + \operatorname{div} \left(\rho \mathbf{v} \otimes \mathbf{v} - \mathbf{T} \right) = \mathbf{J} \times \mathbf{B} + \rho \mathbf{b},$$

$$\frac{\partial}{\partial t} \left(\rho \left(\varepsilon + \frac{1}{2} \mathbf{v} \cdot \mathbf{v} \right) \right) + \operatorname{div} \left(\rho \left(\varepsilon + \frac{1}{2} \mathbf{v} \cdot \mathbf{v} \right) \mathbf{v} + \mathbf{q} - \mathbf{T} \mathbf{v} \right) = \mathbf{J} \cdot \mathbf{E} + \rho \mathbf{b} \cdot \mathbf{v} + \rho r.$$

$$(11.94)$$

Certainly, they have the same form as the corresponding conservation laws of the single component continuum with two source terms following from the interaction with the electromagnetic field: the reduced Lorentz force $\mathbf{J} \times \mathbf{B}$ (the contribution of the electric field $\mathbf{E}$ does not appear due to the quasi-neutrality!) and the Joule heating $\mathbf{J} \cdot \mathbf{E}$. We return further to these equations when we discuss the equations of the classical MHD-theory.

11.4.4　Two-component model

We proceed to present some details of the simplest multi-component model. For simplicity we assume that it is a fully ionized hydrogen plasma. More general cases can also be easily constructed. Then the quasi-neutrality condition yields

$$n^e = n^i =: n, \quad \hat{\rho}^e = \hat{\rho}^i = 0, \quad \hat{\mathbf{p}}^e = -\hat{\mathbf{p}}^i =: \hat{\mathbf{p}}, \quad \hat{\varepsilon}^e = -\hat{\varepsilon}^i =: \hat{\varepsilon}. \tag{11.95}$$

Obviously, the indices e and i refer to electrons and ions. The fields of this model are

$$\mathcal{V}_2 = \left\{ n, \mathbf{v}^e, \mathbf{v}^i, T^e, T^i, \mathbf{E}, \mathbf{H} \right\}. \tag{11.96}$$

It is convenient to transform the partial velocities in this model in the following way $\{\mathbf{v}^e, \mathbf{v}^i\} \to \{\mathbf{v}, \mathbf{J}\}$ according to the obvious relations

$$\rho = \mu^e n + \mu^i n \approx \mu^i n,$$

$$\rho\mathbf{v} = \mu^e n \mathbf{v}^e + \mu^i n \mathbf{v}^i \quad \Rightarrow \quad \mathbf{v} = \mathbf{v}^i + \frac{\mu^e}{\mu^i}\mathbf{v}^e,$$

$$\mathbf{J} = q^e n \mathbf{v}^e + q^i n \mathbf{v}^i \quad \Rightarrow \quad \frac{1}{en}\mathbf{J} = \mathbf{v}^i - \mathbf{v}^e, \tag{11.97}$$

where $q^e = -q^i = -e$ is the charge of the electron.[12] Hence,

$$\mathbf{v}^e = \mathbf{v} - \frac{1}{en}\mathbf{J}, \quad \mathbf{v}^i = \mathbf{v} + \frac{1}{en}\mathbf{J}\frac{\mu^e}{\mu^i} \quad \Rightarrow \quad \mathbf{u}^e = -\frac{1}{en}\mathbf{J}, \quad \mathbf{u}^i = \frac{1}{en}\mathbf{J}\frac{\mu^e}{\mu^i}. \tag{11.98}$$

With this transformation in mind, we can replace the partial mass balance equations with the bulk mass conservation. The second equation is identical with the condition for the electric current, i.e.,

$$\frac{\partial n}{\partial t} + \operatorname{div}(n\mathbf{v}) = 0, \quad \rho = (\mu^e + \mu^i)\, n \approx \mu^i n,$$

$$\operatorname{div} \mathbf{J} = 0. \tag{11.99}$$

Simultaneously, the partial momentum balance equations can be replaced by the bulk momentum conservation and the combination of the partial equations which yields the equation for the current. We have then

$$\frac{\partial(n\mathbf{v})}{\partial t} + \operatorname{div}\left(n\mathbf{v} \otimes \mathbf{v} - \frac{1}{\mu^i}\mathbf{T}\right) = \frac{1}{\mu^i}\mathbf{J} \times \mathbf{B} + n\mathbf{b},$$

$$\frac{\partial \mathbf{J}}{\partial t} + \operatorname{div}\left(\mathbf{J} \otimes \mathbf{v} + \mathbf{v} \otimes \mathbf{J} - \frac{1}{en}\mathbf{J} \otimes \mathbf{J} + \frac{e}{\mu^e}\mathbf{T}^e - \frac{e}{\mu^i}\mathbf{T}^i\right) \tag{11.100}$$

$$= \frac{e^2 n}{\mu^e}(\mathbf{E} + \mathbf{v} \times \mathbf{B}) - \frac{e}{\mu^e}\mathbf{J} \times \mathbf{B} - \frac{e}{\mu^e}\hat{\mathbf{p}} + en(\mathbf{b}^i - \mathbf{b}^e),$$

where

$$\mathbf{T} = \mathbf{T}^e + \mathbf{T}^i - \frac{\mu^e}{e^2 n}\mathbf{J} \otimes \mathbf{J}, \tag{11.101}$$

and the contributions of the order $(\mu^e/\mu^i)^k$, $k = 1, 2$, were neglected. The derivation of the second equation is as follows. In Equations $(11.88)_2$, we multiply the equation for

---

[12] $1[e] = 1.60218 \cdot 10^{-19}$ [C]$= 4.80320 \cdot 10^{-10}$ [statCoulomb]

$\alpha = e$ by $-e$ and the equation for $\alpha = i$ by e and add. Bearing Relations (11.98) in mind and neglecting higher order terms we obtain $(11.100)_2$.

Now we make the following operation on the partial energy balance equations. We replace one of the partial equations by the bulk energy conservation law and combine the two partial equations in a way similar to this for partial momenta. The result is

$$\frac{\partial}{\partial t}\left(\rho\left(\varepsilon+\frac{1}{2}\mathbf{v}\cdot\mathbf{v}\right)\right)+\operatorname{div}\left(\rho\left(\varepsilon+\frac{1}{2}\mathbf{v}\cdot\mathbf{v}\right)\mathbf{v}+\mathbf{q}-\mathbf{Tv}\right)=\mathbf{J}\cdot\mathbf{E}+\rho\mathbf{b}\cdot\mathbf{v}+\rho r,$$

$$\mu^e n\left[\frac{\partial}{\partial t}\left(\varepsilon^i-\varepsilon^e\right)+\mathbf{v}\cdot\operatorname{grad}\left(\varepsilon^i-\varepsilon^e\right)\right]+\frac{\mu^e}{e}\mathbf{J}\cdot\operatorname{grad}\left(\varepsilon^e+\frac{\mu^e}{\mu^i}\varepsilon^i\right)$$

$$=\operatorname{tr}\left[\left(\frac{\mu^e}{\mu^i}\mathbf{T}^i-\mathbf{T}^e\right)\operatorname{grad}\mathbf{v}\right] \tag{11.102}$$

$$+\operatorname{tr}\left[\left(\mathbf{T}^e+\left(\frac{\mu^e}{\mu^i}\right)^2\mathbf{T}^i\right)\operatorname{grad}\left(\frac{1}{en}\mathbf{J}\right)\right]$$

$$+\operatorname{div}\left(\mathbf{q}^e-\frac{\mu^e}{\mu^i}\mathbf{q}^i\right)+\mu^e n\left(r^i-r^e\right)-\hat{\varepsilon}+\mathbf{v}\cdot\hat{\mathbf{p}}-\frac{1}{en}\mathbf{J}\cdot\hat{\mathbf{p}}.$$

Equations $(11.99)_1$, $(11.100)_1$ and $(11.103)_1$ have the form of bulk balance equations which, roughly speaking, govern the behavior of the fields $\{n,\mathbf{v},T\}$ where T is some average of two partial temperatures T^e,T^i. As already mentioned Equation $(11.99)_2$ is a constraint which is directly connected with the condition of quasi-neutrality. The bulk quantities of Equation $(11.103)_1$ are defined as follows

$$\rho\varepsilon\approx\mu^i n\varepsilon=\mu^i n\left(\varepsilon^i+\frac{1}{2}\frac{1}{(en)^2}\mathbf{J}\cdot\mathbf{J}\left(\frac{\mu^e}{\mu^i}\right)^2\right)+\mu^e n\left(\varepsilon^e+\frac{1}{2}\frac{1}{(en)^2}\mathbf{J}\cdot\mathbf{J}\right)$$

$$\Rightarrow\quad \varepsilon=\varepsilon^i+\frac{\mu^e}{\mu^i}\varepsilon^e+\frac{1}{2}\frac{\mu^e}{\mu^i}\frac{1}{(en)^2}\mathbf{J}\cdot\mathbf{J}. \tag{11.103}$$

$$\mathbf{q}=\mathbf{q}^e+\mathbf{q}^i-\frac{\mu^e}{e}\left(\varepsilon^e-\varepsilon^i\right)-\frac{\mu^e}{e}\frac{1}{2}\frac{\mu^e}{\mu^i}\frac{1}{(en)^2}\mathbf{J}\cdot\mathbf{JJ}+\frac{1}{en}\left(\mathbf{T}^e-\mathbf{T}^i\right)\mathbf{J},$$

$$r=r^e+r^i-\frac{\mu^e}{e}\left(\mathbf{b}^e-\mathbf{b}^i\right)\cdot\mathbf{J}.$$

The remaining two equations $(11.100)_2$ and $(11.103)_2$ have a particularly interesting interpretation which we proceed to discuss. Namely, Equation $(11.100)_2$ is the forrunner of Ohm's law relating the electric current $\mathbf{J}$ and the electric field $\mathbf{E}$. Simultaneously, Equation $(11.103)_2$ specifies the conditions of relaxation of two partial temperatures to the common temperature of both components in the local thermodynamical equilibrium. In order to appreciate these properties we have to specify constitutive assumptions. For the two-component plasma which we discuss in this example it is sufficient to assume that both components are ideal gases (see Section 6.1 of Part I, in particular, Relations (6.50),

(6.51) and (6.52))

$$\mathbf{T}^{e,i} = -p^{e,i}\mathbf{1}, \quad \mu^{e,i}n\varepsilon^{e,i} = \frac{3}{2}p^{e,i}, \quad \varepsilon^{e,i} = z\frac{kT^{e,i}}{\mu^{e,i}} + \alpha, \tag{11.104}$$

where the constant $z = 3/2$ and the constant α is immaterial. The justification of this constitutive assumption as well as the form of other constitutive relations, which we use further, follow in the modeling of plasma from the kinetic theory. In particular, the relations of transport coefficients and relaxation times following from the Grad method and from the Chapman-Enskog method result from this approach (see Chapter 19 of the book of S. Chapman and T. G. Cowling [72] and the book of V. M. Zhdanov [457], where the comparison of the methods and many numerical data are presented).

Thermodynamics does not offer any general method of defining the temperature T for mixtures. In the case of ideal gases which we consider in this section we may make the assumption that the constitutive relation for a mixture should have the form of the constitutive relation for the ideal gas. Such an assumption was proposed in a few recent papers of Tommaso Ruggeri and his coworkers (e.g. [328]-[331]). In those papers the intrinsic part of the specific internal energy serves the purpose of the definition of an average temperature. This is indeed a solution provided that we can neglect the contribution of the electric current $\mathbf{J}$ to the specific bulk internal energy $(11.103)_1$. Then we have

$$\varepsilon = 2z\frac{kT}{\mu} + \alpha = z\frac{kT^i}{\mu^i} + \alpha + \frac{\mu^e}{\mu^i}\left(z\frac{kT^e}{\mu^e} + \alpha\right), \quad \mu = \mu^i + \mu^e$$
$$\Rightarrow \quad T = \frac{1}{2}\left(T^i + T^e\right). \tag{11.105}$$

With this definition in mind we can transform Equation $(11.102)_1$ into the heat transfer equation provided we add a constitutive relation for the heat flux $\mathbf{q}$ or the constitutive relations for both partial heat flux vectors $\mathbf{q}^{e,i}$ and then use the definition $(11.103)_2$. Neither of these possibilities seems to be simple. In addition, the contribution of the electric current (i.e., equivalently, the influence of diffusion velocities) can be assumed to be small for such multi-component materials as soils. It is not small at all for plasmas. For instance, in the case of tokamak conditions the temperature of electrons is of the order of 10^8 K. Then, bearing the data of Footnote 9 of this chapter in mind, the correction term (the thermal velocity of electrons is in this case approximately $0.1741 \cdot 10^7$ m/s) yields approximately $0.33 \cdot 10^8$ K. This means that there would be a considerable heat flux between the regions of different electric currents even if the partial temperatures were equal in those regions. Even in the case of the assumption on a linear dependence on the gradient of the temperature T the problem of the heat conductivity coefficient κ is not solved very satisfactorily. Some hints can be found in the book of J. O. Hirschfelder, C. F. Curtiss and R. B. Bird [156] (Section 8.2c).

We return now to the remaining two equations containing the sources of momentum and energy. The simplest constitutive relation for the vector of the momentum source is the linear dependence on the diffusion velocity (see Relation (11.26) of Part I) while the energy source is linearly dependent on the difference of the partial temperatures. Macroscopically these forms are motivated by the dissipation inequality which consists of the heat flux, momentum source and energy source contributions. As they must be positive definite in the simplest case the heat flux is linearly dependent on the temperature gradients and the sources have the above described structure.

In our case, this means that the source $\hat{\mathbf{p}}$ must be proportional to $\mathbf{J}$ (compare (11.98)) and the energy source must be proportional to $(T^e - T^i)$. In the kinetic theory of the multi-component plasma these sources are related to the collision frequencies $\nu_{\alpha\beta}$, $\alpha, \beta = 1, ..., A$, of particles of different components (e.g. see [457]).

$$\hat{\mathbf{p}}^\alpha = -n^\alpha \mu^\alpha \sum_{\alpha=1}^{A} \frac{\mu^\beta}{\mu^\alpha + \mu^\beta} \nu_{\alpha\beta} \left(\mathbf{v}^\alpha - \mathbf{v}^\beta\right), \quad \hat{\varepsilon}^\alpha = -n^\alpha \mu^\alpha \sum_{\alpha=1}^{A} \frac{\mu^\beta}{(\mu^\alpha + \mu^\beta)^2} \nu_{\alpha\beta} 3k \left(T^\alpha - T^\beta\right).$$

(11.106)

For two components, it is only the electron-ion collision frequency $\nu = \nu_{ei}$ which remains in these relations and, bearing the definitions (11.98) in mind, we obtain

$$\hat{\mathbf{p}} = \frac{\mu^e}{e}\nu\mathbf{J}, \quad \hat{\varepsilon} = -\frac{\mu^e}{\mu^i}n\nu 3k \left(T^e - T^i\right).$$

(11.107)

The sources (11.106) may also be linearly dependent on the gradients of the temperatures and the particle density n. We ignore in the first approximation these contributions.

Now, bearing $(11.107)_1$ in mind, Relation $(11.100)_2$ obtains the form

$$\frac{1}{\nu}\left[\frac{\partial \mathbf{J}}{\partial t} + \text{div}\left(\mathbf{J} \otimes \mathbf{v} + \mathbf{v} \otimes \mathbf{J} - \frac{1}{en}\mathbf{J} \otimes \mathbf{J}\right) - e\,\text{grad}\left(nk\left(\frac{T^e}{\mu^e} - \frac{T^i}{\mu^i}\right)\right)\right] + \mathbf{J}$$

$$= \sigma\left(\mathbf{E} + \mathbf{v} \times \mathbf{B}\right) - \frac{\beta}{B}\mathbf{J} \times \mathbf{B}, \quad \sigma = \frac{e^2 n}{\nu\mu^e} = \varepsilon_0\frac{\omega_P^2}{\nu}, \quad \beta = \frac{eB}{\nu\mu^e} = \frac{\omega_L}{\nu}, \quad B = \sqrt{\mathbf{B} \cdot \mathbf{B}}.$$

(11.108)

Without loss of generality we have assumed that $\mathbf{b}^e = \mathbf{b}^i$. The coefficient σ is called the longitudinal electrical conductivity of plasma and the coefficient β is called the Hall coefficient. If we ignore the evolution effects in this equation it follows

$$\mathbf{J} \approx \sigma\left(\mathbf{E} + \mathbf{v} \times \mathbf{B}\right) - \beta\frac{\mathbf{J} \times \mathbf{B}}{B}.$$

(11.109)

This is the Ohm law commonly used in the MHD considerations for plasmas. The first contribution is identical with the original Ohm law without the magnetic field. The omission of the terms at the left-hand side of (11.111) is not always justifiable (compare p. 235 of the book [51]).

The influence of the magnetic field yields anisotropic effects. In order to see it clearly, let us consider a simple example of the law for the special case: $\mathbf{v} = 0$ and $\mathbf{B} = (0, 0, B)^T$. Then follows immediately

$$J_1 = \frac{\sigma}{1 + \beta^2}\left(E_1 - \beta E_2\right), \quad J_2 = \frac{\sigma}{1 + \beta^2}\left(E_2 + \beta E_1\right), \quad J_3 = \sigma E_3.$$

(11.110)

Hence, the Hall effect yields indeed electric anisotropy. The current $\mathbf{J}$ is not parallel to the electric field $\mathbf{E}$ anymore. Moreover, it can be easily shown that the Hall effect induces also the anisotropy of other transport phenomena – heat conduction, diffusion and viscosity. The proportionality of the coefficient β to the strength of the magnetic flux $\mathbf{B}$ shows that the anisotropy effect grows with the magnitude of the magnetic flux.

We proceed to discuss the influence of the energy source $\hat{\varepsilon}$. We substitute the constitutive relations (11.104) and (11.107) in Equation $(11.102)_2$. In order to see the leading

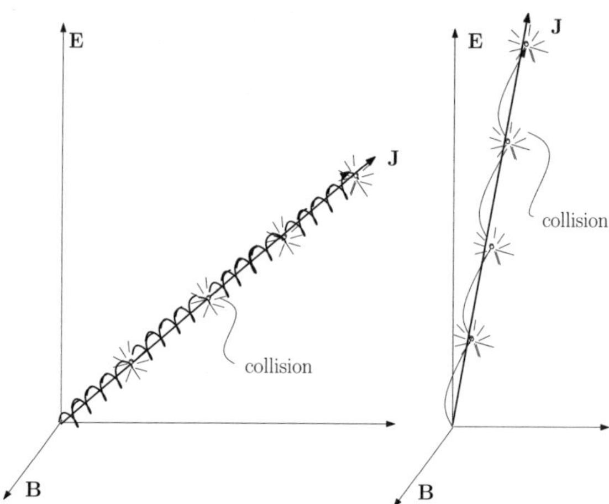

Figure 11.5: *Schematic to the Hall effect. On the left-hand side the gyro- (Larmor) frequency ω_L is much larger than the collision frequency ν, i.e., $\beta \gg 1$. In this case the Hall effect is negligible. The other extreme case is shown on the right-hand side.*

features of the resulting equation we leave out the nonlinear contributions $\mathbf{J} \cdot \operatorname{grad} T^{e,i}$, $\mathbf{J} \cdot \operatorname{grad} n$ and $\mathbf{J} \cdot \mathbf{J}$. It follows that

$$\left(\frac{\partial}{\partial t} + \mathbf{v} \cdot \operatorname{grad}\right)\left(T^e - T^i\right) + 4\frac{\mu^e}{\mu^i}\nu\left(T^e - T^i\right)$$

$$= -\frac{2}{3}\frac{1}{nk}\operatorname{div}\left(\mathbf{q}^e - \mathbf{q}^i\right) + \frac{2}{3}\left(T^e - T^i\right)\left(\frac{\partial}{\partial t} + \mathbf{v} \cdot \operatorname{grad}\right)\ln n. \tag{11.111}$$

Independently from the conduction properties as well as the heterogeneity of the particle density, this equation indicates the relaxation of the partial temperatures to their common value with

$$t_{\text{relaxation}} = \frac{1}{4}\frac{\mu^i}{\mu^e}\frac{1}{\nu}. \tag{11.112}$$

This formula justifies our earlier remark that the large difference in the atomic weight of electrons and ions yields long relaxation times. For hydrogen plasma of the particle density $\sim 10^{18}$ the collision frequency is $\nu = 10^6$ [1/s] which means that $t_{\text{relaxation}} \approx 2 \cdot 10^{-3}$ [s].

11.5   Magnetohydrodynamics of a single component fluid

11.5.1   Dimensionless notation, approximations

The considerations presented for plasma in the previous section motivate the construction of a model in which the thermomechanical part is based on thermomechanical equations of a single component fluid interacting with the electromagnetic field. We present such a model in this section. However, as the main task of those considerations is the description of plasma we still assume the quasi-neutrality and we use the characteristic notions defined for plasmas.

In many practical, particularly astrophysical problems, the following conditions are satisfied

1. The time scale t_0 is of the same order of magnitude as L/V where L is the characteristic length and V is the characteristic velocity of the flow field, i.e.,

$$R_t := \frac{t_0 V}{L} = O(1).$$ (11.113)

   In such problems the characteristic frequencies should be relatively small. For instance, if we take $V = 10^4$ m/s which is a typical high speed of a satellite, and $L = 1$ m, we have $L/V = 10^{-4}$ s. We shall not consider any high frequency phenomena with frequencies much higher than 10^4 Hz.

2. The electrical field which may be characterized by the value E_0 is of the same order of magnitude as the induced electric field $\mathbf{v} \times \mathbf{B}$. In other words, the non-dimensional parameter

$$R_E := \frac{E_0}{V B_0} = O(1),$$ (11.114)

   where B_0 is the strength of the characteristic magnetic flux. Bearing the simplified Ohm law (11.109) without Hall's effect ($\beta \approx 0$) in mind we see that it is a good approximation for very large electric conductivity σ, because as $\sigma \to \infty$ we would expect that $\mathbf{E} \approx -\mathbf{v} \times \mathbf{B}$. Otherwise the conduction current $\mathbf{J}$ will become very large for a slight motion of the plasma.

3. The velocity of the flow is much smaller than the velocity of light c (non-relativistic approximation), i.e.,

$$R_c := \frac{V^2}{c^2} = V^2 \varepsilon_0 \mu_0 = o(1).$$ (11.115)

   Combination of conditions 2 and 3 yields then the following relation for the energy

$$\frac{1}{2}\mathbf{D} \cdot \mathbf{E} \approx \frac{1}{2}\varepsilon_0 E_0^2 \approx \frac{1}{2}\varepsilon_0 V^2 B_0^2 \approx \frac{1}{2}\varepsilon_0 \mu_0^2 V^2 H_0^2 = \frac{1}{2}\mu_0 H_0^2 R_c$$ (11.116)

$$\Rightarrow \quad \frac{1}{2}\mathbf{D} \cdot \mathbf{E} \ll \frac{1}{2}\mathbf{B} \cdot \mathbf{H},$$

   (compare the energy balance equation (11.34)).

4. The radiation terms are negligible. In the case of interactions of the matter and radiation the constitutive relations for the stresses, the energy and the energy flux should be corrected by terms reflecting those interactions. The simplest model (the black-body radiation) consists of the following relations for the radiation pressure, the radiation energy and the radiation energy flux

$$p_R = \frac{1}{3}\sigma_R T^4, \quad \sigma_R = 5.66 \cdot 10^{-8} \ [\text{W/m}^2 \cdot \text{K}^4] - \begin{array}{l} \text{Stefan-Boltzmann} \\ \text{constant} \end{array}$$ (11.117)

$$\varepsilon_R = \sigma_R T^4, \quad \mathbf{q}_R = D_R \operatorname{grad} \varepsilon_R, \quad D_R = \frac{c l_R}{3}, \quad l_R = \frac{1}{K_R \varrho},$$

where l_R is the so-called Rosseland mean free path of radiation and K_R is the opacity.[13] Black-body radiation may appear if the characteristic length $L > R_{BB} = 10^{23} T^{7/2} / (z^2 n^i n^e)$. This may happen for the Bremsstrahlung, i.e., the radiation due to collisions which is changing the momentum of the particles. The influence of dirt is small but it may be important for the diagnosis. The influence of the so-called Cherenkov radiation is small.[14] However, it may be large in cyclotrons.

5. The non-electric body forces are negligible.

We are now in the position to introduce dimensionless variables which are defined as follows

$$\widetilde{\mathbf{x}} = \frac{\mathbf{x}}{L}, \quad \widetilde{\mathrm{grad}} = L\,\mathrm{grad} \quad \widetilde{t} = \frac{t}{t_0}, \quad \widetilde{\mathbf{v}} = \frac{\mathbf{v}}{V},$$

$$\widetilde{\mathbf{E}} = \frac{\mathbf{E}}{E_0}, \quad \widetilde{\mathbf{B}} = \frac{B}{B_0}, \quad \widetilde{\mathbf{J}} = \frac{J}{\sigma V B_0}. \tag{11.118}$$

Then the Maxwell equations have the form

$$\frac{\varepsilon_0 E_0}{t_0} \frac{\partial \widetilde{\mathbf{E}}}{\partial \widetilde{t}} + \sigma V B_0 \widetilde{\mathbf{J}} = \frac{B_0}{L\mu_0} \widetilde{\mathrm{rot}}\,\widetilde{\mathbf{B}},$$

$$\frac{B_0}{t_0} \frac{\partial \widetilde{\mathbf{B}}}{\partial \widetilde{t}} + \frac{E_0}{L} \widetilde{\mathrm{rot}}\,\widetilde{\mathbf{E}} = 0. \tag{11.119}$$

Bearing the notation introduced above in mind we have

$$-R_c \frac{R_E}{R_t} \frac{\partial \widetilde{\mathbf{E}}}{\partial \widetilde{t}} + \widetilde{\mathrm{rot}}\,\widetilde{\mathbf{B}} = R_\sigma \widetilde{\mathbf{J}}, \quad R_\sigma = \sigma \mu_0 L V = \frac{LV}{\nu_H}, \quad \nu_H = \frac{1}{\sigma \mu_0}, \tag{11.120}$$

where R_σ is called the magnetic Reynolds number and ν_H the magnetic viscosity.

It is easy to check that all dimensionless quantities are of the order of magnitude of unity. Consequently, within the non-relativistic MHD, the contribution $\varepsilon_0 \partial \mathbf{E}/\partial t$ is much smaller than the contribution $\widetilde{\mathrm{rot}}\,\widetilde{\mathbf{B}}$ and we can use the approximate relation

$$\widetilde{\mathrm{rot}}\,\widetilde{\mathbf{B}} = R_\sigma \widetilde{\mathbf{J}} \quad \Rightarrow \quad \mathbf{J} = \frac{1}{\mu_0} \mathrm{rot}\,\mathbf{B}, \tag{11.121}$$

instead of (11.120).

On the other hand, the second equation of (11.119) has the form

$$\frac{1}{R_t R_E} \frac{\partial \widetilde{\mathbf{B}}}{\partial \widetilde{t}} + \widetilde{\mathrm{rot}}\,\widetilde{\mathbf{E}} = 0. \tag{11.122}$$

Hence, both terms are of the same order of magnitude.

---

[13]Opacity is the measure of impenetrability to radiation. In radiative transfer, it describes the absorption and scattering of radiation in a medium.

[14]Cherenkov radiation is the electromagnetic radiation emitted when a charged particle (e.g. an electron) passes through a dielectric medium at a speed greater than the phase velocity of light in that medium.

Collecting Maxwell's equations simplified for the non-relativistic model we obtain

$$\mathbf{J} = \frac{1}{\mu_0} \operatorname{rot} \mathbf{B} \quad \Rightarrow \quad \operatorname{div} \mathbf{J} = 0,$$

$$\frac{\partial \mathbf{B}}{\partial t} = -\operatorname{rot} \mathbf{E}, \tag{11.123}$$

$$\mathbf{E} = \frac{1}{\sigma} \mathbf{J} - \mathbf{v} \times \mathbf{B}.$$

It is now easy to eliminate the electric field $\mathbf{E}$ and the electric current $\mathbf{J}$ from the model. From the above equations follows

$$\mathbf{E} = \frac{1}{\sigma} \left[\frac{1}{\mu_0} \operatorname{rot} \mathbf{B} \right] - \mathbf{v} \times \mathbf{B} = \nu_H \operatorname{rot} \mathbf{B} - \mathbf{v} \times \mathbf{B}.$$

Hence,

$$\frac{\partial \mathbf{B}}{\partial t} = \operatorname{rot} (\mathbf{v} \times \mathbf{B}) - \operatorname{rot} (\nu_H \operatorname{rot} \mathbf{B}). \tag{11.124}$$

The magnetic viscosity entering this equation can also be written in the following form

$$\nu_H = \frac{1}{\sigma \mu_0} = \frac{1}{\varepsilon_0 \mu_0} \frac{\nu}{\omega_P^2} = c^2 \frac{\nu}{\omega_P^2} = \frac{\mu^e \nu}{e^2 n \mu_0} = \frac{(\mu^e)^2 \nu}{e^2 \mu_0} \frac{1}{\rho^e}, \tag{11.125}$$

where ν is the electron-ion collision frequency and the plasma frequency ω_P is given by Relation (11.75). Hence, in general, the magnetic viscosity depends on the partial mass density ρ^e. However, σ is usually so large that ν_H can be considered to be small and constant.

Let us now return to the thermomechanical part of the model. As already shown the electromagnetic terms can be transformed into the divergent form. In the particular case under consideration

$$\mathbf{J} \times \mathbf{B} = \frac{1}{\mu_0} (\operatorname{rot} \mathbf{B}) \times \mathbf{B} = \frac{1}{\mu_0} \operatorname{div} (\mathbf{B} \otimes \mathbf{B}) - \frac{1}{2} \frac{1}{\mu_0} \operatorname{grad} B^2,$$

$$\mathbf{E} \cdot \mathbf{J} = (\nu_H \operatorname{rot} \mathbf{B} - \mathbf{v} \times \mathbf{B}) \cdot \frac{1}{\mu_H} (\operatorname{rot} \mathbf{B}).$$

Hence, the full set of equations has the following form

$$\dot{\rho} + \rho \operatorname{div} \mathbf{v} = 0,$$

$$\rho \dot{\mathbf{v}} = -\operatorname{grad} \left(p + \frac{B^2}{2\mu_0} \right) + \operatorname{div} \left(\mathbf{T}_{dev} + \frac{1}{\mu_0} \mathbf{B} \otimes \mathbf{B} \right) + \rho \mathbf{b},$$

$$\rho \dot{\varepsilon} = \operatorname{tr} (\mathbf{T} \operatorname{grad} \mathbf{v}) - \operatorname{div} \mathbf{q} + r + \frac{1}{\sigma} \mathbf{J} \cdot \mathbf{J} \tag{11.126}$$

$$= \operatorname{tr} (\mathbf{T} \operatorname{grad} \mathbf{v}) - \operatorname{div} \mathbf{q} + r + \frac{\nu_H}{\mu_0} |\operatorname{rot} \mathbf{B}|^2,$$

$$\frac{\partial \mathbf{B}}{\partial t} = \operatorname{rot} (\mathbf{v} \times \mathbf{B}) - \operatorname{rot} (\nu_H \operatorname{rot} \mathbf{B}),$$

where

$$\mathbf{T}_{dev} = \mathbf{T} + p\mathbf{1}, \quad p = -\frac{1}{3}\mathrm{tr}\mathbf{T},$$

$$\dot{\rho} = \frac{\partial \rho}{\partial t} + \mathbf{v} \cdot \mathrm{grad}\, \rho,$$

$$\dot{\mathbf{v}} = \frac{\partial \mathbf{v}}{\partial t} + (\mathbf{v} \cdot \mathrm{grad})\, \mathbf{v}, \tag{11.127}$$

$$\dot{\varepsilon} = \frac{\partial \varepsilon}{\partial t} + \mathbf{v} \cdot \mathrm{grad}\, \varepsilon.$$

Sometimes it is convenient to use the pressure as an independent field. Then, the energy balance equation is transformed by use of the following identity

$$\mathrm{tr}\,(\mathbf{T}\,\mathrm{grad}\,\mathbf{v}) = \mathrm{tr}\,(\mathbf{T}_{dev}\,\mathrm{grad}\,\mathbf{v}) - p\,\mathrm{tr}\,\mathrm{div}\,\mathbf{v}$$

$$= \mathrm{tr}\,(\mathbf{T}_{dev}\,\mathrm{grad}\,\mathbf{v}) + \frac{p\dot{\rho}}{\rho}$$

$$= \mathrm{tr}\,(\mathbf{T}_{dev}\,\mathrm{grad}\,\mathbf{v}) - \rho\left(\frac{p}{\rho}\right)^{\bullet} + \dot{p}.$$

Consequently, the energy balance assumes the following form

$$\rho\dot{h} = \dot{p} + \mathrm{tr}\,(\mathbf{T}_{dev}\mathbf{D}_v) - \mathrm{div}\,\mathbf{q} + \rho r + \frac{\nu_H}{\mu_0}\,|\mathrm{rot}\,\mathbf{B}|^2, \tag{11.128}$$

where h is the enthalpy and $\mathbf{D}_v$ the symmetric part of the velocity gradient (the stretching tensor, page 33 of Part I or 25 here)

$$h = \varepsilon + \frac{p}{\rho}, \quad \mathbf{D}_v = \frac{1}{2}\left(\mathrm{grad}\,\mathbf{v} + (\mathrm{grad}\,\mathbf{v})^T\right). \tag{11.129}$$

We are now in the position to perform the analysis of the order of magnitude of terms in the above formulated MHD equations. We limit the attention to such systems satisfying the following constitutive relations

$$p = nkT, \quad \mathbf{T}^D = \eta_0\mathbf{D}_v^D, \quad \mathbf{T}^D = \mathbf{T} + p\mathbf{1}, \quad \mathbf{D}_v^D = \mathbf{D}_v - \frac{1}{3}\mathrm{tr}\mathbf{D}_v\mathbf{1},$$

$$\tag{11.130}$$

$$h = c_p T, \quad \mathbf{q} = -K\,\mathrm{grad}\,T, \quad r = 0,$$

where η_0 is the shear (dynamic) viscosity, c_p is the specific heat under constant pressure and K is the heat (thermal) conductivity. These relations may often be oversimplified. For instance, it is claimed that in plasma the thermal conductivity should by anisotropic and that $K_{\parallel} \gg K_{\perp}$, i.e., this conductivity is much higher in the direction of the magnetic flux than in the perpendicular direction. For the estimates which we are proceeding to make, these more complex constitutive relations are not needed.

In addition to the dimensionless variables (11.118) we introduce the following quantities

$$\widetilde{\rho} = \frac{\rho}{\rho_0}, \quad \widetilde{\mathbf{v}} = \frac{\mathbf{v}}{V}, \quad \widetilde{p} = \frac{p}{\rho_0 V^2}, \quad \widetilde{\mathbf{H}} = \frac{\mathbf{H}}{H_0}, \quad \widetilde{\mathbf{T}}^D = \frac{\mathbf{T}^D}{\eta_0 V/L},$$

$$\tag{11.131}$$

$$\widetilde{h} = \frac{h}{c_P T_0}, \quad \widetilde{\mathbf{q}} = \frac{\mathbf{q}}{\kappa T_0}L, \quad \widetilde{\nu_H} = \frac{\nu_H}{\nu_{H0}},$$

where the parameters η_0, c_P, κ and $\nu_{H0} = 1/\sigma_0\mu_0$ are assumed to be constant. σ_0 denotes the reference value of the electrical conductivity. This may be easily generalized but we do not need here such an extension. Further on, except of the permeabilities of the vacuum ε_0 and μ_0, the quantities with the index "zero" refer to their reference values.

The governing set of equations can now be written in the following dimensionless form

$$\frac{\partial\widetilde{\rho}}{\partial\widetilde{t}} + R_t\,\widetilde{\mathrm{div}}\,(\widetilde{\rho\mathbf{v}}) = 0,$$

$$\widetilde{\rho}\left(\frac{\partial\widetilde{\mathbf{v}}}{\partial\widetilde{t}} + R_t\widetilde{\mathbf{v}}\cdot\widetilde{\mathrm{grad}}\,\widetilde{\mathbf{v}}\right) = -R_t\,\widetilde{\mathrm{grad}}\left(\widetilde{p} + R_H\widetilde{H}^2\right)$$
$$+ \frac{R_t}{Re}\,\widetilde{\mathrm{div}}\left(\widetilde{\mathbf{T}^D} + R_H Re\widetilde{\mathbf{H}}\otimes\widetilde{\mathbf{H}}\right),$$

$$\widetilde{\rho}\left(\frac{\partial\widetilde{h}}{\partial\widetilde{t}} + R_t\widetilde{\mathbf{v}}\cdot\widetilde{\mathrm{grad}\,h}\right) = \left(\frac{V^2}{c_pT_0}\right)\left(\frac{\partial\widetilde{p}}{\partial\widetilde{t}} + R_t\widetilde{\mathbf{v}}\cdot\widetilde{\mathrm{grad}\,p}\right)$$
$$+\frac{R_t}{Re}\left(\frac{V^2}{c_pT_0}\right)\mathrm{tr}\left(\widetilde{\mathbf{T}^D\mathbf{D}_v}\right) - \frac{R_t}{RePr}\,\widetilde{\mathrm{div}}\,\widetilde{\mathbf{q}} + R_t\frac{R_H}{Re}\left(\frac{V^2}{c_pT_0}\right)\widetilde{\nu_H}\left|\widetilde{\mathrm{rot}\,\widetilde{\mathbf{H}}}\right|^2,$$

$$\frac{\partial\widetilde{\mathbf{H}}}{\partial\widetilde{t}} = R_t\,\widetilde{\mathrm{rot}}\left(\widetilde{\mathbf{v}}\times\widetilde{\mathbf{H}}\right) - \frac{R_t}{R_\sigma}\,\widetilde{\mathrm{rot}}\left(\widetilde{\nu_H}\,\widetilde{\mathrm{rot}}\,\widetilde{\mathbf{H}}\right),$$

$$(11.132)$$

where

$$Re = \frac{\rho_0 VL}{\eta_0}, \quad R_\sigma - \frac{VL}{\nu_{H0}}, \quad R_H = \frac{\mu_0 H_0^2}{\rho_0 V^2}, \quad Pr = \frac{c_\mu\eta_0}{K}, \qquad (11.133)$$

and these are: Reynolds number, magnetic Reynolds number, magnetic pressure number and Prandtl number, respectively. Before we explain the physical meaning of these numbers, let us make a few remarks on the ideal gas law (compare Section 6.1 in Part I). The enthalpy and the specific heats have within this model the form

$$h = \varepsilon + \frac{p}{\rho} = (z+1)\frac{R}{M}T + \alpha, \quad c_V = \frac{\partial\varepsilon}{\partial T} = z\frac{R}{M}, \quad c_P = \frac{\partial h}{\partial T} = (z+1)\frac{R}{M},$$

$$\gamma = \frac{c_P}{c_V} = \frac{z+1}{z}, \quad c_{\mathrm{adiab}} = \sqrt{\gamma\frac{R}{M}T}, \quad M_0^2 = \frac{V^2}{c_{\mathrm{adiab}}^2} \quad\Rightarrow\quad \frac{V^2}{c_PT_0} = (\gamma-1)M_0^2,$$

$$(11.134)$$

where $R = 8.3143\cdot 10^3$ [J/kg·K] – the universal gas constant, M – the relative molecular mass, c_{adiab} – speed of sound in adiabatic conditions, M_0 – Mach number.[15]

We are now in the position to present a physical insight of the above introduced parameters.

1. γ – *adiabatic exponent*;

    it is a measure of the relative internal complexity of the molecules of plasma.

2. M_0 – *Mach number*;

---

[15]γ as an adiabatic exponent should be distinguished from the ratio of speeds appearing in the Lorentz transformation!

it is a measure of the compressibility of the fluid due to a high flow velocity. For very small Mach numbers the variation of the density, i.e., the compressibility effect, due to the variation of the velocity of the flow field is negligibly small and the fluid may be considered as incompressible. For large Mach numbers the effect of compressibility must be accounted for.

3. *Re – Reynolds number*;

it is a measure of the ratio of the inertial force to the viscous force (compare Section 5.2.2). When the Reynolds number of a system is small, the viscous force is predominant and the effect of viscosity is important in the whole flow field. When the Reynolds number is large, the inertial force is predominant and the effect of viscosity is important only in the narrow boundary layer region near the solid boundary or in any region of large variations of viscosity such as interiors of shock waves.

4. *Pr – Prandtl number*;

it is a measure of the relative importance of the viscosity and the heat conduction. It can be written in the form[16]

$$Pr = \frac{\eta_0/\rho_0}{K/\rho_0 c_p} = \frac{\nu}{\kappa} = \frac{\text{kinematic viscosity}}{\text{thermal diffusivity}}. \tag{11.135}$$

The value of the kinematic viscosity alludes boundary layer regions when it is small. The Prandtl number relates two dissipative mechanisms of the viscous heat conducting fluid.

5. *R_H – magnetic pressure number*;

it is the ratio of the magnetic pressure $\mu_0 H_0^2/2$ over the dynamical pressure $\rho_0 V^2/2$. Only when R_H is of the order of unity or larger, the fluid flow will be affected noticeably by the magnetic field. If R_H is much smaller than unity, the terms due to the magnetic field in both equations of motion and of energy can be neglected. The velocity

$$V_H = H_0 \sqrt{\frac{\mu_0}{\rho_0}}, \tag{11.136}$$

is known as the characteristic speed of Alfvén's wave.

6. *R_σ – magnetic Reynolds number*;

it has a similar form as the ordinary Reynolds number with the magnetic viscosity ν_{H0} in place of the kinematic viscosity η_0/ρ_0. The magnetic Reynolds number is obtained from kinematic considerations on the influence of the flow motion on the magnetic field. This number may also be regarded as the ratio of the characteristic dimension of the flow field L to the characteristic length L_σ

$$R_\sigma = \frac{L}{L_\sigma}, \quad L_\sigma = \frac{\nu_{H0}}{V} = \frac{1}{\sigma_0 \mu_0 V}, \tag{11.137}$$

---

[16]The thermal diffusivity is also called the coefficient of thermometric conductivity.

or as the ratio of the flow velocity V to a characteristic velocity V_σ

$$R_\sigma = \frac{V}{V_\sigma}, \quad V_\sigma = \frac{\nu_{H0}}{L} = \frac{1}{\sigma_0 \mu_0 L}. \tag{11.138}$$

The length L_σ may be regarded as a characteristic distance on which the magnetic field is moving through the conductor. It is the well-known characteristic length in the skin effect of the ordinary electromagnetic theory. If $L \gg L_\sigma$, i.e., $R_\sigma \gg 1$ the magnetic field will stay with the flow in the form of the so-called frozen-in field, and it will be greatly influenced by the motion of the fluid.

Let us make a brief remark on the frozen-in fields. It is rather easy to prove (e.g. see Section 2.3 in [31]) that any field $\mathbf{B}$ satisfying the Gauss-Faraday law (11.14) can be written in the form

$$\mathbf{B} = \operatorname{grad} \alpha \times \operatorname{grad} \beta \quad \Rightarrow \quad \operatorname{div} \mathbf{B} \equiv 0, \tag{11.139}$$

where α and β are arbitrary, sufficiently smooth scalar functions. For any given magnetic field this representation is not unique. For instance, without changing $\mathbf{B}$ one can add an arbitrary function of β to α or an arbitrary function of α to β

$$\begin{aligned}
\mathbf{B} &= \operatorname{grad} \alpha \times \operatorname{grad} \beta \\
&= \operatorname{grad}\left(\alpha + f\left(\beta\right)\right) \times \operatorname{grad} \beta \\
&= \operatorname{grad} \alpha \times \operatorname{grad}\left(\beta + f\left(\alpha\right)\right).
\end{aligned} \tag{11.140}$$

The evolution equations for α and β can be easily obtained from Equation $(11.126)_4$ under the condition $R_\sigma \gg 1$ (i.e., $\nu_{H0} \ll LV$; this condition defines a perfectly conducting fluid)

$$\frac{\partial \mathbf{B}}{\partial t} = \operatorname{rot}\left(\mathbf{v} \times \mathbf{B}\right). \tag{11.141}$$

Substitution of (11.139) in (11.141) yields immediately

$$\frac{d\alpha}{dt} \operatorname{grad} \beta - \frac{d\beta}{dt} \operatorname{grad} \alpha = \operatorname{grad} \phi, \tag{11.142}$$

$$\frac{d\alpha}{dt} = \frac{\partial \alpha}{\partial t} + \mathbf{v} \cdot \operatorname{grad} \alpha, \quad \frac{d\beta}{dt} = \frac{\partial \beta}{\partial t} + \mathbf{v} \cdot \operatorname{grad} \beta,$$

where ϕ is an arbitrary function of space and time. We can, certainly, choose $\phi = 0$ and then

$$\frac{d\alpha}{dt} = 0, \quad \frac{d\beta}{dt} = 0, \tag{11.143}$$

which means that α and β are constant along the trajectories of the fluid particles. This means that the magnetic field lines must move with the fluid. Nonuniqueness of the above representation is a subtle problem which we shall not discuss here.

In another limit of the small magnetic Reynolds number, i.e., for $L \ll L_\sigma$ the magnetic field will not be influenced noticeably by the motion of the fluid.

It should be noticed that even though the parameters

$$\left\{\gamma, M_0, Re, Pr, R_H, R_\sigma\right\}, \tag{11.144}$$

are sufficient to describe the magnetogasdynamic motion they are not unique. For instance, the following two parameters have been used by many authors:

7. R_h – *Hartmann number*

$$R_h = \sqrt{Re R_H R_\sigma} = \mu_0 H_0 L \sqrt{\frac{\sigma_0}{\eta_0}} = \sqrt{\frac{\sigma_0 \mu_0^2 H_0^2 V}{\eta_0 \frac{V}{L^2}}} = \sqrt{\frac{\text{magnetic force}}{\text{viscous force}}}. \qquad (11.145)$$

It has been first used by J. Hartmann in 1937 [154] in the study of the flow in channels where the important forces are the magnetic force and the viscous force.

8. R_M – *magnetic parameter*

$$R_M = \sqrt{R_H R_\sigma} = \mu_0 H_0 \sqrt{\frac{\sigma_0 L}{\rho_0 V}} = \sqrt{\frac{\sigma_0 \mu_0^2 H_0^2 V}{\rho_0 \frac{V^2}{L}}} = \sqrt{\frac{\text{magnetic force}}{\text{inertial force}}}. \qquad (11.146)$$

As an illustration of the applicability of those numbers we present a few special cases.

1. $R_\sigma \ll 1$: then the magnetic field $\mathbf{H}$ is independent of the motion and the thermo-mechanical equations

$$\dot{\rho} + \rho \operatorname{div} \mathbf{v} = 0,$$

$$\rho \dot{\mathbf{v}} = -\operatorname{grad}\left(p + \mu_0 \frac{1}{2} H^2\right) + \operatorname{div}\left(\mathbf{T}^D + \mu_0 \mathbf{H} \otimes \mathbf{H}\right),$$

$$\rho \dot{h} = \dot{p} + \operatorname{tr}\left(\mathbf{T}^D \mathbf{D}\right) - \operatorname{div} \mathbf{q} + \mu_0 \nu_H \left|\operatorname{rot} \mathbf{H}\right|^2,$$

$$p = \frac{R}{M} \rho T, \quad \mathbf{T}^D = \eta_0 \mathbf{D}_v^D,$$

$$(11.147)$$

(compare (11.130), (11.134)) contain the given contributions of the magnetic field.

2. $R_H \ll 1$: the motion is independent of the magnetic field $\mathbf{H}$, i.e., after solving the thermomechanical equations

$$\dot{\rho} + \rho \operatorname{div} \mathbf{v} = 0,$$

$$\rho \dot{\mathbf{v}} = -\operatorname{grad} p + \operatorname{div} \mathbf{T}^D,$$

$$\rho \dot{h} = \dot{p} + \operatorname{tr}\left(\mathbf{T}^D \mathbf{D}_v\right) - \operatorname{div} \mathbf{q},$$

$$p = \frac{R}{M} \rho T, \quad \mathbf{T}^D = \eta_0 \mathbf{D}_v^D,$$

$$(11.148)$$

one has to solve the equation for the magnetic field

$$\frac{\partial \mathbf{H}}{\partial t} = \operatorname{rot}\left(\mathbf{v} \times \mathbf{H}\right) - \operatorname{rot}\left(\nu_H \operatorname{rot} \mathbf{H}\right). \qquad (11.149)$$

3. In the case of low Mach numbers (incompressible fluids) $\rho = $ const. and the unknown fields are

$$\{\mathbf{H}, \mathbf{v}, p, T\}. \qquad (11.150)$$

The governing equations for this case

$$\operatorname{div} \mathbf{v} = 0,$$

$$\rho \dot{\mathbf{v}} = -\operatorname{grad}\left(p + \mu_0 \frac{1}{2}H^2\right) + \operatorname{div}\left(\mathbf{T}^D + \mu_0 \mathbf{H} \otimes \mathbf{H}\right),$$

$$\frac{\partial \mathbf{H}}{\partial t} + \mathbf{v}\cdot(\operatorname{grad})\mathbf{H} - \mathbf{H}\cdot(\operatorname{grad})\mathbf{H} = \nu_H \Delta^2 \mathbf{H}, \quad \Delta^2(\ldots) = \operatorname{div}\operatorname{grad}(\ldots),$$

$$\tag{11.151}$$

and in the energy balance equation we take the limit $M_0 \to 0$ but leave the term $M_0^2 \dfrac{R_H}{R_\sigma}$ because the latter is not negligible. Hence,

$$\rho \dot{h} = -\operatorname{div}\mathbf{q} + \mu_0 \nu_H \left|\operatorname{rot}\mathbf{H}\right|^2, \tag{11.152}$$

with the constitutive relations

$$h = c_p T, \quad \mathbf{q} = -\kappa \operatorname{grad} T, \quad p = \frac{R}{M}\rho T, \quad \mathbf{T}^D = \eta_0 \mathbf{D}_v^D. \tag{11.153}$$

If the contribution $M_0^2 \dfrac{R_H}{R_\sigma}$ is also small, the last term in (11.152) does not appear and the problem of heat conduction can be solved separately.

4. *Ideal plasma*: this term was introduced by S. Lundquist in 1952 and it concerns the model of the inviscid, non-heat conducting and infinitely electrically conducting gas. In this case

$$Re \to \infty, \quad R_\sigma \to \infty, \quad Pr \quad \text{finite.} \tag{11.154}$$

The governing equations are hyperbolic and they have the form

$$\dot{\rho} + \rho \operatorname{div}\mathbf{v} = 0,$$

$$\rho \dot{\mathbf{v}} - \mu_0\left(\mathbf{H}\cdot\operatorname{grad}\right)\mathbf{H} = -\operatorname{grad}\left(p + \mu_0 \frac{1}{2}H^2\right),$$

$$\tag{11.155}$$

$$\frac{\partial \mathbf{H}}{\partial t} = \operatorname{rot}\left(\mathbf{v}\times\mathbf{H}\right),$$

$$\dot{\eta} = 0,$$

where η is the specific entropy.

11.5.2   Some steady state flows

In order to illustrate deviations of the solutions of some problems of electrically conducting fluids from these for neutral fluids (compare Chapter 5) in the following two subsections we present two simple examples of flows.

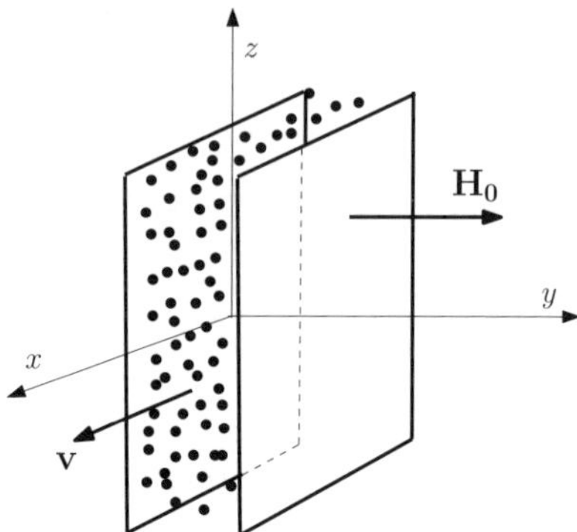

Figure 11.6: *Configuration of the Poiseuille-Hartmann flow.*

11.5.2.1   Poiseuille-Hartmann flow

Let us consider the two-dimensional steady laminar flow of an incompressible and electrically conducting fluid of constant viscosity and constant electrical conductivity between two parallel straight insulated walls. The velocity of the flow is parallel to the walls which lie in the distance $2L$ apart in the direction of the y-axis (see Figure 11.6). There is an external magnetic field of constant strength H_0 in the y-direction. All fields are independent of the z-coordinate and of time t.

In general, such a problem is described by the set of equations (11.151). We assume that the coupling in the equation for the enthalpy (11.152) can be ignored. Then the distribution of the temperature is given by an independent heat conduction equation. We shall not consider it here. For the particular flow under consideration, the remaining equations have the following dimensionless form

$$-\frac{\partial \widetilde{p}}{\partial \widetilde{x}} + \frac{1}{Re}\frac{\partial}{\partial \widetilde{y}}\left(\frac{\partial \widetilde{v}_x}{\partial \widetilde{y}} + R_H Re \widetilde{H}_x\right) = 0,$$

$$-\frac{\partial}{\partial \widetilde{y}}\left(\widetilde{p} + \frac{1}{2}R_H \widetilde{H}_x^2\right) = 0, \tag{11.156}$$

$$-\frac{\partial \widetilde{v}_x}{\partial \widetilde{y}} = \frac{1}{R_\sigma}\frac{\partial^2 \widetilde{H}_x}{\partial \widetilde{y}^2},$$

where

$$\widetilde{x} = \frac{x}{L}, \quad \widetilde{y} = \frac{y}{L}, \quad Re = \frac{\varrho_0 V L}{\eta_0}, \quad R_H = \frac{\mu_0 H_0^2}{\varrho_0 V^2}, \quad R_\sigma = \frac{VL}{\nu_{H0}},$$
$$\widetilde{v}_x = \frac{v_x}{V}, \quad \widetilde{p} = \frac{p}{\varrho_0 V^2}, \quad \widetilde{H}_x = \frac{H_x}{H_0}. \tag{11.157}$$

Obviously, the incompressibility condition $(11.151)_1$ is identically satisfied. Let us eliminate $\widetilde{p}$ and $\widetilde{H}_x$ from the above equations. We obtain immediately

$$\frac{\partial^3 \widetilde{v}_x}{\partial \widetilde{y}^3} + R_h^2\frac{\partial \widetilde{v}_x}{\partial \widetilde{y}} = 0, \quad R_h^2 = R_H Re R_\sigma. \tag{11.158}$$

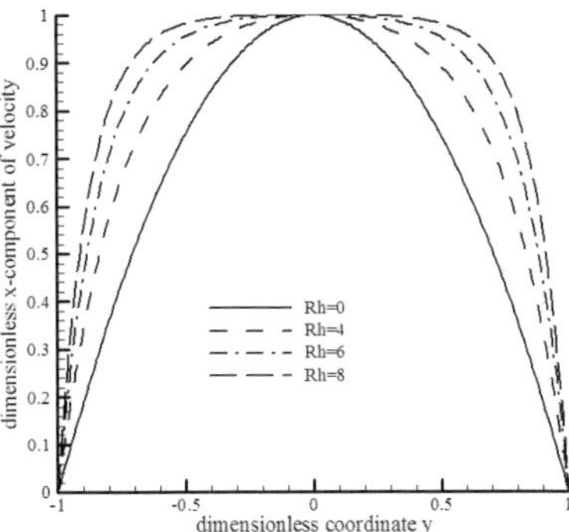

Figure 11.7: *The x-component of the velocity distribution in y-direction for the plane Poiseuille-Hartmann flow.*

Hence, the general solution has the form

$$\widetilde{v}_x = \begin{cases} A_0\left(1-\widetilde{y}\right)\left(1+\widetilde{y}\right) + B_0\widetilde{y} + C_0 & \text{for} \quad R_h = 0, \\[2mm] \dfrac{A}{R_h}\cosh\left(R_h\widetilde{y}\right) + \dfrac{B}{R_h}\sinh\left(R_h\widetilde{y}\right) + C & \text{for} \quad R_h \neq 0, \end{cases} \tag{11.159}$$

where A_0, B_0, C_0, A, B and C are constants. The problem is symmetric and the boundary conditions for the velocity have the classical form for viscous fluids

$$\widetilde{v}_x\left(\widetilde{y} = \pm 1\right) = 0, \quad \widetilde{v}_x\left(y = 0\right) = 1. \tag{11.160}$$

Consequently, we obtain the solution

$$\widetilde{v}_x = \begin{cases} \left(1-\widetilde{y}\right)\left(1+\widetilde{y}\right) & \text{for} \quad R_h = 0, \\[3mm] \dfrac{\cosh R_h - \cosh\left(R_h\widetilde{y}\right)}{\cosh R_h - 1} & \text{for} \quad R_h \neq 0. \end{cases} \tag{11.161}$$

This is shown in Figure 11.7. The distribution of the velocity for $R_h = 0$ is a parabola identical with this for the usual Poiseuille flow (see Chapter 5). For intermediate values of the Hartmann number R_h, i.e., for intermediate values of the external magnetic field the distribution approaches the square form which it has for $R_h \to \infty$

$$\lim_{R_h \to \infty} \widetilde{v}_x = \begin{cases} 1 & \text{for} \quad |\widetilde{y}| < 1 \\ 0 & \text{for} \quad |\widetilde{y}| = 1 \end{cases}. \tag{11.162}$$

For very large values of R_h the velocity distribution is almost constant in the central portion between the two plates and then drops very rapidly to zero near the plates. The reason for this is the fact that the magnetic field increases the total shearing stress of the flow field. Since the shearing stress due to the magnetic field is proportional to the magnetic field H_x which will be zero on the wall, near the wall the total shearing stress will be largely produced by the viscous force which results in a very large velocity gradient

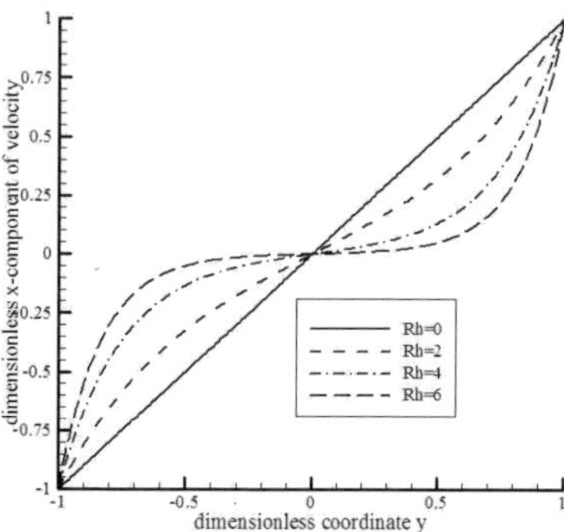

Figure 11.8: *The x-component of the velocity distribution in y-direction for the plane Couette flow.*

near the wall. This effect of the magnetic stress is qualitatively similar to that of the turbulent stress which produces uniform flow in the central portion between two plates and the large velocity gradient near the walls.

Now the distribution of the pressure $\widetilde{p}$ and of the magnetic field component $\widetilde{H}_x$ can be easily obtained from the relations

$$\frac{\partial^2 \widetilde{H}_x}{\partial \widetilde{y}^2} = R_\sigma R_h \frac{\sinh\left(R_h \widetilde{y}\right)}{\cosh R_h - 1}, \quad \widetilde{p} = -\frac{1}{2} R_H \left(\widetilde{H}_x\right)^2 + \widetilde{p}_0. \tag{11.163}$$

11.5.2.2   Couette flow

We consider the case in which the two plates are situated at $\widetilde{y} = \pm 1$, the relative velocity between the two plates is $2V$ and there is no pressure gradient in the flow field. Then, the solution still has the form (11.159) while the boundary conditions are as follows

$$\widetilde{v}_x\left(\widetilde{y} = 1\right) = 1, \quad \widetilde{v}_x\left(\widetilde{y} = -1\right) = -1, \quad \widetilde{v}_x\left(\widetilde{y} = 0\right) = 0. \tag{11.164}$$

Hence, the solution for $R_h \neq 0$ has the form

$$\widetilde{v}_x = \frac{\sinh\left(R_h \widetilde{y}\right)}{\sinh R_h}. \tag{11.165}$$

This function is shown in Figure 11.8. As R_h tends to zero we have the velocity distribution of the Couette flow of the ordinary hydrodynamics, i.e., a straight line. In the limit $R_h \to \infty$ we have again a uniform flow between the two plates and a very large velocity gradient near the walls.

As before, the magnetic field and the pressure can be obtained by quadrature.

To complete the solution let us find the electric current **J** and the electric field **E**. To this aim we use Relations (11.123), i.e.,

$$\mathbf{J} = \mathrm{rot}\,\mathbf{H}, \quad \mathbf{E} = \frac{1}{\sigma}\mathbf{J} - \mu_0 \mathbf{v} \times \mathbf{H}. \tag{11.166}$$

In our particular case

$$J_x = J_y = 0, \quad J_z = -\frac{\partial H_x}{\partial y} = -\frac{H_0}{L}\frac{\partial \widetilde{H}_x}{\partial \widetilde{y}},$$

$$E_x = E_y = 0, \quad E_z = -\frac{1}{\sigma}\frac{\partial H_x}{\partial y} - \mu_0 v_x H_0 = -\frac{1}{\sigma}\frac{H_0}{L}\left[\frac{\partial \widetilde{h}_x}{\partial \widetilde{y}} + R_\sigma \widetilde{v}_x\right]. \tag{11.167}$$

As expected, the non-zero components of both those fields are pointing in the z-directions.

It should be mentioned, that if the external magnetic field is oriented in the direction of the plates instead of being perpendicular to the plates, the velocity distribution of both plane Poiseuille flow and plane Couette flow will not be affected by the magnetic field and their distributions are exactly the same as the distributions of ordinary hydrodynamics.

11.6 A few remarks on the stability of plasmas

11.6.1 Linear stability analysis of the ideal plasma model

One of the main reasons for the construction of various models of plasma is the analysis of its stability. The development of tokamaks, one of many confinements of plasma, has shown as early as in the '60s of the last century that all these devices are limited in the range of their operation by instabilities. Many features of these instabilities can be predicted, at least qualitatively, by MHD models. An excellent presentation of the subject can be found in the book of G. Bateman [31], some newer results and relations to practical problems are presented in the CISM book [63] (in particular, articles of E. K. Maschke and J. J. Rasmussen in this book) but the subject still develops. In this section we demonstrate an example of the linear stability analysis of the model described by Equations (11.155) (compare the original work of B. B. Kadomtsev [195]). The advantage of the linear analysis follows from the fact that any disturbance can be written as the sum of eigenfunctions which are unique for any given equilibrium and boundary conditions. Since the coefficients of the disturbance are constant in time when the static equilibrium is linearized, all eigenfunctions possess a simple exponential time dependence.

For the ideal plasma we use the constitutive relations of the ideal gas (compare (11.134))

$$p = \frac{R}{M}\rho T, \quad \varepsilon = z\frac{R}{M}T + \alpha, \quad d\eta = \frac{1}{T}\left(d\varepsilon - \frac{p}{\rho^2}d\rho\right) \quad \eta = \frac{R}{M}\ln\frac{T^z}{\rho} + \beta. \tag{11.168}$$

Then the last equation of (11.155) yields

$$\rho\dot{p} = \gamma p\dot{\rho}, \quad \gamma = \frac{z+1}{z}. \tag{11.169}$$

As indicated in Chapter 6, we investigate the stability of a static solution which in our case satisfies the following conditions

$$\rho_0 = \rho_0\left(\mathbf{x}\right), \quad \mathbf{v}_0 = 0, \quad \mathbf{B}_0 = \mathbf{B}_0\left(\mathbf{x}\right), \quad T_0 = T_0\left(\mathbf{x}\right), \tag{11.170}$$

fulfilling the field equations

$$\frac{\partial \rho_0}{\partial t} = 0, \quad \frac{1}{\mu_0}\left(\mathbf{B}_0 \cdot \mathrm{grad}\right)\mathbf{B}_0 = \mathrm{grad}\left(p_0 + \frac{1}{2\mu_0}B_0^2\right), \quad \frac{\partial p_0}{\partial t} = 0. \tag{11.171}$$

In order to investigate the linear stability of the solutions of this system, we impose a small disturbance on the static solution

$$\rho = \rho_0 + \rho', \quad \mathbf{v} = \delta v', \quad \mathbf{B} = \mathbf{B}_0 + \mathbf{B}', \quad p = p_0 + p'. \tag{11.172}$$

Substitution of this ansatz in the field equations yields

$$\frac{\partial \rho'}{\partial t} + \operatorname{div}(\rho_0 \mathbf{v}') = 0,$$

$$\rho_0 \frac{\partial \mathbf{v}'}{\partial t} - \frac{1}{\mu_0}\left[(\mathbf{B}' \cdot \operatorname{grad})\mathbf{B}_0 + (\mathbf{B}_0 \cdot \operatorname{grad})\mathbf{B}'\right] = -\operatorname{grad}\left(p' + \frac{1}{\mu_0}\mathbf{B}_0 \cdot \mathbf{B}'\right),$$

$$\frac{\partial \mathbf{B}'}{\partial t} = \operatorname{rot}(\mathbf{v}' \times \mathbf{B}_0),$$

$$\frac{\partial p'}{\partial t} + \mathbf{v}' \cdot \operatorname{grad} p_0 = \frac{\partial \mathbf{B}'}{\partial t} = -\gamma p_0 \operatorname{div} \mathbf{v}'. \tag{11.173}$$

It is customary to introduce a displacement vector ξ to those problems through the integration of the mass balance equation (11.173). Namely

$$\xi(\mathbf{x}, t) = \int_0^t \mathbf{v}'(\mathbf{x}, t)\, dt, \quad \text{i.e.,} \quad \frac{\partial \xi}{\partial t} = \mathbf{v}'. \tag{11.174}$$

Then

$$\begin{aligned}
\rho' &= -\operatorname{div}(\rho_0 \xi), \\
p' &= -\xi \cdot \operatorname{grad} p_0 - \gamma p_0 \operatorname{div}\xi, \\
\mathbf{B}' &= \operatorname{rot}(\xi \times \mathbf{B}_0).
\end{aligned} \tag{11.175}$$

Now, the equation for ξ follows from the momentum balance equation

$$\rho_0 \frac{\partial^2 \xi}{\partial t^2} = \mathbf{F}(\xi), \tag{11.176}$$

where

$$\mathbf{F}(\xi) = \frac{1}{\mu_0}\left[(\mathbf{B}_0 \cdot \operatorname{grad})(\operatorname{rot}(\xi \times \mathbf{B}_0))\right]$$

$$+ \operatorname{grad}\left[(\xi \cdot \operatorname{grad}) p_0 + \gamma p_0 \operatorname{div}\xi - \frac{1}{\mu_0}\mathbf{B}_0 \cdot (\operatorname{rot}(\xi \times \mathbf{B}_0))\right]. \tag{11.177}$$

One can easily prove the following Lemma (e.g. [195]):

The operator $\mathbf{F}$ is linear and self-adjoint, i.e.,

$$\int_{\mathcal{B}_t} \eta \cdot \mathbf{F}(\xi)\, dV = \int_{\mathcal{B}_t} \xi \cdot \mathbf{F}(\eta)\, dV, \tag{11.178}$$

where the volume integration is over the whole domain $\mathcal{B}_t$ of plasma.

This property yields the stability condition. Namely, we investigate properties of normal vibrations

$$\xi = \xi_l(\mathbf{x})\, \mathrm{e}^{i\omega_l t}. \tag{11.179}$$

Then, the substitution in Equation (11.176) yields the eigenvalue problem

$$-\rho_0\omega_l^2\xi_l = \mathbf{F}\left(\xi_l\right).$$

(11.180)

We say that the solution

$$\xi = \sum_l a_l\xi_l e^{i\omega_l t},$$

(11.181)

is *stable* if all imaginary parts of the eigenvalues ω_l vanish: $\forall\, l :$ Im $\omega_l = 0$. Clearly, this means that in such a case the amplitudes of the normal vibrations remain bounded for all times.

For self-adjoint operators one can prove the following Lemma

1. Im $\omega_l^2 = 0 \quad \Rightarrow \quad$ *ω_l are either real or imaginary,*

2. $\{\xi_l\}$ *(the eigenvectors) are orthogonal, i.e., they form the basis in the ξ-space,*

3. *one can formulate the variation principle for the eigenvalues $\delta\left(\omega^2\right) = 0$, where*

$$\omega^2 = \frac{-\int_{\mathcal{B}_t}\xi\cdot\mathbf{F}\left(\xi\right)dV}{\int_{\mathcal{B}_t}\rho_0\xi^2 dV}.$$

(11.182)

The above properties give rise to the formulation of a method of stability analysis which is called the energy principle for stability. Namely, by means of the above mentioned variational principle we can easily prove two theorems.

Theorem 1: *The static solution is stable if and only if, for all values of the index l, $\omega_l^2 > 0$.*

Theorem 2: *The static solution is stable if and only if $\int_{\mathcal{B}_t}\xi\cdot\mathbf{F}\left(\xi\right)dV < 0$.*

Bearing this in mind we have

$$\int_{\mathcal{B}_t}\xi\cdot\left(\rho_0\ddot{\xi}\right)dV = \tfrac{1}{2}\int_{\mathcal{B}_t}\left[\dot{\xi}\cdot\mathbf{F}\left(\xi\right) + \xi\cdot\mathbf{F}\left(\dot{\xi}\right)\right]dV$$

$$= \frac{\partial}{\partial t}\int_{\mathcal{B}_t}\tfrac{1}{2}\xi\cdot\mathbf{F}\left(\xi\right)dV \quad \Rightarrow \quad \frac{\partial}{\partial t}\int_{\mathcal{B}_t}\left[\frac{1}{2}\rho_0\dot{\xi}^2 - \frac{1}{2}\xi\cdot\mathbf{F}\left(\xi\right)\right]dV = 0.$$

(11.183)

Hence, the second contribution is the potential energy of the disturbance. From the constancy of the energy of the disturbance, it follows that any perturbation ξ that decreases the potential energy produces an increase of the kinetic energy which indicates that the system is linearly unstable. Now follows

Theorem 3: *(energy principle) Let*

$$\delta W = -\int_{\mathcal{B}_t}\frac{1}{2}\xi\cdot\mathbf{F}\left(\xi\right)dV.$$

(11.184)

The static solution is stable if and only if

$$\forall\xi : \ \delta W > 0.$$

(11.185)

It is instructive to expand this relation using the definition (11.177) of the operator $\mathbf{F}(\xi)$ and Relations (11.175). Then, the potential energy of the disturbance can be written in the following form

$$\delta W = -\frac{1}{2} \int_{\mathcal{B}_t} \xi \cdot \{\operatorname{grad}\left[\xi \cdot \operatorname{grad} p_0 + \gamma p_0 \operatorname{div} \xi\right]$$
$$+ \mathbf{J}_0 \times \mathbf{B}' + \mathbf{J}' \times \mathbf{B}_0\}, \tag{11.186}$$

$$\mathbf{J}_0 = \operatorname{rot} \mathbf{B}_0, \quad \mathbf{J}' = \operatorname{rot} \mathbf{B}', \quad \mathbf{B}' = \operatorname{rot}(\xi \times \mathbf{B}_0).$$

Now, the integration by parts yields the separation of the boundary contribution

$$\delta W = \delta W_F + \frac{1}{2} \int_{\partial \mathcal{B}_t} (p' + \mathbf{B}_0 \cdot \mathbf{B}') \, \xi \cdot \mathbf{n} dS_t,$$
$$\delta W_F = \frac{1}{2} \int_{\mathcal{B}_t} \left\{ \frac{1}{\mu_0} |\mathbf{B}'|^2 + \mathbf{J}_0 \cdot \left(\xi \times \mathbf{B}' + \gamma p_0 |\operatorname{div} \xi|^2 + (\xi \cdot \operatorname{grad} p_0) \operatorname{div} \xi\right) \right\}. \tag{11.187}$$

The surface term vanishes, for instance, when the plasma is confined within rigid walls. In contact with vacuum there appears an additional surface contribution which is known.

It remains to discuss the contribution δW_F. This has been done in numerous papers and we quote here only a result presented in the book [31] which can be obtained from (11.187) by integrating by parts again. It has the form

$$\delta W_F = \frac{1}{2} \int_{\mathcal{B}_t} \left\{ \underbrace{\frac{1}{\mu_0} |\mathbf{B}'_\perp|^2}_{\text{Alfvén}} + \underbrace{\mu_0 \left|\mathbf{B}'_\parallel - \mathbf{B}_0 \xi \cdot \operatorname{grad} p_0 / |\mathbf{B}_0|^2\right|^2}_{\text{fast magnetoacoustic}} \right.$$

$$\left. + \underbrace{\gamma p_0 |\operatorname{div} \xi|^2}_{\text{acoustic}} + \underbrace{\frac{\mathbf{J}_0 \cdot \mathbf{B}_0}{|\mathbf{B}_0|^2} (\mathbf{B}_0 \times \xi) \cdot \mathbf{B}'}_{\text{kink}} - \underbrace{2\xi \cdot \operatorname{grad} p_0 \, (\xi \cdot \kappa)}_{\text{interchange}} \right\}, \tag{11.188}$$

where κ denotes the normal curvature of the equilibrium magnetic field $\mathbf{B}_0$.

The first term is the magnetic energy of the so-called Alfvén waves. The second term is the potential energy of the fast magnetoacoustic waves. The third term is the potential energy of the ordinary acoustic waves. All these contributions are positive and, consequently, they do not yield any destabilizing effects. This is not the case with the fourth contribution which may drive the so-called kink instabilities. Finally, the last term which is negative may drive the interchange or ballooning instabilities. They correspond to the classical Rayleigh-Taylor instabilities driven by the pressure gradient. In the next section we show a few examples of those instabilities.

11.6.2   A few particular cases of instabilities

As a particular example we consider an infinite cylinder of conducting fluid with an axial current $\mathbf{J}$ and the resulting azimuthal magnetic field $\mathbf{B}_0$. The force $\mathbf{J} \times \mathbf{B}_0$ acting on the

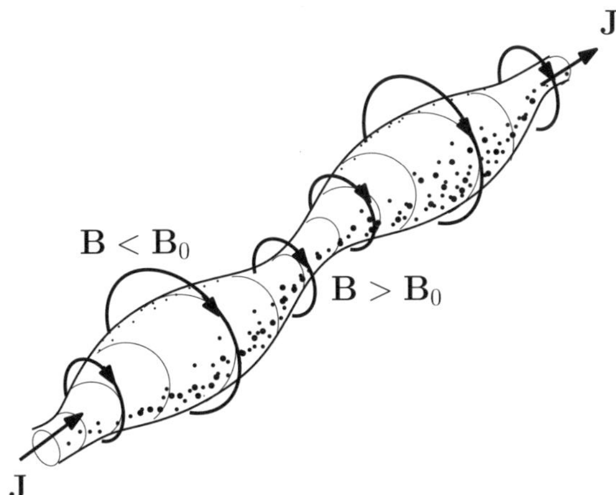

Figure 11.9: *Sausage instability of the pinched plasma column.*

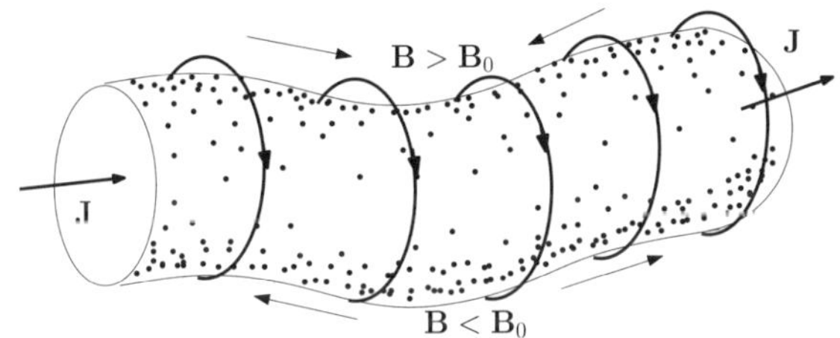

Figure 11.10: *Kink instability of the pinched plasma column.*

fluid yields the radial compression of the column. It is known as the pinch effect. The surfaces of constant pressure are concentric cylinders. These are steady-state conditions. Making use of Equations (11.171) we can easily find the parameters describing this state. The discussion of this solution can be found, for instance, in Chapter 13 of the book of J. A. Bittencourt [51]. Now, if we impose a dynamical disturbance on this steady-state, we find that the steady-state solution is unstable, as indicated by Relation (11.188).

In view of the form of the disturbance (11.179) we expect that the radius of the column will decrease in some places and increase in some other. Then, at locations where the radius has decreased, the magnetic field **B** will be larger than the equilibrium field. This yields the increment of the magnetic pressure and, consequently, the enhancement of the constriction of the plasma. This is an unstable situation and as result one obtains the sausage instability schematically shown in Figure 11.9.

Another type of instability which results from the fact that the dynamical disturbance yields local changes of the curvature is the unstable bending which is called the kink instability. This is schematically shown in Figure 11.10.

Kink instabilities may be much more complicated. In Figure 11.11 we show the kink instability next from the whole sequence specified by the periodicity $\sin\left(\dfrac{m\theta}{2\pi}\right)$, where θ is the angle in the circular cross-section. Then, $m = 1$ corresponds to the first of the

Figure 11.11: *Kink instability with a double twist in the pinched plasma column.*

above kink solutions and $m = 2$ is the kink solution with a double twist.

It should be mentioned, that the above described unstable processes can be stabilized by introducing a magnetic field with a component in the direction of the generating lines of the cylinder.

The instabilities in tokamak and other geometries form an even more complex pattern. For instance, in a toroidal geometry the so-called ballooning appears. It causes the plasma to bulge out most where both the pressure gradient and the curvature are strongest. They require a numerical analysis which is also one of the methods in the analysis of nonlinear theory. In such problems the bifurcation theory, singular perturbation analysis and finally the study of turbulence are used. We shall not go into those problems as the aim of this section was only to point out the role of the thermodynamical MHD-modeling in the stability analysis of plasmas.

Chapter 12

Mechanics of porous materials

12.1 Summary of two-component models

In this book, so far, predominantly continua have been discussed whose kinematics is described by a single mapping – the function of motion. This is not the case for multi-component media, as for instance, fluid mixtures, suspensions, porous or granular media. For such media it is necessary to take into account the microstructure of the body, i.e., to formulate an own function of motion for each component (see Chapters 12 and 13 of Part I about the thermodynamics of immiscible mixtures and poroelastic materials). As was shown by K. Wilmanski in Section 13.5 of Part I on adsorption processes or e.g. by B. Albers [6] some components may move together. A physical adsorption process, e.g. means that at the beginning of the process an adsorbate flows together with a fluid through the channels of a porous medium and then sticks on the surface and moves together with the solid. In each stage of the process two components possess a common kinematics. Otherwise the number of components of a body determines the number of functions of motion needed to describe a process completely.

12.1.1 Immiscible mixtures

As it is common to use the Eulerian description for fluids, it seemed likely that Truesdell proposed an Eulerian continuous description of a body consisting of A different fluid components α. Each of the components possesses its own velocity field

$$\mathbf{v}^\alpha = \mathbf{v}^\alpha\left(\mathbf{x},t\right), \qquad \alpha = 1,...,\mathrm{A}, \qquad \mathbf{x} \in \mathcal{B}_t, \tag{12.1}$$

where $\mathcal{B}_t$ is the current configuration of the body. Partial balance equations for each of the immiscible components follow (see e.g. Truesdell & Toupin [395]). The most important of them are summarized in Table 12.1.

Models for multi-component bodies are a better approximation of reality than classical models for one-component media. However, both classes of models describe with different precision the same processes. This means that for mass, momentum and energy, respectively, the sum of the balance laws of the components (Table 12.1) has to match the classical balance laws (Table 4.2)$_{\mathrm{regular}}$. This results in restrictions for the source terms and for contributions which describe the influence of different kinematics on the process in the one-component model.

Table 12.1: *Eulerian description of partial balance laws for a multi-component body consisting of* A *immiscible fluids.*

mass	$\dfrac{\partial \rho_t^\alpha}{\partial t} + \operatorname{div} \rho_t^\alpha \mathbf{v}^\alpha = \hat{\rho}_t^\alpha,$
momentum	$\dfrac{\partial \rho_t^\alpha \mathbf{v}^\alpha}{\partial t} + \operatorname{div}\left(\rho_t^\alpha \mathbf{v}^\alpha \otimes \mathbf{v}^\alpha - \mathbf{T}^\alpha\right) = \hat{\mathbf{p}}_t^\alpha + \rho_t^\alpha \mathbf{b}^\alpha,$
energy	$\dfrac{\partial}{\partial t}\left[\rho_t^\alpha\left(\varepsilon^\alpha + \tfrac{1}{2}v^{\alpha 2}\right)\right] + \operatorname{div}\left[\rho_t^\alpha\left(\varepsilon^\alpha + \tfrac{1}{2}v^{\alpha 2}\right)\mathbf{v}^\alpha + \mathbf{q}^\alpha - \mathbf{T}^\alpha \mathbf{v}^\alpha\right]$ $= \hat{\varepsilon}_t^\alpha + \rho_t^\alpha \mathbf{b}^\alpha \cdot \mathbf{v}^\alpha + \rho_t^\alpha r^\alpha.$

in which

ρ_t^α - partial mass density of component α per unit volume of the current configuration

$\mathbf{T}^\alpha$ - partial Cauchy stress tensor of component α

$\hat{\mathbf{p}}_t^\alpha$ - partial momentum source of component α

$\mathbf{b}^\alpha$ - partial volume force of component α per unit mass of the current configuration

ε^α - partial inner energy of component α per unit mass of the current configuration

$\mathbf{q}^\alpha$ - partial heat flux of the component α per unit surface of the current configuration

$\hat{\varepsilon}_t^\alpha$ - partial energy source of component α per unit volume of the current configuration

r^α - partial energy radiation of component α per unit mass of the current configuration

A comparison of classical and partial balance equations yields:

$$\rho \equiv \sum_{\alpha=1}^{A} \rho_t^\alpha, \qquad \rho\mathbf{v} \equiv \sum_{\alpha=1}^{A} \rho_t^\alpha \mathbf{v}^\alpha, \tag{12.2}$$

$$\sum_{\alpha=1}^{A} \hat{\rho}_t^\alpha = 0, \qquad \sum_{\alpha=1}^{A} \hat{\mathbf{p}}_t^\alpha, = 0, \qquad \sum_{\alpha=1}^{A} \hat{\varepsilon}_t^\alpha = 0. \tag{12.3}$$

From this follows the stress tensor:

$$\mathbf{T} = \underbrace{\sum_{\alpha=1}^{A} \mathbf{T}^\alpha}_{\text{static part}} + \underbrace{\frac{1}{\displaystyle\sum_{\alpha=1}^{A}\rho_t^\alpha}\left(\sum_{\alpha=1}^{A}\rho_t^\alpha \mathbf{v}^\alpha\right) \otimes \left(\sum_{\alpha=1}^{A}\rho_t^\alpha \mathbf{v}^\alpha\right) - \sum_{\alpha=1}^{A}\left(\rho_t^\alpha \mathbf{v}^\alpha \otimes \mathbf{v}^\alpha\right)}_{\text{kinematic part}}. \tag{12.4}$$

A material surface moves with the velocity $\mathbf{v}$, described by $(12.2)_2$, through the one-component body, in which also the stress tensor $\mathbf{T}$ is defined. However, individual points possess the velocity $\mathbf{v}^\alpha$, not $\mathbf{v}$. Thus, the relative movement yields the transport of a certain amount of momentum and the kinematic part in the stress tensor arises. Similar contributions result in the heat flux and in the energy which contains a contribution of the kinematic energy of relative motion of the components.

12.1.2   Lagrangian description of multi-component porous media

Now, we focus on multi-component bodies which consist of one solid and one or more fluid components. It would be inappropriate to choose different ways of description for the different types of components. Thus, K. Wilmanski [433], [435] proposed a consistent Lagrangian description for all components. It is assumed that the movement of the immiscible fluid components takes place in the channels of the solid phase, the so-called skeleton. Thus, the movement of all components is related to the reference configuration of the solid.

We start with the description of the movement of the solid. Its function of motion is given by

$$\mathbf{x} = \mathbf{f}^S(\mathbf{X},t), \qquad \mathbf{X} \in \mathcal{B}_0, \tag{12.5}$$

in which $\mathbf{X}$ is a material point of the skeleton and $\mathbf{x}$ is its current position. $\mathcal{B}_0$ is the reference configuration of the skeleton which is assumed to be the configuration of the skeleton at an initial instant of time $t = t_0$.

Velocity and deformation gradient are given by the following derivatives

$$\acute{\mathbf{x}}^S = \frac{\partial \mathbf{f}^S}{\partial t}(\mathbf{X},t), \qquad \mathbf{F}^S = \operatorname{Grad} \mathbf{f}^S(\mathbf{X},t). \tag{12.6}$$

In the reference configuration $\mathcal{B}_0$, obviously, $\mathbf{F}^S = \mathbf{1}$ holds.

In the classical theory of mixtures the movement of the fluid components is given by velocity fields in Eulerian description

$$\mathbf{v}^\alpha = \mathbf{v}^\alpha(\mathbf{x},t), \quad \alpha = 1,...,A, \quad \mathbf{x} \in \mathbf{f}^S(\mathcal{B}_0,t). \tag{12.7}$$

The movement of the fluid is hence defined only in the current configuration of the skeleton. However, a relation to the reference configuration has to be found. Therefore, we look at material fluid points which are situated at time t at the same position $\mathbf{x}$ as the material solid point $\mathbf{X}$. After a small increment of time Δt the position of the fluid particle has changed according to

$$\mathbf{x}(t + \Delta t) = \mathbf{x}(t) + \mathbf{v}^\alpha(\mathbf{x}(t),t)\,\Delta t. \tag{12.8}$$

At time $t + \Delta t$ the fluid particle shares its position with another solid point, namely with the material point $\mathbf{X}+\Delta\mathbf{X}$, given by

$$\mathbf{X} \equiv \mathbf{X}(t) = \mathbf{f}^{S-1}(\mathbf{x}(t),t), \qquad \Delta\mathbf{X} = \mathbf{f}^{S-1}(\mathbf{x}(t + \Delta t),t + \Delta t) - \mathbf{f}^{S-1}(\mathbf{x}(t),t). \tag{12.9}$$

From (12.9) one obtains by expansion in a Taylor series

$$\mathbf{x}(t + \Delta t) = \mathbf{f}(\mathbf{X}+\Delta\mathbf{X},t + \Delta t) \approx \mathbf{x}(t) + \mathbf{F}^S(\mathbf{X},t)\Delta\mathbf{X} + \acute{\mathbf{x}}^S(\mathbf{X},t)\Delta t. \tag{12.10}$$

After relating the velocity of the fluid point to the reference configuration of the skeleton

$$\acute{\mathbf{x}}^\alpha = \mathbf{v}^\alpha\left(\mathbf{f}^S(\mathbf{X},t),t\right) \equiv \acute{\mathbf{x}}^\alpha(\mathbf{X},t), \quad \alpha = 1,...,A, \tag{12.11}$$

we obtain relative velocities for the fluid components, i.e., velocities of the fluid components in reference to the skeleton. Comparing (12.8) and (12.10) yields

$$\mathbf{F}^S(\mathbf{X},t)\Delta\mathbf{X} + \acute{\mathbf{x}}^S(\mathbf{X},t)\Delta t = \acute{\mathbf{x}}^\alpha(\mathbf{X},t)\,\Delta t \quad \Rightarrow \quad \Delta\mathbf{X} \approx \acute{\mathbf{X}}^\alpha(\mathbf{X},t)\,\Delta t, \tag{12.12}$$

where

$$\acute{\mathbf{X}}^\alpha(\mathbf{X},t) : \lim_{\Delta t \to 0} \frac{\Delta\mathbf{X}}{\Delta t} = \mathbf{F}^{S-1}\left(\acute{\mathbf{x}}^\alpha - \acute{\mathbf{x}}^S\right), \quad \alpha = 1,...,A. \tag{12.13}$$

These are the *Lagrangian velocity fields of the fluid components*. The classical balance equations for a body consisting of one solid and A fluid components are given in Table 12.2.

For an illustration of the geometrical meaning of the Lagrangian description of relative motion we refer to Section 3.2.1 where we presented a simple example of two bars in extension.

Table 12.2: *Lagrangian description of the classical balance laws for a porous body.*

solid		global	local
mass		$$\dfrac{d}{dt}\int_{\mathcal{P}^S}\rho^S dV = \int_{\mathcal{P}^S}\hat{\rho}^S dV,$$	$$\dfrac{\partial \rho^S}{\partial t} = \hat{\rho}^S, \qquad U\,[\![\rho^S]\!] = 0,$$
momentum		$$\dfrac{d}{dt}\int_{\mathcal{P}^S}\rho^S\acute{\mathbf{x}}^S dV = \oint_{\partial \mathcal{P}^S}\mathbf{P}^S\mathbf{N}dS + \int_{\mathcal{P}^S}\left(\rho^S\mathbf{b}^S + \hat{\mathbf{p}}^S\right)dV,$$	$$\dfrac{\partial}{\partial t}\left(\rho^S\acute{\mathbf{x}}^S\right) = \mathrm{Div}\,\mathbf{P}^S + \rho^S\mathbf{b}^S + \hat{\mathbf{p}}^S,$$ $$\rho^S U\,[\![\acute{\mathbf{x}}^S]\!] + [\![\mathbf{P}^S]\!]\,\mathbf{N} = 0,$$
energy		$$\dfrac{d}{dt}\int_{\mathcal{P}^S}\rho^S\left(\tfrac{1}{2}\acute{x}^{S2} + \varepsilon^S\right)dV =$$ $$= \oint_{\partial \mathcal{P}^s}\left(\mathbf{P}^{ST}\acute{\mathbf{x}}^S - \mathbf{Q}^S\right)\cdot\mathbf{N}dS$$ $$+ \int_{\mathcal{P}^S}\left(\rho^S\mathbf{b}^S\cdot\acute{\mathbf{x}}^S + \rho^S r^S + \hat{\varepsilon}^S\right)dV,$$	$$\dfrac{\partial}{\partial t}\left[\rho^S\left(\tfrac{1}{2}\acute{x}^{S2} + \varepsilon^S\right)\right] + \mathrm{Div}\,\mathbf{Q}^S$$ $$= \mathrm{Div}\left(\mathbf{P}^{ST}\acute{\mathbf{x}}^S\right) + \rho^S\mathbf{b}^S\cdot\acute{\mathbf{x}}^S + \rho^S r^S + \hat{\varepsilon}^S,$$ $$\rho^S U\,[\![\tfrac{1}{2}\acute{x}^{S2} + \varepsilon]\!]$$ $$+ [\![\mathbf{P}^{ST}\acute{\mathbf{x}}^S - \mathbf{Q}^S]\!]\cdot\mathbf{N} = 0.$$

fluids		global	local
mass		$$\dfrac{d}{dt}\int_{\mathcal{P}^\alpha}\rho^\alpha dV = \int_{\mathcal{P}^\alpha}\hat{\rho}^\alpha dV,$$	$$\dfrac{\partial \rho^\alpha}{\partial t} + \mathrm{Div}\left(\rho^\alpha\acute{\mathbf{X}}^\alpha\right) = \hat{\rho}^\alpha,$$ $$\left[\!\left[\rho^\alpha\left(\acute{\mathbf{X}}^\alpha\cdot\mathbf{N} - U\right)\right]\!\right] = 0,$$
momentum		$$\dfrac{d}{dt}\int_{\mathcal{P}^\alpha}\rho^\alpha\acute{\mathbf{x}}^\alpha dV = \oint_{\partial \mathcal{P}^\alpha}\mathbf{P}^\alpha\mathbf{N}dS + \int_{\mathcal{P}^\alpha}\left(\rho^\alpha\mathbf{b}^\alpha + \hat{\mathbf{p}}^\alpha\right)dV,$$	$$\dfrac{\partial}{\partial t}\left(\rho^\alpha\acute{\mathbf{x}}^\alpha\right) + \mathrm{Div}\left(\rho^\alpha\acute{\mathbf{x}}^\alpha \otimes \acute{\mathbf{X}}^\alpha - \mathbf{P}^\alpha\right)$$ $$= \rho^\alpha\mathbf{b}^\alpha + \hat{\mathbf{p}}^\alpha,$$ $$\rho^\alpha\left(\acute{\mathbf{X}}^\alpha\cdot\mathbf{N} - U\right)[\![\acute{\mathbf{x}}^\alpha]\!] - [\![\mathbf{P}^\alpha]\!]\cdot\mathbf{N} = 0,$$
energy		$$\dfrac{d}{dt}\int_{\mathcal{P}^\alpha}\rho^\alpha\left(\tfrac{1}{2}\acute{x}^{\alpha 2} + \varepsilon^\alpha\right)dV$$ $$= \oint_{\partial \mathcal{P}^\alpha}\left(\mathbf{P}^{\alpha T}\acute{\mathbf{x}}^\alpha - \mathbf{Q}^\alpha\right)\cdot\mathbf{N}dS$$ $$+ \int_{\mathcal{P}^\alpha}\left(\rho^\alpha\mathbf{b}^\alpha\cdot\acute{\mathbf{x}}^\alpha + \rho^\alpha r^\alpha + \hat{\varepsilon}^\alpha\right)dV,$$	$$\dfrac{\partial}{\partial t}\left[\rho^\alpha\left(\tfrac{1}{2}\acute{x}^{\alpha 2} + \varepsilon^\alpha\right)\right] +$$ $$+ \mathrm{Div}\left[\rho^\alpha\left(\tfrac{1}{2}\acute{x}^{\alpha 2} + \varepsilon^\alpha\right)\acute{\mathbf{X}}^\alpha + \mathbf{Q}^\alpha\right]$$ $$= \mathrm{Div}\left(\mathbf{P}^{\alpha T}\acute{\mathbf{x}}^\alpha\right) + \rho^\alpha\mathbf{b}^\alpha\cdot\acute{\mathbf{x}}^\alpha + \rho^\alpha r^\alpha + \hat{\varepsilon}^\alpha,$$ $$\rho^\alpha\left(\acute{\mathbf{X}}^\alpha\cdot\mathbf{N} - U\right)[\![\tfrac{1}{2}\acute{x}^{\alpha 2} + \varepsilon^\alpha]\!]$$ $$- [\![\mathbf{P}^{\alpha T}\acute{\mathbf{x}}^\alpha - \mathbf{Q}^\alpha]\!]\cdot\mathbf{N} = 0.$$

in which

ρ^S - partial mass density of the solid per unit volume of the reference configuration

$\hat{\rho}^S$ - mass source of the solid per unit volume of the reference configuration

$\mathbf{P}^S$ - partial Piola-Kirchhoff stress tensor of the solid

$\hat{\mathbf{p}}^S$ - momentum source of the solid

$\mathbf{b}^S$ - partial volume force of the solid per unit mass of the solid

ε^S - specific inner energy of the solid per unit mass of the solid

$\mathbf{Q}^S$ - partial heat flux in the solid per unit surface of the solid

$\hat{\varepsilon}^S$ - partial energy source of the solid per unit volume of the reference configuration

r^S - energy radiation in the solid per unit mass of the solid

ρ^α - partial mass density of the α-component per unit volume
of the reference configuration of the solid

$\hat{\rho}^\alpha$ - mass source of the α-component per unit volume
of the reference configuration of the solid

$\mathbf{P}^\alpha$ - partial Piola-Kirchhoff stress tensor of the α-component related
to the reference configuration of the solid

$\hat{\mathbf{p}}^\alpha$ - momentum source in the α-component

$\mathbf{b}^\alpha$ - partial volume force of the α-component per unit mass
in the reference configuration of the solid

ε^α - specific inner energy of the α-component per unit mass
in the reference configuration of the solid

$\mathbf{Q}^\alpha$ - partial heat flux in the α-component per unit surface
of the reference configuration of the solid

$\hat{\varepsilon}^\alpha$ - partial energy source of the α-component per unit volume
of the reference configuration of the solid

r^α - energy radiation in the α-component per unit mass
in the reference configuration of the solid.

The global equations for the components α in Table 12.2 hold for each subbody $\mathcal{P}^\alpha \subset \mathcal{B}_0$ and they are material in relation to the respective kinematics of the α-component.

12.1.3   Eulerian description of multi-component porous media

The equations of Table 12.2 can be easily transformed into the balance equations in Eulerian form. We show this exemplary for the mass balance of the fluid component.

For the partial time derivative in Table 12.2 we assumed $\mathbf{X} = $ const. Thus, it is the material time derivative in relation to the solid. We conduct the following transformation in which the star indicates the transformation of the variables $(\mathbf{X},t) \to (\mathbf{x},t)$ by means of (12.5)

$$\frac{\partial \rho^\alpha}{\partial t}(\mathbf{X},t) = \frac{\partial J^S \rho^\alpha J^{S-1}}{\partial t} = J^S \frac{\partial \rho^\alpha J^{S-1}}{\partial t} + \rho^\alpha \mathbf{F}^{S-T} \cdot \frac{\partial \mathbf{F}^S}{\partial t}$$

$$= J^S \left(\frac{\partial \rho^\alpha J^{S-1}}{\partial t} + \rho^\alpha J^{S-1} \mathbf{F}^{S-T} \cdot \operatorname{Grad} \acute{\mathbf{x}}^S \right)$$

$$\overset{*}{=} J^S \left(\frac{\partial \rho^\alpha J^{S-1}}{\partial t} + \mathbf{v}^S \cdot \operatorname{grad} \left(\rho^\alpha J^{S-1} \right) + \rho^\alpha J^{S-1} \operatorname{div} \mathbf{v}^S \right)$$

$$= J^S \left(\frac{\partial \rho^\alpha J^{S-1}}{\partial t} + \operatorname{div} \left(\rho^\alpha J^{S-1} \mathbf{v}^S \right) \right). \tag{12.14}$$

In this equation $J^S \equiv \det \mathbf{F}^S \neq 0$ is the Jacobi determinant and the following identity holds: $\mathbf{v}^S \equiv \acute{\mathbf{x}}^S \left(\mathbf{f}^{S-1} \left(\mathbf{x},t \right), t \right)$.

Similarly, the divergence operator can be transformed

$$\operatorname{Div} \left(\rho^\alpha \acute{\mathbf{X}}^\alpha \right) (\mathbf{X},t) = \operatorname{Div} \left[\rho^\alpha \mathbf{F}^{S-1} \left(\acute{\mathbf{x}}^\alpha - \acute{\mathbf{x}}^S \right) \right]$$

$$= J^S \mathbf{F}^{S-T} \cdot \operatorname{Grad} \left[\rho^\alpha J^{S-1} \left(\acute{\mathbf{x}}^\alpha - \acute{\mathbf{x}}^S \right) \right]$$

$$\overset{*}{=} J^S \operatorname{div} \left[\rho^\alpha J^{S-1} \left(\mathbf{v}^\alpha - \mathbf{v}^S \right) \right]. \tag{12.15}$$

Inserting this into the mass balance (Table 12.2) we obtain

$$\frac{\partial \rho^\alpha J^{S-1}}{\partial t} + \operatorname{div} \left(\rho^\alpha J^{S-1} \mathbf{v}^S \right) + \operatorname{div} \left[\rho^\alpha J^{S-1} \left(\mathbf{v}^\alpha - \mathbf{v}^S \right) \right] = \hat{\rho}^\alpha J^{S-1}$$

$$\tag{12.16}$$

$$\Leftrightarrow \frac{\partial \rho_t^\alpha}{\partial t} + \operatorname{div} \rho_t^\alpha \mathbf{v}^\alpha = \hat{\rho}_t^\alpha \qquad \text{with} \qquad \rho_t^\alpha = \rho^\alpha J^{S-1}, \quad \hat{\rho}_t^\alpha = \hat{\rho}^\alpha J^{S-1}.$$

For the sake of completeness, in the following table the classical partial balance equations for porous media in Eulerian form are also summarized.

Table 12.3: *Eulerian description of the classical balance laws for a porous body.*

solid	regular	singular
mass	$\dfrac{\partial \rho_t^S}{\partial t} + \operatorname{div} \rho_t^S \mathbf{v}^S = \hat{\rho}_t^S,$	$\left[\!\left[\rho_t^S \left(c - \mathbf{v}^S \!\cdot\! \mathbf{n} \right) \right]\!\right] = 0,$
momentum	$\dfrac{\partial}{\partial t} \left(\rho_t^S \mathbf{v}^S \right) + \operatorname{div} \left(\rho_t^S \mathbf{v}^S \otimes \mathbf{v}^S - \mathbf{T}^S \right)$ $= \rho_t^S \mathbf{b}^S + \hat{\mathbf{p}}_t^S,$	$\left[\!\left[\rho_t^S \mathbf{v}^S \left(c - \mathbf{v}^S \!\cdot\! \mathbf{n} \right) \right]\!\right]$ $+ \left[\!\left[\mathbf{T}^S \right]\!\right] \mathbf{n} = 0,$
energy	$\dfrac{\partial}{\partial t} \left[\rho_t^S \left(\tfrac{1}{2} v^{S2} + \varepsilon^S \right) \right]$ $+ \operatorname{div} \left[\rho_t^S \left(\tfrac{1}{2} v^{S2} + \varepsilon^S \right) \mathbf{v}^S + \mathbf{q}^S \right]$ $= \operatorname{div} \left(\mathbf{T}^{ST} \mathbf{v}^S \right) + \rho_t^S \mathbf{b}^S \!\cdot\! \mathbf{v}^S + \rho_t^S r^S + \hat{\varepsilon}_t^S.$	$\left[\!\left[\rho_t^S \left(\tfrac{1}{2} v^{S2} + \varepsilon^S \right) \left(c - \mathbf{v}^S \!\cdot\! \mathbf{n} \right) \right]\!\right]$ $= \left[\!\left[\left(\mathbf{q}^S - \mathbf{T}^S \mathbf{v}^S \right) \right]\!\right] \cdot \mathbf{n}.$

Table 12.3 (cont'd)

fluids	regular	singular
mass	$\dfrac{\partial \rho_t^\alpha}{\partial t} + \operatorname{div} \rho_t^\alpha \mathbf{v}^\alpha = \hat{\rho}_t^\alpha,$	$[\![\rho_t^\alpha \left(c - \mathbf{v}^\alpha \cdot \mathbf{n} \right)]\!] = 0,$
momentum	$\dfrac{\partial}{\partial t} \left(\rho_t^\alpha \mathbf{v}^\alpha \right) + \operatorname{div} \left(\rho_t^\alpha \mathbf{v}^\alpha \otimes \mathbf{v}^\alpha - \mathbf{T}^\alpha \right)$ $= \rho_t^\alpha \mathbf{b}^\alpha + \hat{\mathbf{p}}_t^\alpha,$	$[\![\rho_t^\alpha \mathbf{v}^\alpha \left(c - \mathbf{v}^\alpha \cdot \mathbf{n} \right)]\!]$ $+ [\![\mathbf{T}^\alpha]\!]\, \mathbf{n} = 0,$
energy	$\dfrac{\partial}{\partial t} \left[\rho_t^\alpha \left(\tfrac{1}{2} v^{\alpha 2} + \varepsilon^\alpha \right) \right]$ $+ \operatorname{div} \left[\rho_t^\alpha \left(\tfrac{1}{2} v^{\alpha 2} + \varepsilon^\alpha \right) \mathbf{v}^\alpha + \mathbf{q}^\alpha \right]$ $= \operatorname{div} \left(\mathbf{T}^{\alpha T} \mathbf{v}^\alpha \right) + \rho_t^\alpha \mathbf{b}^\alpha \cdot \mathbf{v}^\alpha + \rho_t^\alpha r^\alpha + \hat{\varepsilon}_t^\alpha.$	$[\![\rho_t^\alpha \left(\tfrac{1}{2} v^{\alpha 2} + \varepsilon^\alpha \right) \left(c - \mathbf{v}^\alpha \cdot \mathbf{n} \right)]\!]$ $= [\![\left(\mathbf{q}^\alpha - \mathbf{T}^\alpha \mathbf{v}^\alpha \right)]\!] \cdot \mathbf{n}.$

with

ρ_t^S - partial mass density of the solid per unit volume of the current configuration

$\hat{\rho}_t^S$ - mass source of the solid per unit volume of the current configuration

$\mathbf{T}^S$ - partial Cauchy-Green stress tensor of the solid, $\mathbf{T}^S = J^{S-1} \mathbf{P}^S \mathbf{F}^{ST}$

$\hat{\mathbf{p}}_t^S$ - momentum source of the solid

$\mathbf{b}^S$ - partial volume force of the solid per unit mass of the solid

ε^S - specific inner energy of the solid per unit mass of the solid

$\mathbf{q}^S$ - partial heat flux in the solid per unit surface of the solid, $\mathbf{q}^S = J^{S-1} \mathbf{F}^S \mathbf{Q}^S$

$\hat{\varepsilon}_t^S$ - partial energy source of the solid per unit volume of the current configuration

r^S - energy radiation in the solid per unit mass of the solid

ρ_t^α - partial mass density of the α-component per unit volume of the current configuration

$\hat{\rho}_t^\alpha$ - mass source of the α-component per unit volume of the current configuration

$\mathbf{T}^\alpha$ - partial Cauchy-Green stress tensor of the α-component, $\mathbf{T}^\alpha = J^{S-1} \mathbf{P}^\alpha \mathbf{F}^{ST}$

$\hat{\mathbf{p}}_t^\alpha$ - momentum source in the α-component

$\mathbf{b}^\alpha$ - partial volume force of the α-component per unit mass in the current configuration

ε^α - specific inner energy of the α-component per unit mass in the current configuration

$\mathbf{q}^\alpha$ - partial heat flux in the α-component per unit surface of the current configuration, $\mathbf{q}^\alpha = J^{S-1} \mathbf{F}^S \mathbf{Q}^\alpha$

$\hat{\varepsilon}_t^\alpha$ - partial energy source of the α-component per unit volume of the current configuration

r^α - energy radiation in the α-component per unit mass.

12.2   Two-component models with constitutive relations for the porosity

In Chapter 12 of Part I (pages 268-270) it has been shown that in some theories of porous and granular materials incompressibility of both components is assumed, i.e., both realistic mass densities are constant

$$\rho^{SR} = \text{const.}, \qquad \rho^{FR} = \text{const.} \qquad (12.17)$$

In this case the current mass densities ρ_t^F and ρ_t^S can be reduced to the single field of the porosity n and the partial mass balance equations reduce to a field equation for the porosity and a constraint condition of the model. It was mentioned that this condition yields certain limitations on the constitutive relations appearing in the model which are not always physically and mathematically acceptable. These problems can be avoided if the constitutive relations depend on the gradient of the porosity, grad n.

This class of models may have importance in the description of some soil mechanical problems but it is not appropriate to deal with wave propagation processes. Namely, it is not hyperbolic and this leads to a reduced number of real eigenvalues corresponding to the speeds of propagation. Consequently, the analysis does not reveal all modes of propagation of weak discontinuity waves. In particular, the $P2$-wave in saturated porous media (see Section 13.4 of Part I or e.g. [449]) cannot be described.

Here, we consider two further classes of models with constitutive relations for the porosity [442]. In the first class labeled (I) we assume that the real fluid in the pores is incompressible. This is a weaker assumption than this of the full incompressibility discussed in Part I. This means that changes of the geometry of the medium are controlled by two independent processes. On the one hand the skeleton may macroscopically deform and the porosity remains constant. Only the deformation gradient $\mathbf{F}^S$ changes. We say that the process proceeds in *undrained conditions*. On the other hand the porosity may change by the drainage and still the macroscopic geometry of the skeleton may remain unchanged: $\mathbf{F}^S = 1$. Then, changes of mass density of the fluid are determined by changes of the porosity alone

$$\rho_t^F = n\rho^{FR}, \qquad \rho^{FR} = \text{const.} \tag{12.18}$$

In the second class of models labeled (C) we allow for arbitrary changes of the mass densities ρ_t^F and ρ_t^S but the porosity is assumed to be given by a constitutive relation. In a general case, this relation is assumed to be of the form $n = n(\rho_t^F/\rho_t^S)$ which is the consequence of the dimensional analysis for isotropic materials. Here, we assume the simplest form of this relation

$$n = n_0 \frac{\rho_t^F \, \rho_0^S}{\rho_0^F \, \rho_t^S}, \tag{12.19}$$

where ρ_t^F, ρ_t^S and n_0 denote reference constant values of the partial mass densities and of the porosity.

Let us note that, in general, the porosity n may change as well without accompanying changes of the macroscopic mass densities ρ_t^F, ρ_t^S. This happens when changes of n are compensated by changes of the real mass densities ρ^{FR} and ρ^{SR} in such a way that their products $n\rho^{FR}$ and $(1-n)\rho^{SR}$ remain constant. Such changes are not controllable on the macroscopic level. They must proceed spontaneously. Consequently, they yield a relaxation of the porosity characteristic for microstructural variables. Relaxation properties are, of course, dissipative. They require a source term in the relation for the changes of porosity (see Chapter 13 of Part I).

12.2.1 Fields and field equations

The fields which describe mechanical processes in the model discussed in this section are as follows

1. the reference partial mass density of the fluid component

$$\rho^F = \rho^F(\mathbf{X}, t) = \rho_t^F J^S \equiv n J^S \rho^{FR}, \quad J^S := \det \mathbf{F}^S, \quad \mathbf{X} \in \mathcal{B}_0, \tag{12.20}$$

2. the field of partial velocity of the solid $\acute{\mathbf{x}}^S\left(\mathbf{X},t\right)$ on the macroscopic level of description,

3. the field of partial velocity of the fluid $\acute{\mathbf{x}}^F\left(\mathbf{X},t\right)$ on the macroscopic level of description,

4. the macroscopic deformation gradient of the solid $\mathbf{F}^S\left(\mathbf{X},t\right)$.

By means of the velocity fields one can define the macroscopic filter velocity $\mathbf{w}\left(\mathbf{X},t\right)$, and its corresponding Lagrangian image $\acute{\mathbf{X}}^F\left(\mathbf{X},t\right)$ appearing in the balance equations

$$\mathbf{w} := \acute{\mathbf{x}}^F - \acute{\mathbf{x}}^S, \quad \acute{\mathbf{X}}^F := \mathbf{F}^{S-1}\mathbf{w}. \tag{12.21}$$

The reference partial mass density of the skeleton $\rho^S \equiv \rho_0^S$ does not appear among those fields because it is constant in time if we assume that there is no mass exchange between the components. Its current value is given by the relation

$$\rho_t^S = \rho_0^S J^{S-1}, \tag{12.22}$$

which satisfies identically the partial mass conservation law in Eulerian description.

Summing up we can write the following relations for the porosity in the above mentioned two classes of models

$$(I)\text{-models} \quad : \quad n = J^{S-1}\frac{\rho^F}{\rho^{FR}}, \quad \rho^{FR} = \text{const.,}$$

$$(C)\text{-models} \quad : \quad n = \frac{\rho^F}{\rho_0^{FR}}, \quad \rho_0^{FR} = \frac{\rho_0^F}{n_0} = \text{const.} \tag{12.23}$$

The fields must fulfill the following balance equations in Lagrangian description (see last section; volume forces equal to zero)

- mass conservation of the fluid component

$$\frac{\partial \rho^F}{\partial t} + \text{Div}\left(\rho^F \acute{\mathbf{X}}^F\right) = 0, \tag{12.24}$$

- momentum balance for the skeleton

$$\rho^S\frac{\partial \acute{\mathbf{x}}^S}{\partial t} = \text{Div}\,\mathbf{P}^S + \hat{\mathbf{p}}, \tag{12.25}$$

- momentum balance for the fluid

$$\rho^F\left(\frac{\partial \acute{\mathbf{x}}^F}{\partial t} + \acute{\mathbf{X}}^F\cdot\text{Grad}\,\acute{\mathbf{x}}^F\right) = \text{Div}\,\mathbf{P}^F - \hat{\mathbf{p}}. \tag{12.26}$$

In addition, the deformation gradient $\mathbf{F}^S$ must fulfill the *integrability conditions* yielding the existence of the field of motion of the skeleton. They consist of two parts.

The kinematic compatibility condition relates the time derivative and the gradient of velocity

$$\frac{\partial \mathbf{F}^S}{\partial t} = \text{Grad}\,\acute{\mathbf{x}}^S. \tag{12.27}$$

The second part, a geometrical compatibility condition, is a symmetry relation

$$\operatorname{Grad} \mathbf{F}^S = \left(\operatorname{Grad} \mathbf{F}^S\right)^{\overset{23}{T}}, \tag{12.28}$$

or, in Cartesian coordinates

$$\frac{\partial F^S_{kK}}{\partial X^L} = \frac{\partial F^S_{kL}}{\partial X^K}, \tag{12.29}$$

where the small index refers to Eulerian coordinates, and the capital index to Lagrangian coordinates.

Obviously, all operators $\operatorname{Grad}, \operatorname{Div}, \frac{\partial}{\partial X^K}$ refer to Lagrangian variables.

As we have mentioned, the conditions (12.27) and (12.28) yield the existence of the field of motion of the solid, say $\mathbf{f}^S(\mathbf{X}, t)$, whose derivatives give the deformation gradient and the partial velocity, respectively

$$\mathbf{F}^S = \operatorname{Grad} \mathbf{f}^S, \quad \acute{\mathbf{x}}^S = \frac{\partial \mathbf{f}^S}{\partial t}. \tag{12.30}$$

The momentum balance equations contain the partial Piola-Kirchhoff stresses $\mathbf{P}^S$ and $\mathbf{P}^F$ which are related to the usual Cauchy stresses by the following transformation rules (compare (4.10))

$$\mathbf{T}^S = J^{S-1}\mathbf{P}^S\mathbf{F}^{ST}, \quad \mathbf{T}^F = J^{S-1}\mathbf{P}^F\mathbf{F}^{ST}. \tag{12.31}$$

The momentum equations contain as well the source $\hat{\mathbf{p}}$ which is the diffusion force.

We have to perform a closure in order to obtain field equations from the above balance relations. For poroelastic materials we consider further two models following from two choices of the constitutive variables. Namely we choose either

$$\mathcal{C}^{(1)} = \left\{\rho^F, \mathbf{F}^S, \acute{\mathbf{X}}^F\right\}, \tag{12.32}$$

or

$$\mathcal{C}^{(2)} = \left\{\rho^F, \mathbf{F}^S, \acute{\mathbf{X}}^F, \operatorname{Grad} n\right\}. \tag{12.33}$$

The porosity n does not appear among these variables because it is either given by Relation $(12.23)_1$ or by Relation $(12.23)_2$. Consequently, it can be eliminated from the set of independent constitutive variables.

The following functions must be given in terms of the constitutive relations

$$\mathcal{F} = \left\{\mathbf{P}^S, \mathbf{P}^F, \hat{\mathbf{p}}, \psi^S, \psi^F\right\}, \tag{12.34}$$

where ψ^S and ψ^F denote partial Helmholtz free energies appearing further in the second law of thermodynamics.

For reasons of material objectivity we should choose not only the *relative* velocity as the variable but also one of the *objective measures* of deformation. We shall do so further. However, the exploitation of the second law of thermodynamics is easier if we impose the objectivity after the exploitation of the entropy inequality.

For any of the choices of constitutive variables the constitutive relations are assumed to have the form

$$\mathcal{F} = \mathcal{F}\left(\mathcal{C}^{(\alpha)}\right), \quad \alpha = 1, 2, \tag{12.35}$$

which is sufficiently smooth for all operations which we perform in the sequel.

Let us make two methodological remarks.

In geophysical applications the Biot model is widely used. It is a linear model as is also the simplified model used for the analysis of acoustic waves (see Chapter 13). In the present section we show under which circumstances the linear Biot model follows from the nonlinear model for immiscible mixtures. We start the considerations with nonlinear models because the exploitation of the second law of thermodynamics for a model with linear constitutive laws cannot be made consistent with explicit nonlinear contributions to the field equations. This yields serious flaws of the thermodynamical analysis known in all nonlinear field theories.

Secondly, we should point out that the restriction to incompressible real fluids in the class (I) of models *does not lead to any constraints*. This may simply be interpreted as a change of variables: changes of the partial mass density of the fluid are replaced by corresponding changes of the porosity. There is no reaction force on such a "constraint".

Let us mention in passing that the above change of variables may lead to some mathematical problems due to the restriction $0 < n < 1$. In a linear model changes of porosity with respect to its initial value n_0 are small. The choice of a real value of n_0, say between 0.1 and 0.6, guarantees that this restriction is indeed fulfilled. In nonlinear models we have to take care of this constraint. If thermodynamic restrictions are derived by means of the partial mass density, which we do here, such a restriction does not appear.

12.2.2 Thermodynamic admissibility

12.2.2.1 Second law of thermodynamics

We present here only the second law of thermodynamics for two-component systems for which the temperature is constant. If this is the case, it may be formulated as follows (e.g. [442]). For any solution of the field equations (i.e., for any *thermodynamical process*) the following inequality

$$\rho^S \frac{\partial \psi^S}{\partial t} + \rho^F \left(\frac{\partial \psi^F}{\partial t} + \acute{\mathbf{X}}^F \cdot \operatorname{Grad} \psi^F \right)$$
$$-\mathbf{P}^S \cdot \frac{\partial \mathbf{F}^S}{\partial t} - \mathbf{P}^F \cdot \operatorname{Grad} \acute{\mathbf{x}}^F - \hat{\mathbf{p}} \cdot \mathbf{w} \le 0, \tag{12.36}$$

must be satisfied identically.

The main technical problem by the exploitation of this inequality is the limitation to the solutions of the field equations. This can be eliminated by means of Lagrange multipliers introduced to thermodynamics by I-Shih Liu (e.g. [437]). Namely, it can be shown that the following inequality

$$\rho^S \frac{\partial \psi^S}{\partial t} + \rho^F \left(\frac{\partial \psi^F}{\partial t} + \acute{\mathbf{X}}^F \cdot \operatorname{Grad} \psi^F \right) - \mathbf{P}^S \cdot \frac{\partial \mathbf{F}^S}{\partial t} - \mathbf{P}^F \cdot \operatorname{Grad} \acute{\mathbf{x}}^F - \hat{\mathbf{p}} \cdot \mathbf{w}$$

$$- \Lambda^n \left\{ \frac{\partial \rho^F}{\partial t} + \operatorname{Div} \left(\rho^F \acute{\mathbf{X}}^F \right) \right\} - \Lambda^F \cdot \left\{ \frac{\partial \mathbf{F}^S}{\partial t} - \operatorname{Grad} \acute{\mathbf{x}}^S \right\}$$

$$- \Lambda^{vS} \cdot \left\{ \rho^S \frac{\partial \acute{\mathbf{x}}^S}{\partial t} - \operatorname{Div} \mathbf{P}^S - \hat{\mathbf{p}} \right\} \tag{12.37}$$

$$- \Lambda^{vF} \cdot \left\{ \rho^F \left(\frac{\partial \acute{\mathbf{x}}^F}{\partial t} + \acute{\mathbf{X}}^F \cdot \operatorname{Grad} \acute{\mathbf{x}}^F \right) - \operatorname{Div} \mathbf{P}^F + \hat{\mathbf{p}} \right\} \le 0,$$

must hold for any fields and not only for the solutions of the field equations. The multipliers $\Lambda^n, \Lambda^F, \Lambda^{vS}, \Lambda^{vF}$ are functions of the constitutive variables $\mathcal{C}^{(\alpha)}$ for α either equal to 1 or to 2. We proceed to discuss the consequences of the above condition for the two different models.

12.2.2.2 $\mathcal{C}^{(1)}$-models

We consider the model with the constitutive variables given by Relation (12.32). It is not necessary to distinguish between (I)-models and (C)-models because the only difference appears in the final results due to the substitution of either $(12.23)_1$ or $(12.23)_2$.

It is easy to check that the chain rule of differentiation in inequality (12.37) yields the linearity of this inequality with respect to the following derivatives

$$\left\{ \frac{\partial \rho^F}{\partial t}, \frac{\partial \mathbf{F}^S}{\partial t}, \frac{\partial \acute{\mathbf{x}}^S}{\partial t}, \frac{\partial \acute{\mathbf{x}}^F}{\partial t} \right\}, \tag{12.38}$$

as well as

$$\left\{ \operatorname{Grad} \rho^F, \operatorname{Grad} \mathbf{F}^S, \operatorname{Grad} \acute{\mathbf{x}}^S, \operatorname{Grad} \acute{\mathbf{x}}^F \right\}. \tag{12.39}$$

Consequently, as the inequality must hold for *all fields*, the coefficients of these derivatives have to vanish. We obtain from the contributions of the time derivatives (12.38) the following relations

$$\Lambda^n = \rho^S \frac{\partial \psi^S}{\partial \rho^F} + \rho^F \frac{\partial \psi^F}{\partial \rho^F},$$

$$\mathbf{P}^S + \Lambda^F = \rho^S \frac{\partial \psi^S}{\partial \mathbf{F}^S} + \rho^F \frac{\partial \psi^F}{\partial \mathbf{F}^S} - \rho^S \left(\mathbf{F}^{S-T} \frac{\partial \psi^S}{\partial \acute{\mathbf{X}}^F} \right) \otimes \acute{\mathbf{X}}^F - \rho^F \left(\mathbf{F}^{S-T} \frac{\partial \psi^F}{\partial \acute{\mathbf{X}}^F} \right) \otimes \acute{\mathbf{X}}^F,$$

$$\rho^S \Lambda^{vS} = -\rho^F \Lambda^{vF} = -\rho^S \frac{\partial \psi^S}{\partial \acute{\mathbf{X}}^F} - \rho^F \frac{\partial \psi^F}{\partial \acute{\mathbf{X}}^F}. \tag{12.40}$$

On the other hand the coefficients of the spatial derivatives (12.39) lead to the following identities

$$\left(\rho^F \frac{\partial \psi^F}{\partial \rho^F} - \Lambda^n \right) \acute{\mathbf{X}}^F + \frac{\partial \mathbf{P}^{ST}}{\partial \rho^F} \Lambda^{vS} + \frac{\partial \mathbf{P}^{FT}}{\partial \rho^F} \Lambda^{vF} = 0,$$

$$\overset{23}{\operatorname{sym}} \left(\frac{\partial \psi^F}{\partial \mathbf{F}^S} + \Lambda^n \mathbf{F}^{S-T} \right) \otimes \acute{\mathbf{X}}^F = 0,$$

$$\Lambda^F = \mathbf{P}^F = -\rho^F \Lambda^n \mathbf{F}^{S-T}. \tag{12.41}$$

The residual inequality

$$\hat{\mathbf{p}} \cdot \mathbf{w} \geq 0, \tag{12.42}$$

remains, which defines the dissipation density of the system.

For technical reasons we make the following simplifying assumption

$$\frac{\partial \psi^S}{\partial \acute{\mathbf{X}}^F} = \frac{\partial \psi^F}{\partial \acute{\mathbf{X}}^F} = 0. \tag{12.43}$$

We consider further a model in which this assumption is not being made. It has consequences on the nonlinear contributions of the diffusion velocity.

Combination of Relations $(12.40)_1$ and $(12.41)_1$ yields now

$$\Lambda^n = \rho^F \frac{\partial \psi^F}{\partial \rho^F}, \quad \frac{\partial \psi^S}{\partial \rho^F} = 0. \tag{12.44}$$

The second part of this relation has the most important bearing on the structure of the interactions described by the model. Namely, the partial free energy of the skeleton does not react on changes of the porosity. We see in a moment what is the reaction of the partial stresses on this property. Such a conclusion would be impossible if we performed the exploitation of the second law for a linear model.

Relation $(12.41)_2$ leads after easy calculations to the relation

$$\frac{\partial \psi^F}{\partial \mathbf{F}^S} = -\Lambda^n \mathbf{F}^{S-T}. \tag{12.45}$$

Hence, Relations $(12.40)_2$ and $(12.41)_2$ yield the following relations for the partial Piola-Kirchhoff stresses

$$\mathbf{P}^S = \rho^S \frac{\partial \psi^S}{\partial \mathbf{F}^S}, \quad \mathbf{P}^F = -\rho^{F2} \frac{\partial \psi^F}{\partial \rho^F} \mathbf{F}^{S-T}, \tag{12.46}$$

or, after the transformation to partial Cauchy stresses, described by (12.31),

$$\mathbf{T}^S = \rho_t^S \frac{\partial \psi^S}{\partial \mathbf{F}^S} \mathbf{F}^{ST}, \quad \mathbf{T}^F = -p^F \mathbf{1}, \quad p^F := \rho_t^{F2} \frac{\partial \psi^F}{\partial \rho_t^F}. \tag{12.47}$$

These are classical thermodynamical relations for elastic materials and ideal fluids, respectively. The most important property of these relations is the fact that the identities $(12.44)_2$ and (12.45) yield

$$\mathbf{T}^S = \mathbf{T}^S \left(\mathbf{F}^S \right), \quad p^F = p^F \left(\rho_t^F \right). \tag{12.48}$$

The latter requires an assumption on isotropy and shall be proven in the next subsection.

Consequently, we obtain an analogon of the simple mixture model which we have considered for mixtures of fluids (see Part I). This means that one *cannot obtain Biot's model* by linearization of the above model. Couplings between the partial stresses which are appearing in the original Biot model with the material constant Q would have to violate the second law of thermodynamics. This property seems to be common for all multi-component models of poroelastic materials which do not contain higher gradients among the constitutive variables.

Obviously, the model contains a coupling due to the relative motion of the components which is described by the source $\hat{\mathbf{p}}$. For this reason, the solutions of boundary/initial value problems and, consequently, the local values of the partial stresses follow from the *coupled* field equations.

We should also mention that the similarity of the relation for p^F for incompressible real fluids ((I)-models) to the relation for compressible fluids ((C)-models) is misleading. This relation has an entirely different physical interpretation. For compressible fluids Relation $(12.48)_2$ yields the following *linear form* of the constitutive relation for the partial pressure

$$p^F = p_0^F + \kappa \left(\rho^F - \rho_0^F \right), \tag{12.49}$$

where p_0^F is a reference pressure and ρ_0^F the corresponding reference partial mass density. In such a case, the compressibility coefficient κ describes elastic properties of the fluid. Simultaneously, its square root specifies the speed of the longitudinal wave in the fluid. This is not the case for the model with the incompressibility assumption. As the real fluid in this case is incompressible it has no elastic properties. Consequently, the relation for p^F describes its dependence on changes of the porosity which are due to microscopical morphological changes of the skeleton such as a redistribution of the grains. Hence, acoustic properties related to such a constitutive relation for the pressure cannot be extracted from the microscopic properties of the real fluid component.

Incidentally, such a model supports views advocated by W. G. Gray (e.g. see: [141]) that macroscopic constitutive relations of the components in a macroscopic model cannot be directly related to the constitutive properties of the real components, and even less they can be derived by any averaging procedure for a single real microscopic component. The macroscopic constitutive properties reflect for each component microscopic properties of both real components as well as microscopic interactions between them.

In the next subsection we consider a higher gradient model which allows for interactions in the constitutive relations for the partial stresses.

12.2.2.3 $\mathcal{C}^{(2)}$-models

We proceed to consider the model based on the constitutive variables (12.33). However, we limit the attention to the simplest case in which the model is *linear* with respect to the gradient of the porosity. In such a case it may appear only in the constitutive relation for the source $\hat{\mathbf{p}}$ because this is the only vectorial constitutive function. We simplify the model even further and assume linearity of the source with respect to both filter velocity $\mathbf{w}$ and $\operatorname{grad} n$. Consequently,

$$\hat{\mathbf{p}} = \pi\mathbf{w} - N \operatorname{grad} n, \tag{12.50}$$

where the material parameters π and N may still depend on ρ^F and $\mathbf{F}^S$. We have left out a possible nonlinear contribution proportional to the vector product $\mathbf{w} \times \operatorname{grad} n$ which would appear in a general nonlinear isotropic model. In contrast to linear contributions such a term would be non-dissipative. The minus sign in (12.50) is related to the property of incompressible models in which N coincides with the pore pressure. In general this may not be the case. This parameter seems to be always positive. However, such a property does not follow from the second law of thermodynamics.

We assume also Relation (12.43) to hold.

Bearing Relation (12.23) in mind, we obtain the following contribution of the source to the inequality (12.37)

$$(I)\text{-models:} \quad \hat{\mathbf{p}} \cdot \mathbf{w} = \pi\mathbf{w} \cdot \mathbf{w} - N \frac{1}{\rho^{FR} J^S} \left\{ \operatorname{Grad} \rho^F - \rho^F \mathbf{F}^{ST} \operatorname{Div} \mathbf{F}^{S-1} \right\} \cdot \acute{\mathbf{X}}^F,$$

$$\tag{12.51}$$

$$(C)\text{-models:} \quad \hat{\mathbf{p}} \cdot \mathbf{w} = \pi\mathbf{w} \cdot \mathbf{w} - N \frac{1}{\rho_0^{FR}} \operatorname{Grad} \rho^F \cdot \acute{\mathbf{X}}^F, \quad \mathbf{w} \equiv \mathbf{F}^S \acute{\mathbf{X}}^S.$$

Contributions appearing with the parameter N in these relations change the identities of the previous subsection following from coefficients of spatial derivatives. Namely, Relations $(12.41)_1$ and $(12.41)_2$ will be influenced. We collect all these results of the second law of thermodynamics in the juxtaposition (Table 12.4).

After easy manipulations we obtain the following relations for the partial Piola-Kirchhoff

Table 12.4: *Results of the second law of thermodynamics for (I)-models and (C)-models.*

(I)-models	(C)-models
$\Lambda^n = \rho^S \dfrac{\partial \psi^S}{\partial \rho^F} + \rho^F \dfrac{\partial \psi^F}{\partial \rho^F},$	$\Lambda^n = \rho^S \dfrac{\partial \psi^S}{\partial \rho^F} + \rho^F \dfrac{\partial \psi^F}{\partial \rho^F},$
$\mathbf{P}^S = \rho^S \dfrac{\partial \psi^S}{\partial \mathbf{F}^S} + \rho^F \dfrac{\partial \psi^F}{\partial \mathbf{F}^S} - \mathbf{P}^F,$	$\mathbf{P}^S = \rho^S \dfrac{\partial \psi^S}{\partial \mathbf{F}^S} + \rho^F \dfrac{\partial \psi^F}{\partial \mathbf{F}^S} - \mathbf{P}^F,$
$\mathbf{P}^F = -\rho^F \Lambda^n \mathbf{F}^{S-T},$	$\mathbf{P}^F = -\rho^F \Lambda^n \mathbf{F}^{S-T},$
$\rho^S \dfrac{\partial \psi^S}{\partial \rho^F} - \dfrac{N}{\rho^{FR}} J^{S-1} = 0, \quad \rho^{FR} = \text{const.}$	$\rho^S \dfrac{\partial \psi^S}{\partial \rho^F} - \dfrac{N}{\rho_0^{FR}} = 0,$
$\dfrac{\partial \psi^F}{\partial \mathbf{F}^S} + \rho^F \dfrac{\partial \psi^F}{\partial \rho^F} \mathbf{F}^{S-T} = 0,$	$\dfrac{\partial \psi^F}{\partial \mathbf{F}^S} + \left(\rho^F \dfrac{\partial \psi^F}{\partial \rho^F} + \dfrac{N}{\rho_0^{FR}} \right) \mathbf{F}^{S-T} = 0.$

stresses in both classes of models

$$(I)\text{-models:} \qquad \mathbf{P}^S = \rho^S \frac{\partial \psi^S}{\partial \mathbf{F}^S} + nN\mathbf{F}^{S-T},$$

$$(C)\text{-models:} \qquad \mathbf{P}^S = \rho^S \frac{\partial \psi^S}{\partial \mathbf{F}^S},$$

(12.52)

$$(I)\text{-models \& } (C)\text{-models: } \mathbf{P}^F = - \left(\rho^{F2} \frac{\partial \psi^F}{\partial \rho^F} + nN \right) \mathbf{F}^{S-T}, \qquad (12.53)$$

as well as the following identities for the partial free energy functions

$$(I)\text{-models} \quad : \quad \frac{\partial \psi^F}{\partial \mathbf{F}^S} = -\rho^F \frac{\partial \psi^F}{\partial \rho^F} \mathbf{F}^{S-T}, \quad \rho^S \frac{\partial \psi^S}{\partial \rho^F} = \frac{N}{\rho^{FR} J^S},$$

(12.54)

$$(C)\text{-models} \quad : \quad \frac{\partial \psi^F}{\partial \mathbf{F}^S} = - \left(\rho^F \frac{\partial \psi^F}{\partial \rho^F} + \frac{N}{\rho_0^{FR}} \right) \mathbf{F}^{S-T}, \quad \rho^S \frac{\partial \psi^S}{\partial \rho^F} = \frac{N}{\rho_0^{FR}}.$$

Then, for the partial Cauchy stresses follow relations of the form

(I)-models:

$$\mathbf{T}^S = \rho^S J^{S-1}\frac{\partial\psi^S}{\partial\mathbf{F}^S}\mathbf{F}^{ST} + nNJ^{S-1}\mathbf{1}, \quad \mathbf{T}^F = -p^F\mathbf{1},$$

$$p^F := \rho^{F2}J^{S-1}\frac{\partial\psi^F}{\partial\rho^F} + nNJ^{S-1},$$

$$(12.55)$$

(C)-models:

$$\mathbf{T}^S = \rho^S J^{S-1}\frac{\partial\psi^S}{\partial\mathbf{F}^S}\mathbf{F}^{ST}, \quad \mathbf{T}^F = -p^F\mathbf{1},$$

$$p^F := \rho^{F2}J^{S-1}\frac{\partial\psi^F}{\partial\rho^F} + nNJ^{S-1}.$$

It can be seen that both models contain couplings of stresses which may lead to Biot's constitutive relations for the stresses (see Part I or e.g. [49], [381]) of the linear model. We proceed to investigate this question.

In order to simplify the construction of the linear model we evaluate the above nonlinear relations for isotropic materials. In such a case the free energies ψ^S, ψ^F satisfying the principle of material objectivity depend on the deformation gradient $\mathbf{F}^S$ only through the invariants of its symmetric part. For instance, we can choose the invariants of the right Cauchy-Green deformation tensor $\mathbf{C}^S$ (compare (2.16))

$$\mathbf{C}^S := \mathbf{F}^{ST}\mathbf{F}^S, \quad I := \operatorname{tr}\mathbf{C}^S, \quad II := \frac{1}{2}\left(I^2 - \operatorname{tr}\mathbf{C}^{S2}\right), \quad III \equiv J^{S2} := \det\mathbf{C}^S. \quad (12.56)$$

Then,

$$\psi^S = \psi^S\left(I, II, III, \rho^F\right), \quad \psi^F = \psi^F\left(I, II, III, \rho^F\right). \quad (12.57)$$

Let us exploit the identity $(12.54)_1$ under the above assumption. Bearing the following relation in mind

$$\frac{\partial\psi^F}{\partial\mathbf{F}^S} = 2\mathbf{F}^S\frac{\partial\psi^F}{\partial\mathbf{C}^S}, \quad (12.58)$$

we obtain

$$\frac{\partial\psi^F}{\partial\mathbf{F}^S}\mathbf{F}^{ST} = 2\left(\frac{\partial\psi^F}{\partial I}\mathbf{B}^S + \frac{\partial\psi^F}{\partial II}I\mathbf{B}^S - \frac{\partial\psi^F}{\partial II}\mathbf{B}^{S2} + \frac{\partial\psi^F}{\partial III}III\mathbf{1}\right), \quad (12.59)$$

$$\mathbf{B}^S := \mathbf{F}^S\mathbf{F}^{ST}, \quad III \equiv J^{S2},$$

where the symmetric tensor $\mathbf{B}^S$ is the left Cauchy-Green deformation tensor. Now we can write $(12.54)_1$ in the form

(I)-models :

$$\left(2III\frac{\partial\psi^F}{\partial III} + \rho^F\frac{\partial\psi^F}{\partial\rho^F}\right)\mathbf{1} + 2\left(\frac{\partial\psi^F}{\partial I} + \frac{\partial\psi^F}{\partial II}I\right)\mathbf{B}^S - 2\frac{\partial\psi^F}{\partial II}\mathbf{B}^{S2} = 0,$$

$$(12.60)$$

(C)-models :

$$\left(2III\frac{\partial\psi^F}{\partial III} + \rho^F\frac{\partial\psi^F}{\partial\rho^F} + \frac{N}{\rho_0^{FR}}\right)\mathbf{1} + 2\left(\frac{\partial\psi^F}{\partial I} + \frac{\partial\psi^F}{\partial II}I\right)\mathbf{B}^S - 2\frac{\partial\psi^F}{\partial II}\mathbf{B}^{S2} = 0.$$

These relations must hold for arbitrary deformations $\mathbf{B}^S$. Consequently, according to the corollaries of the Cayley-Hamilton Theorem (5.137), the coefficients of the tensors $\mathbf{1}$, $\mathbf{B}^S$ and $\mathbf{B}^{S2}$ have to vanish independently. Hence, in both cases the free energy ψ^F must be independent of the invariants I, II, and, consequently (compare the relations for N in Table 12.4), N must be independent of these two invariants in the (C)-models but not necessarily in the (I)-models. In addition we have

$$(I)\text{-models:} \qquad \psi^F = \psi^F\left(J^S, \rho^F\right), \quad J^S\frac{\partial \psi^F}{\partial J^S} + \rho^F\frac{\partial \psi^F}{\partial \rho^F} = 0; \tag{12.61}$$

$$(C)\text{-models:} \qquad \psi^F = \psi^F\left(J^S, \rho^F\right), \quad J^S\frac{\partial \psi^F}{\partial J^S} + \rho^F\frac{\partial \psi^F}{\partial \rho^F} = -\frac{N}{\rho_0^{FR}}, \tag{12.62}$$

$$N = N\left(J^S, \rho^F\right).$$

The differential equation $(12.61)_2$ for (I)-models can be easily solved by the method of characteristics. If we denote the variable along the characteristic by ξ we can write this equation in the characteristic form

$$\frac{dJ^S}{d\xi} = J^S, \quad \frac{d\rho^F}{d\xi} = \rho^F, \quad \frac{d\psi^F}{d\xi} = 0. \tag{12.63}$$

Consequently, ψ^F is constant along the characteristics, and these are labeled by the following initial values

$$\rho^F J^{S-1} = \text{const.}, \tag{12.64}$$

which follows from $(12.63)_{1,2}$. This means that the solution of the identity $(12.54)_1$ for isotropic materials with an incompressible real fluid is of the form

$$(I)\text{-models: } \psi^F = \psi^F\left(\rho_t^F\right). \tag{12.65}$$

The problem is more complicated for compressible materials. However, we can simplify Equation $(12.62)_2$ if we change the variables in a way suggested by the above solution for the (I)-model. Namely, we obtain

$$(C)\text{-models:} \qquad \left(\rho^F, J^S\right) \to \left(\rho_t^F, J^S\right) \quad \Longrightarrow \quad \psi^F = \psi^F\left(\rho_t^F, J^S\right), \tag{12.66}$$

and

$$J^S\frac{\partial \psi^F}{\partial J^S} = -\frac{N}{\rho_0^{FR}} \quad \Longrightarrow \quad \psi^F = \psi_{\text{ideal}}^F\left(\rho_t^F\right) - \frac{1}{\rho_0^{FR}}\int_1^{J^S}\frac{N\left(\rho_t^F, \xi\right)}{\xi}d\xi. \tag{12.67}$$

It is instructive to apply Relation $(12.54)_2$ in the formulae (12.55) for the Cauchy stresses. For isotropic materials we obtain immediately

(I)-models:

$$\mathbf{T}^S = \rho^S J^{S-1}\left\{\left(J^S\frac{\partial \psi^S}{\partial J^S} + \rho^F\frac{\partial \psi^S}{\partial \rho^F}\right)\mathbf{1} + 2\left(\frac{\partial \psi^S}{\partial I} + I\frac{\partial \psi^S}{\partial II}\right)\mathbf{B}^S - 2\frac{\partial \psi^S}{\partial II}\mathbf{B}^{S2}\right\},$$

$$\mathbf{T}^F = -\rho^F J^{S-1}\left(\rho^F\frac{\partial \psi^F}{\partial \rho^F} + \rho^S\frac{\partial \psi^S}{\partial \rho^F}\right)\mathbf{1}, \tag{12.68}$$

(C)-models:

$$\mathbf{T}^S = \rho^S J^{S-1}\left\{J^S\frac{\partial \psi^S}{\partial J^S}\mathbf{1} + 2\left(\frac{\partial \psi^S}{\partial I} + I\frac{\partial \psi^S}{\partial II}\right)\mathbf{B}^S - 2\frac{\partial \psi^S}{\partial II}\mathbf{B}^{S2}\right\}, \tag{12.69}$$

$$\mathbf{T}^F = -\rho^F J^{S-1}\left(\rho^F\frac{\partial \psi^F}{\partial \rho^F} + \rho^S\frac{\partial \psi^S}{\partial \rho^F}\right)\mathbf{1}.$$

Let us discuss first the structure of the stress relations for the (I)-models. As indicated by (12.65) the first contribution to the partial stress $\mathbf{T}^F$ cannot contain any coupling to the deformation of the skeleton. Consequently, it is the derivative of the partial Helmholtz free energy ψ^S with respect to ρ_t^F which relates $\mathbf{T}^F$ to the deformation of the skeleton. According to Relation $(12.54)_2$ for the existence of this coupling, the coefficient N must be different from zero. For the symmetry required in Biot's model it is necessary to introduce a rather complicated dependence of the partial free energy function ψ^S on the mass density ρ^F which would create a term in $\rho^S \frac{\partial \psi^S}{\partial J^S}$ canceling out the contribution $\rho^S \rho^F J^{S-1} \frac{\partial \psi^S}{\partial \rho^F}$ to the stress $\mathbf{T}^S$ (the wrong sign!) and simultaneously produces another one introducing the coupling to the deformation of the fluid. Even though it is possible in principle, such a model does not seem to be very plausible and we do not investigate it any further.

Let us mention in passing that in the particular case of the constant partial free energy ψ^F Relation $(12.55)_2$ for incompressible real fluids yields $N = \frac{p^F}{n}$ which is the pore pressure of classical models. Hence, in this particular case, the coefficient in the diffusion force (12.50) coincides with that of classical models of consolidation. This structure has been indicated in the work [429] on nonlinear sources.

The structure of the stress relations for (C)-models is simpler. As thermodynamical requirements do not lead to any restrictions of the free energy ψ^S on ρ_t^F, and the free energy ψ^F on J^S, we may produce as a particular case the desired dependence and symmetry.

Concluding the above thermodynamical considerations we see that Biot's constitutive relations for the stresses can be derived from nonlinear $\mathcal{C}^{(2)}$-models by a specific choice of partial free energies. This means that such a transition requires a higher gradient model as a background.

12.3   Double- and multi-porosity models

12.3.1   Fissured rocks – example of a model with double porosity

Most theories of seepage are based on the concept of a porous medium consisting of a solid and the pore space which is filled by one or more fluids. It is well known that the ratio of the volume of the pore space to the entire volume of the porous medium is the porosity (in the last section and in Chapter 13 denoted by n). One of the first approaches stating that such a concept with a single porosity is inadequate for strata under natural conditions is the work [30] published in 1960 by G. I. Barenblatt, I. P. Zheltov and I. N. Kochina. They pointed out that in all natural strata, the development of fissuring is a characteristic feature which has to be taken into account in the description of seepage processes.

Because simple models seem only to be appropriate for a qualitative investigation of seepage phenomena in such media, a more evolved model is presented in [30]. In this paper the mean characteristics of fissured media and of the flow (porosities, permeabilities, pressures, seepage velocity, etc.) are described and basic laws in terms of these mean characteristics are formulated. The ideas of the paper are repeated here.

At each point in the space two liquid pressures are introduced, namely 1) the liquid pressure in the pores and 2) the pressure of the liquid in the fissures α. The transfer of liquid between the fissures and the pores is taken into account. General equations for the seepage of a liquid in a porous medium with double porosity are derived.

Figure 12.1: *Rock texture, Widemouth Bay, North Cornwall, figure from http://themagicofcornwall.photoshelter.com/.*

12.3.1.1 Basic physical concepts

A fissured rock consists of pores and permeable blocks – this means that the blocks are separated from each other by a system of fissures (see Figure 12.1). The dimensions of the blocks vary for different types of rocks within wide limits, depending on the extent to which fissures are developed. The widths of the fissures are considerably greater than the characteristic dimensions of the pores, so that the permeability of the fissure system considerably exceeds the permeability of the system of pores in the individual blocks. A characteristic feature of fissured rocks is that the fissures occupy a much smaller volume than the pores, so that the coefficient of fissuring of the rock n_1 – the ratio of the volume of the cavity space occupied by the fissures to the total volume of the rock – is considerably smaller than the porosity of the individual blocks n_2. There exists a wide literature on fissured rocks, e.g. [4, 270, 295].

If the system of fissures is sufficiently well developed, according to [30] the motion of the liquid in fissured rocks can be investigated by the following method. For each point in the space, two liquid pressures, p_1 and p_2, are introduced. The pressure p_1 represents the average pressure of the liquid in the fissures in the neighborhood of the given point, while the pressure p_2 is the average pressure of the liquid in the pores in the neighborhood of the given point. In order to obtain reliable averages, the representative elementary volume in which averaging takes place (see Chapter 12 of Part I) should include a sufficiently large number of blocks. Thus, not only a sufficient number of pores in any infinitely small volume has to be accounted for, as is done in the classical theory, but also for a sufficient number of blocks.

Similarly, two seepage velocities can be defined at each point in space. The vector $\mathbf{v}_1$ is the seepage velocity of the liquid along the fissures. I.e., that the projection of this vector in a particular direction is equal to the flow of the liquid through the cross section of the fissures of a small zone passing through the given point in a direction perpendicular to the given direction, divided by the density of the liquid and the total area of this zone. In the same way, the projection of vector $\mathbf{v}_2$, the seepage velocity of the liquid through the pores in a given direction, equals the flow of the liquid through the cross section of the blocks of the small zone mentioned, also divided by the density of the liquid and the total area of the zone.

According to [30] a characteristic feature of fissured rocks is that the flow of the liquid proceeds essentially along the fissures. Thus, the velocity of the liquid through the blocks is negligibly small compared to the seepage along the fissures. If the boundary between the fissures and the blocks is assumed to be impermeable, the fissured rock can be considered as a coarse-grained porous medium in which the fissures play the role of pores and the blocks play the role of grains. If, furthermore, the fissures are sufficiently narrow and the velocity of the liquid is sufficiently small, the motion of the liquid along the fissures will be inertialess and Darcy's law is fulfilled

$$\mathbf{v}_1 = -\frac{k_1}{\mu} \operatorname{grad} p_1, \tag{12.70}$$

where k_1 is the permeability of the system of fissures and μ is the viscosity of the liquid. If inertia of the motion would be taken into account, instead of Darcy's law a more complicated nonlinear law had to be used.

During non-steady motion of a liquid in fissured rocks liquid is transferred between blocks and fissures. Therefore, it is necessary to take into account the outflow of liquid from the blocks into the pores. The transfer of liquid from the pores and the blocks takes place essentially under a sufficiently smooth change of pressure, and, therefore, it can be assumed that this pressure is quasi-stationary, i.e., it is independent of time explicitly. It is obvious then that during motion of a homogeneous liquid in the fissures of the rock, the volume of the liquid V, which flows from the blocks into the fissures per unit time and unit volume of the rock, depends on (1) the viscosity of the liquid μ; (2) the pressure drop between the pores and the fissures $p_2 - p_1$; and (3) on certain characteristics of the rock. On the basis of dimensional analysis (see e.g. [346] by L. I. Sedov), we obtain the expression

$$V = \frac{\alpha}{\mu} \left(p_2 - p_1 \right), \tag{12.71}$$

where α is a dimensionless parameter of the fissured rock.

Thus, for the mass q of the liquid which flows from the pores into the fissures per unit of time and per unit of rock volume, the following equation is valid

$$q = \frac{\rho \alpha}{\mu} \left(p_2 - p_1 \right), \tag{12.72}$$

where ρ is the density of the liquid.

12.3.1.2   Equation of motion of a uniform liquid in fissured rocks

The conservation of mass of liquid in the presence of fissures can be written as

$$\underbrace{\frac{\partial n_1 \rho}{\partial t}}_{\text{small}} + \operatorname{div} \rho \mathbf{v}_1 - q = 0. \tag{12.73}$$

Since the volume of the fissures is assumed to be small, the first term, the change in the mass of the liquid due to compression in the fissures (change of the volume of the fissures), is small compared to the second term, the changes in the mass of the liquid caused by the inflow of the liquid along the fissures through the boundary of this element. Therefore,

Relation (12.73) can be disregarded. Inserting Equation (12.70) into Equation (12.73), taking into account the fact that the liquid is slightly compressible, so that $\rho = \rho_0 + \beta \delta p$ (ρ_0 is the density of the liquid at some standard pressure, β is the coefficient of compressibility of the liquid,[1] δp is the change in the pressure relative to the standard pressure), assuming that the medium is homogeneous and neglecting the small higher-order terms, we obtain

$$k_1 \Delta^2 p_1 + \alpha \left(p_2 - p_1 \right) = 0, \tag{12.74}$$

where Δ^2 is the Laplace operator. The equations of mass conservation of the liquid which is present in the pores can be written analogously

$$\frac{\partial n_2 \rho}{\partial t} + \underbrace{\operatorname{div} \rho \mathbf{v}_2}_{\text{small}} + q = 0, \tag{12.75}$$

so that the quantity of liquid which flows into the fissures equals the quantity of the liquid which flows out of the blocks.

In view of the low permeability of the blocks, the second term of Equation (12.75) can be disregarded. It expresses the changes in the mass of the liquid within the pores in some element of the rock, due to the inflow of liquid along the pores through the boundaries of the element. It is small compared to the first term which represents changes in the mass of the liquid in the pores due to expansion, and also to changes in the volume of the pores.

Furthermore, the porosity of the blocks n_2 in the case of a constant pressure upon the stratum depends, generally speaking, on the pressure of the liquid in the fissures p_1 and the pressure of the liquid in the pores p_2. However, the volume of the fissures in the rock is considerably smaller than the volume of the pores. Therefore, the influence of the pressure of the liquid in the fissures p_1 on the porosity of the blocks can be disregarded compared to the influence of the pressure of the liquid in the pores p_2. It can be assumed that $dn_2 = \beta_{c2} dp_2$, where β_{c2} is the coefficient of compressibility of the blocks. Neglecting small terms of higher order, we obtain then

$$\left(\beta_{c2} + n_0 \beta \right) \frac{\partial p_2}{\partial t} + \frac{\alpha}{\mu} \left(p_2 - p_1 \right) = 0, \tag{12.76}$$

where n_0 is the porosity of the blocks at standard pressure. Equations (12.74) and (12.76) describe, according to [30], the motion of the liquid in fissured rocks. Eliminating from these equations p_2, we obtain for the pressure of the liquid in the fissures p_1 the equation

$$\frac{\partial p_1}{\partial t} - \eta \frac{\partial \Delta^2 p_1}{\partial t} - \kappa \Delta^2 p_1, \tag{12.77}$$

where the coefficient $\kappa = k_1 / \left[\mu \left(\beta_{c2} + n_0 \beta \right) \right]$ represents the coefficient of piezo-conductivity of the fissured rock. Note that this does not correspond to the permeability of the system of fissures k_1, but to the porosity and compressibility of the blocks. The coefficient $\eta = k_1/\alpha$ represents a further characteristic of fissured rocks. If η tends to zero, it corresponds to a reduction of the block dimensions and an increase in the degree of fissuring, and Equation (12.77) will, obviously, tend to coincide with the ordinary equation for seepage of a liquid under elastic conditions. In [30] it is mentioned that for various rocks the parameter η will assume values within wide limits – from a few cm^2 to values of the order of 10^{10} cm^2.

---

[1]In the last section, in the model introduced in Chapter 13 and in Part I the compressibility is denoted by κ (compare Equation (12.49)) which represents here the coefficient of piezo-conductivity of the fissured rock.

12.3.1.3 Equations of motion of a homogeneous liquid in a medium with double porosity

The system of equations (12.74) and (12.76) represents a particular case of the system of equations of motion of a homogeneous liquid in a medium with double porosity. Since in some cases the latter equations may be of interest, we repeat here briefly their derivation [30].

The motion of a uniform liquid in a porous medium with double porosity is considered. Thus, two different systems are investigated: the first one consists of relatively wide pores – fissures and blocks; the porosity is n_1. The blocks in themselves are porous, consisting of grains which are separated by fine pores; they form the second porous medium. The second porosity is denoted by n_2. Both porosities, n_1 and n_2, will depend on the pressures of the liquid in the blocks and in the pores, p_1 and p_2, so that

$$dn_1 = \beta_{c1}dp_1 - \beta_* dp_2, \qquad dn_2 = \beta_{c2}dp_2 - \beta_{**}dp_1, \qquad (12.78)$$

where β_{c1}, β_{c2}, β_* and β_{**} are positive constant coefficients.

The equations for conservation of mass of the liquid for both media are of the form (12.73) and (12.75), respectively. Assuming that the flow of the liquid is inertialess, the Darcy law for the two media can be written as

$$\mathbf{v}_1 = -\frac{k_1}{\mu}\,\mathrm{grad}\,p_1, \qquad \mathbf{v}_2 = -\frac{k_2}{\mu}\,\mathrm{grad}\,p_2, \qquad (12.79)$$

where k_1 is the porosity of the system of pores of the first order (the fissures) and k_2 the porosity of the system of pores of the second order (the pores of the blocks).

Instead of (12.76) the following system of equations is obtained

$$\frac{k_1}{\mu}\Delta^2 p_1 = (\beta_{c1} + n_{10}\beta)\,\frac{\partial p_1}{\partial t} - \beta_*\frac{\partial p_2}{\partial t} - \frac{\alpha}{\mu}\,(p_2 - p_1)\,,$$

$$\frac{k_2}{\mu}\Delta^2 p_2 = (\beta_{c2} + n_{20}\beta)\,\frac{\partial p_2}{\partial t} - \beta_{**}\frac{\partial p_1}{\partial t} + \frac{\alpha}{\mu}\,(p_2 - p_1)\,, \qquad (12.80)$$

where n_{10} and n_{20} are the values of the first and second order porosity at standard pressure.

If the pressure p_2 changes, say decreases, at a constant pressure of the upper strata, on the one hand, n_1 will increase as a result of the compression of the blocks and, on the other hand, it will decrease as a result of compression by the overlying strata. These effects will apparently compensate each other to some extent. The situation is similar for n_2 for a change of the pressure p_1. Therefore, we should consider a model of double porosity for which both porosities depend only on the appropriate pressure, so that the coefficients β_* and β_{**} in Equation (12.78) can be considered small and the appropriate terms in this equation can be disregarded.

Equations (12.80) for such a model of a porous medium with double porosity are of the form similar to the equations for heat transfer in a heterogeneous medium considered by Rubinstein [326] who was according to [23] one of the first to consider heat conduction in a dual component heterogeneous media, where an average temperature is considered for each component and Fourier's law is obeyed in each of the components. His equations have the form

$$\frac{k_1}{\mu}\Delta^2 p_1 = (\beta_{c1} + n_{10}\beta)\,\frac{\partial p_1}{\partial t} - \frac{\alpha}{\mu}\,(p_2 - p_1)\,,$$

$$\frac{k_2}{\mu}\Delta^2 p_2 = (\beta_{c2} + n_{20}\beta)\,\frac{\partial p_2}{\partial t} + \frac{\alpha}{\mu}\,(p_2 - p_1)\,. \qquad (12.81)$$

Thus, disregarding in Equations (12.80) the terms representing a change in the mass of the liquid due to the compressibility of the first medium and the compression of the liquid in the pores of the first order, and moreover, the changes in the mass of the liquid as a result of the seepage inflow along the pores of the second order, again the equations of motion of a liquid in a fissured porous medium (12.74) and (12.76) are obtained.

Finally, Equation (12.77) to which corresponds the pressure distribution of the liquid in the pores p_1, can be written as

$$\beta_0 \frac{\partial p_1}{\partial t} - \text{div}\left[\frac{k_1}{\mu} \operatorname{grad} p_1 + \eta \beta_0 \frac{\partial}{\partial t} \operatorname{grad} p_1\right] = 0, \tag{12.82}$$

where β_0 is the total effect of compressibility equal to $\beta_{c2} + n_0\beta$. This form of writing the basic equation indicates, that motion in the system of fissures can be considered as the motion of a liquid in a porous medium with a total compressibility coefficient β_0, and the expression for the velocity of seepage of the liquid can be written as

$$\mathbf{v} = -\frac{k_1}{\mu} \operatorname{grad} p_1 - \eta \beta_0 \frac{\partial}{\partial t} \operatorname{grad} p_1. \tag{12.83}$$

Initial and boundary conditions have to be added to solve the problem. Among the possible types of boundary conditions the most important are the following:

1. The pressure p_1 at the boundary of the rock of volume s under consideration is given (boundary condition of first type or Dirichlet boundary condition)

$$p_1|_S = f(S, t). \tag{12.84}$$

2. At the boundary s of the surface S the flow of the liquid is given (boundary condition of second type or Neumann boundary condition)

$$\left\{\frac{k_1}{\mu} \frac{\partial p_1}{\partial n} + \eta \beta_0 \frac{\partial}{\partial t}\left(\frac{\partial p_1}{\partial n}\right)\right\}\Bigg|_S = f(S, t), \tag{12.85}$$

($\partial/\partial n$ is the derivative along the normal to the surface S), and,

3. At the boundary a linear combination of the pressure and the flow of the liquid, generally speaking with variable coefficients A and B, is given (boundary condition of third type or mixed boundary condition, often called Robin boundary condition)

$$\left\{Ap_1 + B\left[\frac{k_1}{\mu} \frac{\partial p_1}{\partial n} + \eta \beta_0 \frac{\partial}{\partial t}\left(\frac{\partial p_1}{\partial n}\right)\right]\right\}\Bigg|_S = f(S, t). \tag{12.86}$$

These types of boundary conditions have been discussed for porous media and in analogy to heat conduction in [7].

12.3.1.4 Non-steady-state flow of liquid in a gallery

One special boundary value problem is picked from the paper [30]. It concerns the non-steady-state seepage in drainage galleries and it is formulated as follows: at the initial instant, the pressure of the liquid P_0 in a semi-infinite stratum $(0 \leq x < \infty)$ is constant; the pressure at the boundary $x = 0$ suddenly takes the value P_1, differing from P_0, and then remains constant. The problem of determining seepage flow requires the solution of the equation

$$\frac{\partial p_1}{\partial t} - \eta\frac{\partial^3 p_1}{\partial x^2 \partial t} = \kappa\frac{\partial^2 p_1}{\partial x^2}, \tag{12.87}$$

for the initial and boundary conditions

$$p_1(x,0) = P_0, \qquad p_1(-0,t) = P_1, \tag{12.88}$$

(the boundary pressure is given immediately to the left of the boundary $x = 0$). At the initial instant there will be a pressure jump at the boundary equal to $(P_0 - P_1)$. According to [30], at time t this jump will equal $(P_0 - P_1)e^{-\kappa t/\eta}$, so that the pressure of the liquid immediately to the right of the boundary will equal

$$p_1(+0,t) = P_1 + (P_0 - P_1)e^{-\kappa t/\eta}. \tag{12.89}$$

To find the distribution at any desired instant of time t, the Laplace transform (see Appendix B.2) with respect to time t is applied. We set

$$p_1(x,t) = P_0 + -(P_0 - P_1)u(x,t). \tag{12.90}$$

Then, for determining $u(x,t)$ ($t \geq 0$, $0 \leq x < \infty$), we obtain the boundary-value problem

$$\frac{\partial u}{\partial t} - \eta\frac{\partial^3 u}{\partial x^2 \partial t} = \kappa\frac{\partial^2 u}{\partial x^2}, \qquad u(+0,t) = 1 - e^{-\kappa t/\eta}, \quad u(x,0) = 0. \tag{12.91}$$

Let

$$U(x,\lambda) = \int_0^\infty e^{-\lambda t}u(x,t)dt. \tag{12.92}$$

Applying the Laplace transform to Equation (12.91) we obtain

$$\frac{d^2 U}{dx^2} - \frac{\lambda}{\kappa + \lambda\eta}U = 0, \qquad U(0,\lambda) = \frac{1}{\lambda(\kappa + \lambda\eta)}, \qquad U(\infty,\lambda) = 0,$$

$$U = \frac{1}{\lambda(\kappa + \lambda\eta)}\exp\left(-\sqrt{\frac{\lambda}{\kappa + \lambda\eta}}x\right), \tag{12.93}$$

whence, on the basis of the rule of inversion [105], we obtain

$$u(x,t) = \frac{1}{2\pi i}\int_{\gamma-i\infty}^{\gamma+i\infty} e^{\lambda t}\exp\left(-\sqrt{\frac{\lambda}{\kappa + \lambda\eta}}x\right)\frac{d\lambda}{\lambda(\kappa + \lambda\eta)}. \tag{12.94}$$

The evolution of the integral on the right-hand side of the previous equation yields

$$2\pi i - 2i\int_0^1 e^{-\sigma t}\sin\left(\sqrt{\frac{\sigma}{1-\sigma}}x\right)\frac{d\sigma}{\sigma(1-\sigma)}. \tag{12.95}$$

Inserting this expression into Equation (12.94) and substituting the variables $\sigma\,(1-\sigma)=\nu^2$ we obtain

$$u(x,t)=1-\frac{2}{\pi}\int_0^\infty \frac{\sin\nu x}{\nu}\exp\left(-\frac{\nu^2\kappa t}{1+\nu^2\eta}\right)d\nu=1-\exp\left(-\frac{\kappa t}{\eta}\right)$$

$$\tag{12.96}$$

$$-\frac{2}{\pi}\int_0^\infty \frac{\sin\nu x}{\nu}\left\{\exp\left(-\frac{\nu^2\kappa t}{1+\nu^2\eta}\right)-\exp\left(-\frac{\kappa t}{\eta}\right)\right\}d\nu.$$

From this and from Equation (12.90) we finally obtain

$$p_1(x,t)=P_1+\frac{2\,(P_0-P_1)}{\pi}\int_0^\infty \frac{\sin\nu x}{\nu}\left\{\exp\left(-\frac{\nu^2\kappa t}{1+\nu^2\eta}\right)\right.$$

$$\tag{12.97}$$

$$\left.-\exp\left(-\frac{\kappa t}{\eta}\right)\right\}d\nu+(P_0-P_1)\exp\left(-\frac{\kappa t}{\eta}\right).$$

Since the integral in Equation (12.97) and the integral obtained after differentiation with respect to x under the integral sign are both uniformly convergent, the expression given by Equation (12.97) can be differentiated with respect to x. Hence, from Equation (12.97), we obtain the expression for the flow of the liquid through the boundary of the stratum $x=0$

$$q=\frac{k_1}{\mu}\left(\frac{\partial p_1}{\partial x}\right)_{x=+0}=-\frac{2\,(P_0-P_1)\,k_1}{\pi\mu}\int_0^\infty\left\{\exp\left(-\frac{\nu^2\kappa t}{1+\nu^2\eta}\right)\right.$$

$$\left.-\exp\left(-\frac{\kappa t}{\eta}\right)\right\}d\nu=-\frac{2\,(P_0-P_1)\,k_1}{\pi\mu\sqrt{\eta}}\exp\left(-\frac{\kappa t}{\eta}\right) \tag{12.98}$$

$$\times\int_0^\infty\left\{\exp\left(\frac{\kappa t}{\eta\,(1+\zeta^2)}\right)-1\right\}d\zeta.$$

It is interesting to compare the derived solutions with the appropriate solution of the problem of the theory of seepage in a porous medium with single porosity. This well-known self-similar solution is obtained from Equation (12.97) for the case when $\eta=0$

$$p_1(x,t)=P_1+\frac{2\,(P_0-P_1)}{\pi}\int_0^{1/2\xi}e^{-\beta}d\beta=P_1+(P_0-P_1)\,\Phi\left(\frac{1}{2}\xi\right),$$

$$\tag{12.99}$$

$$\xi=\frac{x}{\sqrt{\kappa t}}.$$

where Φ is the symbol of the Kramp function. For comparing the pressures, we look at Equations (12.97) and (12.99), which correspond to fissured and ordinary porous media, respectively. For both equations the distribution quantities $u(x,t)=(p_1-P_1)/(P_0-P_1)$

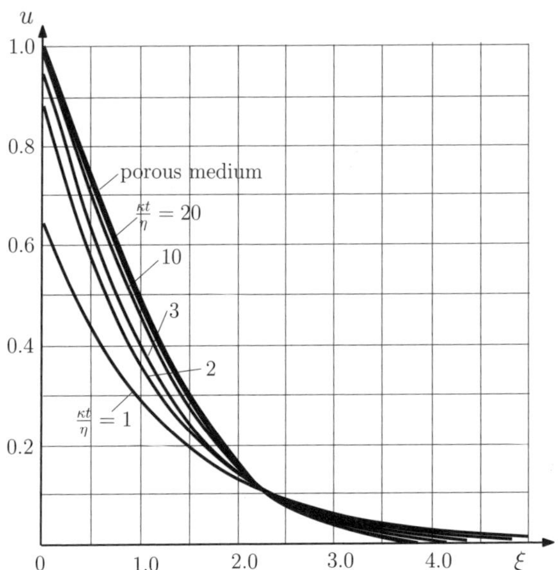

Figure 12.2: $u(x,t) = (p_1 - P_1)/(P_0 - P_1)$ *for various values of the parameter* $\kappa t/\eta$ *have been plotted as a function of the self-similar variable* ξ, *i.e., comparison of fissured and ordinary porous media (after Figure 2 of [30]).*

for various values of the parameter $\kappa t/\eta$ have been plotted as a function of the self-similar variable ξ in Figure 12.2.

It can be seen that with increasing $\kappa t/\eta$ the pressure distribution in a fissured rock tends asymptotically to the self-similar distribution which is obtained in an ordinary porous medium. It can also be seen from the example that the most characteristic property of the non-steady-state flow of liquids in fissured rocks is the occurrence of some delay in the transient processes: the characteristic time of this delay is $\tau = \eta/\kappa$. Thus, the following general conclusion can be advanced: in considering processes of non-steady-state infiltration in fissured rocks, the ordinary equations of non-steady-state flow in a porous medium can be applied only if the characteristic times of the process under consideration are long compared to the delay times. However, if the characteristic times of the process are comparable to the delay time, it is necessary to apply the model and the basic equations presented in this subsection which is based on [30]. The estimates which have been carried out have shown that fissuring must be taken into account in many cases when investigating such processes as the restoration of pressure in shut-down wells and, generally, transient processes during changes of the operating conditions of a well.

The pioneering work of Barenblatt et al.[30] which was presented above has been widely discussed and extended since its publication in 1960. For instance, in [254] it is compared to two other early approaches of double-porosity [418, 198] which also deal with fissured groundwater reservoirs. Multi-porosity or multi-permeability systems have been used extensively to model various types of composite media (see e.g. [34]). The mathematical theory of such systems is obtained as an application of the theory of degenerate evolution equations e.g.in [353] or of the homogenization theory in [24]. Such processes in deforming composite media including a finite element formulation are addressed in a series of papers [199].

12.3.2 Multi-porosity models in bioengineering – swelling media

Besides the above described original application of multi-porosity models to reservoir engineering another application area arose, namely bioengineering. In [181] J. Huyghe et al. describe in a short form the benefit of multi-porosity models in bioengineering; for a detailed presentation see e.g. [176]. Here, we reproduce the introduction briefly to suggest a second type of application of multi-porosity models.

Human tissues are porous media. In order to understand the etiology of diseases and to design adequate solutions, accurate quantitative understanding of the poromechanics of these tissues is necessary. In this section a multi-porosity model which was applied before to fractured rocks is applied in its finite deformation form to cardiovascular disease. On the other hand, a swelling model developed in the context of intervertebral disc mechanics, is presently assessed by petroleum engineers in the context of bore hole stability in shale formations. The ionization of shales, clays, hydrogels and tissues result in an intimate coupling between electrical and mechanical behavior. We are concerned with ionized porous structures imbibed with electrolyte solutions, in which interfacial phenomena often determine the macroscopic behavior (compare the chapter on interactions with electromagnetic fields and the sections on boundary layers).

The role of porous media mechanics is evident: most biology involves fluid flow, convection of solutes, deformation of a solid skeleton. Even on the level of a single cell, the number of molecules involved is so high that modeling has necessarily to involve continuum mechanics. Here, two applications, blood perfusion and swelling, are discussed.

12.3.2.1 Cardiovascular disease

Coronary artery disease is believed to be the number one cause of mortality of human beings. The coronary vascular system is a pore structure inside a deforming solid which is called the heart muscle. It is subject to large deformations. The pore structure is highly organized to deliver each heart cell from waste materials and to supply them with nutrients. This is done through a multi-porosity structure (see Figure 12.3). The different porosities are identified as arteries, arterioles, capillaries, venules and veins. They have each a characteristic pore size, a characteristic blood velocity, characteristic wall properties and characteristic pathology. A well-known quantity in multi-porosity flow is the perfusion. This is the flow of fluid from one porosity to another. Flow has the dimension m^3s^{-1} and, thus, is defined per unit volume while the perfusion has the dimension of s^{-1}. There is evidence that blood perfusion is affected by the deformation of the heart muscle [362]. Furthermore, blood perfusion affects the deformation, both metabolically as mechanically [283]. These findings demonstrate that the coronary system should be modeled using a multi-porosity, finite deformation approach. The essential difference between the perfusion model [179, 180] and classical multi-porosity models of fractured rocks is the coupling between the interporosity flow and the intraporosity flow by the coupling factor $\underline{K}_0$ (see Figure 12.3):

$$-\left[\begin{array}{cc} K_{00} & \underline{K}_0 \\ \underline{K}_0^T & \underline{K} \end{array}\right] \left[\begin{array}{c} \dfrac{\partial p}{\partial x_0} \\ \nabla \underline{p} \end{array}\right] = \left[\begin{array}{c} q_0 \\ \underline{q} \end{array}\right]. \tag{12.100}$$

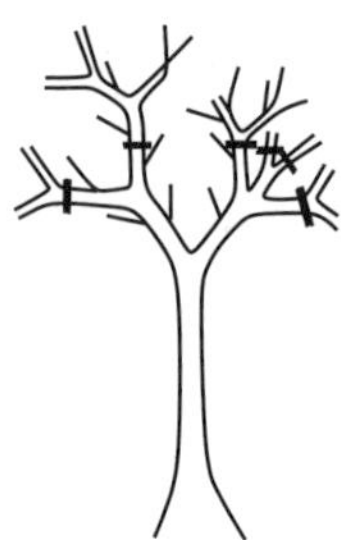

Figure 12.3: *A dense tree structure can be modeled as a multi-porosity continuum. Each generation of branches – the stem, larger branches, intermediate branches, small branches and twigs – is one porosity. Here, the interface between two subsequent porosities is illustrated. The normal to this interface, pointing from large to smaller branches, is directed upwards on average. A gradient of the chemical potential from smaller towards larger branches, therefore, induces an upward flux of sap. Conversely, a gradient in the chemical potential between the upper branches and lower branches induces a flux of sap from larger branches to smaller branches. This example illustrates the coupling factor between interporosity flow and intraporosity flow in the Darcy-type equation (12.100); after [181].*

The multi-porosity model was implemented in 2D, 3D and axisymmetric finite elements [406]. Animal experiments were undertaken to verify the concept [402]. The same equations were derived twice: using mixture theory [405] and using averaging theorems [177, 178]. A micro-macro transformation was derived to compute hydraulic permeabilities from the microstructure [179, 180].

12.3.2.2 Swelling

Swelling is known as a symptom of disease since antiquity. It is associated with ionization of a porous medium. This is true for clays as well as for living tissues. In this subsection the governing equations, as derived by [255], are given for the special case of infinitesimal quadriphasic mechanics of compressible charged porous media. Finite deformations are investigated e.g. in [175]. Specific constitutive material behavior is considered and corresponding relations for the osmotic pressure and electrical potential at equilibrium are given. A porous medium saturated with a monovalent ionic solution is considered. The current volume fraction of the solid is φ^s, the current volume fraction of the ionic solution is φ^f. The corresponding initial volume fractions are φ_0^s and φ_0^f. The solution is a molecular mixture of water (w), cations ($+$) and anions ($-$). The partial densities of water, cations and anions are in the current state ρ^w, ρ^+ and ρ^-, and in the initial state ρ_0^w, ρ_0^+ and ρ_0^-. The unstrained volume change of constituent ζ^β of constituent $\beta = w, +, -$ is

$$\zeta^\beta = \frac{\rho^\beta - \rho_0^\beta}{\rho_i^\beta}, \tag{12.101}$$

in which ρ_i^β is an intrinsic reference density such that $\sum_{\beta=w,+,-} \frac{\rho_0^\beta}{\rho_i^\beta} = \varphi_0^f$.

Balance laws The momentum balance, neglecting inertia, is given by

$$\nabla \cdot \boldsymbol{\sigma} = 0, \tag{12.102}$$

where $\boldsymbol{\sigma}$ is the Cauchy stress tensor. The mass balance of constituent α requires

$$\frac{\partial \zeta^{\beta}}{\partial t} + \nabla \cdot \left(\mathbf{v}^{\beta s} \right) = 0, \tag{12.103}$$

where $\mathbf{v}^{\beta s}$ is the unstrained volume flux of constituent β with respect to the solid. Index α denotes all constituents, i.e., the electrically charged solid (s), water (w), cations $(+)$ and anions $(-)$; while β, γ are all constituents except the charged solid matrix. As the mixture is assumed to be fully saturated, the following relation for the saturation condition holds

$$\alpha \nabla \cdot \mathbf{v}^{s} + \frac{1}{M} \frac{\partial p}{\partial t} - \sum_{\beta = w,+,-} \frac{\partial \zeta^{\beta}}{\partial t} = 0, \tag{12.104}$$

in which $\mathbf{v}^{s}$ is the solid velocity, α and M are Biot coefficients (see e.g. [50]). Electroneutrality requires

$$\sum_{\beta = w,+,-} F \frac{z^{\beta}}{\bar{V}^{\beta}} \frac{\partial \zeta^{\beta}}{\partial t} = 0, \tag{12.105}$$

with F Faraday's constant, $z^{\beta}, \bar{V}^{\beta}$ the valence and molar volume of constituent β.

Constitutive behavior The fluxes obey a coupled form of Darcy's, Fick's and Ohm's law

$$\mathbf{v}^{\beta s} = - \sum_{\gamma = w,+,-} \mathbf{K}^{\beta \gamma} \nabla \mu^{\gamma} \quad \text{for} \quad \beta = w,+,-, \tag{12.106}$$

with $\mathbf{K}^{\beta \gamma}$ a positive definite symmetric permeability matrix. In Relation (12.106) the electro-chemical potentials μ^{γ} are defined as

$$\mu^{\gamma} = \frac{\partial W}{\partial \zeta^{\beta}} + \frac{F z^{\beta}}{\bar{V}^{\beta}} \xi, \tag{12.107}$$

with ξ and W the electrical potential and the energy function, respectively.

The stress appearing in the momentum balance (12.102) is the partial derivative of the energy function W with respect to the infinitesimal strain tensor $\varepsilon = 1/2(\nabla \mathbf{u} + (\nabla \mathbf{u})^{T})$, with $\mathbf{u}$ the displacement vector of the solid

$$\boldsymbol{\sigma} = \frac{\partial W}{\partial \varepsilon}. \tag{12.108}$$

The energy function is the sum of the poroelastic strain energy and a mixing energy

$$W = -\alpha M \operatorname{tr} \varepsilon \sum_{\beta = w,+,-} \zeta^{\beta} + \frac{1}{2} M \left(\sum_{\beta = w,+,-\zeta} \zeta^{\beta} \right)^{2} + \frac{1}{2} \left(\frac{2\nu G}{1 - 2\nu} + \alpha^{2} M \right) \operatorname{tr}^{2}\varepsilon$$

$$+ G\varepsilon : \varepsilon + \sum_{\beta = w,+,-} \mu_{0}^{\beta} \zeta^{\beta} \tag{12.109}$$

$$- RTT \left(\frac{\zeta^{+}}{\bar{V}^{+}} + \frac{\zeta^{-}}{\bar{V}^{-}} \right) \ln \zeta^{w} + RT \frac{\zeta^{+}}{\bar{V}^{+}} \left(\ln \frac{\zeta^{+}}{\bar{V}^{+}} - 1 \right) RT \frac{\zeta^{-}}{\bar{V}^{-}} \left(\ln \frac{\zeta^{-}}{\bar{V}^{-}} - 1 \right).$$

The form (12.109) of W assumes linear isotropic elasticity and Donnan osmosis as the swelling mechanism of the shale. G is the shear modulus, ν is Poisson's ratio, K_f is the volumetric modulus of the solution, μ_0^β is the reference potential of constituent β, R is the gas constant, T the absolute temperature and Γ the osmotic coefficient. Given this expression for the energy function, the electrochemical potentials (12.107) take the form

$$\mu^w = \mu_0^w + p - RT\Gamma\left(\frac{\zeta^+}{\bar{V}^+ + \zeta^w} + \frac{\zeta^-}{\bar{V}^- - \zeta^w}\right), \tag{12.110}$$

$$\mu^+ = \mu_0^+ + p + \frac{RT}{\bar{V}^+}\ln\frac{\zeta^+}{(\zeta^w)^\Gamma} + \frac{F}{\bar{V}^+}\xi, \tag{12.111}$$

$$\mu^- = \mu_0^- + p + \frac{RT}{\bar{V}^-}\ln\frac{\zeta^-}{(\zeta^w)^\Gamma} + \frac{F}{\bar{V}^+}\xi, \tag{12.112}$$

considering that

$$p = -\alpha M \operatorname{tr}\boldsymbol{\varepsilon} + M \sum_{\beta=w,+,-} \zeta^\beta. \tag{12.113}$$

The differential form of Equations (12.110)-(12.112) is

$$\frac{\partial}{\partial t}\left(\mu^\beta - \frac{Fz^\beta}{\bar{V}^\beta}\xi - p\right) = \sum_{\gamma=w,+,-} \mathbf{W}^{\beta\gamma}\frac{\partial\zeta^\gamma}{\partial t}, \qquad \beta = w,+,-, \tag{12.114}$$

in which

$$\mathbf{W}^{\beta\gamma} = \partial^2 W/\partial\zeta^\beta\partial\zeta^\gamma. \tag{12.115}$$

For the numerical implementation of the above equations see [404]. The displacements, the pressures, the electrochemical potentials of water, cations and anions, and the electrical potentials are the degrees of freedom.

$$\begin{bmatrix} \underline{S} & -\underline{L} & \underline{0} & \underline{0} \\ -\underline{L}^T & -\sum_{\beta,\gamma}\underline{C}_{\alpha\beta} - \frac{1}{M} & \sum_\beta\underline{C}_{\alpha\beta} & -\sum_{\beta,\gamma}\underline{C}_{\alpha\beta}\frac{z^\beta F}{\bar{V}^\beta} \\ \underline{0} & \sum_\gamma\underline{C}_{\alpha\beta} & -\underline{K}_{\beta\gamma} - \underline{C}_{\alpha\beta} & \sum_\gamma\underline{C}_{\alpha\beta}\frac{z^\gamma F}{\bar{V}^\gamma} \\ \underline{0} & -\sum_{\beta,\gamma}\underline{C}_{\alpha\beta}\frac{z^\beta F}{\bar{V}^\beta} & \sum_{\beta,\gamma}\underline{C}_{\alpha\beta}\frac{z^\beta F}{\bar{V}^\beta} & -\sum_{\beta,\gamma}\frac{z^\beta F}{\bar{V}^\beta}\underline{C}_{\alpha\beta}\frac{z^\gamma F}{\bar{V}^\gamma} \end{bmatrix} \begin{bmatrix} \delta\underline{u} \\ \delta\underline{p} \\ \delta\underline{\mu}^\gamma \\ \delta\underline{\xi} \end{bmatrix} = \begin{bmatrix} \underline{U} - \sum_\beta\underline{Q}_\beta \\ \underline{Q}_\beta + \sum_\gamma(\underline{T}_1 + \underline{T}_2) \\ -\sum_\alpha\frac{z^\beta F}{\bar{V}^\beta}\underline{Q}_\alpha \end{bmatrix}$$

$$\underline{S} = \int_\Omega \left[(\nabla\Phi\mathbf{e})^C : \frac{\partial^2 W}{\partial\boldsymbol{\varepsilon}} : \nabla\Phi\mathbf{e}\right] d\Omega, \qquad \underline{L} = \int_\Omega \alpha\nabla\cdot\Phi\mathbf{e}p\, d\Omega,$$

$$\underline{C}_{\alpha\beta} = \int_\Omega \Psi\left(W^{\alpha\beta}\right)^{-1}\Psi^T\, d\Omega, \qquad \underline{K}_{\alpha\beta} = \theta\int_\Omega \nabla\Psi^C\cdot K^*_{\alpha\beta}\cdot\nabla\Psi\Delta t\, d\Omega,$$

$$\underline{R} = -\int_\Gamma \Phi\mathbf{e}\cdot\boldsymbol{\sigma}\cdot\mathbf{n}\, d\Gamma - \int_\Omega \boldsymbol{\sigma}_e\cdot\nabla\Phi^I\mathbf{e}\, d\Omega + \int_\Omega \nabla\cdot\Phi\mathbf{e}p\, d\Omega, \tag{12.116}$$

$$\underline{U} = \int_\Omega \Psi\left(\operatorname{tr}\boldsymbol{\varepsilon} - \operatorname{tr}\boldsymbol{\varepsilon}_n + \frac{p - p_n}{M}\right) d\Omega, \qquad \underline{T}_1 = \theta\int_\Omega (K_{\alpha\beta}\cdot\nabla\mu^\alpha)\cdot\nabla\Psi\Delta t\, d\Omega,$$

$$\underline{T}_2 = (1-\theta)\int_\Omega (K_{\alpha\beta}\cdot\nabla\mu_n^\alpha)\cdot\nabla\Psi\Delta t\, d\Omega, \qquad \underline{Q}_\beta = \int_\Omega \Psi\left(\zeta^\beta - \zeta_n^\beta\right) d\Omega.$$

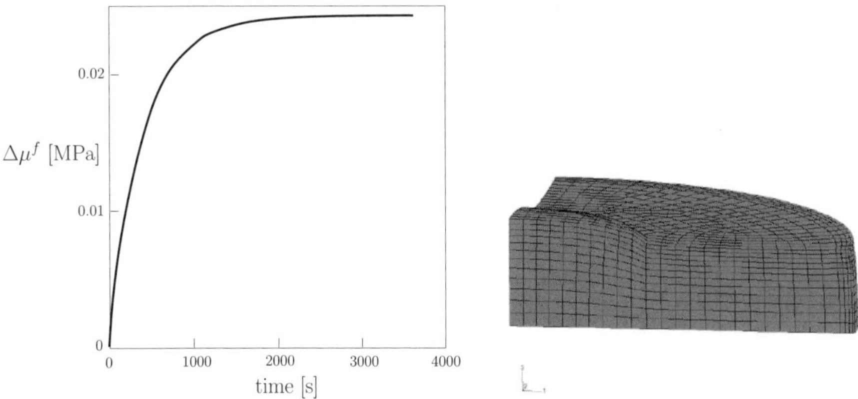

Figure 12.4: *Left: qualitative behavior of the chemical potential of the fluid as a function of time during swelling of a one-dimensional ionized medium (after [181]); right: free swelling of an intervertebral disc in physiological salt solution, after excision from its neighboring vertebrae. The mesh covers one eight of an axisymmetric intervertebral disc, which is cylindrical in the undeformed state. The scale of the displacements is identical to the scale of the geometry (from [181]).*

Continuity of the displacements and the electrochemical potentials is enforced both within one element and between different elements. As the displacements are continuous, but non-differentiable across the element interfaces, the strains, unstrained volumes and concentrations are discontinuous across the element boundaries. From Equations (12.110), (12.111) and (12.112), one infers that the continuity of the electrochemical potentials is incompatible with the continuity of pressure and electrical potential, if the unstrained volumes are discontinuous. Hence, it is necessary not to enforce continuity of the pressure and the electrical potential. Figure 12.4 (left) shows the qualitative form of the chemical potential of the fluid as function of time during swelling of a one-dimensional ionized medium, the right-hand side represents the deformed mesh of swelling simulation of an intervertebral disc after excision. The swelling response is observed experimentally and indicates that in the in vivo state (before excision) swelling pressures the disc fibers even in its unloaded state. This prestressing might have a protecting function against hernia (fissuring of the disc).

12.4 Biomechanics of soft tissues

In this section we repeat the contents of the chapter of the same title in volume three of the Handbook of Material Behavior [223] by Gerhard A. Holzapfel [159] (for a detailed version see, e.g. [162]). This introduction into the topic refers to the classical introductory book by Yuan-Cheng Fung [131]. One of the pioneers in the investigation of this highly topical field in which large deformations, and thus, a nonlinear behavior of the material play a leading part, is surely Van C. Mow. E.g. in his highly cited papers [257, 258, 259] he introduces biphasic and triphasic theories for biological tissues. One of the most recent books in this field is by Stephen C. Cowin and Stephen B. Doty [99].

The presented general model is a fully three-dimensional material description of soft tissues for which nonlinear continuum mechanics is used as the fundamental basis [158], [281]. It is based partly on histological information, i.e., the microscopic structure of organs and tissues. With as few material parameters as possible, an efficient constitutive formulation is searched for which approximates all types of soft tissues with a reasonable accuracy over a large strain range. The general model describes the highly nonlinear and anisotropic behavior of soft tissues as composites reinforced by two families of collagen fibers. The constitutive framework is based on the theory of the mechanics of fiber-reinforced composites [363]. The performance and the physical mechanism of the model is presented in [160]. As a representative example, the general model for soft tissues is specified to predict the mechanical response of healthy and young arteries under physiological loading conditions [161]. The model neglects active components, i.e., contracting elements with biochemical energy supply which are controlled by biological mechanisms, and is concerned with the description of the passive state of arteries. Furthermore, it does not consider acute and long-term changes in the geometry and/or the mechanical response of tissues due to, for example, drugs, aging and disease. In clinical procedures tissues may undergo irreversible (plastic) deformations [161] which are of medical importance. Constitutive equations for describing plastic deformations of, for example, arteries are proposed in [374].

12.4.1   Structure of soft tissues – collagen and elastin

According to [159] a primary group of tissue which binds, supports and protects our human body and structures such as organs is *soft connective tissue*. In contrary to other tissues, it is a biological material in which the cells are separated by extracellular material. Connective tissues may be distinguished from hard (mineralized) tissues such as bones for their high flexibility and their soft mechanical properties (for details see, for example, [131]). Examples for soft tissues are tendons, ligaments, blood vessels, skins or articular cartilages. Tendons are muscle-to-bone linkages to stabilize the bony skeleton (or to produce motion), while ligaments are bone-to-bone linkages to restrict relative motion. Blood vessels have to distend in response to pulse waves and the skin is the largest single organ (16% of the human adult weight). Articular cartilages form the surface of body joints, distribute loads across joints and minimize contact stresses and friction. Soft connective tissues of our body are complex fiber-reinforced composite structures. Their mechanical behavior is strongly influenced by the concentration and structural arrangement of constituents such as *collagen* and *elastin*, the hydrated matrix of proteoglycans, and the topographical site and respective function in the organism.

Collagen is a protein which is a major constituent of the extracellular matrix of connective tissue. It is the main load carrying element in a wide variety of soft tissues and is very important to human physiology. Collagen molecules are linked to each other by covalent bonds building collagen fibrils. In the structure of tendons and ligaments, for example, collagen appears as parallel oriented fibers [45], while many other tissues have an intricate disordered network of collagen fibers embedded in a gelatinous matrix of proteoglycans. The intramolecular crosslinks of collagen gives the connective tissues the strength which varies with age, pathology, etc.

Elastin, like collagen, is a protein which is a major constituent of the extracellular matrix of connective tissue. It is present as thin strands in soft tissues such as skin, lung, ligamenta flava of the spine and ligamentum nuchae (the elastin content of the

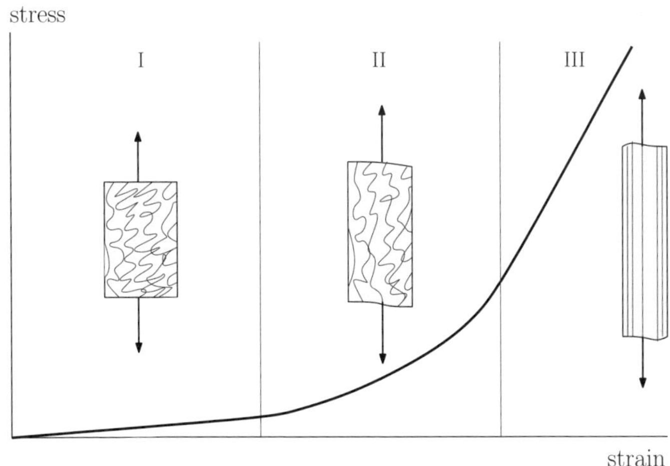

Figure 12.5: *Schematic diagram of a typical (tensile) stress-strain curve for skin showing the associated collagen fiber morphology (after [159]).*

latter is about 5 times that of collagen). The long flexible elastin molecules build up a three-dimensional (rubber-like) network, which may be stretched to about 2.5 of the initial length of the unloaded configuration. The mechanical behaviour of elastin may be explained within the concept of entropic elasticity. As for rubber, the random molecular conformations, and hence the entropy, change with deformation. Elasticity arises through entropic straightening of the chains, i.e., a decrease of entropy, or an increase of internal energy (see, for example, [157] or [158]). Elastin is essentially a linearly elastic material. It displays very small relaxation effects (they are larger for collagen).

12.4.2 General mechanical characteristic of soft tissues

Soft tissues behave anisotropically because of their fibers which tend to have preferred directions. In a microscopic sense they are non-homogeneous materials because of their composition. The tensile response of soft tissue is nonlinear stiffening and tensile strength depends on the strain rate. In contrast to hard tissues, soft tissues may undergo large deformations. Some soft tissues show viscoelastic behavior (relaxation and/or creep).

We repeat here from [159] the explanation of the tensile stress-strain behavior for skin, an organ consisting mainly of connective tissues. Such a mechanical behavior is representative of many (collagenous) soft connective tissues. For the connective tissue parts of the skin the three-dimensional network of fibers appears to have preferred directions parallel to the surface. However, in order to prevent out-of-plane shearing, some fiber orientations also have components out-of-plane.

Figure 12.5, which is a reproduction of Figure 1 of [159], shows a schematic diagram of a typical J-shaped (tensile) stress-strain curve for skin. This form, representative for many soft tissues, differs significantly from stress-strain curves of hard tissues or from other types of (engineering) materials. In addition, Figure 12.5 shows how the collagen fibers straighten with increasing stress.

The deformation behavior for skin may be studied in three phases I, II and III:

Phase I In the absence of load the collagen fibers, which are woven into rhombic-shaped pattern, are in relaxed conditions and appear wavy and crimped. Unstretched skin behaves approximately isotropically. Initially low stress is required to achieve large deformations of the individual collagen fibers without requiring stretch of the fibers. In phase I the tissue behaves like a very soft (isotropic) rubber sheet, and the elastin fibers (which keep the skin smooth) are mainly responsible for the stretching mechanism. The stress-strain relation is approximately linear, the elastic modulus of skin in phase I is low (0.1-2 MPa).

Phase II In phase II, as the load is increased, the collagen fibers tend to line up with the load direction and bear loads. The crimped collagen fibers gradually elongate and they interact with the hydrated matrix. With deformation the crimp angle in collagen fibrils leads to a sequential uncrimping of fibrils. Note that the skin is normally under tension in vivo.

Phase III In phase III, at high tensile stresses, the crimp patterns disappear and the collagen fibers become straighter. They are primarily aligned with one another in the direction in which the load is applied. The straightened collagen fibers resist the load strongly and the tissue becomes stiff at higher stresses. The stress-strain relation becomes linear again. Beyond the third phase the ultimate tensile strength is reached and fibers begin to break.

The mechanical properties of soft tissues depend strongly on the topography, risk factors, age, species, physical and chemical environmental factors such as temperature, osmotic pressure, pH, and on the strain rate.

12.4.3 Model

As pointed out above, we repeat here the model presented by G. Holzapfel in [159]. For details see his original works.

At any referential position $\mathbf{X}$ of the tissue the existence of a Helmholtz free-energy function Ψ is postulated. The decoupled form is assumed

$$\Psi = U(\mathbf{X}; J) + \bar{\Psi}(\mathbf{X}; \bar{\mathbf{C}}, \mathbf{A}_1, \mathbf{A}_2), \qquad (12.117)$$

where U is a purely volumetric (dilatational) contribution and $\bar{\Psi}$ is a purely isochoric (volume-preserving) contribution to the free energy Ψ. The modified right Cauchy-Green tensor is denoted by $\bar{\mathbf{C}} = \bar{\mathbf{F}}^T \bar{\mathbf{F}}$ and $\bar{\mathbf{F}} = J^{-1/3}\mathbf{F}$ is the unimodular (distortional) part of the deformation gradient $\mathbf{F}$, with $J = \det \mathbf{F} < 0$ denoting the local volume ratio. In addition, in Equation (12.117), $\{\mathbf{A}_1, \mathbf{A}_2\}$ is a set of two (second-order) tensors which characterize the anisotropic properties of the tissue at any $\mathbf{X}$. The structure tensors $\mathbf{A}_1$ and $\mathbf{A}_2$ are defined as the tensor products $\mathbf{a}_{0i} \otimes \mathbf{a}_{0i}$, where $\mathbf{a}_{0i}$, $i = 1, 2$, are two unit vectors characterizing the orientations of the families of collagen fibers in the (undeformed) reference configuration of the tissue (see Figure 12.6).

Since most types of soft tissues are regarded as incompressible (for example, arteries do not change their volume within the physiological range of deformation [64]), now attention is focused on the description of their isochoric deformation behavior characterized by the energy function $\bar{\Psi}$. The simple additive split

$$\bar{\Psi} = \bar{\Psi}_{\text{iso}}(\mathbf{X}; \bar{I}_1) + \bar{\Psi}_{\text{aniso}}(\mathbf{X}; \bar{I}_4, \bar{I}_6), \qquad (12.118)$$

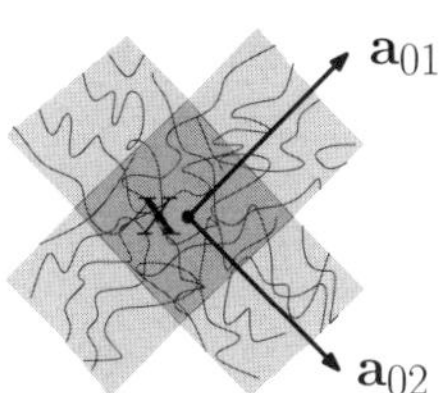

Figure 12.6: *Arrangement of collagen fibers in the reference configuration characterized by two unit vectors $\mathbf{a}_{01}, \mathbf{a}_{02}$ at position $\mathbf{X}$ (after [159]).*

of $\bar{\Psi}$ into a part $\bar{\Psi}_{\mathrm{iso}}$ associated with isotropic deformations and a part $\bar{\Psi}_{\mathrm{aniso}}$ associated with anisotropic deformations is suggested. This is sufficiently general to capture the salient mechanical feature of soft tissue elasticity, a more general constitutive framework is presented e.g. in [160, 161]. In Relation (12.118) $\bar{I}_1 = \bar{\mathbf{C}} : \mathbf{I}$ is the first invariant of tensor $\bar{\mathbf{C}}$ ($\mathbf{I}$ is the second-order unit tensor), and the following definitions of the invariants, which are stretch measures for the two families of collagen fibers (see, for example, [363], [158]) have been used

$$\bar{I}_4(\bar{\mathbf{C}}, \mathbf{a}_{01}) = \bar{\mathbf{C}}{:}\mathbf{A}_1, \qquad \bar{I}_6(\bar{\mathbf{C}}, \mathbf{a}_{02}) = \bar{\mathbf{C}}{:}\mathbf{A}_1. \tag{12.119}$$

The invariants $\bar{I}_4$ and $\bar{I}_6$ are squares of the stretches in the directions of $\mathbf{a}_{01}$ and $\mathbf{a}_{02}$, respectively. Isotropy is described through the invariant $\bar{I}_1$ and anisotropy through $\bar{I}_4$ and $\bar{I}_6$.

Since the (wavy) collagenous structure of tissues is not active at low stresses (it does not store strain energy) $\bar{\Psi}_{\mathrm{iso}}$ is associated with the mechanical response of the non-collagenous matrix of the material (which is less stiff than its elastin fiber constituent). To determine the non-collagenous matrix response, the isotropic neo-Hookean model according to

$$\bar{\Psi}_{iso} = \frac{c}{2}(\bar{I}_1 - 3), \tag{12.120}$$

is proposed to use, where $c > 0$ is a stress-like material parameter. However, to model the (isotropic) non-collagenous matrix material any Ogden-type elastic material may be applied [281].

According to morphological findings at highly-loaded tissues the families of collagen fibers become straighter and the resistance to stretch is almost entirely due to collagen fibers (the tissue becomes stiff). Hence, the strain energy stored in the collagen fibers is taken to be governed by the polyconvex (anisotropic) function

$$\bar{\Psi}_{\mathrm{aniso}} = \frac{k_1}{2k_2}\left\{\exp\left[k_2(\bar{I}_4 - 1)^2\right] - 1\right\} + \frac{k_3}{2k_4}\left\{\exp\left[k_4(\bar{I}_6 - 1)^2\right] - 1\right\}, \tag{12.121}$$

where $k_1 > 0$, $k_3 > 0$, are stress-like material parameters and $k_2 > 0$, $k_4 > 0$, are dimensionless parameters. According to relations (12.118), (12.120), (12.121), the collagen fibers do not influence the mechanical response of the tissue in the low stress domain. Due to the crimp structure of collagen fibers it is assumed that they do not support compressive stresses which implies that they are inactive in compression. Hence, the relevant part of the anisotropic function (12.121) is omitted in this case. If, for example, $\bar{I}_4 \leq 1$ and $\bar{I}_6 \leq 1$, then the soft tissue responds similarly to a rubber-like (purely isotropic) material described by the energy function (12.120). However, in extension, that is when $\bar{I}_4 > 1$ or $\bar{I}_6 > 1$, the collagen fibers are active and energy is stored in the fibers.

Function (12.117) enables the Cauchy stress tensor, denoted by $\mathbf{T}$, to be derived in the decoupled form

$$\mathbf{T} = \mathbf{T}_{\text{vol}} + \bar{\mathbf{T}}, \quad \text{with} \quad \mathbf{T}_{\text{vol}} = p\mathbf{I}, \quad \bar{\mathbf{T}} = 2J^{-1}\text{dev}\left(\bar{\mathbf{F}}\frac{\partial\bar{\Psi}}{\partial\bar{\mathbf{C}}}\bar{\mathbf{F}}^T\right), \quad (12.122)$$

with the volumetric contribution $\mathbf{T}_{\text{vol}}$ and the isochoric contribution $\bar{\mathbf{T}}$ to the Cauchy stresses. In the stress relation (12.122), $p = dU/dJ$ denotes the hydrostatic pressure and $\text{dev}(\cdot)$ furnishes the deviatoric operator in Eulerian description. The operator is defined as $\text{dev}(\cdot) = (\cdot) - \frac{1}{3}[(\cdot) : \mathbf{I}]\mathbf{I}$, so that $\text{dev}(\cdot): \mathbf{I} = 0$.

Using the additive split (12.118) and particularizations (12.120), (12.121), Equation (12.122)$_3$ builds an explicit constitutive expression for the isochoric behavior of soft connective tissues in Eulerian description, i.e.,

$$\bar{\mathbf{T}} = c\,\text{dev}\,\bar{\mathbf{B}} + \sum_{i=4,6} 2\bar{\Psi}_i\,\text{dev}\,(\mathbf{a}_i \otimes \mathbf{a}_i), \quad (12.123)$$

where $\bar{\mathbf{B}} = \bar{\mathbf{F}}\bar{\mathbf{F}}^T$ denotes the modified left Cauchy-Green tensor,

$$\bar{\Psi}_4 = \frac{\partial\bar{\Psi}_{\text{aniso}}}{\partial\bar{I}_4} = k_1(\bar{I}_4 - 1)\left\{\exp\left[k_2(\bar{I}_4 - 1)^2\right] - 1\right\}, \quad (12.124)$$

$$\bar{\Psi}_6 = \frac{\partial\bar{\Psi}_{\text{aniso}}}{\partial\bar{I}_6} = k_3(\bar{I}_6 - 1)\left\{\exp\left[k_4(\bar{I}_6 - 1)^2\right] - 1\right\}, \quad (12.125)$$

are (scalar) response functions and $\mathbf{a}_i = \bar{\mathbf{F}}\mathbf{a}_{0i}$, $i = 1, 2$, are the Eulerian counterparts of the unit vectors $\mathbf{a}_{0i}$.

The specific form of the proposed constitutive equation (12.123) requires the five material parameters c, k_1, k_2, k_3, k_4 whose interpretations can be partly based on the underlying histological structure, i.e., matrix and collagen of the tissue. Note that in (12.123), orthotropic ($k_1 = k_3, k_2 = k_4$), transversely isotropic ($k_1 = 0$ or $k_3 = 0$) and isotropic hyperelastic descriptions ($k_1 = k_3 = 0$) at finite strains are included as special cases.

12.4.4   Example: A model for the artery

In order to specify the constitutive framework for soft tissues introduced in previous subsection, in this subsection a model for the passive state of the healthy and young artery is adopted from [159] (no pathological changes in the intima, which is the innermost arterial layer frequently affected by atherosclerosis). It is suitable for predicting three-dimensional distributions of stresses and strains under physiological loading conditions with reasonable accuracy. For an adequate model of arteries incorporating the active state (contraction of smooth muscles) see [300]. For a detailed study of the mechanics of arterial walls see the extensive review [168].

Experimental tests show that the elastic properties of the media (middle layer of the artery) and adventitia (outermost layer of the artery) are significantly different [453]. The media is much stiffer than the adventitia. In addition, the arterial layers have different physiological tasks, and hence the artery is modeled as a thick-walled elastic circular tube consisting of two layers corresponding to the media and adventitia. In a young non-diseased artery the intima (innermost layer of the artery) exhibits negligible wall-thickness and mechanical strength. Each tissue layer is treated as a composite reinforced by two

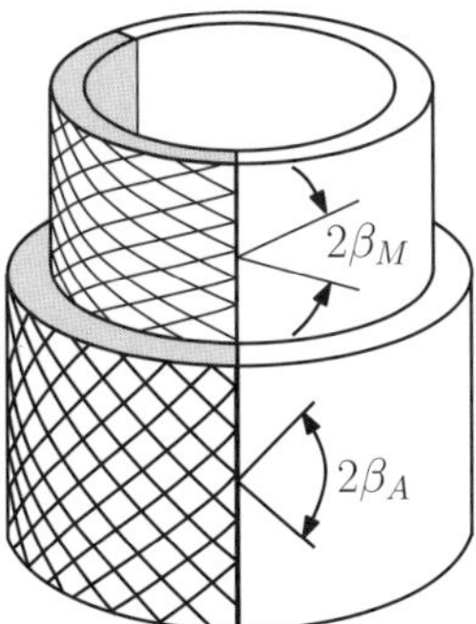

Figure 12.7: *Load-free configuration of an idealized artery modeled as a thick-walled circular tube consisting of two layers, i.e., the media and adventitia (after [159]).*

families of collagen fibers which are symmetrically disposed with respect to the cylinder axis. Hence, each tissue layer is considered as cylindrically orthotropic so that a tissue layer behaves like a so-called balanced angle-ply laminate. The same forms of the strain-energy functions (12.120), (12.121) are used for each tissue layer (each layer responds with similar mechanical characteristics) but a different set of material parameters. Hence, Equation (12.118) takes on the specified form

$$\bar{\Psi}_M = \frac{C_M}{2}(\bar{I}_1 - 3) + \frac{k_{1M}}{2k_{2M}} \sum_{i=4,6} \left\{ \exp\left[k_{2M}(\bar{I}_{iM} - 1)^2\right] - 1 \right\}, \tag{12.126}$$

$$\bar{\Psi}_A - \frac{C_A}{2}(\bar{I}_1 - 3) + \frac{k_{1A}}{2k_{2A}} \sum_{i=4,6} \left\{ \exp\left[k_{2A}(I_{iA} - 1)^2\right] - 1 \right\}. \tag{12.127}$$

The result is a two-layer model incorporating six material parameters, three for the media M, i.e., c_M, k_{1M}, k_{2M}, and three for the adventitia A, i.e., c_A, k_{1A}, k_{2A}.

The invariants, associated with the anisotropic parts of the two tissue layers are defined by $\bar{I}_{4j} = \bar{\mathbf{C}} : \mathbf{A}_{1j}$ and $\bar{I}_{6j} = \bar{\mathbf{C}} : \mathbf{A}_{2j}$, $j = M, A$. The structure tensors $\mathbf{A}_{1j}$ and $\mathbf{A}_{2j}$ are given by

$$\mathbf{A}_{1j} = \mathbf{a}_{01j} \otimes \mathbf{a}_{01j}, \qquad \mathbf{A}_{2j} = \mathbf{a}_{02j} \otimes \mathbf{a}_{02j}, \qquad j = M, A. \tag{12.128}$$

Employing a cylindrical coordinate system, the components of the unit (direction) vectors $\mathbf{a}_{01j}$ and $\mathbf{a}_{02j}$ read in matrix notation

$$[\mathbf{a}_{01j}] = \begin{bmatrix} 0 \\ \cos\beta_j \\ \sin\beta_j \end{bmatrix}, \qquad [\mathbf{a}_{02j}] = \begin{bmatrix} 0 \\ \cos\beta_j \\ -\sin\beta_j \end{bmatrix}, \qquad j = M, A, \tag{12.129}$$

and β_j, $j = M, A$, are the angles between the collagen fibers and the circumferential direction in the media and adventitia (see Figure 12.7). Small components of the (collagen) fiber orientation in the radial direction, as, for example, reported for human brain arteries [125], are neglected.

Residual stresses It has been known for some years that arteries which are excised from the body and not subjected to any loads are not stress-free (or strain-free) [399].

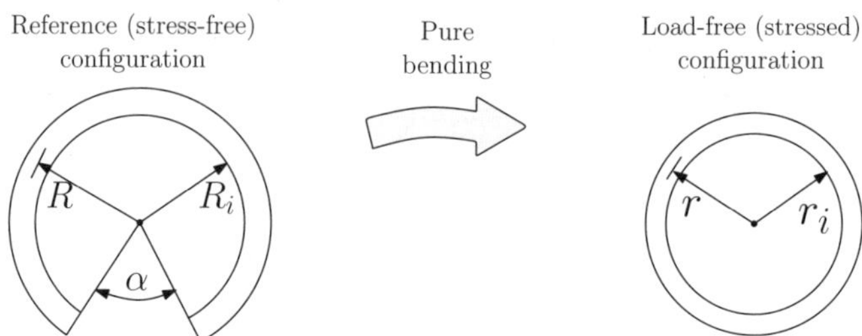

Figure 12.8: *Cross-sectional representation of one arterial layer at the reference (stress-free) and load-free (stressed) configurations (after [159]).*

If, for example, the media and adventitia are separated and cut in a radial direction the two arterial layers will spring open to form open (stress-free) sectors, which, in general, have different opening angles. In general, the residual stress-state is very complex, and residual stresses (strains) in the axial direction may also occur.

By considering the arterial layers as circular cylindrical tubes we may characterize the reference (stress-free) configuration of one arterial layer as a circular sector, as shown in Figure 12.8. For each arterial layer of the blood vessel a certain opening angle α can be found by experimental methods. The importance of incorporating residual stresses associated with the load-free (but stressed) configuration into the computation has been emphasized in, for example, [77], [161]. Consideration of residual strains has a strong influence on the global pressure/radius response of arteries and also on the stress and strain distributions across the deformed arterial wall. For analytical studies of residual stresses see, for example, [300], which contains further references.

Therefore, it is essential to incorporate the residual stresses inherent in many biological tissues. One possible approach to consider of the influence of residual stresses on the overall three-dimensional stress behavior is to measure the strain energy from the load-free (stressed) configuration and to include the residual stresses (see the contributions of G. A. Holzapfel and R. W. Ogden in [163] or [164]). Another approach is to start with the energy function relative to the stress-free (and fixed) configuration, as assumed in the presented models, and determine the deformation required to reach the load (stressed) configuration. Figure 12.8 shows the cross-sectional representation of one arterial layer at the load-free configuration obtained from the reference configuration by pure bending.

With the condition of incompressibility, the radius r of an arterial layer in the load-free configuration may be computed from the radius R of the associated reference configuration as [161]

$$r = \sqrt{\frac{R^2 - R_i^2}{k\lambda_z} + r_i^2}, \qquad k = \frac{2\pi}{2\pi - \alpha}, \tag{12.130}$$

where r_i, R_i are the internal radii associated with the two configurations. The (constant) axial stretch is denoted by λ_z and the parameter k is a convenient measure of the tube opening angle in the stress-free configuration.

12.4.5 Material parameters, numerical analysis and results

Preferred directions in soft tissues are well specified by the orientation of prolate cell nuclei. They can be identified in microphotographs of appropriately stained histological sections. By visual inspection there exists a high directional correlation between smooth muscle cells and collagen fibers. Hence, the (bell-shaped distribution of) collagen fiber orientations may be obtained from an image processing analysis of stained histological sections. The angle β (and thus the unit vectors $\mathbf{a}_{01}$ and $\mathbf{a}_{02}$) may be identified as the mean value of the corresponding statistical distribution.

Values of the material parameters associated with the model for soft tissues are then obtained by fitting the equations to the experimental data of the soft tissue of interest by using standard nonlinear fitting algorithms, such as the Levenberg-Marquardt algorithm. For the case that the mean values of the orientation of cell nuclei (collagen fiber) may not be identified experimentally, it is suggested to treat the collagen fiber orientations as additional (phenomenological) 'material' parameters.

The energy functions are well-suited for use in nonlinear finite element software, which enables complex boundary-value problems to be solved. Aspects of finite element implementation and numerical analysis of the model are presented in [160]. Furthermore, computations may be carried out with some of the commercially available mathematical software-packages which allow symbolic computation. In [161] such a numerical technique for solving the bending, axial extension, inflation and torsion problem of an artery is described.

Values of the parameters correspond to the functions (12.126), (12.127) and are given for a representative carotid artery from a rabbit. The material parameters c, k_1, k_2 and the angles of collagen fibers β are

$$
\begin{aligned}
c_M &= 3.0 \text{ kPa}, & k_{1M} &= 2.3632 \text{ kPa}, & k_{2M} &= 0.8393, & \beta_M &= 29°, \\
c_A &= 0.3 \text{ kPa}, & k_{1A} &= 0.5620 \text{ kPa}, & k_{2A} &= 0.7112, & \beta_A &= 62°.
\end{aligned}
\tag{12.131}
$$

In the adventitia many collagen fibers run closer to the axial direction of the artery, while in the media the collagen fibers tend to run around the circumference. The fiber angles are meant to be associated with the reference (stress-free) configuration. Note that the change of the through-thickness mean value of the angle due to bending to the load-free (stressed) configuration (see Figure 12.8) is small so that it has a negligible influence on the stress-strain analysis of arteries. By using a wall of thickness 0.39 mm (adopted from [77]) and making the assumption that the media occupies 2/3 of the arterial wall thickness, the parameters (12.131) predict the characteristic orthotropic behavior of a carotid artery under combined bending, inflation, axial extension and torsion, as documented in [161].

Chapter 13

Thermodynamics of porous materials with the balance equation of porosity

In the '90s Krzysztof Wilmanski introduced a model for the description of multi-component porous media with the porosity as a field and with an own balance equation (see e.g. Chapter 13 of Part I or the simplified version for saturated porous media in Subsection 13.1.2 of this book). Since this time the model has been used for several applications (diffusion processes, wave propagation, adsorption, piping and other problems). Some fields of application have already been shown in Part I, others will be shown in this section. But before, shortly the derivation of the balance equation of porosity and the model are shown (for details see Part I). Originally, it is a nonlinear, multi-component thermodynamic model. However, for applications it is mostly sufficient to consider simplified linearized versions for a small number of components, with common constant temperature for all components and without considering the gradient of porosity (then the model is called Simple Mixture Model instead of Full Model). In the following only these linearized versions of the model are revisited.

13.1 Summary: Balance equation of porosity and associated models

13.1.1 Balance equation of porosity

The smeared-out partial mass densities, ρ_t^S and ρ_t^F, and the porosity n arise from averaging over a Representative Element Volume (REV)

$$n := \frac{1}{V} \int_{REV} \chi^F dV, \qquad \rho_t^S := \frac{1}{V} \int_{REV} \rho^{SR} \left(1 - \chi^F\right) dV = (1 - n) \rho_t^{SR},$$

$$\rho_t^F := \frac{1}{V} \int_{REV} \rho^{FR} \chi^F dV = n \rho_t^{FR}, \tag{13.1}$$

where V is the volume of the REV, χ^F the distribution function of the fluid which is one if at a position fluid appears and zero if solid exists. ρ^{SR} and ρ^{FR} are true mass densities and ρ_t^{SR} and ρ_t^{SR} are current true mass densities. These realistic (true) mass densities are defined in each point of a domain and not in points of the skeleton or of the fluid, respectively.

Changes of porosity may be described by a kind of balance equation with nonexistent source if the components are incompressible. If $\rho_t^{SR} = $ const. and $\rho_t^{FR} = $ const. the current mass densities ρ_t^F, ρ_t^S can be reduced to the single field of porosity n. The partial mass balance equations reduce to the following form

$$\frac{1}{\rho_t^{SR}}\left\{\frac{\partial \rho_t^S}{\partial t} + \operatorname{div}\left(\rho_t^S \mathbf{v}^S\right)\right\} \equiv -\frac{\partial n}{\partial t} + \operatorname{div}\left((1-n)\,\mathbf{v}^S\right) = 0,$$

$$\frac{1}{\rho_t^{FR}}\left\{\frac{\partial \rho_t^F}{\partial t} + \operatorname{div}\left(\rho_t^F \mathbf{v}^F\right)\right\} \equiv \frac{\partial n}{\partial t} + \operatorname{div}\left(n\mathbf{v}^F\right) = 0,$$

(13.2)

which can be combined to the following one

$$\operatorname{div}\left(n\mathbf{v}^F + (1-n)\,\mathbf{v}^S\right) = 0. \tag{13.3}$$

This was the consequence of incompressibility of the components (compare Part I or Section 12.3). If the components are *compressible,* which is assumed in the following, we are missing this equation. The compressibility of the components is an important feature in the wave analysis and many other problems of practical bearing. Linear models yield for compressible components a relation for the porosity which does not coincide with this derived for incompressible materials. As mentioned above, the porosity equation following from the mass conservation law does not contain a source. However, such a source would describe a spontaneous relaxation of porosity as well as a production of porosity by, for instance, damage. We know from experience with other microstructural variables that this is an important property yielding evolution equations for such variables.

As in the case of incompressible components also for compressible constituents we motivate an equation for the porosity by a transition from a semimacroscopic model (derivation see Part I). Again, this is done by means of averaging over the Representative Elementary Volume. Finally, this results in the *balance equation of porosity* in regular points and in Lagrangian description

$$\frac{\partial \Delta_n}{\partial t} + \operatorname{Div}\mathbf{J} = \hat{n}, \quad \Delta_n = n - n_E, \tag{13.4}$$

where Δ_n is the deviation of porosity from the equilibrium value n_E, the latter together with the macroscopic *flux of porosity* $\mathbf{J}$, and the source of porosity $\hat{n}$ must be given by constitutive relations. The *source* of porosity is caused, for instance, by microstructural relaxation processes or cracking. The flux of porosity results from the relative motion of the components. We expect that n tends to an equilibrium under constant external conditions. The equilibrium value of porosity n_E satisfies Equation (13.4) with the flux and source equal to zero. The *equilibrium changes of the porosity, $\partial n_E/\partial t$*, are related to internal deformations of the skeleton. As we consider porous materials with elastic skeleton, this contribution, in contrast to the other two, must appear in equilibrium as well as in non-equilibrium processes.

In Eulerian description the Div-operator has to be exchanged by the div-operator, i.e.,

$$\frac{\partial \Delta_n}{\partial t} + \operatorname{div}\mathbf{J} = \hat{n}, \quad \Delta_n = n - n_E. \tag{13.5}$$

13.1.2 Full and simplified models with the balance equation of porosity

The *Full Model* and the *Simple Mixture Model,* which results by some simplifications from the former, are used in this chapter for the description of saturated porous media. For the investigation of the various problems, covered in this book, the choice of the model depends on the level of difficulty of the problem, and keeping in mind the minimization of the numerical effort. For complex problems, it is convenient to use a model which is as simple as possible. Therefore, for most of the topics the Simple Mixture Model is used.

Full Model

The *Full Model* (see Wilmanski [442]) is based on the assumption that a porous material is described by a two-component mixture-like continuum. The two components are the grains which are called the solid or the skeleton, S, and the pore fluid, F. We consider only isothermal processes, i.e., the temperature of the system stays constant during the whole process. From the limitation that the temperature does not belong to the fields follows that only the following set of fields is chosen for the description of the porous medium:

- partial mass densities of the solid and of the fluid, respectively: ρ^S, ρ^F,

- partial velocities of the solid and of the fluid, respectively: $\mathbf{v}^S, \mathbf{v}^F$ and

- the porosity, n.

These fields are defined on a common domain $\mathcal{B}$ of the porous medium and are functions of time. Sufficient smoothness assumptions are made which we do not discuss here. The field equations for these fields follow from the balance equations whose linear form is as follows

- partial mass balance

$$\frac{\partial \rho^S}{\partial t} + \rho_0^S \mathrm{div}\, \mathbf{v}^S = 0, \quad \frac{\partial \rho^F}{\partial t} + \rho_0^F \mathrm{div}\, \mathbf{v}^F = 0, \tag{13.6}$$

- partial momentum balance equations

$$\rho_0^S \frac{\partial \mathbf{v}^S}{\partial t} = \mathrm{div}\, \mathbf{T}^S + \hat{\mathbf{p}}, \quad \rho_0^F \frac{\partial \mathbf{v}^F}{\partial t} = \mathrm{div}\, \mathbf{T}^F - \hat{\mathbf{p}}, \tag{13.7}$$

- porosity balance equation

$$\frac{\partial \Delta_n}{\partial t} + \Phi\, \mathrm{div}\left(\mathbf{v}^F - \mathbf{v}^S\right) = \hat{n}, \quad \Delta_n = n - n_E. \tag{13.8}$$

In these equations ρ_0^S and ρ_0^F are initial (constant) partial mass densities, $\mathbf{T}^S$ and $\mathbf{T}^F$ are the partial Cauchy stress tensors of the solid and of the fluid, respectively, $\hat{\mathbf{p}}$ is a momentum source while $\hat{n}$ is a porosity source. The material parameter of the porosity flux is denoted by Φ and the porosity in thermodynamical equilibrium by n_E. As we neglect temperature changes, the latter appears when there is no diffusion, i.e., if the partial velocities of both components are the same: $\mathbf{v}^S = \mathbf{v}^F$.

We assume the medium to be linearly isotropic which means that the macroscopic constitutive relations have the following form

$$\mathbf{T}^S = \mathbf{T}_0^S + \lambda^S e \mathbf{1} + 2\mu^S \mathbf{e}^S + Q\varepsilon \mathbf{1} + \beta \Delta_n \mathbf{1} - N\left(n - n_0\right)\mathbf{1},$$

$$\mathbf{T}^F = -p^F \mathbf{1}, \quad p^F = p_0^F - \rho_0^F \kappa \varepsilon - Qe + \beta \Delta_n - N\left(n - n_0\right), \tag{13.9}$$

$$n_E = n_0\left(1 + \delta e\right), \quad \hat{\mathbf{p}} = \pi\left(\mathbf{v}^F - \mathbf{v}^S\right), \quad \hat{n} = -\frac{\Delta_n}{\tau_n}.$$

In these relations, $\mathbf{T}_0^S, p_0^F$ and n_0 are the initial partial stress in the solid, the initial partial pressure in the fluid and the initial porosity, respectively. The quantities

$$e = \operatorname{tr} \mathbf{e}^S = \frac{\rho_0^S - \rho^S}{\rho_0^S} \quad \text{and} \quad \varepsilon = \frac{\rho_0^F - \rho^F}{\rho_0^F}, \tag{13.10}$$

describe partial volume changes of the skeleton and of the fluid, respectively. Hence, the model contains the following set of macroscopic material parameters

$$\left\{ \lambda^S, \mu^S, \kappa, Q, N, \delta, \beta, \pi, \tau_n, \Phi \right\}. \tag{13.11}$$

Obviously, the first two parameters are Lamé constants of the two-component model, i.e., elastic constants of the solid and κ is the macroscopic compressibility of the fluid. Q and N are coupling constants whose influence mainly on the propagation of sound waves in saturated porous media is described in detail in [18]. The first one quantifies the static coupling effect between the partial stresses. The second one describes the influence of the porosity gradient. Equilibrium changes of the porosity are described by δ, while Φ, β, π and τ_n are non-equilibrium parameters related to the porosity flux, the coupling of the stresses to non-equilibrium porosity changes, to diffusion and to relaxation properties of the porosity. All these parameters are functions of the initial porosity n_0 and some other initial quantities.

Simple Mixture Model

In some of the following applications we rely on a simpler model than the Full Model. Some effects are neglected in the *Simple Mixture Model* (for the model proposed in a fully nonlinear and multi-component form, see e.g., K. Wilmanski Part I, [434, 435, 437, 438]) but it was deduced taking into account thermodynamic fundamental laws. Macroscopic modeling of saturated porous materials leads inevitably to a theory of immiscible mixtures considered as an extension of the classical Truesdell theory of fluid mixtures (e.g. [392]).

Certainly, the two mentioned models have to be compared to the Biot Model which became famous for the description of saturated porous media. It does not belong to the class of models which incorporates the porosity as a field but it takes into account three effects which are neglected by the Simple Mixture Model. These are:

- an added mass effect reflected in Biot's model by off-diagonal contributions to the matrix of partial mass densities

- a static coupling effect between partial stresses (the parameter Q which also appears in the Full Model) and

- a dependence of the permeability coefficient π (a constant in both the Full and the Simple Mixture Model) on the frequency (viscous effects).

The added mass effect is neglected because it yields a non-objectivity of Biot's equations [440, 445]. Even though, it could be shown in these works that Biot's model follows by linearization from a nonlinear thermodynamically unobjectionable and objective model, the fact persists that the relative accelerations which are part of the Biot Model, are non-objective.

The coupling of partial stresses Q is neglected in the Simple Mixture Model because it is assumed to yield only quantitative corrections without changing the qualitative behavior of the system, at least in the range of a relatively high stiffness of the skeleton in comparison with the fluid. This has been analyzed for bulk waves in Albers & Wilmanski [17]. It has been shown within the classical thermodynamic theory of fluid mixtures by Müller (e.g. see [261]) that such couplings are eliminated in macroscopic models by the second law of thermodynamics if one does not account for gradients of partial mass densities. Such models without couplings are called *simple mixtures*. K. Wilmanski could show by the direct exploitation of the second law of thermodynamics [442] that also the Biot coupling follows by linearization of a thermodynamical model in which a dependence on the porosity gradient is included.

The third difference to the Biot Model – π assumed to be constant – has for small frequencies only a small influence on the results for waves. For high frequencies, on the other hand, the results for the wave speeds which have been achieved with the constant parameter are slightly smaller and a difference exists in the attenuation for large frequencies. However, this difference between the results is not qualitative.

We present here the linear form of the Simple Mixture Model for a two-component poroelastic saturated medium. Within this model the process is described by the macroscopic fields $\rho^F(\mathbf{x},t)$ – the partial mass density of the fluid, $\mathbf{v}^F(\mathbf{x},t)$ – the velocity of the fluid, $\mathbf{v}^S(\mathbf{x},t)$ – the velocity of the skeleton, $\mathbf{e}^S(\mathbf{x},t)$ – the symmetric tensor of small deformations of the skeleton and the porosity n.

As it is the case for the Full Model – but in contrast to the Biot Model – in the Simple Mixture Model the porosity belongs to the class of fields and thus the model contains additionally to the classical balance equations the balance of porosity.

The set of linear equations is given by

$$\frac{\partial \rho^F}{\partial t} + \rho_0^F \operatorname{div} \mathbf{v}^F = 0, \qquad \left| \frac{\rho^F - \rho_0^F}{\rho_0^F} \right| \ll 1,$$

$$\rho_0^F \frac{\partial \mathbf{v}^F}{\partial t} + \kappa \operatorname{grad} \rho^F + \beta \operatorname{grad}(n - n_E) + \hat{\mathbf{p}} = 0, \quad \hat{\mathbf{p}} := \pi \left(\mathbf{v}^F - \mathbf{v}^S \right),$$

$$\rho_0^S \frac{\partial \mathbf{v}^S}{\partial t} - \operatorname{div} \left[\lambda^S \left(\operatorname{tr} \mathbf{e}^S \right) \mathbf{1} + 2\mu^S \mathbf{e}^S + \beta (n - n_E) \mathbf{1} \right] - \hat{\mathbf{p}} = 0, \qquad (13.12)$$

$$\frac{\partial \mathbf{e}^S}{\partial t} = \operatorname{sym} \operatorname{grad} \mathbf{v}^S, \quad \left\| \mathbf{e}^S \right\| \ll 1, \quad n_E := n_0 \left(1 + \delta \operatorname{tr} \mathbf{e}^S \right),$$

$$\frac{\partial (n - n_E)}{\partial t} + \Phi \operatorname{div} \left(\mathbf{v}^F - \mathbf{v}^S \right) + \frac{n - n_E}{\tau_n} = 0, \qquad \left| \frac{n - n_0}{n_0} \right| \ll 1.$$

As before, ρ_0^F, ρ_0^S and n_0 denote constant reference values of the partial mass densities and porosity, respectively, and $\kappa, \lambda^S, \mu^S, \beta, \pi, \tau_n, \delta, \Phi$ are constant material parameters. The first one describes the macroscopic compressibility of the fluid component, the next two are macroscopic elastic constants of the skeleton, β is the coupling constant with the porosity, π is the coefficient of bulk permeability, τ_n is the relaxation time of porosity and δ, Φ describe equilibrium and non-equilibrium changes of porosity, respectively.

For linear analyses, e.g. linear wave analyses, coupling effects through β can be neglected. In some papers on the subject (e.g. Wilmanski [438, 439, 441]) the case $\beta = 0$ has been investigated and the statement has already been checked. Finally, it could be verified by Albers & Wilmanski in [18].

If the constant β is equal to zero the partial stresses $\mathbf{T}^S$ and $\mathbf{T}^F$ in the skeleton and in the fluid, respectively, are not coupled. Such a coupling, even though of a different –

static – nature, is required in Biot's model in which the partial stresses (compare Equations (13.9)) have the form

$$\mathbf{T}^S = \lambda^S \left(\operatorname{tr} \mathbf{e}^S\right) \mathbf{1} + 2\mu^S \mathbf{e}^S - Q \frac{\rho^F - \rho_0^F}{\rho_0^F} \mathbf{1},$$

$$\mathbf{T}^F = -\left[\kappa \left(\rho^F - \rho_0^F\right) - Q \operatorname{tr} \mathbf{e}^S\right] \mathbf{1}. \tag{13.13}$$

It can be easily shown that Biot's coupling constant Q (also appearing in the Full Model) has the order of magnitude of the pore pressure, i.e., 10^5 Pa in air-saturated soils and rocks. This must be compared with the elastic constants λ^S, μ^S and $\kappa\rho_0^F$ which are at least of order 10^8 Pa. Hence, similarly to the assumption that $\beta = 0$, the coupling constant Q can be left out in linear analyses as well.

13.2 Freezing and thawing

13.2.1 Introduction

In this section the modeling of freezing and thawing processes in two-component saturated porous materials is demonstrated. An iterative procedure for the calculation of the mechanical properties is shown. It incorporates two different stages of the process according to the actual temperature (above or below the freezing point). First, isothermal diffusion in the poroelastic range without freezing is considered while the second range contains the process of freezing. The model for the latter range is based on the Gurson-Tvergaard-Needleman theory for plastic deformations. The measure of damage is described by the extent of the porosity changes caused by freezing. The material parameters which depend on the porosity follow from a micro-macro procedure.

Freezing processes in porous materials cause damage through large volume changes by cryo-swelling, i.e., during the transition from water to ice. Other causes for damage are the high pressure of water which does not freeze due to the shift of the melting point of ice and the flow of water from unfrozen regions through shrinking channels. Owing to the Gibbs-Thomson effect it may happen that the water, confined in the pores, forms ice lenses at a temperature lower than the bulk freezing point. Damage caused by these freezing processes yields changes in the porosity and the creation of microcracks in the skeleton. In the course of the process these microcracks coalesce and yield a global destruction of the material. If diffusion processes can be neglected, such deformations can usually be described by various models arising from the fundamental Gurson model (1977) [147]. An extension of this model in which the evolution of the damage parameter is identified with porosity changes was proposed by Tvergaard and Needleman (1981-1987) (see [151], [225] for a detailed presentation or directly e.g. [269, 397]). Some details of this model and a modification are presented in this book. The section on freezing and thawing gives an account of the paper of K. Wilmanski and A. Alsabry [22].

Models for freezing and thawing processes usually follow from micromechanical considerations. On the one hand, they can be based on a model of microcracks in cavities (e.g. [166], [249]) – in this case diffusion processes are neglected. An alternative is to build a macroscopic model for which the skeleton is assumed to be elastic (e.g. [94] where an excellent analysis of different processes coupled to freezing is presented but the analysis is based on elastic potentials, or [312] where, apart from the erroneous nonlinear part, a construction of mass sources due to freezing for poroelastic materials is presented).

Also here a simplified macroscopic model is presented which is based on the continuous theory of immiscible mixtures. It is assumed that the porous material is fully saturated and, thus, a two-component model is used. Two ranges of processes are distinguished: the first, called PE-range (poroelastic range), contains isothermal diffusive processes without freezing. It is modeled by a modified Biot theory presented in Subsection 13.2.2. The second range, F-range (range of freezing) reflects the process of freezing and, again, it is assumed that it is isothermal and diffusion-free. Processes belonging to this range are described in Subsection 13.2.3. The whole long-term thermomechanical process of deformation, diffusion, freezing and damage is assumed to be a sequence of these ranges. Thus, the current approach is called an iterative description of freezing.

Such a procedure neglects some important processes such as heat transfer or cryosuction in the range of freezing. However, it enables to solve engineering boundary value problems by simple and easily available numerical methods. Within the range of damage the model is almost classical and within the range of two-component diffusion it goes beyond known problems of porous media only due to heterogeneity of the material.

13.2.2 Modeling of the diffusion range (PE-range) without freezing

The continuous model of isothermal processes in saturated porous materials in the PE-range is described by the fields

$$\left\{ \rho^S, \rho^F, n, \mathbf{v}^S, \mathbf{v}^F, \mathbf{e}^S \right\}, \tag{13.14}$$

defined on a common domain $\mathcal{B} \subset \Re^3$. The first two are the partial mass densities of the skeleton and water, respectively, n denotes the current porosity. $\mathbf{v}^S$ and $\mathbf{v}^F$ are partial velocities and $\mathbf{e}^S$ is the Almansi-Hamel deformation tensor. Since we consider a linear model it satisfies the following conditions

$$\left\| \mathbf{e}^S \right\| \ll 1, \quad |\varepsilon| \ll 1, \quad \left(\mathbf{e}^S - \lambda^{(\alpha)} \mathbf{1} \right) \mathbf{n}^{(\alpha)} = \mathbf{0}, \quad \alpha = 1, 2, 3,$$

$$\left\| \mathbf{e}^S \right\| = \max \left| \lambda^{(\alpha)} \right|, \quad \varepsilon = \frac{\rho_0^F - \rho^F}{\rho_0^F}, \tag{13.15}$$

where the eigenvalues $\lambda^{(\alpha)}$ are principal stretches of the skeleton. They define the volume changes of the solid for small deformations

$$e = \operatorname{tr} \mathbf{e}^S = \sum_{\alpha=1}^{3} \lambda^{(\alpha)}. \tag{13.16}$$

The Simple Mixture Model with $\beta = 0$ but under consideration of body forces is used. The field equations which follow from partial balance laws of mass and momentum for the linear theory have the following form

$$\frac{\partial \rho^S}{\partial t} + \rho_0^S \operatorname{div} \mathbf{v}^S = 0, \quad \frac{\partial \rho^F}{\partial t} + \rho_0^F \operatorname{div} \mathbf{v}^F = 0, \quad \rho_0^S, \quad \rho_0^F = \text{const.},$$

$$\rho_0^S \frac{\partial \mathbf{v}^S}{\partial t} = \operatorname{div} \mathbf{T}^S + \hat{\mathbf{p}}^S + \rho_0^S \mathbf{b}^S, \tag{13.17}$$

$$\rho_0^F \frac{\partial \mathbf{v}^F}{\partial t} = \operatorname{div} \mathbf{T}^F + \hat{\mathbf{p}}^F + \rho_0^F \mathbf{b}^F.$$

All quantities with index zero denote initial values. $\mathbf{T}^S$ and $\mathbf{T}^F$ are partial Cauchy stress tensors and the momentum sources $\hat{\mathbf{p}}^S$ and $\hat{\mathbf{p}}^F$ satisfy the conservation law $\hat{\mathbf{p}}^S = -\hat{\mathbf{p}}^F$. The body forces $\mathbf{b}^S$ and $\mathbf{b}^F$ may contain noninertial forces.

Since for freezing processes the fluid is assumed to be water, the fluid component is supposed to be ideal, i.e.,

$$\mathbf{T}^F = \sigma_{kl}^F \mathbf{e}_k \otimes \mathbf{e}_l = -p^F \mathbf{1}, \qquad \text{i.e.,} \qquad \sigma_{kk}^F = -3p^F, \tag{13.18}$$

where $\mathbf{e}_k$ are unit basis vectors of Cartesian coordinates.

In addition to the mass and momentum balances, the nonequilibrium deviation of the porosity from its equilibrium value is assumed to satisfy a balance law, the balance equation of porosity (see 13.1.1 or for more details [436]),

$$\frac{\partial \Delta_n}{\partial t} + \Phi_0 \mathrm{div}\left(\mathbf{v}^F - \mathbf{v}^S\right) = \hat{n}, \qquad \Delta_n = n - n_E. \tag{13.19}$$

The scalar character of the material parameter Φ_0 implies that isotropy of the medium is assumed. The constitutive quantities $\left\{\mathbf{T}^S, p^F, \hat{\mathbf{p}}^S, n_E, \hat{n}\right\}$ appearing in the balance laws must be specified by constitutive relations. For the Biot-like model, used here, these are (see [93] or Part I of this book)

$$\mathbf{T}^S = \mathbf{T}_0^S + \lambda^S e \mathbf{1} + 2\mu^S \mathbf{e}^S + Q\varepsilon\mathbf{1}, \qquad p^F = p_0^F - Qe - \rho_0^F \kappa\varepsilon,$$

$$\hat{\mathbf{p}}^S = -\hat{\mathbf{p}}^F = \pi\left(\mathbf{v}^F - \mathbf{v}^S\right), \qquad \hat{n} = 0, \qquad n_E = n_0\left(1 + \delta e\right), \tag{13.20}$$

where

$$\frac{\partial \mathbf{e}^S}{\partial t} = \mathrm{sym\,grad}\,\mathbf{v}^S \quad \Rightarrow \quad \frac{\partial e}{\partial t} = \mathrm{div}\,\mathbf{v}^S. \tag{13.21}$$

The model contains the elastic constants of the skeleton λ^S and μ^S, the compressibility coefficient of water κ, the static Biot coupling coefficient Q, the material parameter for elastic changes of porosity δ and the permeability coefficient π. The latter is a time independent scalar which means that hereditary effects connected with the diffusion and anisotropy effects connected with the tortuosity are neglected (compare [447] or for the influence of anisotropic tortuosity Section 13.4 of this book).

A porosity gradient is not considered in the model but could be easily included [443], [447]. However, this would require an extension of the micro-macro relations (dependence not only on the porosity but also on its gradient) and the solution of this problem cannot be presented in a closed form [443].

The assumption of the source of porosity being zero ($\hat{n} = 0$) means that the above equations reflect the range without freezing. This assumption will be modified in the F-range. In the present case the balance equation of porosity can be easily solved and the following relation for current values of the porosity is obtained (e.g. [443])

$$n = n_0\left(1 + \delta e + \frac{\Phi_0}{n_0}\left(\varepsilon - e\right)\right). \tag{13.22}$$

13.2.2.1   Specification of the material parameters

One of the material parameters which have to be specified, the shear modulus μ^S, cannot be derived by a micro-macro transition. For this reason, it is either assumed to be given or, as often is done for soils, it is related to other material constants and to the Poisson ratio ν which is assumed to be a given constant (for soils $\nu \approx 0.33$).

In engineering applications it is more convenient to work with the bulk stress $\mathbf{T}$ rather than with partial stresses $\mathbf{T}^S$ and $\mathbf{T}^F$. Addition of the partial stresses (see (13.20)) yields the bulk stress

$$\mathbf{T} = \mathbf{T}^S + \mathbf{T}^F = \mathbf{T}_0 + Ke\mathbf{1} + 2\mu^S \operatorname{dev} \mathbf{e}^S - C\zeta\mathbf{1}, \qquad \operatorname{dev} \mathbf{e}^S = \mathbf{e}^S - \frac{1}{3}\operatorname{tr} \mathbf{e}^S \mathbf{1},$$

(13.23)

$$p = \frac{p^F}{n_0} = p_0 - Ce + M\zeta,$$

where $\mathbf{T}_0$ is the initial bulk stress and p_0 the initial value of the pore pressure p. The so-called Biot material parameters (since they are often used in analyses based on Biot's model) are defined by the relations

$$K = \lambda^S + \frac{2}{3}\mu^S + \rho_0^F \kappa + 2Q, \qquad C = \frac{Q + \rho_0^F \kappa}{n_0}, \qquad M = \frac{\rho_0^F \kappa}{n_0^2}.$$

(13.24)

This set replaces the parameters λ^S, Q and κ. In (13.23) appears a new field, the increment of fluid content ζ, which can be expressed by a relation between the initial porosity and the volume changes of the components: $\zeta = n_0 (e - \varepsilon)$. It is convenient to work with this variable because, according to the mass balance equations $(13.17)_1$, it satisfies the relation

$$\frac{\partial \zeta}{\partial t} = \operatorname{div} \left(\mathbf{v}^F - \mathbf{v}^S \right),$$

(13.25)

which means that it is identically zero when diffusion does not arise, i.e., when the velocities of the two components are the same. Using this quantity, the following form of the porosity balance equation (13.19) is obtained

$$\frac{\partial}{\partial t} \left(\Delta_n - \frac{\Phi_0}{n_0}\zeta \right) = \hat{n}.$$

(13.26)

Consequently, the balance equation of porosity reduces in this linear model to a pure evolution equation and this means, in turn, that it does not require boundary conditions.

The transformed balance equations of momentum in which inertial forces are neglected, since they are immaterial in processes of freezing, have the following form

$$\operatorname{div} \mathbf{T} + \rho_0 \mathbf{b} = 0, \qquad \rho_0 = \rho_0^S + \rho_0^F, \qquad \rho_0 \mathbf{b} = \rho_0^S \mathbf{b}^S + \rho_0^F \mathbf{b}^F,$$

(13.27)

$$\operatorname{grad}\left(n_0 p \right) + \pi \left(\mathbf{v}^F - \mathbf{v}^S \right) - \rho_0^F \mathbf{b}^F = 0.$$

The deformation measure must satisfy the compatibility condition

$$\frac{\partial \mathbf{e}^S}{\partial t} = \operatorname{sym} \operatorname{grad} \mathbf{v}^S.$$

(13.28)

The second row of (13.27) describes a paradox related to the porosity varying in space. Namely, for the case that body forces are absent and the pore pressure is constant, the variation of the porosity, which is a pure geometrical property, yields a difference in the velocities of the components. El Tani discusses this hydrostatic paradox in [111] and explains that it can be removed by the assumption that the momentum source depends on the porosity gradient [445]. However, it was already mentioned above, that the porosity gradient is ignored in the present model. Actually, if this would not be the case, one would have to correct the constitutive relation $(13.23)_2$ by addition of a term proportional to the increment of the porosity.

For the description of processes in the PE-range the set (13.23) and (13.25)–(13.28) is the full set of field equations. Sometimes, it is convenient to replace the increment of fluid contents by the pore pressure as unknown field but we shall not do so here.

We return to the estimation of Biot's material parameters (13.24). This can be done by means of a micro-macro transition (see [443] for two-component media or [13] for three-component materials). In these approaches it is assumed that the material parameters of the components, i.e., the material properties of the system in the limits of the initial porosity, $n_0 = 0$ and $n_0 = 1$, are known. As a result of the so-called gedankenexperiments (compression tests; for the original idea see [50]), we obtain for the two-component medium, considered here, the Gassmann relations (e.g. see [443])

$$K = \frac{(K_s - K_d)^2}{\frac{K_s^2}{K_W} - K_d} + K_d, \qquad C = \frac{K_s(K_s - K_d)}{\frac{K_s^2}{K_W} - K_d}, \qquad M = \frac{K_s^2}{\frac{K_s^2}{K_W} - K_d}, \tag{13.29}$$

where

$$K_W = \left(\frac{1 - n_0}{K_s} + \frac{n_0}{K_f}\right)^{-1}, \tag{13.30}$$

and K_s and K_f are the compressibility moduli of the skeleton and of water. K_d is the so-called drained compressibility modulus. The Gassmann relations specify the dependence of the macroscopic material parameters on the initial porosity n_0. With the same procedure also the dependence of the parameters δ and Φ_0 appearing in the porosity balance equation can be obtained. However, they are not needed here and therefore we refer in this regard to [446] or [13].

The specification of the permeability coefficient π remains. In most applications, and also for freezing/thawing processes treated in this section, we assume that the random structure of the porous material yields isotropy, which means that this material parameter is a scalar. Only in Section 13.4 anisotropy of the permeability is considered. Then, the permeability is the product of the inverse of the tortuosity tensor and the permeability coefficient π. In 1927 J. Kozeny proposed a relation for this parameter, connecting it with the viscosity of the fluid in the channels [206]. However, the relation contained a flaw concerning the dependence on the tortuosity which has been corrected by N. Epstein [113] in 1989. The tortuosity τ is in the isotropic case a quantity, which is equal to the ratio of the length of streamlines between two points lying close to each other, say A and B (see Figure 13.1), and the distance of these points. Consequently, $\tau \geq 1$. While in the original Kozeny relation the dependence on the tortuosity was proposed to be linear, in Epstein's corrected version of the permeability coefficient the tortuosity is a squared value

$$\pi = \frac{9}{4}\frac{\gamma b \mu_\nu}{d^2}\tau^2\left(\frac{1 - n_0}{n_0}\right)^2 \qquad \text{with} \qquad \gamma = \frac{g \rho_0^F}{n_0}, \tag{13.31}$$

where g is the gravity acceleration, b is the so-called capillary shape factor (e.g. 32 for circular pores and 48 for parallel slits), μ_ν is the true dynamic viscosity of the fluid and d is the characteristic diameter. The coefficient $\frac{9}{4}b\tau^2$ is often assumed to be equal to 180. This value corresponds for $b = 33.3$ to the tortuosity $\tau = 1.55 \approx \frac{\pi}{2}$. For a detailed discussion of different formulations of the permeability we refer to paper [447].

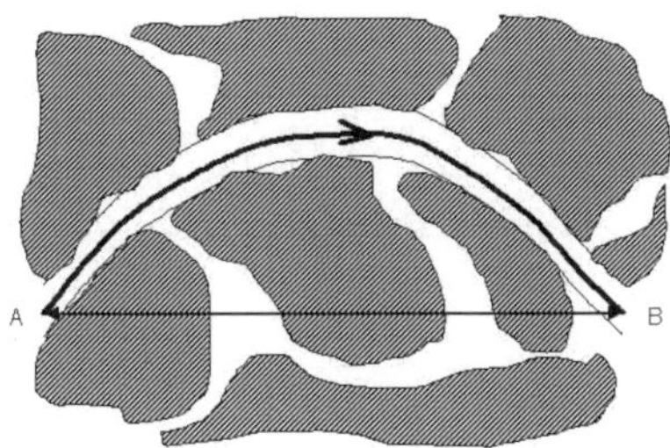

Figure 13.1: *The tortuosity τ is the ratio of the distance between points* A *and* B *to the length of the channels indicated by the line with arrow, from [19].*

Figure 13.2: *Photo of a phenomenon – most likely palsas – taken in Berlin in 2010.*

13.2.3 Governing equations for material damaged by freezing (F-range)

Creation of ice lenses in porous materials yields a number of accompanying effects which substantially change the properties of the medium. First, in the domain of freezing $\mathcal{B}_f \subset \mathcal{B} \subset \mathfrak{R}^3$ diffusion is hampered. This means that the increment of the fluid content ζ in $\mathcal{B}_f$ is identically zero in the model presented in the last subsection. Hence, a one-component, composite-like, description of the medium can be used in the freezing range (F-range).

The inhibition of the diffusion and the volume change of water due to freezing raise the pore pressure to very high values. This has been found in early experiments on frost heaving in soils (e.g. [291]) and was confirmed in numerous modern experiments (e.g. [292]). It is also the reason for the creation of palsas, which are frost heaves occurring mostly in polar and subpolar climates containing permanently frozen ice lenses. They occur particularly in bogs – most likely also the phenomenon shown in the picture taken in Berlin (Figure 13.2) can be attributed to the heaving of soil due to freezing and can be identified as a kind of palsas. Not only experimental results are available: In a few recent works O. Coussy and coworkers proposed a theoretical description of freezing and thawing processes in porous materials which yield relations for the pore pressure and the shift of the melting point [94], [95], [121]. The theory is based on a microscopic approach in which freezing in the pores is described by the classical Gibbs-Thomson and Young-Laplace equations. Thereby, a simple pore structure (spherical voids with cylindrical connections) is considered. Freezing and thawing are described as slow time evolution processes in which changes of the pore pressure behave according to the evolution equation

$$\tau_f \frac{dp}{dt} + p = S_f \left(T_f - T \right), \tag{13.32}$$

where $\tau_f = \mu_\nu \left(\frac{3}{4} \mu^S + \frac{1}{K_{ice}} \right)$ is the relaxation time and K_{ice} is the compressibility modulus of ice, T_f is a uniform initial temperature and S_f is the entropy of fusion. The pore pressure p is now identical to the ice crystal pressure. For the purpose of the iterative method presented here, this evolution process (for details see [95]) is neglected. For frozen soils this is justified because the capillary motion of water is slow and the corresponding relaxation time very small. For Cairo sand, for instance, the order of magnitude of the relaxation time of $\tau_f = 0.55 \times 10^{-11}$ s follows from the typical values $\mu^S = 161$ MPa and $K_{ice} = 1200$ MPa given in [342] and the viscosity of water under normal conditions $\mu_\nu = 1.002 \times 10^{-3}$ Pa s.

Consequently, we can consider freezing processes as rate independent quasistatic processes of damage: cavities are nucleated, grow by plastic deformation and yield mesocracks by coalescence. These three mechanisms, characteristic for geomaterials, are described by the extension of Gurson's model proposed by Tvergaard and Needleman [269], [225]. According to the Gurson-Tvergaard-Needleman (GTN) model there exists a critical value of porosity n_C, created by the microdamage, at which microcracks coalesce and yield a macroscopic damage of the material. This would be the last F-range in the iteration process beyond which the model does not work anymore. Hence, the porosity considered in the iteration must be smaller than this critical value. Then, the GTN model is based on the following yield function [225]

$$f_{GTN} = \frac{\sigma_{eq}^2}{\sigma_s^2} + 2q_1 n \cosh \left(\frac{1}{2} q_2 \frac{\sigma_{kk}}{\sigma_s} \right) - 1 - (q_1 n)^2, \tag{13.33}$$

$$\text{where} \quad \sigma_{eq} = \sqrt{\frac{3}{2} \sigma_{ij}^D \sigma_{ij}^D}, \quad \sigma_{ij}^D = \sigma_{ij} - \frac{1}{3} \sigma_{kk} \delta_{ij}.$$

The values 1.5 and 1.0 for the material parameters q_1 and q_2, respectively, seem to be confirmed for metals by a numerical analysis of the so-called cell model of damage [151]. Most likely, the same values can also be accepted for other materials. As before, n denotes the current porosity. σ_s is the yield stress which may consist of the sum of two parts: $\sigma_s = \sigma_Y + \sigma_H$, where σ_Y is the yield stress at the beginning of the damage process and σ_H the isotropic hardening. As already mentioned, Hao and Brocks [151] justified the above yield function by averaging the microscopic cell model over a simple Representative Elementary Volume.

The GTN model is based on the normality rule which yields the following relation for plastic strain rate

$$\dot{e}_{ij}^p = \dot{\lambda} \left[\frac{3\sigma_{ij}^D}{\sigma_s^2} + \frac{q_1 q_2 n}{\sigma_s} \sinh \left(\frac{1}{2} q_2 \frac{\sigma_{kk}}{\sigma_s} \right) \delta_{ij} \right], \tag{13.34}$$

where, as usual, the plastic multiplier $\dot{\lambda}$ follows from the consistency conditions at yield: $f_{GTN} = 0$, $\dot{f}_{GTN} = 0$. In contrast to the classical plasticity theory, the rate of equivalent plastic strain is defined by the relation

$$\dot{e}^p_{eq} = \frac{\sigma_{ij}\dot{e}^p_{ij}}{(1-n)\,\sigma_s}. \tag{13.35}$$

The most important issue of damage by freezing is the evolution of the porosity due to damage. As changes of the damage parameter and of the porosity are assumed to be identical, we can write for the source of porosity in the porosity balance equation (13.19)

$$\hat{n} = (1-n)\,\dot{e}^p_{kk} + \dot{n}_n, \tag{13.36}$$

where the first part is, obviously, the contribution of growth of the voids due to plastic deformation while the second contribution describes nucleation. This process is controlled by the growth of the ice lenses and, therefore, it seems to be reasonable to assume that the rate is proportional to the equivalent plastic strain rate

$$\dot{n}_n = N_n \dot{e}^p_{eq}, \tag{13.37}$$

where N_n is a material parameter dependent on the latent heat and the temperature of the transformation.

Now, bearing in mind the assumption that in the range of freezing diffusion does not occur, we can write the porosity balance equation (13.19) in the following form

$$\frac{\partial n}{\partial t} = n_{in}\delta\frac{\partial e}{\partial t} + (1-n)\frac{\partial e^p_{kk}}{\partial t} + N_n\frac{\partial e^p_{eq}}{\partial t}, \tag{13.38}$$

where n_{in} replaces the initial porosity n_0 of the first PE-range. It is the value of porosity in the PE-range preceding the F-range under consideration. It is simultaneously the initial value for the above evolution equation.

The lack of diffusion yields easily the relation between elastic strain and bulk stress. This relation is needed if we want to formulate equations analogous to the Prandtl-Reuss equations of the classical plasticity. Namely, for $\zeta = 0$ from (13.23) follows

$$e_{el} = \operatorname{tr}\mathbf{e}^S_{el} = \frac{1}{3K}\operatorname{tr}\left(\mathbf{T} - \mathbf{T}_0\right), \qquad \operatorname{dev}\mathbf{e}^S_{el} = \frac{1}{2\mu^S}\operatorname{dev}\left(\mathbf{T} - \mathbf{T}_0\right), \tag{13.39}$$

where $\mathbf{e}_{el}$ is the elastic part of the deformation, $\dot{\mathbf{e}}^S = \dot{\mathbf{e}}^S_{el} + \dot{\mathbf{e}}^S_p$, $\mathbf{e}^S_p = e^p_{ij}\mathbf{e}_i \otimes \mathbf{e}_j$. The Gassmann relation for the material parameter K and the corresponding parameter K_W have already been introduced in (13.29) and (13.30), respectively. As explained above, the shear modulus μ^S is related to the constant Poisson number ν and to other material constants, in the present case to K:

$$\mu^S = \frac{3}{2}\frac{1-2\nu}{1+\nu}K. \tag{13.40}$$

These relations, referring to the current porosity n, replace the classical relations between the material parameters and the damage parameter [225].

The set of equations (13.33)-(13.38), together with the equilibrium condition for bulk stresses (13.27)$_1$, forms the complete set of field equations for the F-range. As there is no diffusion in the domain $\mathcal{B}_f$, the boundary conditions are simple because they do not have to describe flows through the boundary of this domain which would be the case for the two-component model with diffusion.

13.2.4   Iterative procedure

Instead of presenting numerical solutions of boundary value problems, we indicate here some properties of the model, presented above, in relation to the dependence of the field equations on the damage parameter.

The procedure begins with the solution of a boundary value problem for the two-component poroelastic material (PE-range) as described in Subsection 13.2.2. It is assumed that the temperature T in this range is higher than the freezing temperature T_f, and the initial porosity n_0 is assumed to be constant in time. Among the results of the calculations one obtains the field $n = n_{(1)}(\mathbf{x})$ described by Relation (13.22). As the model is linear, all material parameters are evaluated for the initial porosity n_0. They are given by Relations (13.29) and (13.31). In the second step it is assumed that the temperature is lowered to a value smaller than the temperature of freezing T_f. Solution of the corresponding boundary value problem in the F-range is made by means of equations presented in Subsection 13.2.3 with the initial porosity equal to the porosity $n_{(1)}(\mathbf{x})$ of the preceding PE-range. Among the other results new values of porosity $n = n_{(2)}(\mathbf{x})$ follow from these calculations as solution of Equation (13.38). This field measures the damage of the material in the F-range due to freezing. Now, we may proceed to the next step which is again the PE-range for the new value of the temperature, higher than the temperature of freezing. The initial value of the porosity is now $n_{(2)}(\mathbf{x})$ which means that, in contrast to the first PE-range, the boundary value problem in this range is heterogeneous. The new values of the material parameters, calculated by means of Relations (13.29) and (13.31), are now dependent on the position $\mathbf{x}$. We continue this iteration until, after a certain number of freezing-thawing cycles, the porosity exceeds at some place a critical value (coalescence of microcracks!) and the material is treated as destroyed.

Let us note that the material parameters depend in this iteration procedure on the damage in a different way in the two ranges. The elastic parameters such as the bulk compressibility K are at any given point $\mathbf{x}$ constant within one PE-range but they change in the process of loading in the F-range. This is different from standard damage mechanics.

In the next two Figures 13.3 and 13.4 we present the dependence of the bulk compressibility parameter and the permeability coefficient on the porosity. The following values for the material parameters are used

$$K_s = 48 \times 10^9 \text{ Pa}, \quad K_f = 2.25 \times 10^9 \text{ Pa}, \quad K_d = \frac{K_s}{1 + 50n}, \quad \pi\,(n = 0.1) = 10^{10}\,\frac{\text{kg}}{\text{m}^3\text{s}}.$$
$$(13.41)$$

The relation for the drained modulus K_d, the so-called Geertsma formula, is often used in models for soils.

In Figure 13.3 the compressibility modulus K is presented. It is shown for a wide range of porosities from $n = 0.1$ to $n = 0.5$ in order to show the shape of the function. The range from 0.3 to 0.35 is plotted as a solid line because this is approximately the range of changes of the porosity due to damage if the initial porosity is chosen to be $n_0 = 0.3$. Clearly, the compressibility modulus is decaying in a similar manner as for the classical laws of damage mechanics.

In Figure 13.4 the dependence of the permeability coefficient π on the porosity is illustrated. For $n = 0.1$ the permeability coefficient is chosen to be $\pi = 10^{10}\,\frac{\text{kg}}{\text{m}^3\text{s}}$ which corresponds approximately to 0.01 darcy $= 10^{-14}$ m^2 (so-called intrinsic permeability, for details see: e.g. [447]). The coefficient is plotted for the same range of porosities, again, in order to show the shape of the function. And, again, in the range $n = 0.3$ to $n = 0.35$ it

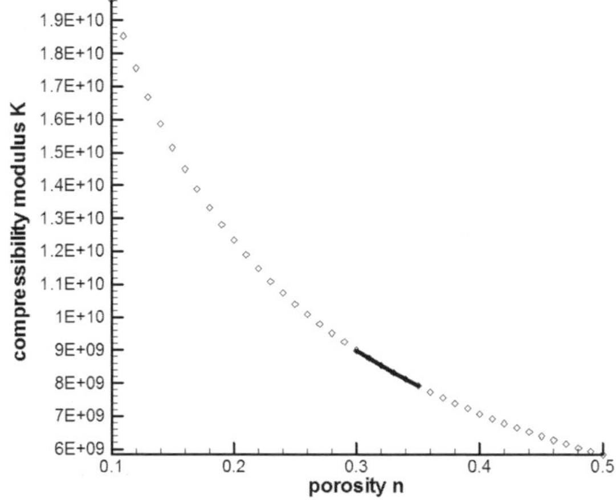

Figure 13.3: *Compressibility modulus as a function of porosity.*

is drawn as a solid line. As expected, the damage-increasing porosity yields a decreasing permeability coefficient which means that the diffusive force – a resistance to the relative motion – decays with growing damage.

This example confirms the expectation that the micro-macro relations of the mechanics of porous materials can replace speculative dependencies on the damage parameter within the classical damage mechanics.

13.2.5   Some remarks

The iterative description of freezing processes of porous materials seems to be sufficiently simple for practical applications in such engineering problems as frost heaving of soils or freezing techniques in the construction of tunnels and pits. It is also capable of extensions on nonisothermal processes with diffusion in the range of freezing. In particular, this latter problem seems to be important from the physical point of view. Damage done by ice lenses appears primarily due to the increment of pore pressure which, in turn, is a consequence of cryosuction, i.e., diffusive motion in the medium. However, from the point of view of numerical evaluation such extensions may be very difficult.

13.3   Linear stability of a 1D flow under transversal disturbance with adsorption

13.3.1   Introduction

In Chapter 6 several types of instabilities have been introduced. Since porous media consist of two and more components including both solid and fluid components, at least two kinds of instabilities may appear in such media: structural instabilities and flow instabilities. Examples for structural instabilities are piping (the brakedown of a granular medium) or shear band formation. The appearance of Bénard cells (compare Section 6.3)

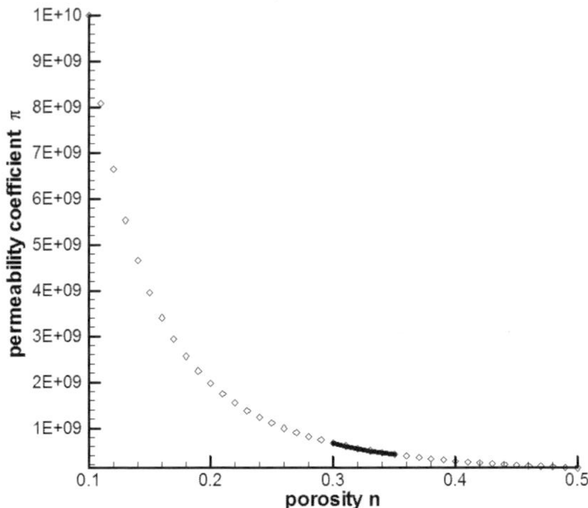

Figure 13.4: *Permeability coefficient as a function of porosity.*

or the change from laminar to turbulent flows belong to the class of flow instabilities. All kinds of instability have in common that they appear due to competition of at least two mechanisms. In the case of structural instabilities e.g. a threshold nonlinearity competes with the permeability of the material. A flow instability arises if a kinematic nonlinearity acts against, for example, viscosity.

Here, we consider one special type of flow instability: a steady state 1D flow in a porous medium is the base flow on which we superpose a small disturbance with adsorption. The disturbances satisfy the equations of the model for multi-component systems with adsorption (summarized in Subsection 13.3.2 and introduced in [6], compare also Section 13.5 of Part I). It is considered that a fluid/adsorbate mixture flows through the channels of a skeleton. In this case a kinematic nonlinearity acts against the permeability (diffusion) of the medium. Adsorption processes contribute in a nonlinear way to the field equations and essentially influence the stability properties.

Small disturbances are inevitably present in any real system but their effect on stable systems is mostly ephemeral. Consider, for example, the influence of wavelike disturbances on the circulation of the atmosphere. Also the superposition of the base flow considered in this section with disturbances without mass exchange leads to completely stable situations [9]. But the disturbance with adsorption yields the existence of an unstable region which means that adsorption slows down the relaxation process tremendously or even controls the loss of stability of the base flow (see [8, 9]) for a certain region of permeability coefficients. If the base flow is unstable, the disturbances will grow in amplitude with time and space. However, the values of the permeability coefficient for which the base flow is unstable are much smaller than those used in other applications shown in this chapter.

The region of instability for a transversal disturbance of the base flow with adsorption will be shown in dependence on three important model parameters:

- The bulk permeability coefficient π enters the field equations and describes the effective resistance of the skeleton to the flow of the fluid.

- The surface permeability α enters the model through boundary conditions of third type and accounts for the properties of the surface. It is one of the material parameters which determine the fluid velocity.

- The third parameter, the mass density of the adsorbate on the internal surface ρ_{ad}^A, enters the model through the mass source. It is proportional to the size of the internal surface which plays an enormous role for the global rate and amount of adsorption. For different soils it may vary some orders of magnitude.

The paper [8] is the source of the contents of this section.

13.3.2　Adsorption/diffusion model

We investigate the stability behavior of a 1D base flow under perturbations with mass exchange. The disturbance satisfies the following equations of the adsorption/diffusion model introduced in e.g. B. Albers [6]. This model is based on the above introduced model with balance equation of porosity by K. Wilmanski (for details see e.g. [435, 436] or Part I). However, in this section we assume the porosity to be constant so that the model presented here is simpler than the version introduced in [6].

We consider a process of physical adsorption in a three-component porous medium. A fluid-adsorbate mixture flows through the channels of the skeleton. The model takes into account three components: the solid, the fluid and an adsorbate which either flows with the same velocity as the fluid through the channels of the skeleton or settles down on the inner surface of the porous body. Under the assumption of a small concentration of the adsorbate we can neglect the influence of mass changes of the skeleton. Furthermore, we neglect dynamical disturbances of the skeleton.

Mass balances

The mass balance equation for the liquid (fluid and adsorbate phases together) and the concentration balance have the following form

$$\frac{\partial \rho^L}{\partial t} + \operatorname{div}\left(\rho^L \mathbf{v}^F\right) = \hat{\rho}^A,$$

$$\rho^L \left(\frac{\partial c}{\partial t} + \mathbf{v}^F \cdot \operatorname{grad} c\right) = (1 - c)\,\hat{\rho}^A,$$

(13.42)

where ρ^L is the mass density of the liquid components, i.e., the sum of the mass densities of the fluid and the adsorbate, and c denotes the concentration of the adsorbate in the fluid. The common velocity of the fluid and the adsorbate is $\mathbf{v}^F$. The mass balance for the adsorbate is replaced by the concentration balance.[1] Finally, the intensity of the mass source is denoted by $\hat{\rho}^A$.

---

[1] With

$$\frac{\partial}{\partial t}\left[\rho^L \frac{\rho^A}{\rho^L}\right] = \frac{\partial \rho^L}{\partial t}\frac{\rho^A}{\rho^L} + \rho^L \frac{\partial \frac{\rho^A}{\rho^L}}{\partial t} \quad \text{and} \quad \operatorname{div}\left[\rho^L \frac{\rho^A}{\rho^L}\right]\mathbf{v}^F = \frac{\rho^A}{\rho^L}\operatorname{div}\rho^L \mathbf{v}^F + \rho^L \mathbf{v}^F \cdot \operatorname{grad}\frac{\rho^A}{\rho^L},$$

we get from the mass balance for the adsorbate, i.e., from

$$\frac{\partial \rho^A}{\partial t} + \operatorname{div}\left(\rho^A \mathbf{v}^F\right) = \hat{\rho}^A,$$

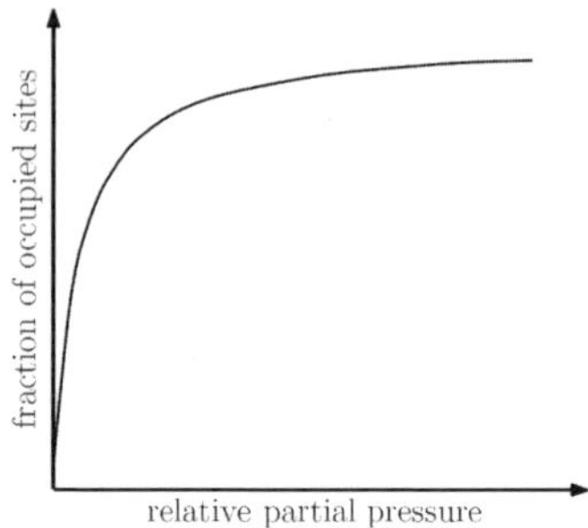

Figure 13.5: *Langmuir adsorption isotherm.*

Mass source

According to the model (see [6]) the mass source is given by the relation

$$
\hat{\rho}^A = -\frac{m^A}{V}\frac{d\left(\xi\, f_{int}\right)}{dt} = -\frac{m^A}{V}\left(f_{int}\frac{d\xi}{dt} + \xi\frac{d\,f_{int}}{dt}\right),
\tag{13.43}
$$

whose derivation is based on the classical Langmuir adsorption theory about occupied (ξ) and bare $(1-\xi)$ sites on the internal surface f_{int} of the solid (for details see e.g. [144]). V is the representative elementary volume REV and m_A denotes the reference mass of adsorbate per unit of the internal surface area.

The first contribution on the right-hand side of (13.43) describes changes in time of the fraction of occupied sites. It is specified by the Langmuir evolution equation which can be written in the form

$$
\frac{\partial \xi}{\partial t} = \left[\frac{cp^L}{p_0}\left(1-\xi\right) - \xi\right]\frac{1}{\tau_{ad}},
\tag{13.44}
$$

where p^L denotes the partial pressure in the liquid (fluid and adsorbate together), p_0 is a Langmuir reference pressure and τ_{ad} is the characteristic time of adsorption. In the case of time changes of ξ equal to zero the well-known Langmuir isotherm of occupied sites, which is illustrated in Figure 13.5, follows

$$
\xi_L = \frac{\frac{cp^L}{p_0}}{1 + \frac{cp^L}{p_0}},
\tag{13.45}
$$

where, according to Dalton's law, for small concentrations of the adsorbate it is assumed that the partial pressure of the adsorbate $p^A \cong cp^L$.

---

the relation:

$$
c\underbrace{\left[\frac{\partial \rho^L}{\partial t} + \operatorname{div}\rho^L \mathbf{v}^F\right]}_{\hat{\rho}^A} + \rho^L\left[\frac{\partial c}{\partial t} + \mathbf{v}^F\cdot\operatorname{grad}c\right] = \hat{\rho}^A.
$$

From this follows

$$
\rho^L\left[\frac{\partial c}{\partial t} + \mathbf{v}^F\cdot\operatorname{grad}c\right] = (1-c)\,\hat{\rho}^A,
$$

which is the concentration balance.

The second part of (13.43) describes the change of the internal surface. In principle, this change is assumed to be coupled with the relaxation of the porosity. However, under the assumption of constant porosity it drops out in the calculations presented in this section. It does not influence essentially the results because the Langmuir part of the mass source dominates on orders of magnitude the part connected with the porosity.

Consequently, we obtain the following form of the mass source

$$\hat{\rho}^A = -\rho_{ad}^A \left\{ \left[\frac{cp^L}{p_0} (1 - \xi) - \xi \right] \frac{1}{\tau_{ad}} \right\}, \tag{13.46}$$

where $\rho_{ad}^A := \frac{m^A f_{int}}{V}$ is the mass density of the adsorbate on the internal surface. This is a very important parameter as we know, for example from J. Bear [33], that the internal surface may change several orders of magnitude. Therefore, ρ_{ad}^A is one of those parameters which is altered to see changes in the stability behavior for various porous materials.

Momentum balance

Due to the same velocity of fluid and adsorbate only one momentum balance jointly for these components is needed. We have

$$\frac{\partial \rho^L \mathbf{v}^F}{\partial t} + \operatorname{div} \left(\rho^L \mathbf{v}^F \otimes \mathbf{v}^F + p^L \mathbf{1} \right) = \left(-\pi + \hat{\rho}^A \right) \mathbf{v}^F \approx -\pi \mathbf{v}^F. \tag{13.47}$$

The order of magnitude of π is much higher than that of $\hat{\rho}^A$ so that on the right-hand side of (13.47) $\hat{\rho}^A$ is negligible. Furthermore, the permeability coefficient π and the compressibility parameter κ are assumed to be constant. We assume the flow to be isochoric which leads to the following linear constitutive relation for the pressure in the liquid phase p^L

$$p^L = \overset{0}{p} + \kappa \overset{1}{\rho}. \tag{13.48}$$

Governing equations in 2D

Summarizing, the 2D model is governed by the following set of equations

$$\frac{\partial \rho^L}{\partial t} + \rho^L \left(\frac{\partial v_x^F}{\partial x} + \frac{\partial v_z^F}{\partial z} \right) + v_x^F \frac{\partial \rho^L}{\partial x} + v_z^F \frac{\partial \rho^L}{\partial z} = -\frac{\rho_{ad}^A}{\tau_{ad}} \left[\frac{cp^L}{p_0} (1 - \xi) - \xi \right],$$

$$\rho^L \left[\frac{\partial c}{\partial t} + v_x^F \frac{\partial c}{\partial x} + v_z^F \frac{\partial c}{\partial z} \right] = -(1 - c) \frac{\rho_{ad}^A}{\tau_{ad}} \left[\frac{cp^L}{p_0} (1 - \xi) - \xi \right],$$

$$\frac{\partial \xi}{\partial t} = \left[\frac{cp^L}{p_0} (1 - \xi) - \xi \right] \frac{1}{\tau_{ad}}, \tag{13.49}$$

$$\rho^L \left[\frac{\partial v_x^F}{\partial t} + v_x^F \frac{\partial v_x^F}{\partial x} + v_z^F \frac{\partial v_x^F}{\partial z} \right] = -\frac{\partial p^L}{\partial x} - \pi v_x^F,$$

$$\rho^L \left[\frac{\partial v_z^F}{\partial t} + v_x^F \frac{\partial v_z^F}{\partial x} + v_z^F \frac{\partial v_z^F}{\partial z} \right] = -\frac{\partial p^L}{\partial z} - \pi v_z^F.$$

13.3.3   Regular perturbation

The stability behavior of the 1D base flow is investigated with respect to transversal (2D) perturbations with mass exchange. For the analysis we use a regular perturbation method restricted to zeroth and first order contributions. This means that we expect the fields to be a superposition of the base solution (indicated by 0) and a small perturbation (indicated by 1). The solutions have the following structure

$$
p^F = \overset{0}{p}(x) + \kappa \overset{1}{\rho}(x,z,t), \qquad c = \overset{0}{c} + \overset{1}{c}(x,z,t), \qquad \xi = \overset{0}{\xi}(x) + \overset{1}{\xi}(x,z,t),
$$

$$
(13.50)
$$

$$
v_x^F = \overset{0}{v} + \overset{1}{v}_x(x,z,t), \qquad v_z^F = \overset{1}{v}_z(x,z,t).
$$

Base solution/zeroth step of perturbation

The base flow satisfies the following set of field equations. It is a steady state one-dimensional flow process in a porous medium

$$
\frac{\partial \rho^F}{\partial t} + \frac{\partial \rho^F v_x^F}{\partial x} = 0, \qquad 0 < x < l
$$

$$
(13.51)
$$

$$
\rho^F \left(\frac{\partial v_x^F}{\partial t} + v_x^F \frac{\partial v_x^F}{\partial x} \right) = -\frac{\partial p^F}{\partial x} - \pi v_x^F.
$$

Equations $(13.51)_{1/2}$ are the mass and momentum balances of the fluid. Here, ρ^F is the mass density of the fluid component, v_x^F is the fluid velocity in x-direction and π is the bulk permeability coefficient. The partial pressure in the fluid is denoted by p^F. Only steady state processes are investigated. Consequently, the base flow does not contain an influence of mass exchange. It is assumed that a deformed skeleton does not contribute to a dynamical disturbance, and therefore relations for the skeleton are not quoted.

The boundary conditions of this problem have the following form

$$
-\rho^F v_x^F \big|_{x=0} = \alpha \left[p^F \big|_{x=0} - n_E p_l \right],
$$

$$
(13.52)
$$

$$
\rho^F v_x^F \big|_{x=l} = \alpha \left[p^F \big|_{x=l} - n_E p_r \right].
$$

They are of third type and express the fact that the flow through the boundary of the body depends on the pressure difference of the partial pressure in the fluid $\left(p^F \right)$ and the external pressure which works on the fluid (on the left-hand side p_l and on the right-hand side p_r). The permeability of the surface, denoted by α, is a material property of the system.

For simplicity we assume the base flow to be isochoric. This means that in the zeroth step of perturbation the mass density is constant: $\overset{0}{\rho} = \text{const.} = \rho_0^F$. Bearing $(13.51)_1$ for the zeroth step in mind, we obtain that also the fluid velocity in x-direction is constant (denoted by $\overset{0}{v}$ without subscribed x): $\overset{0}{v} = \text{const.}$ The stationary form of $(13.51)_2$ allows us to calculate the solution for the partial pressure in the fluid in the zeroth step, $\overset{0}{p}$,

$$
\frac{\partial \overset{0}{p}}{\partial x} = -\pi \overset{0}{v} \quad \Longrightarrow \quad \overset{0}{p} = -\pi \overset{0}{v} x + C, \tag{13.53}
$$

where C is an integration constant which can be determined by use of the boundary condition on the left-hand side of the system ($x = 0$)

$$-\rho_0^F \overset{0}{v} = \alpha \left[p^F \big|_{x=0} - n_E p_l \right],$$
(13.54)

so that we obtain

$$\overset{0}{p} = -\pi \overset{0}{v} x + n_E p_l - \frac{\rho_0^F \overset{0}{v}}{\alpha}.$$
(13.55)

As a second boundary condition for $x = l$ we have

$$\rho_0^F \overset{0}{v} = \alpha \left[p^F \big|_{x=l} - n_E p_r \right],$$
(13.56)

from which the constant velocity of the zeroth step follows in the form

$$\overset{0}{v} = \frac{\alpha p_d}{2\rho_0^F + \alpha \pi l}, \qquad p_d := n_E (p_l - p_r).$$
(13.57)

The base solution forms the zeroth approximation of the above described perturbation. Obviously, it depends alone on the x-variable. In addition to the above fields, we need in the zeroth step with mass exchange the fields of concentration $\overset{0}{c}$ and of the number of occupied sites $\overset{0}{\xi}$ given by equations (13.49)$_{2/3}$.

In the stationary case they reduce to

– concentration balance

$$\rho_0^F \overset{0}{v} \frac{\partial \overset{0}{c}}{\partial x} = -\left(1 - \overset{0}{c}\right) \frac{\rho_{ad}^A}{\tau_{ad}} \left[\overset{0}{c} \frac{\overset{0}{p}}{p_0} \left(1 - \overset{0}{\xi}\right) - \overset{0}{\xi} \right],$$
(13.58)

– evolution equation for fraction of occupied sites

$$0 = \overset{0}{c} \frac{\overset{0}{p}}{p_0} \left(1 - \overset{0}{\xi}\right) - \overset{0}{\xi}.$$
(13.59)

From (13.59) follows the Langmuir adsorption isotherm which depends, of course, on x

$$\overset{0}{\xi} = \frac{\overset{0}{c} \frac{\overset{0}{p}}{p_0}}{1 + \overset{0}{c} \frac{\overset{0}{p}}{p_0}}.$$
(13.60)

On the other hand, (13.58) yields $\overset{0}{c} = \text{const.} = c_0$, where c_0 denotes the initial concentration of the adsorbate in the fluid.

First step of perturbation

For the first step of perturbation (2D) we obtain the following set of equations

$$(x, z) \in \mathcal{B} := (0, l) \times (-b, b)$$

$$\frac{\partial \overset{1}{\rho}}{\partial t} + \rho_0^F \left(\frac{\partial \overset{1}{v}_x}{\partial x} + \frac{\partial \overset{1}{v}_z}{\partial z} \right) + \overset{0}{v} \frac{\partial \overset{1}{\rho}}{\partial x} = \overset{1}{\hat{\rho}}{}^A,$$

$$\rho_0^F \left(\frac{\partial \overset{1}{c}}{\partial t} + \overset{0}{v} \frac{\partial \overset{1}{c}}{\partial x} \right) = (1 - c_0) \overset{1}{\hat{\rho}}{}^A - \overset{1}{c} \frac{\rho_{ad}^A}{\tau_{ad}} \left[c_0 \frac{\overset{0}{p}}{p_0} \left(1 - \overset{0}{\xi} \right) - \overset{0}{\xi} \right],$$

$$\frac{\partial \overset{1}{\xi}}{\partial t} = \left[\left(1 - \overset{0}{\xi} \right) \left(\frac{c_0 \kappa \overset{1}{\rho}}{p_0} + \overset{1}{c} \frac{\overset{0}{p}}{p_0} \right) - \left(1 + c_0 \frac{\overset{0}{p}}{p_0} \right) \overset{1}{\xi} \right] \frac{1}{\tau_{ad}}, \qquad (13.61)$$

$$\rho_0^F \left[\frac{\partial \overset{1}{v}_x}{\partial t} + \overset{0}{v} \frac{\partial \overset{1}{v}_x}{\partial x} \right] = -\kappa \frac{\partial \overset{1}{\rho}}{\partial x} - \pi \overset{1}{v}_x,$$

$$\rho_0^F \left[\frac{\partial \overset{1}{v}_z}{\partial t} + \overset{0}{v} \frac{\partial \overset{1}{v}_z}{\partial x} \right] = -\kappa \frac{\partial \overset{1}{\rho}}{\partial z} - \pi \overset{1}{v}_z,$$

with

$$\overset{1}{\hat{\rho}}{}^A = -\frac{\rho_{ad}^A}{\tau_{ad}} \left[\left(1 - \overset{0}{\xi} \right) \left(\frac{c_0 \kappa \overset{1}{\rho}}{p_0} + \overset{1}{c} \frac{\overset{0}{p}}{p_0} \right) - \overset{1}{\xi} \left(1 + c_0 \frac{\overset{0}{p}}{p_0} \right) \right]. \qquad (13.62)$$

The following boundary conditions are considered

$$-\rho^F v_x^F \big|_{x=0} = \alpha \left[p^F \big|_{x=0} - n_E p_l \right],$$
$$\rho^F v_x^F \big|_{x=l} = \alpha \left[p^F \big|_{x=l} - n_E p_r \right], \qquad \qquad v_z^F \big|_{z=\pm b} = 0. \qquad (13.63)$$

The two conditions on the left-hand side, corresponding to the momentum balance in x-direction, are already introduced in (13.52). They describe in the above explained way the in- and outflow in x-direction of the liquid to the porous body of length l. But we consider also the flow in z-direction. We assume the fields to be periodic in this direction with period $2b$. Application of a Fourier ansatz yields that the two boundary conditions on the right-hand side of (13.63), corresponding to the momentum balance in z-direction, are automatically fulfilled.

In the first step of perturbation the boundary conditions (13.63) have the form

$$-\rho_0^F \overset{1}{v}_x \Big|_{x=0} - \overset{0}{v} \overset{1}{\rho} \Big|_{x=0} = \alpha \kappa \overset{1}{\rho} \Big|_{x=0},$$
$$\rho_0^F \overset{1}{v}_x \Big|_{x=l} + \overset{0}{v} \overset{1}{\rho} \Big|_{x=l} = \alpha \kappa \overset{1}{\rho} \Big|_{x=l}, \qquad \qquad \overset{1}{v}_z \Big|_{z=\pm b} = 0. \qquad (13.64)$$

13.3.4 Wave ansatz for the disturbance

The perturbations in the first step are expressed in terms of the following simple wave ansatz

$$\overset{1}{\rho} = e^{\omega t}\bar{\rho}\left(x\right)\cos kz, \qquad \overset{1}{v}_x = e^{\omega t}\bar{v}_x\left(x\right)\cos kz, \qquad \overset{1}{v}_z = e^{\omega t}\bar{v}_z\left(x\right)\sin kz,$$

$$(13.65)$$

$$\overset{1}{c} = e^{\omega t}\bar{c}\left(x\right)\cos kz, \quad \overset{1}{\xi} = e^{\omega t}\bar{\xi}\left(x\right)\cos kz,$$

where $\bar{\rho}\left(x\right), \bar{v}_x\left(x\right), \bar{v}_z\left(x\right), \bar{c}\left(x\right)$ and $\bar{\xi}\left(x\right)$ are the amplitudes of the disturbances, ω is the frequency – possibly complex, and k is the real wave number defined as

$$k := \frac{\Pi m}{b}, \qquad \Pi = 3.1415...$$

$$(13.66)$$

The Fourier ansatz in z-direction follows from the structure of the boundary conditions. After insertion of this ansatz into the field equations (13.49) we have

$$\omega\bar{\rho} + \rho_0^F\left(\frac{\partial\bar{v}_x}{\partial x} + k\bar{v}_z\right) + \overset{0}{v}\frac{\partial\bar{\rho}}{\partial x}$$

$$= -\frac{\rho_{ad}^A}{\tau_{ad}}\left[\left(\frac{c_0\kappa\bar{\rho}}{p_0} + \bar{c}\frac{\overset{0}{p}}{p_0}\right)\left(1 - \overset{0}{\xi}\right) - \bar{\xi}\left(1 + c_0\frac{\overset{0}{p}}{p_0}\right)\right],$$

$$\rho_0^F\left(\omega\bar{c} + \overset{0}{v}\frac{\partial\bar{c}}{\partial x}\right) = -\left(1 - c_0\right)\frac{\rho_{ad}^A}{\tau_{ad}}\left[\left(\frac{c_0\kappa\bar{\rho}}{p_0} + \bar{c}\frac{\overset{0}{p}}{p_0}\right)\left(1 - \overset{0}{\xi}\right) - \bar{\xi}\left(1 + c_0\frac{\overset{0}{p}}{p_0}\right)\right]$$

$$- \frac{\rho_{ad}^A}{\tau_{ad}}\bar{c}\left[c_0\frac{\overset{0}{p}}{p_0}\left(1 - \overset{0}{\xi}\right) - \overset{0}{\xi}\right],$$

$$(13.67)$$

$$\omega\bar{\xi} = \left[\left(\bar{\rho}\frac{c_0\kappa}{p_0} + \bar{c}\frac{\overset{0}{p}}{p_0}\right)\left(1 - \overset{0}{\xi}\right) - \bar{\xi}\left(1 + c_0\frac{\overset{0}{p}}{p_0}\right)\right]\frac{1}{\tau_{ad}},$$

$$\rho_0^F\left(\omega\bar{v}_x + \overset{0}{v}\frac{\partial\bar{v}_x}{\partial x}\right) = -\kappa\frac{\partial\bar{\rho}}{\partial x} - \pi\bar{v}_x,$$

$$\rho_0^F\left(\omega\bar{v}_z + \overset{0}{v}\frac{\partial\bar{v}_z}{\partial x}\right) = k\kappa\bar{\rho} - \pi\bar{v}_z.$$

Dispersion relation

The following dispersion relation has to be analyzed

$$\left(\omega\mathsf{I} + \mathfrak{A}\right)\mathbf{u} + \mathfrak{B}\mathbf{u}' = 0,$$

$$(13.68)$$

with

$$\mathbf{u} = \left(\bar{\rho}, \bar{c}, \bar{\xi}, \bar{v}_x, \bar{v}_z\right)^T, \quad \mathbf{u}' = \left(\frac{\partial\bar{\rho}}{\partial x}, \frac{\partial\bar{c}}{\partial x}, \frac{\partial\bar{\xi}}{\partial x}, \frac{\partial\bar{v}_x}{\partial x}, \frac{\partial\bar{v}_z}{\partial x}\right)^T,$$

$$
\mathsf{I} = \begin{pmatrix} 1 & 0 & 0 & 0 & 0 \\ 0 & 1 & 0 & 0 & 0 \\ 0 & 0 & 1 & 0 & 0 \\ 0 & 0 & 0 & 1 & 0 \\ 0 & 0 & 0 & 0 & 1 \end{pmatrix}, \quad
\mathfrak{B} := \begin{pmatrix} \overset{0}{v} & 0 & 0 & \rho_0^F & 0 \\ 0 & \overset{0}{v} & 0 & 0 & 0 \\ 0 & 0 & 0 & 0 & 0 \\ \dfrac{\kappa}{\rho_0^F} & 0 & 0 & \overset{0}{v} & 0 \\ 0 & 0 & 0 & 0 & \overset{0}{v} \end{pmatrix}, \tag{13.69}
$$

$$
\mathfrak{A} := \begin{pmatrix}
\dfrac{\rho_{ad}^A}{\tau_{ad}}\dfrac{c_0\kappa}{p_0}\left(1-\overset{0}{\xi}\right) & \dfrac{\rho_{ad}^A}{\tau_{ad}}\dfrac{\overset{0}{p}}{p_0}\left(1-\overset{0}{\xi}\right) & -\dfrac{\rho_{ad}^A}{\tau_{ad}}\left(1+c_0\dfrac{\overset{0}{p}}{p_0}\right) & 0 & \rho_0^F k \\[2ex]
\dfrac{1-c_0}{\rho_0^F}\dfrac{\rho_{ad}^A}{\tau_{ad}}\dfrac{c_0\kappa}{p_0}\left(1-\overset{0}{\xi}\right) & \dfrac{1-c_0}{\rho_0^F}\dfrac{\rho_{ad}^A}{\tau_{ad}}\dfrac{\overset{0}{p}}{p_0}\left(1-\overset{0}{\xi}\right)+\dfrac{1}{\rho_0^F}\dfrac{\rho_{ad}^A}{\tau_{ad}}\left[c_0\dfrac{\overset{0}{p}}{p_0}\left(1-\overset{0}{\xi}\right)-\overset{0}{\xi}\right] & -\dfrac{1-c_0}{\rho_0^F}\dfrac{\rho_{ad}^A}{\tau_{ad}}\left(1+c_0\dfrac{\overset{0}{p}}{p_0}\right) & 0 & 0 \\[2ex]
-\dfrac{c_0\kappa}{p_0}\left(1-\overset{0}{\xi}\right)\dfrac{1}{\tau_{ad}} & -\dfrac{\overset{0}{p}}{p_0}\left(1-\overset{0}{\xi}\right)\dfrac{1}{\tau_{ad}} & \left(1+c_0\dfrac{\overset{0}{p}}{p_0}\right)\dfrac{1}{\tau_{ad}} & 0 & 0 \\[2ex]
0 & 0 & 0 & \dfrac{\pi}{\rho_0^F} & 0 \\[2ex]
-k\dfrac{\kappa}{\rho_0^F} & 0 & 0 & 0 & \dfrac{\pi}{\rho_0^F}
\end{pmatrix},
$$

and boundary conditions

$$
\bar{v}_x\big|_{x=0} = -\frac{\alpha\kappa+\overset{0}{v}}{\rho_0^F}\,\bar{p}\big|_{x=0}\,,
$$
$$
\bar{v}_x\big|_{x=l} = \frac{\alpha\kappa-\overset{0}{v}}{\rho_0^F}\,\bar{p}\big|_{x=l}\,,
$$
$$
\bar{v}_z\big|_{z=\pm b} = 0. \tag{13.70}
$$

The conditions in z-direction are fulfilled automatically due to ansatz (13.65).

13.3.5   Numerical investigation

We solve the eigenvalue problem for ω, using a second order finite difference scheme in an equidistant mesh (length of the body l divided into n parts of length h). The derivatives of the disturbances are written as central differences $\left(\frac{\partial u}{\partial x}=\frac{u(x+h)-u(x-h)}{2h}\right)$ for inner mesh points or as asymmetric ones for the first $\left(\frac{\partial u}{\partial x}=\frac{u(x+h)-u(x)}{h}\right)$ and the last point $\left(\frac{\partial u}{\partial x}=\frac{u(x)-u(x-h)}{h}\right)$. For this linear eigenvalue problem $5n+3$ eigenvalues ω_i are obtained (number of linear equations: $5(n+1)-2$). The exponential form of the ansatz (13.65) yields that the base flow is stable if all real parts of ω_i are negative and unstable if at least one of the $5n+3$ real parts is positive.

For a chosen pair $\left(\alpha,\rho_{ad}^A\right)$ we calculate $\max(\mathrm{Re}\,(\omega))$ in dependence on π. In order to consider the precision of the numerical calculations only those real parts of the eigenvalues are considered, whose absolute value is greater than 10^{-5}. This value has been determined by comparison of results for the same parameters running once a calculation without mass exchange and once a calculation with mass exchange in which the mass exchange was switched off. Many authors solving eigenvalue problems numerically using a finite difference scheme complain slowly converging procedures (e.g. [134]) from which follows that calculations for a large value of elements are necessary. In our case we do not observe such problems: comparison of analytical and numerical results (without mass exchange) shows that the small number of 10 elements is enough to obtain a good agreement of the results in the scope of the just mentioned precision.

Table 13.1: *Typical model parameters for flow processes in soils.*

Length, width of the body l, b	1 m	Equilibrium porosity n_E	0.23
Compressibility κ	$2.25 \cdot 10^6 \ \frac{\text{m}^2}{\text{s}^2}$	Initial mass density ρ_0^L	$2.3 \cdot 10^2 \ \frac{\text{kg}}{\text{m}^3}$
Pressure left-hand side p_l	110 kPa	Pressure right-hand side p_r	100 kPa
Pressure difference working on the fluid		$p_d = n_E \left(p_l - p_r \right)$	2.3 kPa

Table 13.2: *Additional model parameters for adsorption processes in soils.*

Initial concentration c_0	10^{-3}
Langmuir pressure p_0	10 kPa
Characteristic time of adsorp. τ_{ad}	1 s

In the calculations we use the data of Tables 13.1 and 13.2. The data are typical for geotechnical applications. In Table 13.1 parameters for a flow process without adsorption are given, in Table 13.2 quantities are given which enter the model due to the adsorption process.

In Figure 13.6 the results of the stability analysis are shown. The domain of instability is shown in 3D in dependence on the bulk permeability π, the surface permeability α and the coefficient ρ_{ad}^A, characterizing adsorption properties of the internal surface. The last coefficient may change a few orders of magnitude because it is proportional to the internal surface. Below, we see the 2D projections of the instability region.

The transversal disturbance with mass exchange yields an instability of the steady state flow for certain ranges of the parameters. While for small values of π the border of this region strongly varies with the surface permeability parameter α (see third 2D projection), on the side of large π, the unstable region is limited by a value of π approximately $10^{8.548}$. For bigger values of π, the base flow is stable for any α. In the range of extremely large π, not shown anymore in the figure, there appears an instability again for all values of α. However, the model is not applicable in this range anymore. In addition, the numerical procedure is not stable anymore for those very large values of π.

With decreasing α also the region of π decreases for which the flow is unstable until the value $\alpha \approx 10^{-5.690}$ is reached where the base flow is stable on the whole range of π (see the first 2D projection). A similar behavior exists in dependence on ρ_{ad}^A (see second 2D projection): the range of ρ_{ad}^A for which the flow is unstable decreases with decreasing α. For smaller values than $\alpha \approx 10^{-5.690}$ the flow is stable for any ρ_{ad}^A.

13.3.6 Properties of some other disturbances of the base flow

As in many other known cases of flow instabilities a 1D disturbance either without or with mass exchange does not produce an instability of the base flow under consideration. The same is valid for a 2D disturbance without mass exchange. Results for these disturbances are shown in [9]. Although there appears no region of instability for the three mentioned disturbances, the results show important features of the relaxation behavior. For the 1D and 2D disturbances without mass exchange even an analytical investigation was feasible which gave hints for the numerical approach.

From the stable cases we know that the relaxation properties do not change monotonously with the permeability π. They possess rather two different ranges. In the range of smaller

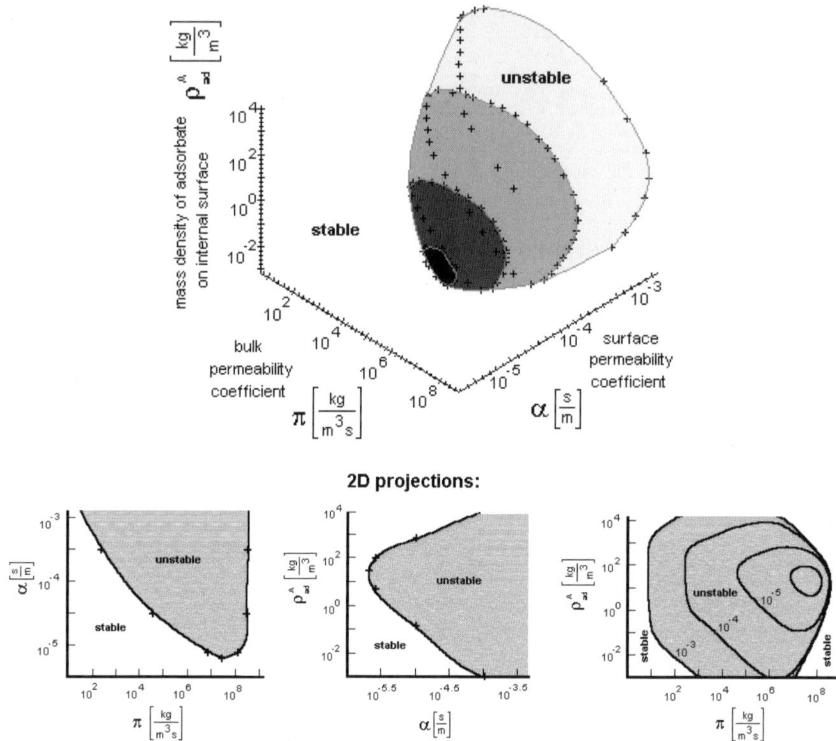

Figure 13.6: *Numerical result for transversal disturbance with mass exchange (from [8]).*

values of π the relaxation is determined by a real part of the complex root, while for larger π this root does not possess an imaginary part. Thus, the perturbation causes vibrations in the range of smaller π whose frequencies cover the whole discrete spectrum. For values of π bigger than this of the turning point, the disturbance is only damped but the damping is smaller than this predicted by the resistance to the diffusion $\left(\text{i.e.,} < \frac{\pi}{\rho_0^F}\right)$. The position of the turning point is determined by the compressibility coefficient of the fluid. We refer to [9] where this property is discussed. In general, the damping is smallest (i.e., the relaxation is slowest) for large and small values of π.

The calculations for different disturbances made evident that the mass exchange slows down the relaxation even a few orders of magnitude. This effect is related to the characteristic time of adsorption τ_{ad}. Little can be said about its experimental values for porous materials because most experiments are conducted under quasistatic conditions. In [9] the influence of τ_{ad} on the relaxation properties is shown. As expected, the disturbances relax faster for smaller characteristic times τ_{ad}, but even for a very short time of adsorption $\tau_{ad} = 10^{-5}$s, this relaxation is considerably longer than entirely without mass exchange. For the present calculations a value of $\tau_{ad} = 1$ s has been used.

The results for the 2D disturbance without mass exchange reveal the same behavior as these for the 1D perturbation. Admittedly, there appear additional solutions but they do neither have influence on the stability behavior nor on the relaxation properties. In general, the relaxation in the 2D case without mass exchange is faster than in the 1D case. All the more it is interesting that the 2D disturbance with mass exchange yields an instability while the 1D disturbance with adsorption does not.

13.3.7 Some remarks

In this section we have shown that for transversal disturbances with mass exchange of a 1D steady state flow through a porous material there appears a region of parameters π, α and ρ_{ad}^A in which the base flow is unstable. This is not the case with respect to a linear longitudinal disturbance without and with mass exchange and to a linear transversal disturbance without mass exchange. For the three latter disturbances the base flow is stable in the whole range of control permeability parameters π and α.

This shows that a disturbance with adsorption has a deep influence on the stability behavior of the base flow. While adsorption already in the case of 1D disturbances decreases the maximum values of real parts about orders of magnitude, for the 2D disturbances it even decides whether the base flow is stable or unstable.

The existing unstable region for the 2D transversal disturbance appears in a region of π where in the case of the other disturbances (1D without and with mass exchange, 2D without mass exchange) appear complex roots and thus vibrations.

The range of instability corresponds to higher velocities for the same pressure difference due to a lower resistance of the boundary. Simultaneously it corresponds to a lower "internal friction" πv_x^F in the momentum balance equation due to lower values of the permeability π.

13.4 Wave propagation in porous media with anisotropic permeability

The influence of anisotropic tortuosity on the propagation of acoustic waves in porous materials has, up to now, not been investigated, while the formulation of the fully isotropic model for both saturated and partially saturated porous materials and the properties of acoustic waves within such a model have a very extensive literature. For instance, these topics are presented in the book by B. Albers [13]. Following up, here we investigate an extended model in which the stress-strain relations are isotropic but the tortuosity is not. The latter sentence implies that in a porous medium two types of anisotropy are conceivable. We present them in the following subsection. A model including anisotropic permeability was recently proposed by K. Wilmanski [448] who investigated its properties for the shear mode which propagates in one of the principal directions of the tortuosity. In [20] B. Albers and K. Wilmanski discussed all modes of acoustic waves without making a simplifying assumption on the direction of the propagation. Results of this investigation are repeated in this section.

13.4.1 Anisotropy of porous materials

Due to the existence of a solid and fluid components, anisotropy of multi-component media can be described in two ways. While in some types, as for instance, composites, the anisotropy of the solid is the decisive factor, for porous media where diffusion takes place, it is the permeability. In the anisotropic case the relative motion of the components in a fully saturated porous material is described by a symmetric tensor of permeability defined by a permeability coefficient and the inverse of a tortuosity tensor introduced by J. Bear and Y. Bachmat [35]. This particular case seems to have an important bearing in description of rocks and some granular materials in which deformations of the anisotropic skeleton are less important than anisotropic diffusion properties. Otherwise, the anisotropy is reflected by more general stress-strain relations.

13.4.1.1　Anisotropy of the stress-strain relations

Linear isotropic elastic materials are described by two material parameters. For such materials the most general form of a stress-strain relation between the Cauchy stress tensor $\mathbf{T} = \sigma_{ij}\mathbf{e}_i \otimes \mathbf{e}_j$ and the strain field $\mathbf{e} = e_{ij}\mathbf{e}_i \otimes \mathbf{e}_j$,

$$\sigma_{kl} = c_{klij}e_{ij}, \tag{13.71}$$

can be written in the form

$$\mathbf{T} = \lambda\,(\mathrm{tr}\,\mathbf{e})\,\mathbf{1} + 2\mu\mathbf{e}, \quad \text{i.e.,} \quad \sigma_{ij} = \lambda e_{kk}\delta_{ij} + 2\mu e_{ij},$$

$$\tag{13.72}$$

$$\text{i.e.,} \quad c_{klij} = \lambda\delta_{ij}\delta_{kl} + \mu\left(\delta_{ik}\delta_{jl} + \delta_{il}\delta_{jk}\right),$$

where λ and μ are the Lamé constants – the two above mentioned material parameters. However, often we are concerned with anisotropic materials which do not possess any particular symmetry properties. This means that samples, cut from such a material, in different directions yield different response to the same external loading. Such materials are, for example, wood, composites (see e.g. [250]), polycrystals or many rock structures. In the general form (13.71) of the stress-strain relation which is not constrained by symmetry and isotropy assumptions c_{ijkl} has 81 ($= 3^4$) coefficients. However, one has to take into account that the Cauchy stresses and the strain field are symmetric, i.e., $\sigma_{ij} = \sigma_{ji}$ and $e_{ij} = e_{ji}$. If this has taken into consideration, the number of independent constants will be reduced to 21. The number can be further decreased if particular cases of material symmetry are considered, for example, monotropy (13 constants), orthotropy (9 constants) or transverse isotropy (5 constants).

S. C. Cowin has presented an elegant modification of the classical Biot model for porous materials which possess anisotropic properties of the stress-strain relations (e.g. [96], [97]). By means of the fabric tensor which describes the microscopic geometry he has extended the Gassmann relations for the coefficients of elasticity. However, his tensor of permeability is assumed to be an isotropic function of the fabric tensor which yields an isotropic dependence on the tortuosity in linear models. This has been indicated by S. C. Cowin and L. Cardoso in their work on acoustic waves in anisotropic porous materials, especially in cancellous bone tissue [98].

13.4.1.2　Anisotropy of the permeability

The most characteristic feature of a porous permeable material is the diffusion, i.e., the relative motion of fluid components with respect to the skeleton. The intensity of the diffusion depends on the microstructure of the material of the skeleton and, in particular, on the shape of the channels and their volume contribution to the total volume of the material (i.e., the porosity). The shape of the channels is described by such parameters as their curvature and the internal surface. They contribute to the macroscopic parameter, called the permeability, which characterizes the intensity of diffusion. A detailed description of the random geometry of the microstructure of porous materials is the subject of two chapters of the book of J. Bear and Y. Bachmat [35]. Among other notions, they introduce a tensor of tortuosity which describes anisotropic properties of the permeability. This notion, in its simplest isotropic form, was introduced at the beginning of XX Century in the description of seepage processes in soils. In formula (20) of [206], which corresponds to the classical Darcy law, J. Kozeny describes the permeability k as inversely proportional to the tortuosity τ, i.e., to the fraction of the length of a streamline between two points to the distance of these two points. However, it turned out that this relation was not correct and, thus, it has been corrected in 1956 by P. C. Carman [65] (formula (1.45); see also N. Epstein [113]) as it should be proportional to τ^2 rather than to τ. It is commonly called the Blake-Kozeny relation (see page 322).

This relation does not yet give evidence on the isotropy of the channels. However, the curvature of the channels may lead to anisotropy. The importance of the anisotropy of the permeability was recognized in a theoretical work of J. Ferrandon in 1948 [124] and confirmed by experimental research of oil-bearing strata carried by W. E. Johnson et al. between 1948 and 1951 (e.g. [185] or [184]). A full theoretical model of anisotropic permeability with experimental identification of coefficients for rocks was presented by A. Scheidegger in 1954 [337]. This subject appeared as well in the investigation of biological tissues (see Section 12.4). Experiments on such materials use nuclear magnetic resonance (NMR) imaging techniques which, in turn, are based on anisotropic diffusion properties contributing to the Bloch-Torrey equation of magnetization (see [447]). It has been shown that these anisotropic properties indicate important conclusions for medical diagnosis.

Permeability tensor of the continuum model

In the continuum model the anisotropy of the material is reflected by a permeability tensor π_{ij}. Its structure is assumed to be as follows

$$\pi_{ij} = \pi_0 T_{ij}^{-1}, \tag{13.73}$$

where

$$\pi_0 = \frac{\mu_v b}{D_h^2} \frac{g \rho_0^F}{n_0}, \tag{13.74}$$

is the previously used scalar permeability or rather resistance parameter (In the continuum model not the Darcy permeability itself but its inverse appears). In (13.74) D_h is the hydraulic diameter, n_0 is the initial porosity, μ_v is the true dynamic viscosity of the fluid in the pores, b is the capillary shape factor (e.g. 32 for circular pores and 48 for parallel slits) and g is the earth gravity (see e.g. [447]). The respective tortuosity tensor $\mathbf{T} = T_{ij}\mathbf{e}_i \otimes \mathbf{e}_j$ is symmetric and various details concerning its origin and structure can be found in the book [35]. If we choose the direction $\mathbf{e}_3$ as the principal direction $\mathbf{t}_3$ of the tortuosity tensor, the latter can be written in the following spectral form

$$\mathbf{T} = T_{ij}\mathbf{e}_i \otimes \mathbf{e}_j = \sum_{\mu=1}^{3} \frac{1}{\tau^{(\mu)^2}} \mathbf{t}_\mu \otimes \mathbf{t}_\mu, \tag{13.75}$$

where

$$
\begin{aligned}
\mathbf{e}_3 &= \mathbf{t}_3, \\
\mathbf{e}_2 \cdot \mathbf{t}_1 &= \sin \alpha, \\
\mathbf{e}_1 \cdot \mathbf{t}_1 &= \mathbf{e}_2 \cdot \mathbf{t}_2 = \cos \alpha, \\
\mathbf{e}_1 \cdot \mathbf{t}_2 &= -\sin \alpha.
\end{aligned}
\tag{13.76}
$$

$\{\mathbf{t}_1, \mathbf{t}_2, \mathbf{t}_3\}$ are unit eigenvectors of the tensor $\mathbf{T}$ and $\{\tau^{(1)}, \tau^{(2)}, \tau^{(3)}\}$ are the square roots of the inverse eigenvalues of the tortuosity tensor. They are called principal tortuosities and, according to Bear and Bachmat [35], they measure the average inverse of the cosines of the angles between a short straight interval in a chosen principal direction and a streamline between the endpoints of this interval. Obviously, in the isotropic case they are all equal to τ appearing in the classical Blake-Kozeny relation.

13.4.2 Governing equations accounting for anisotropic permeability

The governing equations in Eulerian form are given in Table 12.3. The linearized partial momentum balance equations for a two-component porous medium are

$$
\rho^S \frac{\partial \mathbf{v}^S}{\partial t} = \operatorname{div} \mathbf{T}^S + \hat{\mathbf{p}}, \quad \rho^F \frac{\partial \mathbf{v}^F}{\partial t} = -\operatorname{grad} p^F - \hat{\mathbf{p}},
\tag{13.77}
$$

where ρ^S and ρ^F are the initial constant partial mass densities of solid and fluid, respectively, $\mathbf{v}^S$ and $\mathbf{v}^F$ are the partial velocities of these components, $\mathbf{T}^S$ is the partial stress tensor in the skeleton and p^F is the partial pressure in the fluid. $\hat{\mathbf{p}}$ denotes the momentum source.

The constitutive relations for these quantities are similar to these of the Biot model (see Section 13.3 of Part I). The difference of the classical constitutive relations and those accounting for anisotropic permeability appears in the relation for the momentum source. Namely

$$
\mathbf{T}^S = \mathbf{T}_0^S + \lambda^S e \mathbf{1} + 2\mu^S \mathbf{e}^S + Q\varepsilon \mathbf{1}, \quad p^F = p_0^F - Qe - \rho^F \kappa \varepsilon,
\tag{13.78}
$$

$$
\hat{\mathbf{p}} = \hat{p}_i \mathbf{e}_i, \quad \hat{p}_i = \pi_{ij}\left(v_j^F - v_j^S\right), \quad \mathbf{v}^F = v_i^F \mathbf{e}_i, \quad \mathbf{v}^S = v_i^S \mathbf{e}_i,
$$

where $\mathbf{e}_i$ are Cartesian basis vectors. Their scalar product is $\mathbf{e}_i \cdot \mathbf{e}_j = \delta_{ij} = \begin{cases} 0 & \text{if } i \neq j \\ 1 & \text{if } i = j \end{cases}$,

where δ_{ij} is the Kronecker delta. $\mathbf{e}^S$ is the Almansi-Hamel small deformation tensor of the skeleton (for an introduction of different measures of deformation see Subsection 2.1.3), $e = \operatorname{tr} \mathbf{e}^S$ and ε are the volume changes of the solid and of the fluid, respectively. They satisfy the compatibility equations

$$
\frac{\partial \mathbf{e}^S}{\partial t} = \operatorname{sym} \operatorname{grad} \mathbf{v}^S, \quad \frac{\partial \varepsilon}{\partial t} = \operatorname{div} \mathbf{v}^F,
\tag{13.79}
$$

which, in turn, yield identically partial mass conservation laws for the two components.

The macroscopic Lamé constants of the skeleton are denoted by λ^S and μ^S, the compressibility of the fluid by κ and the stress coupling parameter by Q. The latter has been introduced by M. A. Biot (see Section 13.3 of Part I). $\mathbf{T}_0^S$ and p_0^F are the initial values of the partial stress tensor in the skeleton and of the partial pressure in the fluid, respectively. The permeability (or resistance) tensor π_{ij} is given by (13.73).

In contrast to Biot's model in (13.79), the influence of inertial forces attributed to added mass effects, i.e., relative accelerations, are neglected. While in some papers on acoustics of porous media these forces are related to tortuosity effects we feel confirmed that such effects are of higher order of magnitude (e.g. compare [446]).

Changes of the porosity can be easily found a posteriori. They are described by the volume changes of the two components. Appropriate material parameters, describing these changes, are known. However, for the acoustic problems considered in this section they are immaterial (compare Part I).

Further, the material parameters of the model

$$\left\{ \rho^S, \rho^F, \lambda^S, \mu^S, \kappa, Q, \pi_0, \tau^{(1)}, \tau^{(2)}, \tau^{(3)} \right\}, \tag{13.80}$$

are assumed to be given and constant. The permeability (resistance) parameter π_0 can be calculated by use of (13.74).

13.4.3 Monochromatic waves

We consider the propagation of acoustic waves in a two-component poroelastic material. For wave analyses of different models for isotropic poroelastic media (Biot's model, Full Model with the porosity gradient, Simple Model with Biot's coupling parameter $Q = 0$) see either Section 13.4 of Part I or [13]). For the analysis incorporating anisotropic permeability we choose a Cartesian reference system in which Relation (13.76) between the basis vectors and the principal directions of the tortuosity tensor is satisfied. The propagation of monochromatic waves of a given frequency ω is investigated, i.e., we search for the solutions of the governing equations in the following form

$$\mathbf{v}^S = \mathbf{V}^S \mathcal{E}, \quad \mathbf{v}^F = \mathbf{V}^F \mathcal{E}, \quad \mathbf{e}^S = \mathbf{E}^S \mathcal{E}, \quad \varepsilon = E^F \mathcal{E}$$

$$\mathcal{E} = e^{i(\mathbf{k}\cdot\mathbf{x}-\omega t)} \equiv e^{-((\operatorname{Im} k)\mathbf{n}\cdot\mathbf{x})} e^{i\operatorname{Re} k\left(\mathbf{n}\cdot\mathbf{x}-c_{ph}t\right)}, \tag{13.81}$$

$$\mathbf{k} = k\mathbf{n}, \quad \mathbf{n} = \mathbf{e}_1, \quad \mathbf{n}\cdot\mathbf{n} = 1, \quad c_{ph} = \frac{\omega}{\operatorname{Re} k}.$$

In these relations $\mathbf{V}^S, \mathbf{V}^F, \mathbf{E}^S$ and E^F denote constant amplitudes, $\mathbf{k}$ is the wave vector, k is the wave number, $\mathbf{n}$ denotes the propagation direction and c_{ph} is the propagation speed of the monochromatic wave of frequency ω.

Substitution of (13.81) in the compatibility equations (13.79) yields the following auxiliary relations

$$\mathbf{E}^S = -\frac{1}{2\omega}\left(\mathbf{V}^S \otimes \mathbf{k} + \mathbf{k} \otimes \mathbf{V}^S\right), \quad E^F = -\frac{1}{\omega}\mathbf{V}^F \cdot \mathbf{k}. \tag{13.82}$$

By means of the momentum balance equations (13.77) and the constitutive relations (13.78) and – for the present problem of highest interest – (13.73), the following set of algebraic relations is obtained

$$\left(-\rho^S\omega^2\mathbf{1}+\lambda^S\mathbf{k}\otimes\mathbf{k}+\mu^S\left(k^2\mathbf{1}+\mathbf{k}\otimes\mathbf{k}\right)-i\pi_0\omega\mathbf{T}^{-1}\right)\mathbf{V}^S$$
$$+\left(Q\mathbf{k}\otimes\mathbf{k}+i\pi_0\omega\mathbf{T}^{-1}\right)\mathbf{V}^F=0,$$

$$\left(Q\mathbf{k}\otimes\mathbf{k}+i\pi_0\omega\mathbf{T}^{-1}\right)\mathbf{V}^S$$
$$+\left(-\rho^F\omega^2\mathbf{1}+\rho^F\kappa\mathbf{k}\otimes\mathbf{k}-i\pi_0\omega\mathbf{T}^{-1}\right)\mathbf{V}^F=0. \tag{13.83}$$

Obviously, this is an eigenvalue problem for the amplitudes $\mathbf{V}^S$ and $\mathbf{V}^F$ of the waves. Since $\mathbf{n}=\mathbf{e}_1$ this means

$$-\rho^S\omega^2\mathbf{V}^S+\lambda^S k^2\mathbf{e}_1 V_1^S+\mu^S\left(k^2\mathbf{V}^S+k^2\mathbf{e}_1 V_1^S\right)-i\pi_0\omega\mathbf{T}^{-1}\mathbf{V}^S$$
$$+Qk^2\mathbf{e}_1 V_1^F+i\pi_0\omega\mathbf{T}^{-1}\mathbf{V}^F=0,\quad \mathbf{V}^S=V_i^S\mathbf{e}_i,\quad \mathbf{V}^F=V_i^F\mathbf{e}_i,$$

$$Qk^2\mathbf{e}_1 V_1^S+i\pi_0\omega\mathbf{T}^{-1}\mathbf{V}^S$$
$$-\rho^F\omega^2\mathbf{V}^F+\rho^F\kappa k^2\mathbf{e}_1 V_1^F-i\pi_0\omega\mathbf{T}^{-1}\mathbf{V}^F=0. \tag{13.84}$$

As mentioned above, this set was investigated in the work [448] in the special case when the propagation direction $\mathbf{k}$ coincides with the direction $\mathbf{e}_1$ and, simultaneously, with one of the principal directions of the tortuosity tensor, i.e., $\alpha=0$. Under this assumption the properties of shear (transversal) waves were examined. Here, we present the results of [20] and do not assume that the direction of propagation coincides with the principal direction of the tensor $\mathbf{T}$. However, in order to limit tedious calculations, we still assume that one of the principal directions of $\mathbf{T}$ is perpendicular to $\mathbf{k}$: i.e., $\mathbf{k}=k\mathbf{n}=k\mathbf{e}_1$ and $\mathbf{t}_3=\mathbf{e}_3$. Then, as we see in the next subsection, one of the modes of propagation is indeed purely transversal.

13.4.4   Decoupled transversal wave

Multiplication of Equations $(13.83)_1$ and $(13.83)_2$ with $\mathbf{e}_3$ ($\mathbf{e}_3\cdot\mathbf{k}=\mathbf{e}_3\cdot\mathbf{t}_1=\mathbf{e}_3\cdot\mathbf{t}_2=0$) yields two equations for the two components V_3^S and V_3^F

$$-\rho^S\omega^2 V_3^S+\mu^S k^2 V_3^S-i\pi_0\omega\tau^{(3)^2}V_3^S+i\pi_0\omega\tau^{(3)^2}V_3^F=0,$$

$$i\pi_0\omega\tau^{(3)^2}V_3^S-\rho^F\omega^2 V_3^F-i\pi_0\omega\tau^{(3)^2}V_3^F=0, \tag{13.85}$$

where the inverse of the spectral representation (13.75) of the tortuosity tensor has been used. The following dispersion relation can be found

$$\left(\omega^2-\frac{\mu^S}{\rho^S}k^2+\frac{i\pi_0\omega}{\rho^S}\tau^{(3)^2}\right)\left(\omega^2+\frac{i\pi_0\omega}{\rho^F}\tau^{(3)^2}\right)$$

$$+\frac{\rho^F}{\rho^S}\left(\frac{\pi_0\omega}{\rho^F}\tau^{(3)^2}\right)^2=0,\qquad V_3^S\neq 0,\quad V_3^F\neq 0. \tag{13.86}$$

As the remaining components of the vectors $\mathbf{V}^S$ and $\mathbf{V}^F$ are zero, this is clearly a transversal wave, i.e., the amplitude is perpendicular to the propagation direction. The existence of such a purely transversal $S1$-wave follows from the assumption that the eigenvector of

Table 13.3: Material parameters of Alermoehe sandstone.

K_s	K_f	ν	n_0	ρ^{SR}	ρ^S	ρ^F
48 GPa	2.25 GPa	0.2	0.3	$2500\ \frac{\text{kg}}{\text{m}^3}$	$1750\frac{\text{kg}}{\text{m}^3}$	$300\frac{\text{kg}}{\text{m}^3}$

λ^S	μ^S	κ	Q	K_d	K_w	π_0
1.5 GPa	2.25 GPa	$2.05 \cdot 10^6\ \frac{\text{m}^2}{\text{s}^2}$	1.304 GPa	3 GPa	6.76 GPa	$10^{10}\ \frac{\text{kg}}{\text{m}^3\text{s}}$

K_s, K_f – true compressibility moduli of solid and fluid, ν – Poisson number, n_0 – initial porosity, ρ^{SR}, ρ^S – true and partial mass densities of the solid, ρ^F – partial mass density of the fluid, λ^S, μ^S – Lamé parameters, κ – compressibility, Q – coupling parameter, $K_d = \frac{K_s}{1+50n_0}$, $K_w = \left[\frac{1-n_0}{K_s} + \frac{n_0}{K_f}\right]^{-1}$, π_0 – permeability (resistance) coefficient.

the tortuosity tensor $\mathbf{t}_3$ is orthogonal to the propagation direction $\mathbf{n}$. This is the same situation which has been considered in the paper [448]. For this reason we call this $S1$-wave decoupled. For $\tau^{(3)} = 1$ the result (13.86) is identical with the result for isotropic permeability [17]. The propagation speeds of the $S1$-wave in the limits of small and high frequencies are independent of the value of the tortuosity. As for isotropic media the low and high frequency limits are given by

$$c_0 = \lim_{\omega \to 0} \frac{\omega}{\operatorname{Re} k} = \sqrt{\frac{\mu^S}{\rho^S + \rho^F}}, \quad c_\infty = \lim_{\omega \to \infty} \frac{\omega}{\operatorname{Re} k} = \sqrt{\frac{\mu^S}{\rho^S}}, \tag{13.87}$$

respectively (compare e.g. to the well-known results of the wave analysis for soils [13]).

In the following a numerical example will be shown. It is based on data of J. Arnold [25] for Alermoehe sandstone saturated by water. The material parameters, in the notation used in this book, are given in Table 13.3. Arnold conducted experiments by use of a mobile NMR (nuclear magnetic resonance) device. She reported on tortuosities in the interval between 1.06 and 6.50. However, there is no direct information on anisotropy of the structure. This may be only suspected because the samples were extracted from a very similar depth, reported values of the porosity are similar and variations in the permeability seem to be related to variations of the tortuosity. Similar conclusions follow from the work [417] in which the values of the tortuosity for various rocks vary between 3.45 and 7.69.

True and partial mass densities of the skeleton are connected by the relation $\rho^S = (1 - n_0)\,\rho^{SR}$. Then, the Gassmann and Geertsma relations (see e.g. pages 316/317 of Part I or [17]) yield the shear modulus

$$\mu^S = \frac{3}{2}\frac{1 - 2\nu}{1 + \nu}\frac{K_s}{1 + 50n_0}. \tag{13.88}$$

With the material parameters of Table 13.3 the following limit values for high and low frequencies of the speed of propagation result

$$c_0 = 1048\ [\text{m/s}], \quad c_\infty = 1134\ [\text{m/s}].$$

The dispersion relation (13.86) can be easily solved with respect to the wave number. We obtain

$$k^2 = \frac{\omega_0^2}{c_\infty^2}\frac{(\omega/\omega_0)^2}{(\omega/\omega_0)^2 + \tau^{(3)4}}\left[(\omega/\omega_0)^2 + ir\tau^{(3)2}(\omega/\omega_0) + (1 + r)\,\tau^{(3)4}\right], \tag{13.89}$$

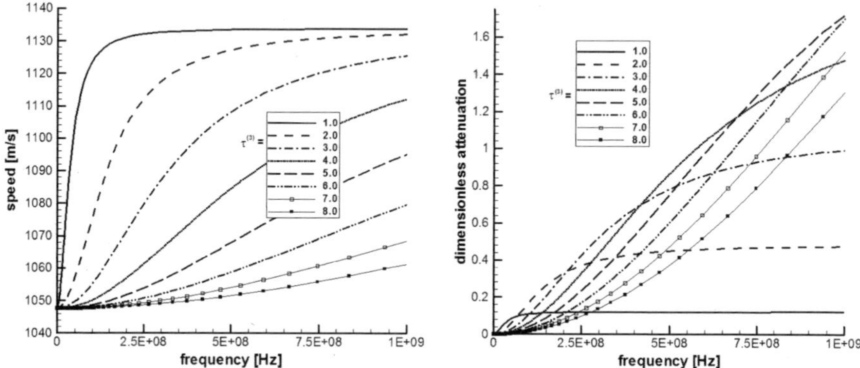

Figure 13.7: *Phase speed c_{ph} [m/s] (left panel) and dimensionless attenuation (right panel) of the decoupled $S1$-wave in function of the frequency for tortuosities $\tau^{(3)} = 1.0$ to $\tau^{(3)} = 8.0$ (from [20]).*

where

$$\omega_0 = \frac{\pi_0}{\rho^F} = 33.3 \cdot 10^6 \ [1/s], \quad c_\infty^2 = \frac{\mu^S}{\rho^S}, \quad r = \frac{\rho^F}{\rho^S}. \tag{13.90}$$

Relation (13.89) yields immediately the phase speed c_{ph} and the attenuation $\operatorname{Im} k$ of the $S1$-wave

$$c_{ph} = \frac{c_\infty \sqrt{2}}{\sqrt{A + \sqrt{A^2 + B^2}}}, \quad \operatorname{Im} k = \frac{\omega_0}{c_\infty \sqrt{2}} \frac{(\omega/\omega_0)\, B}{\sqrt{A + \sqrt{A^2 + B^2}}}, \tag{13.91}$$

where the following abbreviations have been used

$$A = \frac{1}{\left(\omega/\omega_0\right)^2 + \tau^{(3)4}} \left[\left(\omega/\omega_0\right)^2 + (1 + r)\, \tau^{(3)4} \right], \qquad B = \frac{r\tau^{(3)^2} \left(\omega/\omega_0\right)}{\left(\omega/\omega_0\right)^2 + \tau^{(3)4}}. \tag{13.92}$$

Hence, in the limits of low and high frequencies also the limits of the attenuation can be specified

$$\lim_{\omega \to 0} \operatorname{Im} k = 0, \quad \lim_{\omega \to \infty} \operatorname{Im} k = \frac{r\omega_0 \tau^{(3)^2}}{2c_\infty} = \frac{\pi_0 \tau^{(3)^2}}{2\sqrt{\mu^S \rho^S}}. \tag{13.93}$$

The attenuation depends even stronger on the tortuosity than the speed. It grows in square of the tortuosity which indicates the importance of this parameter in the description of waves. In order to point this up more clearly, the attenuation is normalized in the following way

$$\operatorname{Im} k \to \frac{c_\infty \sqrt{2}}{\omega_0} \operatorname{Im} k, \quad \frac{c_\infty \sqrt{2}}{\omega_0} = 4.816 \cdot 10^{-5} \ [m]. \tag{13.94}$$

In Figure 13.7 both the speed and the normalized attenuation of the decoupled $S1$-wave are shown for the material parameters given in Table 13.3.

Indeed, the range of frequencies shown in Figure 13.7 is very large and it is known that such high frequencies do not appear in geophysics. However, as the log- or, respectively, the loglog-presentations of Figure 13.8 show, for small frequencies the speed of propagation remains constant and the high frequency limit is reached only for very high frequencies. The same holds for the attenuation, as can be seen in the right panel of Figure 13.8.

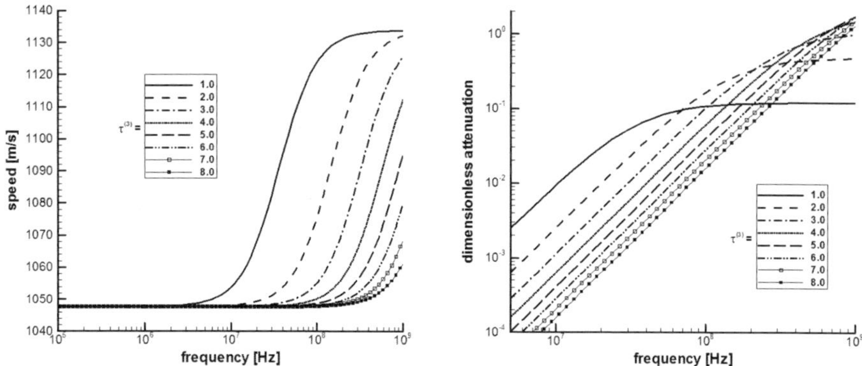

Figure 13.8: *Logarithmic and, respectively, double logarithmic presentation of the phase speed c_{ph} [m/s] (left panel) and the dimensionless attenuation (right panel) of the decoupled S1-wave in function of the frequency for tortuosities $\tau^{(3)} = 1.0$ to $\tau^{(3)} = 8.0$ (from [20]).*

We show also the plots for the remaining waves for a very large range of frequencies. This happens in order to indicate the asymptotic properties of the waves on the fronts of an arbitrary acoustic disturbance which, obviously, propagates with the speed corresponding to the limit of infinite frequencies. Nevertheless, the properties of low frequency monochromatic waves will also be indicated as these are important in practical applications. For the shear S1-waves described in this section it is important to notice that in range of low frequencies they are almost nondispersive, i.e., for a given tortuosity they have nearly the same speed of propagation.

13.4.5 Coupled waves

Multiplication of Equations $(13.83)_1$ and $(13.83)_2$ with $\mathbf{e}_1$ and $\mathbf{e}_2$ yields the following four equations for the four unknown components V_1^S, V_2^S, V_1^F and V_2^F

$$\mathbf{AX} = \mathbf{0} \qquad \text{with} \qquad \mathbf{X} := \left(V_1^S, V_2^S, V_1^F, V_2^F \right)^T. \tag{13.95}$$

The components of the matrix $\mathbf{A}$ are given in the footnote.[2]

This homogeneous set yields the dispersion relation $\det(\mathbf{A}) = 0$ determining the relation between ω and k. For a given real frequency ω three complex solutions for k are

---

[2]The components of matrix $\mathbf{A}$ have the following form

$$A_{11} = -\omega^2 + \frac{\lambda^S + 2\mu^S}{\rho^S} k^2 - \frac{i\pi_0\omega}{\rho^S}\left(\tau^{(1)^2}\cos^2\alpha + \tau^{(2)^2}\sin^2\alpha\right), \quad A_{12} = -\frac{i\pi_0\omega}{\rho^S}\left(\tau^{(1)^2} - \tau^{(2)^2}\right)\sin\alpha\cos\alpha,$$

$$A_{13} = \frac{Q}{\rho^S}k^2 - \frac{i\pi_0\omega}{\rho^S}\left(\tau^{(1)^2}\cos^2\alpha + \tau^{(2)^2}\sin^2\alpha\right), \quad A_{14} = \frac{i\pi_0\omega}{\rho^S}\left(\tau^{(1)^2} - \tau^{(2)^2}\right)\sin\alpha\cos\alpha,$$

$$A_{21} = -\frac{i\pi_0\omega}{\rho^S}\left(\tau^{(1)^2} - \tau^{(2)^2}\right)\sin\alpha\cos\alpha, \quad A_{22} = -\omega^2 + \frac{\mu^S}{\rho^S}k^2 - \frac{i\pi_0\omega}{\rho^S}\left(\tau^{(1)^2}\sin^2\alpha + \tau^{(2)^2}\cos^2\alpha\right),$$

$$A_{23} = \frac{i\pi_0\omega}{\rho^S}\left(\tau^{(1)^2} - \tau^{(2)^2}\right)\sin\alpha\cos\alpha, \quad A_{24} = \frac{i\pi_0\omega}{\rho^S}\left(\tau^{(1)^2}\sin^2\alpha + \tau^{(2)^2}\cos^2\alpha\right),$$

$$A_{31} = \frac{Q}{\rho^S}k^2 + \frac{i\pi_0\omega}{\rho^S}\left(\tau^{(1)^2}\cos^2\alpha + \tau^{(2)^2}\sin^2\alpha\right), \quad A_{32} = \frac{i\pi_0\omega}{\rho^S}\left(\tau^{(1)^2} - \tau^{(2)^2}\right)\sin\alpha\cos\alpha,$$

$$A_{33} = r\left(-\omega^2 + \kappa k^2\right) - \frac{i\pi_0\omega}{\rho^S}\left(\tau^{(1)^2}\cos^2\alpha + \tau^{(2)^2}\sin^2\alpha\right), \quad A_{34} = -\frac{i\pi_0\omega}{\rho^S}\left(\tau^{(1)^2} - \tau^{(2)^2}\right)\sin\alpha\cos\alpha,$$

$$A_{41} = \frac{i\pi_0\omega}{\rho^S}\left(\tau^{(1)^2} - \tau^{(2)^2}\right)\sin\alpha\cos\alpha, \quad A_{42} = \frac{i\pi_0\omega}{\rho^S}\left(\tau^{(1)^2}\sin^2\alpha + \tau^{(2)^2}\cos^2\alpha\right),$$

$$A_{43} = -\frac{i\pi_0\omega}{\rho^S}\left(\tau^{(1)^2} - \tau^{(2)^2}\right)\sin\alpha\cos\alpha, \quad A_{44} = -r\omega^2 - \frac{i\pi_0\omega}{\rho^S}\left(\tau^{(1)^2}\sin^2\alpha + \tau^{(2)^2}\cos^2\alpha\right).$$

obtained. From these follow the phase speeds $c_{ph} = \frac{\omega}{\operatorname{Re} k}$ and the attenuations $\operatorname{Im} k$. Obviously, the solutions are dependent on the angle α which reflects the anisotropy of the tortuosity (see the sketch next to (13.76)).

For the limit values $\alpha = 0$ and $\alpha = \pi/2$ the situation is the same as considered in the work [448]. For these two values the modes of propagation are classical: two longitudinal waves $P1$ and $P2$ and one transversal (shear) mode $S2$ propagate in the medium. Their amplitudes are parallel to the propagation direction for longitudinal waves and perpendicular to this direction for transversal waves. The transversal mode is identical with the decoupled wave considered in the previous section. For the angle α different from the limit values $\alpha = 0$ and $\alpha = \pi/2$ the modes of propagation are neither longitudinal nor transversal. Furtheron we consider this problem in some details.

The properties of the solutions of the set (13.95) will be demonstrated for the data given in Table 13.3 which can be obtained from the basic quantities by use of the following relations

$$\lambda^S = \frac{2\mu^S \nu}{1 - 2\nu}, \qquad \kappa = \frac{n_0^2 \left[K_s^2 / \left(K_s^2 / K_w - K_d \right) \right]}{\rho^F},$$

$$Q = n_0 \frac{K_s(K_s - K_d)}{K_s^2 / K_w - K_d} - \rho^F \kappa, \qquad \rho^F = n_0 \rho^{FR}, \qquad (13.96)$$

$$K_d = \frac{K_s}{1 + 50 n_0}, \qquad K_w = \left[\frac{1 - n_0}{K_s} + \frac{n_0}{K_f} \right]^{-1}.$$

The first three are, again, the Gassmann relations [367] (compare Subsection 13.2.2.1). The expression for the compressibility modulus of the empty matrix (drained compressibility modulus) K_d, has been proposed in this form by Geertsma (see [425]). K_f and K_s are the true compressibilities of the fluid and the solid. K_w follows from Wood's formula.[3] For three different choices of pairs of principal tortuosities the phase speeds and dimensionless attenuations are shown in Figure 13.9. These are: in the left column the pair with the biggest difference between the two values of the principal tortuosities: $\tau^{(1)} = 6.0$ and $\tau^{(2)} = 1.5$, in the middle a medium difference between the tortuosities $\tau^{(1)} = 2.0$ and $\tau^{(2)} = 4.5$ and on the right the nearby values $\tau^{(1)} = 2.8$ and $\tau^{(2)} = 3.2$. The upper row of Figure 13.9 shows the dependence of the phase speeds on the frequency ω while in the bottom row the attenuations as a function of the frequency are illustrated. The five curves of different style represent different values of the angle α, namely $\alpha = 0$, $\alpha = \pi/8$, $\alpha = \pi/4$, $\alpha = 3\pi/8$ and $\alpha = \pi/2$. It is obvious that both for speeds and attenuations a

---

[3]In 1930 A. B. Wood [450] ascertained the influence of the the volume fraction of air in water on the velocity of sound in air-water-mixtures. He proposed to connect the fraction of the constituents to the elastic constants. In the application to porous media the relation is nowadays known as Wood's formula and is not only used for fluid components but – as shown above – also for the solid and the fluid. We show it here in his own notation: Wood assumed that the sound velocity in a mixture of two components is the same as that in a homogeneous medium of mean density ρ and mean elasticity E. The pairs ρ_1, E_1 and ρ_2, E_2 represent the densities and elasticities of the constituents 1 and 2, respectively. The mean density therefore is $\rho = x\rho_1 + (1 - x)\rho_2$, if x is the volume fraction of the first constituent and $1 - x$ of the second one. For the inverse of the mean elasticity then follows $\frac{1}{E} = \frac{x}{E_1} + \frac{1 - x}{E_2}$. From this we obtain a formula for the mean elasticity $E = \frac{E_1 E_2}{x E_2 + (1-x) E_1}$, which is often used to determine the sound speed of a mixture $c = \sqrt{\frac{E}{\rho}} = \sqrt{\frac{E_1 E_2}{[x E_2 + (1-x) E_1][x\rho_1 + (1-x)\rho_2]}}$.

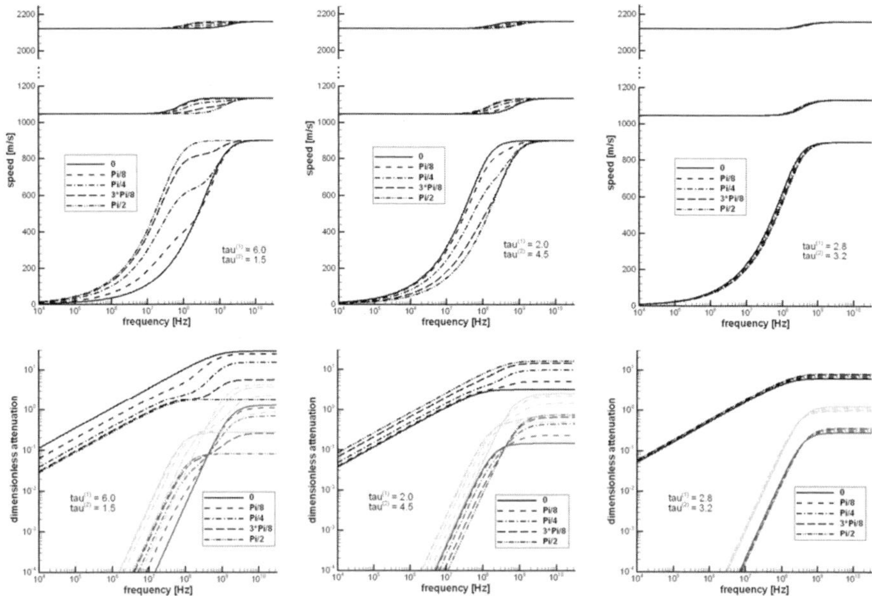

Figure 13.9: *Top row: phase speed in dependence on the logarithmic frequency, fastest wave: pseudo P1. Middle row: pseudo S2-wave, slowest wave: pseudo P2. Bottom row: corresponding dimensionless attenuations, black: pseudo P2, light gray: pseudo S2, dark gray: pseudo P1. The columns represent different pairs of tortuosities: left: $\tau^{(1)} = 6.0$, $\tau^{(2)} = 1.5$, middle: $\tau^{(1)} = 2.0$, $\tau^{(2)} = 4.5$, right: $\tau^{(1)} = 2.8$, $\tau^{(2)} = 3.2$. Speeds and dimensionless attenuations are given for several values of the angle (from [20]).*

bigger difference in the principal tortuosities leads to a wider range for the different values of the angle α. While in the left column the curves are different, in the right column they are nearly lie on top of one another. In the two columns on the right-hand side the smaller value is denoted by $\tau^{(1)}$ and the bigger value by $\tau^{(2)}$. In the first column it is contrary. This is done in order to show that the curves remain principally the same. However, for different order of the tortuosities they belong to different values of the angle α. Inspection of the curves in the left and middle column shows that the curves for different values of the angle α are reverse.

Generally, it becomes obvious, that additionally to the decoupled transversal $S1$-wave three waves are appearing. They are analogous to the $P1$-, $P2$- and S-waves occurring in a saturated porous medium. Thus, they are called pseudo $P1$, pseudo $P2$ and pseudo $S2$ waves. As already mentioned they would coincide with the classical waves if the propagation direction coincided with a principal direction of the tortuosities, in particular in the case of isotropic permeability. The wave with the highest speed and the lowest attenuation (presented in dark gray curves) is the fast longitudinal wave pseudo $P1$. Like the pseudo $S2$ wave it possesses certain high- and low-frequency values which are independent both of the principal tortuosities and of the angle α. The medium speed belongs to the pseudo $S2$ wave as also the medium attenuation (illustrated in light gray).

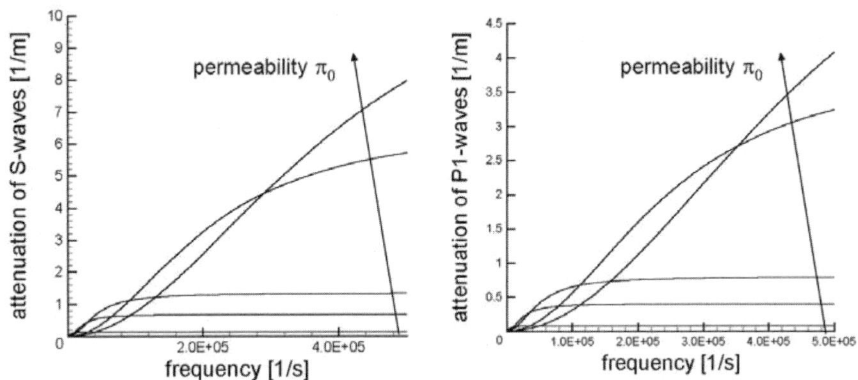

Figure 13.10: *Attenuations of S- and P1-waves in isotropic porous media in dependence on the frequency. Different curves represent values 10^6, $5\cdot10^6$, 10^7, $5\cdot10^7$, 10^8 kg/(m³s) of the permeability parameter π_0 (Figs. 9b and 2a of [449]).*

The smallest speed and the highest attenuation (in black) belong to the slow longitudinal wave pseudo $P2$. This wave starts for zero frequency with zero speed and has also a high-frequency limit which does not depend on the angle α nor on the values of the principal tortuosities. The high and low frequency limits of the speeds of the three pseudo waves are

$$
\begin{aligned}
P1: \quad & \lim_{\omega \to 0} c_{ph} = 2121 \ [\text{m/s}], & \lim_{\omega \to \infty} c_{ph} = 2160 \ [\text{m/s}], \\
S2: \quad & \lim_{\omega \to 0} c_{ph} = 1048 \ [\text{m/s}], & \lim_{\omega \to \infty} c_{ph} = 1134 \ [\text{m/s}], \\
P2: \quad & \lim_{\omega \to 0} c_{ph} = 0 \ [\text{m/s}], & \lim_{\omega \to \infty} c_{ph} = 899 \ [\text{m/s}].
\end{aligned}
\tag{13.97}
$$

Contrary to the limits of the speeds, the frequency limits of the dimensionless attenuation are highly dependent on both the values of the principal tortuosities and the angle α.

In contrast to the propagation speeds for different values of the tortuosity, the attenuation curves intersect. This property of the attenuation may have a bearing in non-destructive testing. For low frequencies the attenuation is lower for higher tortuosities while in the range of high frequencies the attenuation is higher for higher tortuosities. This property was already observed for monochromatic waves in isotropic media. For S-waves a lower permeability led to higher attenuation for low frequencies and lower for high frequencies, for $P1$-waves it is vice versa (see: Figure 13.10 in which Figure 9b and Figure 2a of [449] – the attenuations of S- and $P1$-waves in dependence on the frequency and the permeability parameter π_0 – are reproduced). Consequently, in the low frequency range, where the monochromatic waves are almost nondispersive, for instance transversal waves of the same initial amplitude and different polarization would arrive at the receiver simultaneously but these polarized in the direction of the smaller value of tortuosity with a bigger amplitude than these for a higher tortuosity.

Figure 13.11 illustrates the phase speeds of the three coupled waves dependent on the angle α in the interval $0 \leq \alpha \leq \pi/2$. The curves represent different values of the frequency ω. Again, the top row concerns pseudo $P1$, the middle row pseudo $S2$ and the bottom row the pseudo $P2$-wave. Also the arrangement of the three sets of tortuosities: left – big difference, middle – medium difference, right – small difference is the same as in Figure 13.9.

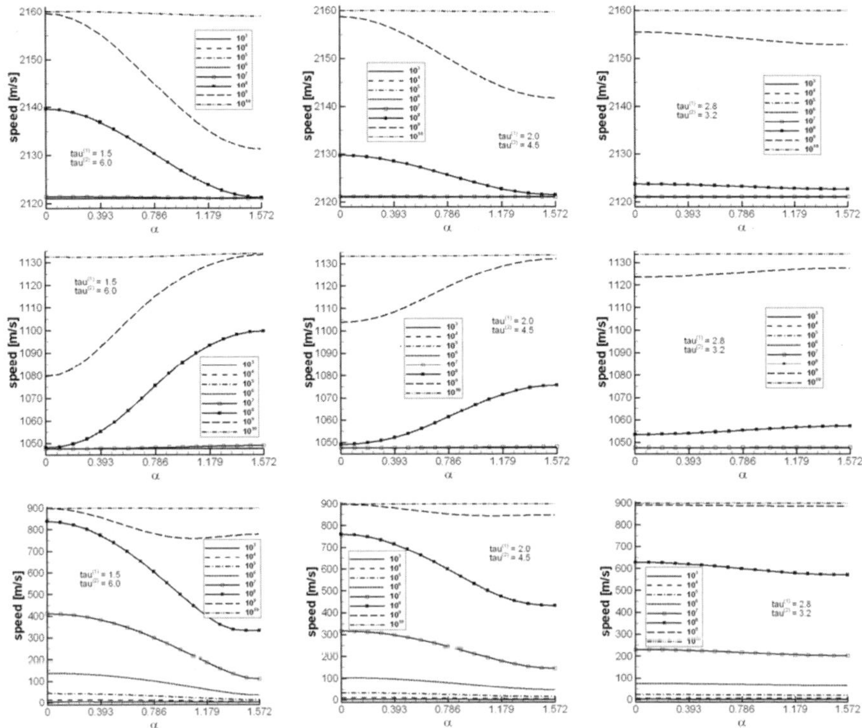

Figure 13.11: *Phase speed in dependence on the angle α, top row: pseudo P1, middle row: pseudo S2-wave, bottom row: pseudo P2. The columns represent different pairs of tortuosities: left: $\tau^{(1)} = 1.5$, $\tau^{(2)} = 6.0$, middle: $\tau^{(1)} = 2.0$, $\tau^{(2)} = 4.5$, right: $\tau^{(1)} = 2.8$, $\tau^{(2)} = 3.2$. The speeds are given for several values of the frequency (from [20]).*

It is obvious that for both waves which exhibit high and low frequency limits different from zero, the pseudo $P1$ and the pseudo $S2$-wave, only at very high frequencies the angle α plays any role. For such frequencies of about 10^9 Hz the difference in the pseudo $P1$-wave speed is around $30/20/10$ m/s depending on the difference in tortuosities. This corresponds to a difference in the speeds of around $75/50/20$ % of the maximal difference between high and low frequency limits. The difference is somewhat smaller for the pseudo $S2$-wave namely $60/35/10$ % which corresponds to around $60/30/10$ m/s. The behavior of the speeds of pseudo $P1$ and pseudo $S2$ are opposite. While the maximum for pseudo $P1$ appears for the angle $\alpha = 0$, for the pseudo $S2$-wave it is apparent for $\alpha = \pi/2$.

The first situation is also the case for the second pseudo longitudinal wave $P2$. However, the maximal difference appears for a somewhat smaller frequency, namely for approximately $\omega = 10^8$ Hz. Quantitatively, the changes in the speed in dependence on α are bigger since this wave starts from the zero speed for zero frequency. However, qualitatively, the changes lie in a comparable range to the other two waves, namely at $55/33/10\%$ $(350/300/100$ m/s$)$. For this wave also curves for lower frequencies are identifiable.

Figure 13.12 shows the corresponding dimensionless attenuations of the three waves.

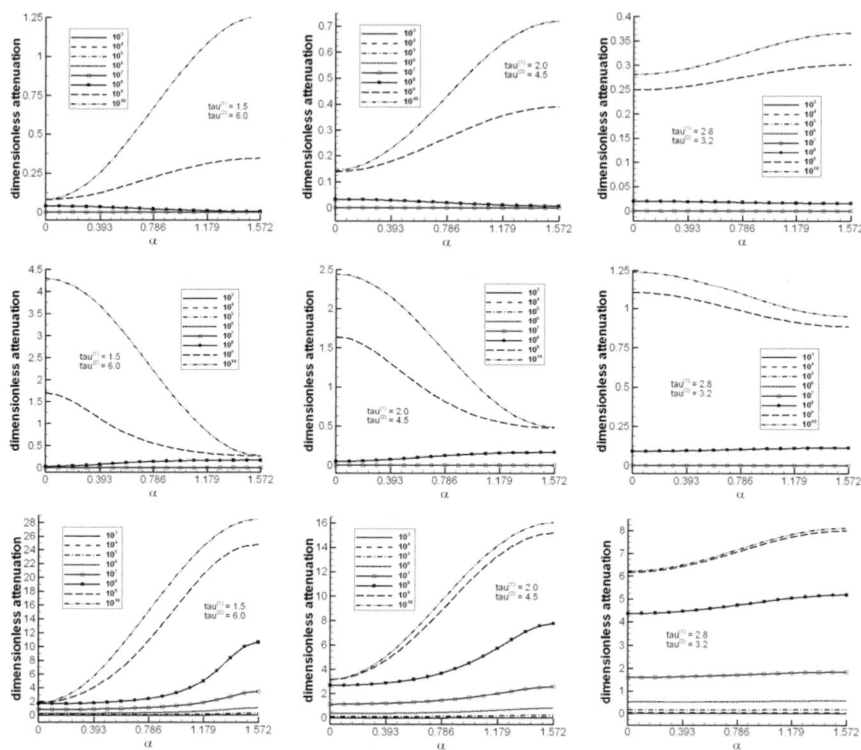

Figure 13.12: *Dimensionless attenuation in dependence on the angle α, top row: pseudo P1, middle row: pseudo S2-wave, bottom row: pseudo P2. The columns represent different pairs of tortuosities: left: $\tau^{(1)} = 1.5$, $\tau^{(2)} = 6.0$, middle: $\tau^{(1)} = 2.0$, $\tau^{(2)} = 4.5$, right: $\tau^{(1)} = 2.8$, $\tau^{(2)} = 3.2$. The attenuations are given for several values of the frequency (from [20]).*

They exhibit a similar strong dependence on the angle α as the speeds and this likewise predominantly for high values of the frequency. As for the classical $P1$- and $S2$-waves the attenuation is also for the pseudo $P1$- and pseudo $S2$-wave rather small. The maximum value for pseudo $P1$ is 1.25, for pseudo $S2$ it is around 4.5. The $P2$-wave is generally strongly damped. This is also the case for the pseudo $P2$-wave. Here, the maximal dimensionless attenuation is 28. The values of the dimensionless attenuations indicate a strong influence of the difference in the principal tortuosities on the attenuation. For all three waves the maximal value of the dimensionless attenuation is for a big difference in the principal tortuosities around the double of this of a medium difference and for a small difference a half ($P1$: 0.4/0.7/1.25, $S2$: 1.25/2.5/4.5, $P2$: 8/16/28).

For a better overview the results of Figures 13.8, 13.9, 13.11 and 13.12 are summarized in 3D plots illustrated in Figures 13.13.

These results indicate that in geophysical applications of low frequency monochromatic waves the orientation of principal directions of tortuosity plays a secondary role in measured speeds and attenuation and most likely cannot be used for any practical purposes.

13.4.6 Shear polarization

Transversal (shear) waves play a particular role in the analysis of anisotropy. This is due to the fact that in isotropic materials the properties of shear waves are independent of the polarization. All waves with an amplitude perpendicular to the direction of propagation have the same speed and the same attenuation. For this reason, we call them S-waves without indicating the direction of the amplitude on the plane perpendicular to the direction of propagation. This is not the case for anisotropic media and, in particular, for porous media with an anisotropic permeability.

In order to investigate the deviation from the plane perpendicular to the direction of propagation of the pseudo transversal (shear) wave $S2$ the shear polarization factor is built. It is defined by

$$d = \left| \operatorname{Re} \frac{V_1^S}{V_2^S} \right| = \left| \operatorname{Re} \frac{N}{D} \right|, \tag{13.98}$$

where N and D follow from the coefficients given in the footnote on page 347

$$N = \frac{i\pi_0\omega}{\rho^S} \left(\tau^{(1)^2} - \tau^{(2)^2} \right) \sin\alpha \cos\alpha \left(1 - \frac{-\omega^2 + \frac{\mu}{\rho^S}k^2}{r\omega^2} \right),$$

$$D - \frac{i\pi_0\omega}{\rho^S} \left(\tau^{(1)^2} \cos^2\alpha + \tau^{(2)^2} \sin^2\alpha \right) \left(1 + \frac{-\omega^2 + \frac{\lambda^S+2\mu^S}{\rho^S}k^2 + \frac{Q}{\rho^S}k^2}{\frac{Q}{\rho^S}k^2 + r\left(-\omega^2 + \kappa k^2\right)} \right) \tag{13.99}$$

$$+ \frac{Q}{\rho^S}k^2 - r\left(-\omega^2 + \kappa k^2\right) \frac{-\omega^2 + \frac{\lambda^S+2\mu^S}{\rho^S}k^2 + \frac{Q}{\rho^S}k^2}{\frac{Q}{\rho^S}k^2 + r\left(-\omega^2 + \kappa k^2\right)}.$$

Here, $k\left(\omega\right)$ is the solution of the dispersion relation corresponding to the pseudo shear wave. The coefficients N and D follow by the elimination of the unknown components V_1^F and V_2^F from the third equation of (13.95). Obviously, the fourth equation is a linear combination of the remaining three equations (13.95).

The limit values of d are

$$d\left(\alpha = 0\right) = d\left(\alpha = \frac{\pi}{2}\right) = 0 \quad \text{and} \quad d\left(\omega = 0\right) = d\left(\omega \to \infty\right) = 0. \tag{13.100}$$

Hence, as expected, the coincidence of the direction of propagation with a principal direction of tortuosity yields pure transversal waves with an amplitude perpendicular to the direction of propagation. Simultaneously, in both limits of the frequency, $\omega = 0$ and $\omega \to \infty$, waves do not feel the anisotropy of tortuosity and become pure transversal waves as well.

Figure 13.15 illustrates the dependence of the shear polarization factor, i.e., of the deviation of the direction of the pseudo shear-wave from the plane perpendicular to the direction of propagation, on the frequency ω and on the angle α. On the left-hand side in a 3D plot both dependences are shown. For clarification in the right column the dependence on the frequency for several values of the angle α is presented in a 2D plot. The limit values of d given in (13.100) are immediately evident. Especially the inspection

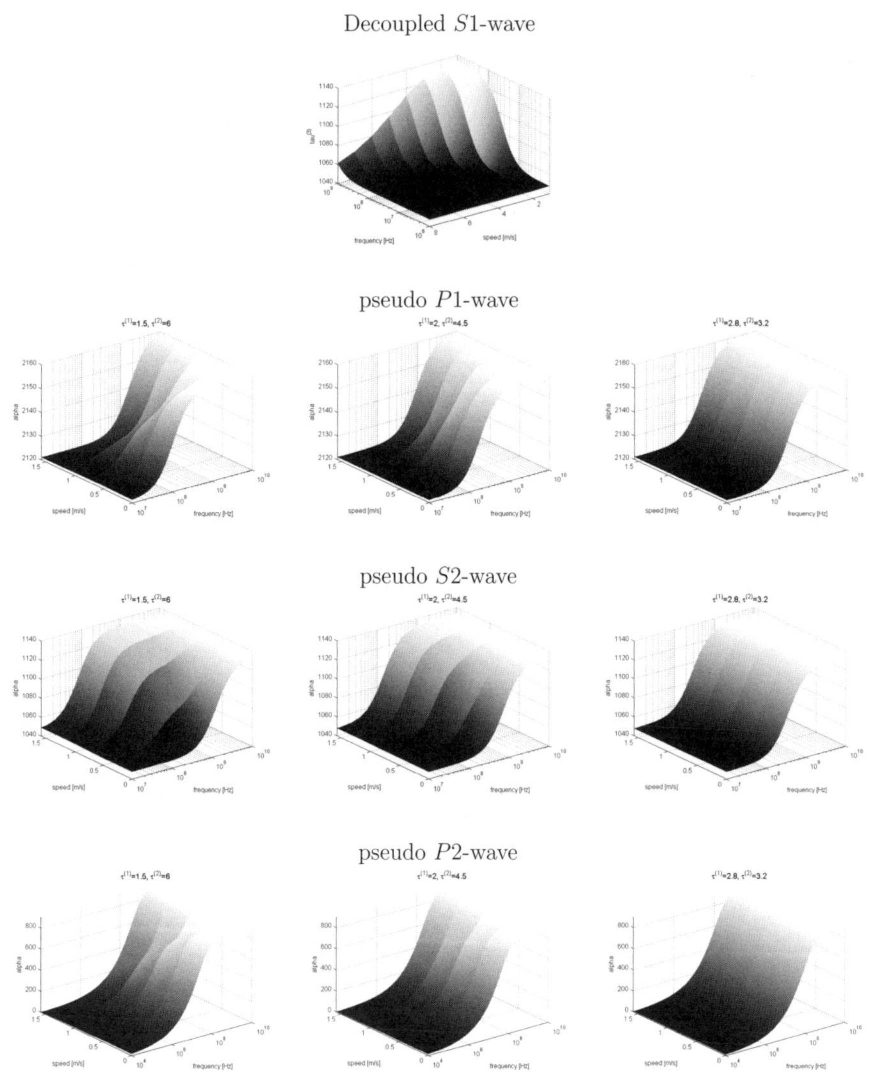

Figure 13.13: *Phase speeds of the four waves appearing in the porous medium with material parameters given in Table 13.3. Top row: decoupled S1-wave as a function of frequency and tortuosity $\tau^{(3)}$, further rows: pseudo P1, S2 and P2-waves in dependence on frequency and angle α for three different pairs of tortuosities $\tau^{(1)}$ and $\tau^{(2)}$ (from [21]).*

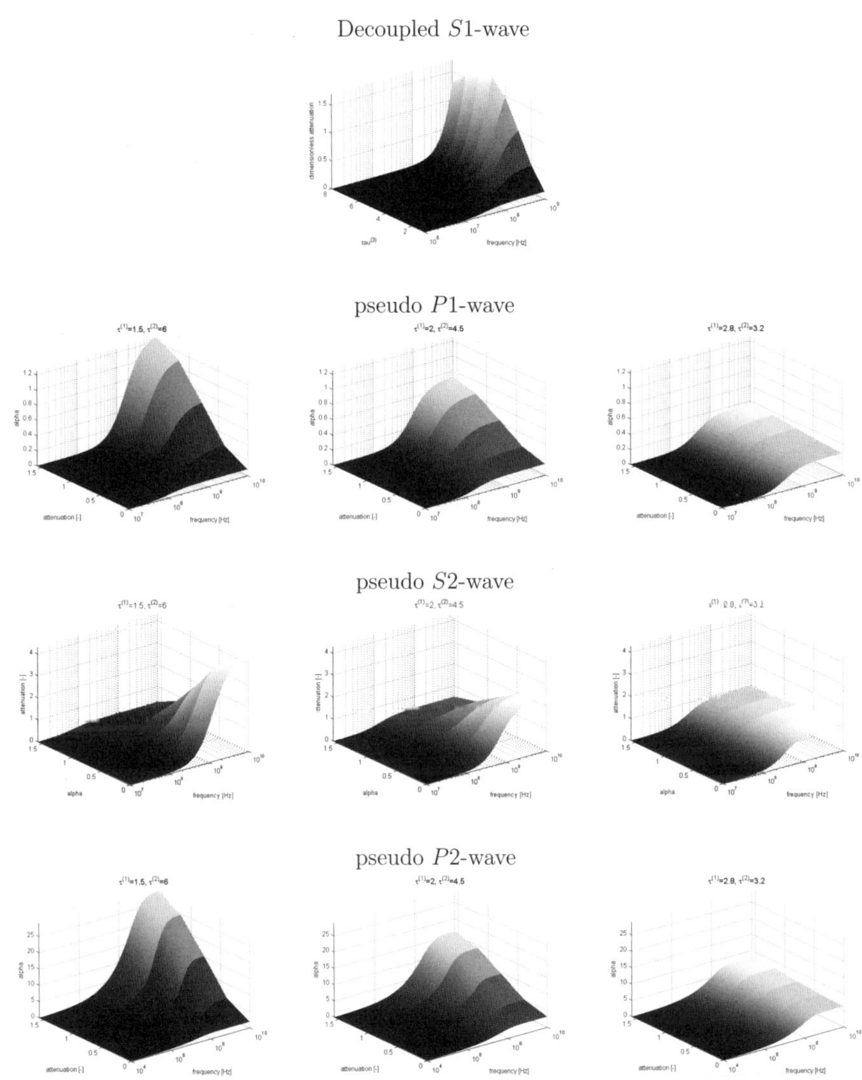

Figure 13.14: *Dimensionless attenuations of the four waves appearing in the porous medium with material parameters given in Table 13.3. Top row: decoupled $S1$-wave as a function of frequency and tortuosity $\tau^{(3)}$, further rows: pseudo $P1$, $S2$ and $P2$-waves in dependence on frequency and angle α for three different pairs of tortuosities $\tau^{(1)}$ and $\tau^{(2)}$ (cf. [21]).*

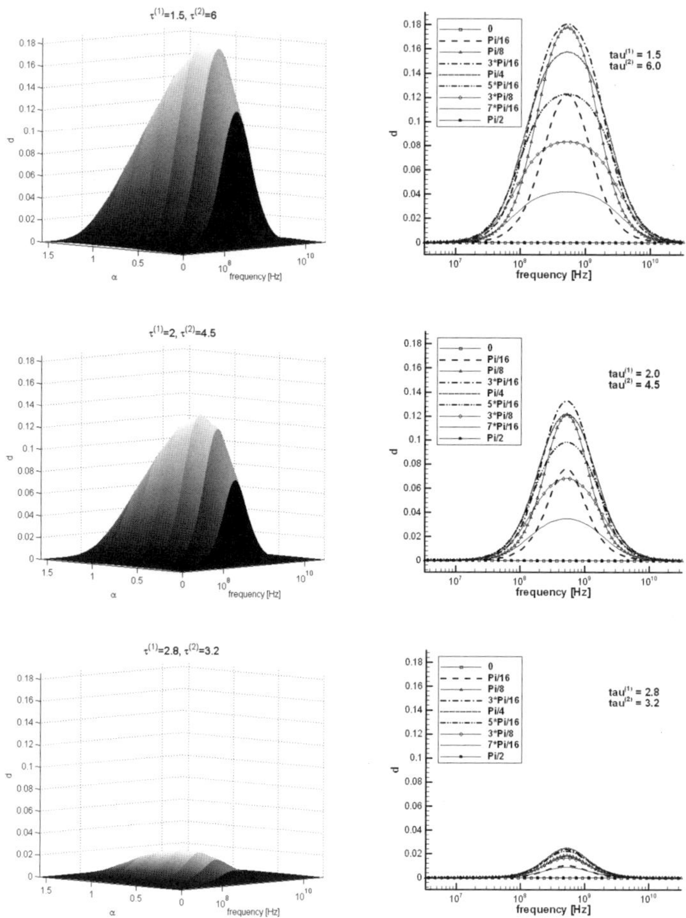

Figure 13.15: *Shear polarization factor, d. Left: 3D plot in dependence on the frequency ω and on the angle α for different pairs of tortuosities: top row: $\tau^{(1)} = 1.5$, $\tau^{(2)} = 6.0$, middle row: $\tau^{(1)} = 2.0$, $\tau^{(2)} = 4.5$, bottom row: $\tau^{(1)} = 2.8$, $\tau^{(2)} = 3.2$; right column: d in dependence on the frequency for the same pairs of principal tortuosities – different curves represent different values of the angle α.*

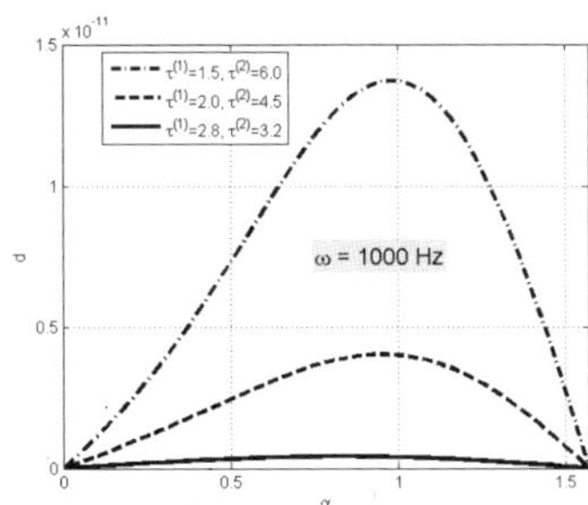

Figure 13.16: *Polarization factor d as a function of the angle α in Alermoehe sandstone for three pairs of principal tortuosities – frequency $\omega = 1000$ Hz.*

of the right column of the figure points up the maximum values of d for the three pairs of principal tortuosities. They vary between around 0.18 for a big difference of the principal tortuosities, 0.13 for a medium difference and 0.02 for a small difference. Additionally to the results in the high frequency region shown in Figure 13.15, in Figure 13.16 for a chosen value of the frequency ($\omega = 1000$ Hz) the dependence on α is presented for frequencies appearing in geotechnical applications. Then, the limits (13.100), again, become obvious and the differences for the three choices of principal tortuosities become evident as well, even if the numeric values are smaller.

The above numerical results seem to indicate, similarly to the speed and attenuation dependencies on the angle α, that a particular orientation of the principal directions of the tortuosity tensor with respect to the direction of propagation may not be very important in geophysical applications (low frequency). This means that in this range pseudo shear waves may be approximately treated as usual pure shear waves. Consequently, the anisotropy is primarily reflected by the influence of the principal tortuosities on the attenuation of monochromatic waves and this gives, in turn, rise to a possibility of a new method of nondestructive acoustic testing of permeability of geomaterials.

13.4.7 Some remarks

The above analysis of monochromatic waves in porous media with an anisotropic permeability reveals the existence of four modes of propagation: two pseudo longitudinal $P1$- and $P2$-waves and two pseudo transversal (shear) modes $S1$ and $S2$. Due to a special choice of the geometry one of the pseudo transversal (shear) modes – $S1$ – becomes in our example a pure transversal wave. In the general case of anisotropy this would, certainly, not be the case. We have found out that the anisotropy of tortuosity yields essential changes of the attenuation of the waves depending on the direction of propagation in relation to the principal directions of tortuosity and on the mode of the wave. Due to the appearance of two shear modes one can construct a device for measuring the anisotropy of the permeability. It would have to induce shear waves of different polarization and different directions of propagation. Then, for the range of low frequencies, one could measure the principal values and the directions of the tortuosity by comparing the amplitudes of the arrivals for different polarization of the signals. This would be an important information for various geophysical applications of the model of porous media. In such problems as seepage processes in road and dam constructions or tunneling in rocks the different permeability in different directions plays an important practical role.

13.5 Wave propagation in three-component porous media

13.5.1 Introduction

In this section we investigate the propagation of sound waves in partially saturated poroelastic media consisting of three components: the solid S, a fluid F and a gas G. This happens by means of a new model [14] (for details see [13]) which is an extension of both the classical Biot model for two-component saturated media (e.g. [381] or Section 13.3 of Part I) and the Simple Mixture Model by K. Wilmanski (e.g. [438], Part I, [13] or Section 13.1.2). The three-component model, which describes linear processes in unsaturated poroelastic materials, contains features of both models: Biot's model incorporates the interaction between the components by partial volume changes through an additional contribution to the partial stresses (the above mentioned parameter Q). While the Biot model incorporates only one coupling (between skeleton and fluid) the present model for unsaturated porous media – which consists of one more component – includes three couplings (solid-fluid Q^F, solid-gas Q^G and fluid-gas Q^{FG}). The latter incorporates the effect of surface tension between the fluid and the gas. The Simple Mixture Model does not include such couplings. Its characteristic is that the porosity belongs to the set of fields and satisfies an own balance equation (presented in Subsection 13.1.1 and in Part I). This is also the case for the three-component model. A systematic method of derivation of the relations between macroscopic (average) material parameters (compressibilities and coupling parameters) and their counterparts for true (microscopic) materials is applied.

It is well known that in a one-component medium there exists a single longitudinal wave P (primary) and one transversal wave S (secondary or shear). The additional component in saturated porous media yields – compared to the one-component body – the existence of an additional bulk wave, the $P2$-wave. In the literature it is also called the Biot slow wave. Due to the existence of a second pore fluid (the gas) in the unsaturated medium besides the two compressional waves ($P1$ and $P2$) and the shear wave (S) which appear in the saturated porous medium an additional compressional wave ($P3$) emerges. The speeds and attenuations of these waves are shown in dependence on the frequency ω and on the initial saturation S_0. The speeds are compared to those of sound waves in air-water-mixures.

13.5.2 Linear model

The linear thermodynamical model without memory effects (for details see [13]) can be described by the essential fields $\left\{\mathbf{v}^S, \mathbf{v}^F, \mathbf{v}^G, \mathbf{e}^S, \varepsilon^F, \varepsilon^G\right\}$ which have to satisfy the field equations

$$\rho_0^S \frac{\partial \mathbf{v}^S}{\partial t} = \operatorname{div}\left\{\lambda^S e\mathbf{1} + 2\mu^S \mathbf{e}^S + Q^F \varepsilon^F \mathbf{1} + Q^G \varepsilon^G \mathbf{1}\right\}$$
$$+ \pi^{FS}\left(\mathbf{v}^F - \mathbf{v}^S\right) + \pi^{GS}\left(\mathbf{v}^G - \mathbf{v}^S\right),$$

$$\rho_0^F \frac{\partial \mathbf{v}^F}{\partial t} = \operatorname{grad}\left\{\rho_0^F \kappa^F \varepsilon^F + Q^F e + Q^{FG}\varepsilon^G\right\} - \pi^{FS}\left(\mathbf{v}^F - \mathbf{v}^S\right), \qquad (13.101)$$

$$\rho_0^G \frac{\partial \mathbf{v}^G}{\partial t} = \operatorname{grad}\left\{\rho_0^G \kappa^G \varepsilon^G + Q^G e + Q^{FG}\varepsilon^F\right\} - \pi^{GS}\left(\mathbf{v}^G - \mathbf{v}^S\right),$$

$$\frac{\partial \mathbf{e}^S}{\partial t} = \operatorname{sym} \operatorname{grad} \mathbf{v}^S, \quad \frac{\partial \varepsilon^F}{\partial t} = \operatorname{div} \mathbf{v}^F, \quad \frac{\partial \varepsilon^G}{\partial t} = \operatorname{div} \mathbf{v}^G, \quad e \equiv \operatorname{tr} \mathbf{e}^S.$$

The set of equations (13.101) coincides with the classical Biot model, if we neglect the third component, i.e., the gas.

The quantities $\mathbf{v}^S, \mathbf{v}^F$ and $\mathbf{v}^G$ are macroscopic velocity fields of the components, $\mathbf{e}^S$ is the macroscopic deformation tensor. Quantities with subindex zero are initial values of the corresponding current quantity.

Instead of the partial mass densities of the components, ρ^S, ρ^F and ρ^G the equations depend on the corresponding volume changes of the components e, ε^F and ε^G for which hold

$$e = \frac{\rho_0^S - \rho^S}{\rho_0^S}, \quad \varepsilon^F = \frac{\rho_0^F - \rho^F}{\rho_0^F}, \quad \varepsilon^G = \frac{\rho_0^G - \rho^G}{\rho_0^G}. \qquad (13.102)$$

In principle, the porosity, n, is also a field and satisfies an own balance equation. However, if we neglect memory effects, the balance equation can be solved and its consideration is no longer necessary to solve the problem.

In the case of multi-component fluid mixtures filling the pores one needs, apart from the porosity, at least one additional microstructural variable – a fraction of the contributions of these components. Here, we consider a three-component medium consisting of a solid and two pore fluids (fluid and gas). Then, the saturation is defined by

$$S = S_f = \frac{\text{volume of the fluid within a } REV}{\text{volume of voids within a } REV},$$

i.e., it is the fraction of the volume of the fluid over the volume of all voids within a representative elementary volume (REV). Instead of the fluid saturation as well one could use the gas saturation: $S_g = 1 - S$, the fraction of the gas over the volume of the voids.

In the present model the current saturation of the fluid S is used but it is not included in the series of fields. Instead, a constitutive law will be introduced for this quantity. In [446] K. Wilmanski discussed that in the literature there appear at least five ways to describe changes of porosity, i.e., of the first microstructural variable. Namely by constitutive assumptions, by an assumption on the incompressibility of true components or by different types of evolution or balance equations for the porosity. While the incompressibility assumption for media consisting of at least one gaseous component is obviously not suitable, the other two options are also conceivable for the incorporation of the second microstructural variable, the saturation, into a model. Here, we decided on a constitutive law which will be described in the next subsection.

13.5.3 Material parameters

In (13.101) the diffusion velocities $\mathbf{v}^F-\mathbf{v}^S$ and $\mathbf{v}^G-\mathbf{v}^S$ appear. They are multiplied by the resistance parameters π^{FS} and π^{GS} which have to be specified according to the material. Both in the Biot model and in the Simple Mixture Model only one of such parameters appears (called π) because only the resistance of the flow of the fluid through the channels of the skeleton is reflected. It is easy to show that this parameter is connected to the classical permeability parameter $\overline{K}$ in the following way

$$\overline{K} = \frac{K}{\rho g} \sim \frac{1}{\pi}, \tag{13.103}$$

where the hydraulic conductivity is denoted by K, the mass density of the fluid by ρ and the earth acceleration by g. Using $\overline{K}_f = k_f\,\overline{K}$ and $\overline{K}_g = k_g\,\overline{K}$, i.e., separating $\overline{K}$ into two parts for the fluid and the gas and proposing similar relations to (13.103) for the components: $\overline{K}_f \sim 1/\pi^{FS}$ and $\overline{K}_g \sim 1/\pi^{GS}$, we obtain

$$\pi^{FS} = \frac{\pi^F}{k_f}, \qquad \pi^{GS} = \frac{\pi^G}{k_g}. \tag{13.104}$$

The parameters π^F and π^G depend on the viscosities of the fluid and the gas, respectively, and thus they are extremely different for the two pore fluids.

For a water-air-mixture in sand the relative permeabilities k_f and k_g for the fluid and the gas have been measured in dependence on the liquid saturation e.g. by Wyckoff and Botset [452] in 1936 (see Figure 13.17 left). In 1980 van Genuchten [403] not only proposed a theoretical relationship between the capillary pressure and the saturation but also formulae for the relative permeabilities

$$k_f = S^{\frac{1}{2}}\left[1-\left(1-S^{\frac{1}{m}}\right)^m\right]^2, \qquad k_g = (1-S)^{\frac{1}{3}}\left(1-S^{\frac{1}{m}}\right)^{2m}, \tag{13.105}$$

where S is the fluid saturation and m a material parameter. From Figure 13.17 (right) in which these permeabilities are plotted for $m = 0.85$ it becomes obvious that the accordance between measured and calculated curves is very good. The corresponding values of π^{FS} and π^{GS} which are needed in the presented model follow from (13.104).

Furthermore, the material parameters $\left\{\lambda^S + \frac{2}{3}\mu^S, \kappa^F, \kappa^G, Q^F, Q^G, Q^{FG}\right\}$ appearing in the Cauchy stress tensors have to be specified according to the material. This is done by applying a transition from the micro- to the macro-scale (for details see [13]). The relation for the capillary pressure which is used in this procedure has been proposed by van Genuchten [403] in 1980

$$p_c(S) = \frac{1}{\alpha_{vG}}\left[S^{(-1/m_{vG})} - 1\right]^{1/n_{vG}}, \tag{13.106}$$

where α_{vG}, m_{vG} and n_{vG} are parameters which depend on the properties of the soil. In the present example we consider an air-water-mixture in sandstone. Other soil types have been investigated in [15]. Sandstones are represented by the values $m_{vG} = 0.5$, $n_{vG} = 2$ and $\alpha_{vG} = 0.00005$. The corresponding calculated capillary pressure curve is shown in Figure 13.18 (left). A comparison to experimental results of K. E.-A. van den Abeele et al. [401] presented on the right-hand side of this figure points up, again, the good agreement calculated and measured values. The choice of the parameters controls

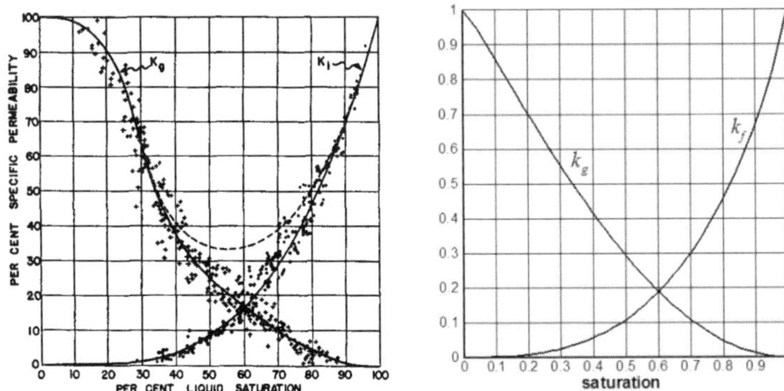

Figure 13.17: *Left: Experimentally obtained relative permeabilities of a water-air-mixture in sand in dependence on the liquid saturation (from [452]), right: relative permeabilities according to formulae (13.105) with m = 0.85.*

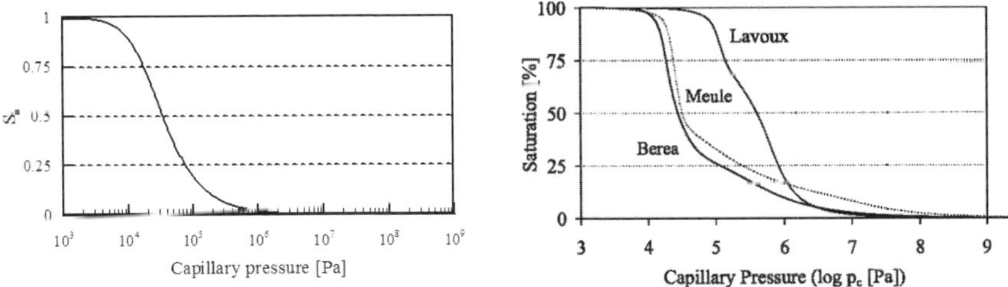

Figure 13.18: *Capillary pressure curves. Left: Calculated using the formula of van Genuchten (13.106) for an air-water-mixture in sandstone, right: experimental curves for Berea- and Meule-sandstone and Lavoux-limestone measured by mercury intrusion (from [401]).*

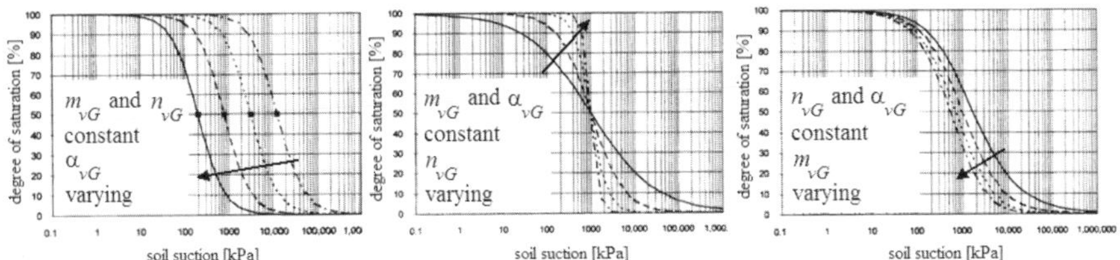

Figure 13.19: *Influence of changes in parameters α_{vG}, m_{vG} and n_{vG} in the van Genuchten equation (13.106). Modified figures from [356].*

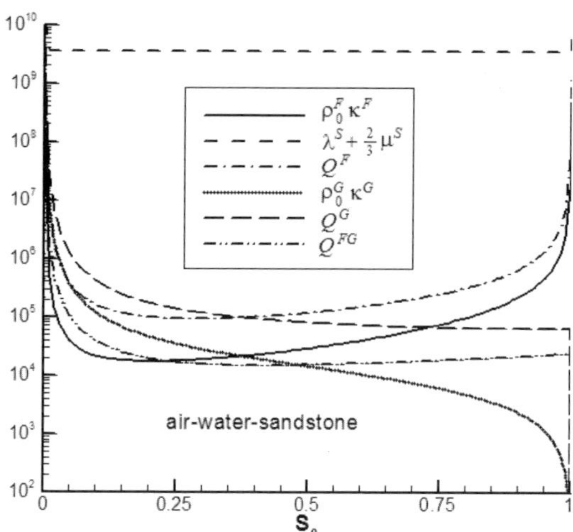

Figure 13.20: *Material parameters* $\left\{\lambda^S + \frac{2}{3}\mu^S, \rho_0^F\kappa^F, \rho_0^G\kappa^G, Q^F, Q^G, Q^{FG}\right\}$ *for an air-water-sandstone system in log-scale in dependence on the initial saturation* S_0.

the shape of the curve and its position in the capillary pressure-saturation-diagram. An illustration of the influence of the change of each of the three parameters can be found in Figure 13.19 which is a modification of a figure of W. Sillers et al. [356].

The numerical values of the microscopic parameters are given in Table 13.4. Therein the initial porosity is denoted by n_0, the microscopic compressibilities of solid, fluid and gas by K_s, K_f and K_g, respectively and the true mass densities of the components by ρ_0^{SR}, ρ_0^{FR} and ρ_0^{GR}. As already mentioned in the last section of this book, the expression for the compressibility modulus of the empty matrix, K_d, has been proposed in this form by J. Geertsma (see [425]). The true initial pressures of the skeleton and the fluid, p_0^{SR} and p_0^{FR}, are negligible in comparison to the compressibilities K_s and K_f. Therefore, they are set equal to zero. This is not the case for the true initial pressure of the gas p_0^{GR} because it is of the same order as K_g. It is defined by the capillary pressure $p_c(S_0)$.

Table 13.4: *Material parameters for an air-water-mixture in sandstone.*

$n_0 = 0.25,$	$K_s = 48 \cdot 10^9$ Pa,	$\rho_0^{SR} = 2650 \; \frac{\text{kg}}{\text{m}^3},$	$p_0^{SR} = 0,$	$m_{vG} = 0.5,$
$K_d = \dfrac{K_s}{1 + 50n_0},$	$K_f = 2.25 \cdot 10^9$ Pa,	$\rho_0^{FR} = 1000 \; \frac{\text{kg}}{\text{m}^3},$	$p_0^{FR} = 0,$	$n_{vG} = 2,$
	$K_g = 1.01 \cdot 10^5$ Pa,	$\rho_0^{GR} = 1.2 \; \frac{\text{kg}}{\text{m}^3},$	$p_0^{GR} = p_c,$	$\alpha_{vG} = 5 \cdot 10^{-5}.$

13.5.4 General propagation condition of monochromatic waves

As in Section 13.4.3 it is assumed that the fields of the model, $\left\{\mathbf{v}^S, \mathbf{v}^F, \mathbf{v}^G, \mathbf{e}^S, \varepsilon^F, \varepsilon^G\right\}$, satisfy the following relations

$$\varepsilon^F = E^F\mathcal{E}, \quad \varepsilon^G = E^G\mathcal{E}, \quad \mathbf{e}^S = \mathbf{E}^S\mathcal{E},$$
$$\mathbf{v}^F = \mathbf{V}^F\mathcal{E}, \quad \mathbf{v}^G = \mathbf{V}^G\mathcal{E}, \quad \mathbf{v}^S = \mathbf{V}^S\mathcal{E}, \tag{13.107}$$

$$n - n_0 = D\mathcal{E}, \qquad \mathcal{E} := \exp i\left(\mathbf{k}\cdot\mathbf{x} - \omega t\right),$$

where $\mathbf{E}^S, E^F, E^G, \mathbf{V}^S, \mathbf{V}^F, \mathbf{V}^G$ and D are constant amplitudes, ω is a given frequency and $\mathbf{k}$ is the, possibly complex, wave vector. Also in this case $\mathbf{k} = k\mathbf{n}$, where k is the complex wave number and $\mathbf{n}$ is a unit vector in the direction of propagation. Such a solution describes the propagation of plane monochromatic waves in an infinite medium whose fronts are perpendicular to $\mathbf{n}$.

Substitution of the above relations in the field equations $(13.101)_4$ yields the following compatibility relations

$$E^F = -\frac{1}{\omega}k\mathbf{n}\cdot\mathbf{V}^F, \quad E^G = -\frac{1}{\omega}k\mathbf{n}\cdot\mathbf{V}^G,$$

$$\mathbf{E}^S = -\frac{1}{2\omega}k\left(\mathbf{n}\otimes\mathbf{V}^S + \mathbf{V}^S\otimes\mathbf{n}\right), \quad \text{i.e.,} \quad e = -\frac{1}{\omega}k\mathbf{n}\cdot\mathbf{V}^S\mathcal{E}.$$

Making use of (13.107) in the remaining field equations leads to the following set

$$
\begin{aligned}
\omega^2\mathbf{V}^S = {}& \frac{\lambda^S}{\rho_0^S}k^2\left(\mathbf{V}^S\cdot\mathbf{n}\right)\mathbf{n} + \frac{\mu^S}{\rho_0^S}k^2\left(\left(\mathbf{V}^S\cdot\mathbf{n}\right)\mathbf{n} + \mathbf{V}^S\right) \\
& + \frac{Q^F}{\rho_0^S}k^2\left(\mathbf{V}^F\cdot\mathbf{n}\right)\mathbf{n} + \frac{Q^G}{\rho_0^S}k^2\left(\mathbf{V}^G\cdot\mathbf{n}\right)\mathbf{n} \\
& + i\frac{\pi^{FS}\omega}{\rho_0^S}\left(\mathbf{V}^F - \mathbf{V}^S\right) + i\frac{\pi^{GS}\omega}{\rho_0^S}\left(\mathbf{V}^G - \mathbf{V}^S\right) = 0,
\end{aligned}
\tag{13.108}
$$

$$
\begin{aligned}
\omega^2\mathbf{V}^F = {}& \kappa^F k^2\left(\mathbf{V}^F\cdot\mathbf{n}\right)\mathbf{n} + \frac{Q^F}{\rho_0^F}k^2\left(\mathbf{V}^S\cdot\mathbf{n}\right)\mathbf{n} \\
& + \frac{Q^{FG}}{\rho_0^F}k^2\left(\mathbf{V}^G\cdot\mathbf{n}\right)\mathbf{n} - i\frac{\pi^{FS}\omega}{\rho_0^F}\left(\mathbf{V}^F - \mathbf{V}^S\right) = 0,
\end{aligned}
\tag{13.109}
$$

$$
\begin{aligned}
\omega^2\mathbf{V}^G = {}& \kappa^G k^2\left(\mathbf{V}^G\cdot\mathbf{n}\right)\mathbf{n} + \frac{Q^G}{\rho_0^G}k^2\left(\mathbf{V}^S\cdot\mathbf{n}\right)\mathbf{n} \\
& + \frac{Q^{FG}}{\rho_0^G}k^2\left(\mathbf{V}^F\cdot\mathbf{n}\right)\mathbf{n} - i\frac{\pi^{GS}\omega}{\rho_0^G}\left(\mathbf{V}^G - \mathbf{V}^S\right) = 0.
\end{aligned}
\tag{13.110}
$$

It is convenient to separate the contributions of the normal and transversal components of the wave vector $k\mathbf{n}$. Let us begin with the transversal component.

13.5.4.1 Transversal wave

The scalar product of (13.108)-(13.110) with an arbitrary unit vector $\mathbf{n}_\perp$ perpendicular to $\mathbf{n}$, i.e., $\mathbf{n}\cdot\mathbf{n}_\perp = 0$, is taken.

$$\omega^2 V_\perp^S = \frac{\mu^S}{\rho_0^S}k^2 V_\perp^S + i\frac{\pi^{FS}\omega}{\rho_0^S}\left(V_\perp^F - V_\perp^S\right) + i\frac{\pi^{GS}\omega}{\rho_0^S}\left(V_\perp^G - V_\perp^S\right),$$

$$\omega^2 V_\perp^F = -i\frac{\pi^{FS}\omega}{\rho_0^F}\left(V_\perp^F - V_\perp^S\right), \qquad \omega^2 V_\perp^G = -i\frac{\pi^{GS}\omega}{\rho_0^G}\left(V_\perp^G - V_\perp^S\right), \tag{13.111}$$

$$V_\perp^S := \mathbf{V}^S\cdot\mathbf{n}_\perp, \quad V_\perp^F := \mathbf{V}^F\cdot\mathbf{n}_\perp, \quad V_\perp^G := \mathbf{V}^G\cdot\mathbf{n}_\perp.$$

The dispersion relation of this wave follows in the form

$$\omega^2 \left[1 - \frac{\mu^S}{\rho_0^S} \left(\frac{k}{\omega} \right)^2 \right] - \pi^{FS} \pi^{GS} \frac{\rho_0^S + \rho_0^F + \rho_0^G}{\rho_0^S \rho_0^F \rho_0^G} \left[1 - \frac{\mu^S}{\rho_0^S + \rho_0^F + \rho_0^G} \left(\frac{k}{\omega} \right)^2 \right]$$

$$+ i\omega \left(\frac{\pi^{FS} + \pi^{GS}}{\rho_0^S} + \frac{\pi^{FS}}{\rho_0^F} + \frac{\pi^{GS}}{\rho_0^G} \right) \left[1 - \frac{\frac{\pi^{FS}}{\rho_0^F} + \frac{\pi^{GS}}{\rho_0^G}}{\frac{\pi^{FS} + \pi^{GS}}{\rho_0^S} + \frac{\pi^{FS}}{\rho_0^F} + \frac{\pi^{GS}}{\rho_0^G}} \frac{\mu^S}{\rho_0^S} \left(\frac{k}{\omega} \right)^2 \right] = 0. \tag{13.112}$$

In the two frequency limits similar phase speeds to those of two-component media are obtained:

$$\omega \to \infty \quad \Rightarrow \quad c_{ph} = c_{S\infty} = \sqrt{\frac{\mu^S}{\rho_0^S}},$$

$$\omega \to 0 \quad \Rightarrow \quad c_{ph} = \sqrt{\frac{\mu^S}{\rho_0^S + \rho_0^F + \rho_0^G}}. \tag{13.113}$$

Thus, since μ^S and ρ_0^S are independent of the saturation, the value for $\omega \to \infty$ is equal to $c_{S\infty}$ also in the limits of the saturation $S_0 = 0$ and $S_0 = 1$. In the limit $\omega \to 0$ the limit values in the limits of the saturation are different:

$$\omega \to 0, \; S_0 = 1 \quad \Rightarrow \quad c_{ph} = \sqrt{\frac{\mu^S}{\rho_0^S + \rho_0^F}},$$

$$\omega \to 0, \; S_0 = 0 \quad \Rightarrow \quad c_{ph} = \sqrt{\frac{\mu^S}{\rho_0^S + \rho_0^G}}. \tag{13.114}$$

13.5.4.2   Longitudinal waves

In order to obtain relations for longitudinal waves the scalar product of equations (13.108)-(13.110) with the vector $\mathbf{n}$ is taken:

$$\left\{ \omega^2 - \frac{\lambda^S + 2\mu^S}{\rho_0^S} k^2 + i \frac{\left(\pi^{FS} + \pi^{GS} \right) \omega}{\rho_0^S} \right\} V_\parallel^S$$

$$- \left\{ \frac{Q^F}{\rho_0^S} k^2 + i \frac{\pi^{FS}\omega}{\rho_0^S} \right\} V_\parallel^F - \left\{ \frac{Q^G}{\rho_0^S} k^2 + i \frac{\pi^{GS}\omega}{\rho_0^S} \right\} V_\parallel^G = 0,$$

$$- \left\{ \frac{Q^F}{\rho_0^F} k^2 + i \frac{\pi^{FS}\omega}{\rho_0^F} \right\} V_\parallel^S + \left\{ \omega^2 - \kappa^F k^2 + i \frac{\pi^{FS}\omega}{\rho_0^F} \right\} V_\parallel^F - \frac{Q^{FG}}{\rho_0^F} k^2 V_\parallel^G = 0,$$

$$- \left\{ \frac{Q^G}{\rho_0^G} k^2 + i \frac{\pi^{GS}\omega}{\rho_0^G} \right\} V_\parallel^S - \frac{Q^{FG}}{\rho_0^G} k^2 V_\parallel^F + \left\{ \omega^2 - \kappa^G k^2 + i \frac{\pi^{GS}\omega}{\rho_0^G} \right\} V_\parallel^G = 0, \tag{13.115}$$

where

$$V_\parallel^S = \mathbf{V}^S \cdot \mathbf{n}, \quad V_\parallel^F = \mathbf{V}^F \cdot \mathbf{n}, \quad V_\parallel^G = \mathbf{V}^G \cdot \mathbf{n}. \tag{13.116}$$

The dispersion relation for the longitudinal waves is very lengthy. Instead of presenting it here, again, the frequency and saturation limits are examined. The problem for the unsaturated porous medium yields the existence of three longitudinal waves: $P1$-, $P2$- and $P3$-waves. In the frequency limits they possess the following phase velocities

$$\omega \to 0 \quad \Rightarrow \quad c_{ph} = \begin{cases} \sqrt{\dfrac{\lambda^S + 2\mu^S + \rho_0^F \kappa^F + \rho_0^G \kappa^G + 2(Q^F + Q^G + Q^{FG})}{\rho_0^S + \rho_0^F + \rho_0^G}} & \text{for } P1\text{-waves} \\ 0 & \text{for } P2\text{-waves} \\ 0 & \text{for } P3\text{-waves} \end{cases} \quad (13.117)$$

For the high frequency limit we distinguish also the limits of the initial saturation

$$\omega \to \infty,\ S_0 = 1 \quad \Rightarrow \quad c_{ph} = \begin{cases} \dfrac{\sqrt{2}}{D^{F+}} \sqrt{\dfrac{\lambda^S + 2\mu^S}{\rho_0^S} \kappa^F - \dfrac{Q^{F2}}{\rho_0^S \rho_0^F}} & \text{for } P1\text{-waves} \\ \dfrac{\sqrt{2}}{D^{F-}} \sqrt{\dfrac{\lambda^S + 2\mu^S}{\rho_0^S} \kappa^F - \dfrac{Q^{F2}}{\rho_0^S \rho_0^F}} & \text{for } P2\text{-waves} \\ 0 & \text{for } P3\text{-waves} \end{cases} \quad ,$$

$$D^{F\pm} := \sqrt{\dfrac{\lambda^S + 2\mu^S}{\rho_0^S} + \kappa^F \mp \sqrt{\left(\dfrac{\lambda^S + 2\mu^S}{\rho_0^S} - \kappa^F\right)^2 + \dfrac{4Q^{F2}}{\rho_0^S \rho_0^F}}},$$

$$(13.118)$$

$$\omega \to \infty,\ S_0 = 0 \quad \Rightarrow \quad c_{ph} = \begin{cases} \dfrac{\sqrt{2}}{D^{G+}} \sqrt{\dfrac{\lambda^S + 2\mu^S}{\rho_0^S} \kappa^G - \dfrac{Q^{G2}}{\rho_0^S \rho_0^G}} & \text{for } P1\text{-waves} \\ \dfrac{\sqrt{2}}{D^{G-}} \sqrt{\dfrac{\lambda^S + 2\mu^S}{\rho_0^S} \kappa^G - \dfrac{Q^{G2}}{\rho_0^S \rho_0^G}} & \text{for } P2\text{-waves} \\ 0 & \text{for } P3\text{-waves} \end{cases} \quad ,$$

$$D^{G\pm} := \sqrt{\dfrac{\lambda^S + 2\mu^S}{\rho_0^S} + \kappa^G \mp \sqrt{\left(\dfrac{\lambda^S + 2\mu^S}{\rho_0^S} - \kappa^G\right)^2 + \dfrac{4Q^{G2}}{\rho_0^S \rho_0^G}}}.$$

These relations shall not be discussed here. Instead, we show the results of the numerical example: air-water-mixture in sandstone (figures and comments from [11]).

13.5.5 Numerical analysis of the wave propagation

The dispersion relations are solved for the complex wave number k. The solution gives rise to the phase velocities $\frac{\omega}{\mathrm{Re}\,k}$ and the attenuations $\mathrm{Im}\,k$ of the four waves. The numerical data of Table 13.4 and the following additional values for the resistances of the two pore fluids water and air are used

$$\pi^F = 10^7 \ \text{kg/m}^3\text{s}, \quad \pi^G = 1.82 \cdot 10^5 \ \text{kg/m}^3\text{s}. \quad (13.119)$$

Further soil textures and other pore fillings are investigated in [13].

Before looking at the numerical results in dependence on the frequency and on the degree of initial saturation, the velocities in the limits of those quantities are numerically determined. We obtain the limit values of the speeds of propagation for the shear wave from (13.113) and (13.114). Thus, for $S_0 = 1$:

$$\lim_{\omega \to \infty} c_{ph} = c_{S\infty} = 908.67 \ \text{m/s},$$

$$(13.120)$$

$$\lim_{\omega \to 0} c_{ph} = c_{S0} = 856.34 \ \text{m/s},$$

and for $S_0 \to 0$

$$\lim_{\omega \to \infty} c_{ph} = \lim_{\omega \to 0} c_{ph} = c_S' = 908.67 \ \text{m/s}. \quad (13.121)$$

Using the material parameters for the saturated porous medium obtained in K. Wilmanski [443]: $\lambda^S + \frac{2}{3}\mu^S = 7.2153 \cdot 10^9$ Pa, $\kappa^F = 2 \cdot 10^6 \frac{\text{m}^2}{\text{s}^2}$ and $Q^F = 1.3493 \cdot 10^9$ Pa, for the $P1$-, $P2$-, and $P3$-waves (13.117), (13.118) we obtain for $S_0 = 1$

$$\lim_{\omega \to \infty} c_{ph} = \begin{cases} c_{P1\infty} = 2391.00 \text{ m/s}, \\ c_{P2\infty} = 1005.61 \text{ m/s}, \\ c_{P3\infty} = 0 \text{ m/s}, \end{cases}$$

$$\lim_{\omega \to 0} c_{ph} = \begin{cases} c_{P10} = 2373.23 \text{ m/s}, \\ c_{P20} = 0 \text{ m/s}, \\ c_{P30} = 0 \text{ m/s}. \end{cases} \tag{13.122}$$

For the limit $S_0 = 0$ limit values of the material parameters obtained in [13] are used: $\lambda^S + \frac{2}{3}\mu^S = 3.56 \cdot 10^9$ Pa, $\kappa^G = 8.42 \cdot 10^4 \frac{\text{m}^2}{\text{s}^2}$ and $Q^G = 6.83 \cdot 10^4$ Pa. Then, the following limit velocities are obtained for $S_0 = 0$, i.e., in the case of air-saturation:

$$\lim_{\omega \to \infty} c_{ph} = \begin{cases} c_{P1\infty} = 1699.88 \text{ m/s}, \\ c_{P2\infty} = 290.11 \text{ m/s}, \\ c_{P3\infty} = 0 \text{ m/s}, \end{cases}$$

$$\lim_{\omega \to 0} c_{ph} = \begin{cases} c_{P10} = 1699.99 \text{ m/s}, \\ c_{P20} = 0 \text{ m/s}, \\ c_{P30} = 0 \text{ m/s}. \end{cases} \tag{13.123}$$

13.5.5.1   Discussion of numerical results

Figures 13.21 and 13.22 show the numerical results of the wave analysis. The wave speeds of the shear wave S and of the three longitudinal waves $P1$, $P2$ and $P3$ are shown on the left-hand sides of the figures, the attenuations on the right-hand sides.

In Figure 13.21 the phase speeds and the attenuations of the four waves are given in dependence on the frequency ω. The different curves represent the values of the initial saturation $S_0 = 0.2, 0.4, 0.6$ and 0.99999. The latter corresponds to a nearly water-saturated sandstone.

As known from the works about waves in saturated porous media and as presented above, the speeds of the transversal wave S and of the $P1$-wave expose a low and a high limit value. For each value of the initial saturation the phase speed of both S- and $P1$-waves grows a little from its initial value to the asymptotic speed for $\omega \to \infty$. In the low frequency range which appears in geotechnical applications the speeds of both waves are nearly constant. The increase occurs in the region of ultrasonic frequencies which may be excited in laboratory experiments. The $P1$-wave is the fastest of the appearing waves. This is related to the fact that it propagates mainly in the skeleton. The transversal wave is slower and also the difference of the limit values is smaller. For very large values of the frequency the speeds of both waves are, again, nearly constant.

The $P2$-wave and the $P3$-wave start from zero speed for zero frequency. Otherwise the frequency dependence is similar to this of the $P1$- and S-wave. At relative high values of the frequency a strong increase of the velocity takes place before an asymptotic value is reached. However, at least near water saturation, the increase is stronger than for the other waves. In the next figure the dependence of the waves on the saturation is shown and it will be obvious that the $P3$-wave only exists if a second pore fluid is present. But here it can be seen already that the velocities for each degree of saturation are much lower than for all other waves. For the highest value of the saturation the velocity of this wave is even so low that it does not appear in the figure anymore.

This is consistent with the attenuation of the $P3$-wave for this degree of saturation. It is some orders of magnitude larger than any other value of the attenuation. Both $P1$- and S-wave have a low attenuation. In the region of high frequencies shown in Figure 13.21 (normal scale of the frequency axis) it is especially low for high degrees of saturation. From the analysis of wave propagation in saturated poroelastic media it is already known that the second longitudinal wave is highly damped and therefore hard to observe especially in non-artificial materials. This problem is even more pronounced for the $P3$-wave because its attenuation is even higher than this of the $P2$-wave. Thus the observation of this wave will be nearly impossible.

Figure 13.22 shows the dependence of the four waves on the initial saturation. Again, on the left-hand side the speeds are given, on the right-hand side the corresponding attenuations are shown. Now, the various lines represent different values of the frequency. Because there result some astonishing curves in the region of very high frequencies (especially if the second pore fluid is oil and not air, see [13]) which have no clear interpretation up to now, velocities and attenuations are given for the relative low frequencies $\omega = 10$, 100, 500, 1000 and 5000 Hz. For geophysical applications these are anyway frequencies which have a practical bearing.

The saturation axis covers the whole region from $0 \leq S_0 \leq 1$, however, observation of the curves shows that the model is not applicable for very low degrees of saturation. That was to be expected because in this region presumptions of the model are not fulfilled anymore. Instead of a continuous fluid with air inclusions in this region of saturation a frothy structure of the pore fluids is encountered. A second reason for acceptance of the limitation is that smaller values of the saturation do not appear in reality (a residual amount of fluid is trapped in the channels).

The shear modulus, μ^S, is for a given Poisson number constant and the wave velocity depends only on the mass density of the skeleton. Thus, it decreases linearly with increasing saturation. In nearly the whole range of saturations also the other elastic constant of the skeleton, λ^S, changes only marginally. However, for a degree of saturation which is closely to the state of water saturation it increases abruptly and reaches approximately the double of the value before. Thus, the $P1$-wave proceeds nearly constant for almost all values of the saturation and increases only for very large values of S_0. Again, it is obvious that the changes in attenuation of both waves are small.

For the other waves, $P2$ and $P3$, the influence of the degree of saturation is higher. These waves are effected by the existence of the fluid and the gas. The speed of the $P2$-wave decreases as the degree of saturation increases up to approximately 75% and then increases rapidly until saturation is reached. It is opposite for the $P3$-wave: it increases with increasing S_0 up to 75% and then decreases. Both for gas saturation and for water saturation this wave disappears. This shows that it is driven by the capillary pressure and thus only exists if a second pore fluid is existent (see the work [396] whose authors were most likely the first in predicting this form of $P3$-waves in such media). Its velocity is much smaller than this of the other waves, however, its attenuation is much higher. Due to the high values of attenuation it is not astonishing that we are not aware of any attempt to measure this third longitudinal wave, either in artificial media or even less in non-artificial ones.

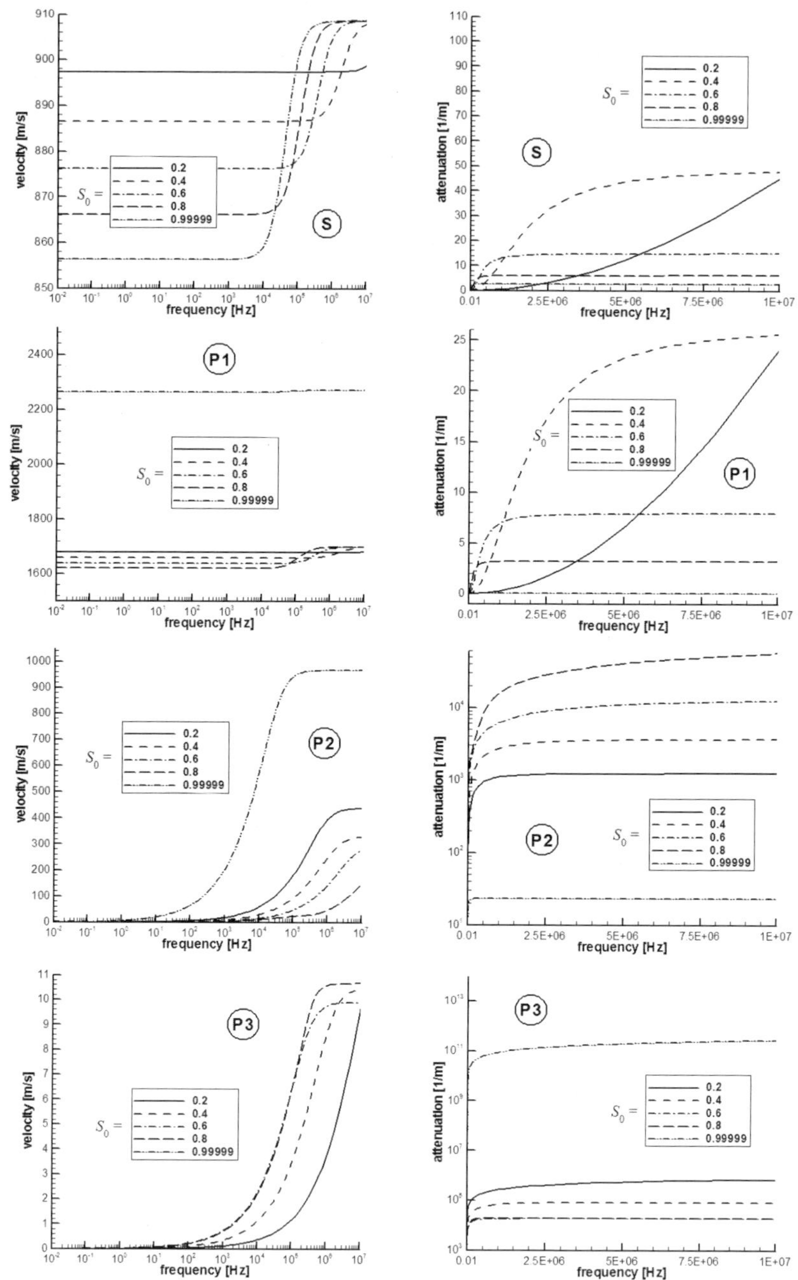

Figure 13.21: *Speeds (left) and attenuations (right) of the four waves* $P1, P2, P3$ *and* S *in dependence on the frequency* ω *(from [11]).*

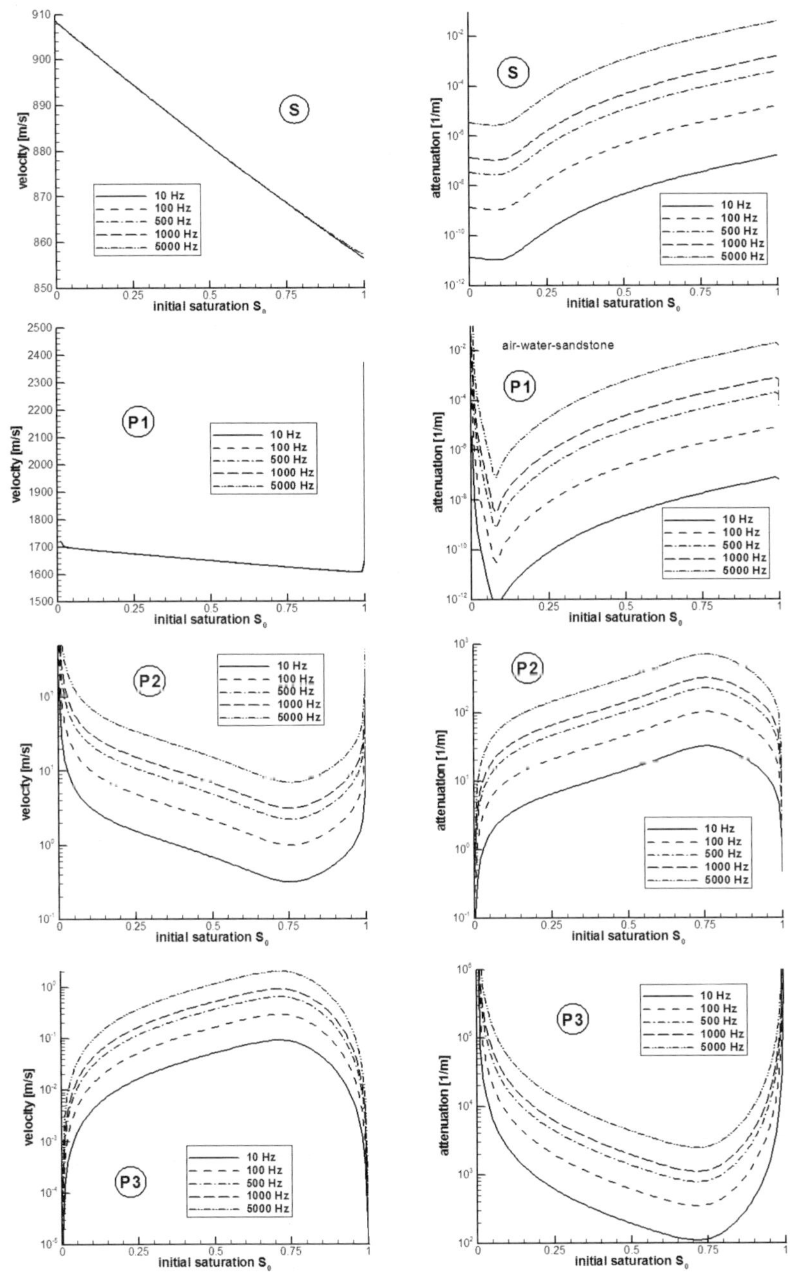

Figure 13.22: *Speeds (left) and attenuations (right) of the four waves $P1, P2, P3$ and S in dependence on the initial saturation S_0 (from [11]).*

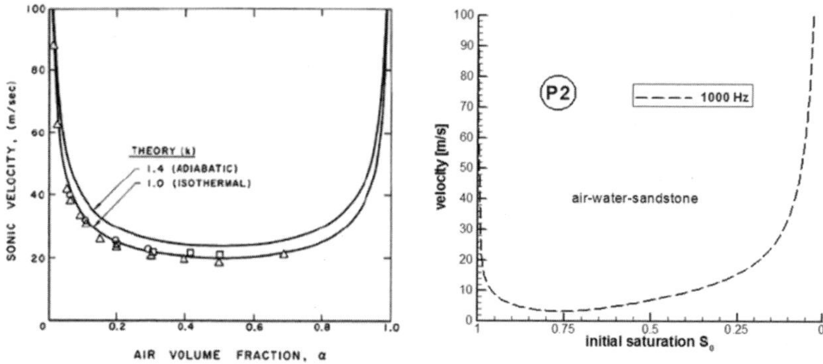

Figure 13.23: *Left: velocity of sound in air-water mixtures [58] in dependence on the air volume fraction, right: speed of the P2-wave in dependence on the air fraction (initial saturation in reverse axis).*

13.5.6   Comparison with suspensions and experiments

The behavior of the wave velocities in the air-water-mixture in sandstone, of course, shows similarities to the propagation of sound in suspensions, especially in an air-water-mixture. It is well known that the existence of air bubbles in water reveals a minimum in the sonic velocity. Figure 13.23 (left) is taken from the book [58] and shows results of the velocity of sound in an air-water-mixture for a frequency of 1000 Hz in dependence on the air volume fraction. As shown in right panel of the figure such a behavior also appears in the partially saturated porous medium since the air-water-mixture is only jacketed by the solid particles.

Besides the similarity to the propagation of sound in suspensions there exists also qualitative agreement of the calculated wave velocities with experimental results. The reproduction of one figure of [268], Figure 13.24 (left), shows the measured wave velocities of the S- and $P1$-waves in Massilon sandstone (porosity $n_0 = 0.23$). Due to the high attenuation of the $P2$- and $P3$-wave these waves could not be observed. However, in spite of somewhat different conditions (porosity, frequency, temperature etc.) the accordance of the experimentally observed S- and $P1$-wave velocities and their calculated counterparts shown in Figure 13.24 (right) is quite well. Both $P1$-velocities show the strong increase of this velocity for high degrees of saturation and both S-wave velocities behave linearly and nearly constant in the whole range of saturations. Also the order of magnitude of the wave velocities matches closely.

13.5.7   Some remarks

The wave analysis of a model for partially saturated soils containing three elastic constants and three coupling constants has been accomplished. For the special case of an air-water-mixture in sandstones four body waves are predicted: three longitudinal waves, $P1$, $P2$, $P3$, and one shear wave, S.

The fastest longitudinal wave, $P1$, which propagates mainly in the skeleton, and the shear wave, S, are relatively unaffected by the saturation except for a small region near

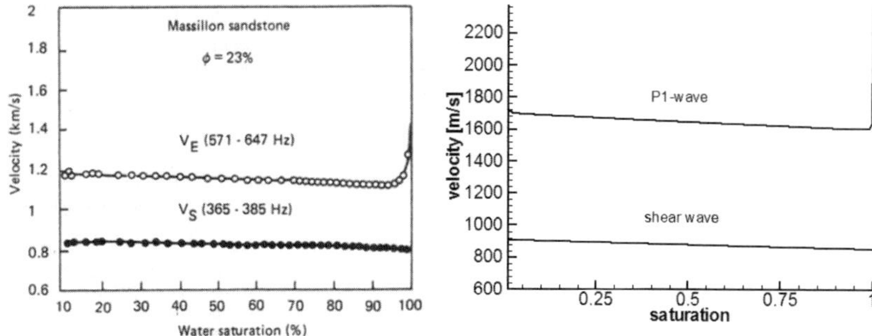

Figure 13.24: *Comparison of experimental results (left) by Murphy and numerical results for the velocities of the fast longitudinal and the shear wave.*

water saturation. In this region the velocity of the $P1$-wave increases abruptly to nearly the double of its value for other saturations. Experimental results found in the literature support the occurrence of this effect. This may be an important feature for applications in geotechnics. It provides the hope for the development of a non-destructive testing method to warn against land slides. The latter occur if the degree of saturation exceeds a certain value. It is favorable for such a method that this feature occurs for the $P1$-wave which is the first arrival on an oscillogram. However, it is disadvantageous that it appears in a very narrow range of saturation. Further applications are imaginable as for example in piping, oil industry, landfilled waste management or for the remediation of compacted residual soils. The $P2$- and $P3$-waves, which are effected by the existence of the fluid and the gas, are much more affected by the degree of saturation. They behave in the opposite way. While the velocity of the $P2$-wave has a strong minimum for medium values of saturation (as it is the case for suspensions) the $P3$-wave only exists in this region of saturation. Both for the water- and for the gas-saturated medium it does not emerge since it is evoked by the capillary pressure between the pore fluids. Both these waves are strongly damped, for the $P3$-wave the attenuation is so high that an observation of this wave in the field is nearly impossible.

Appendices

Appendix A

Basic notions

A.1 Mathematical basics shown on the example of polar decomposition

In Section 2.1.4 the procedure for the polar decomposition of a tensor is described. In this appendix a numerical example for the procedure is demonstrated which, simultaneously, illustrates the basic notions related to matrix manipulations.

<u>matrices</u>   deformation gradient – 3-rowed, quadratic matrix

$$\mathbf{F} = \begin{pmatrix} 1 & 1 & 0 \\ 2 & 2 & 1 \\ 0 & 2 & 1 \end{pmatrix} \leftarrow i\text{-th row} \qquad (A.1)$$

$$\uparrow$$
$$k\text{-th column}$$

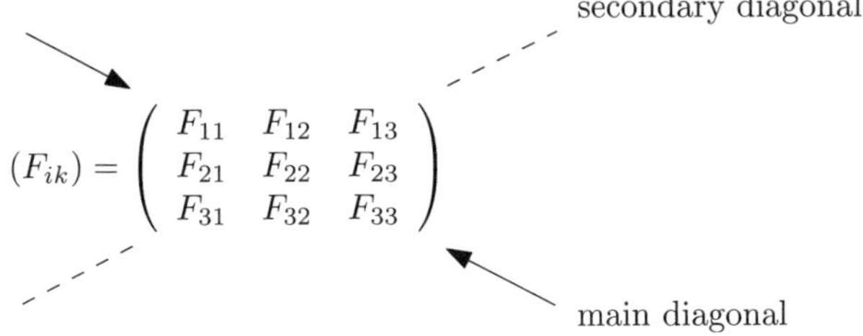

<u>transpose</u>

The result of the interchange of the rows and columns of a matrix is the transposed matrix

$$\mathbf{F}^T = \begin{pmatrix} 1 & 2 & 0 \\ 1 & 2 & 2 \\ 0 & 1 & 1 \end{pmatrix}. \qquad (A.2)$$

<u>addition</u>

Addition of matrices is performed element by element.

matrix multiplication

"row times column"

$$\left(\begin{array}{c}\boxed{\text{row}}\end{array}\right)\cdot\left(\begin{array}{c}\begin{smallmatrix}c\\o\\l\\u\\m\\n\end{smallmatrix}\end{array}\right)=\left(\begin{array}{c}\blacksquare\end{array}\right)$$

Right Cauchy-Green deformation tensor

$$\mathbf{C} = \mathbf{F}^T\mathbf{F} = \begin{pmatrix} 1 & 2 & 0 \\ 1 & 2 & 2 \\ 0 & 1 & 1 \end{pmatrix}\begin{pmatrix} 1 & 1 & 0 \\ 2 & 2 & 1 \\ 0 & 2 & 1 \end{pmatrix} = \begin{pmatrix} 5 & 5 & 2 \\ 5 & 9 & 4 \\ 2 & 4 & 2 \end{pmatrix} \tag{A.3}$$

trace
Sum of the elements of the main diagonal

$$\operatorname{tr}(\mathbf{C}) = 5 + 9 + 2 = 16 \tag{A.4}$$

determinant

1. Laplace expansion: calculation by expansion according to a chess board scheme

	+	−	+		+	−	+
+	C_{11}	C_{12}	C_{13}	+	5	5	2
−	C_{21}	C_{22}	C_{23}	−	5	9	4
+	C_{31}	C_{32}	C_{33}	+	2	4	2

Expansion along the first column:

$$\begin{vmatrix} 5 & 5 & 2 \\ 5 & 9 & 4 \\ 2 & 4 & 2 \end{vmatrix} = 5\cdot\begin{vmatrix} 9 & 4 \\ 4 & 2 \end{vmatrix} - 5\cdot\begin{vmatrix} 5 & 2 \\ 4 & 2 \end{vmatrix} + 2\cdot\begin{vmatrix} 5 & 2 \\ 9 & 4 \end{vmatrix}$$

subdeterminant (minor)

guideline for 2-rowed determinants:

"main diagonal minus secondary diagonal"

$$\det(\mathbf{C}) = 5\cdot(9\cdot2 - 4\cdot4) - 5\cdot(5\cdot2 - 4\cdot2) + 2\cdot(5\cdot4 - 9\cdot2) = 4. \tag{A.5}$$

Using the expansion rule of Laplace one is able to expand the determinant of a $n \times n$-matrix "along a row or along a column". The formulae are

$$\det(\mathbf{C}) = \sum_{i=1}^{n}(-1)^{i+j}\cdot C_{ij}\cdot\det(M_{ij}) \quad \text{(expansion along the } j\text{-th column)},$$

$$\det(\mathbf{C}) = \sum_{j=1}^{n}(-1)^{i+j}\cdot C_{ij}\cdot\det(M_{ij}) \quad \text{(expansion along the } i\text{-th row)},$$

where M_{ij} is the $(n-1) \times (n-1)$-submatrix of $\mathbf{C}$, which arises by deletion of the i-th row and the j-th column.

2. calculation according to the rule of Sarrus

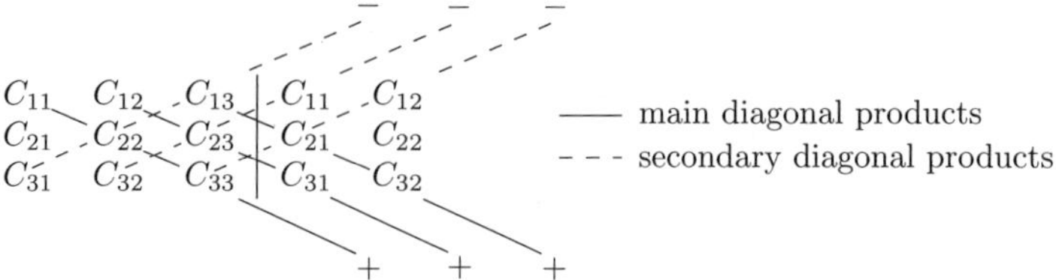

Columns 1 and 2 of the determinant are put again on the right-hand side beside the determinant. The value of the determinant is obtained by summing up the three main diagonal products and subtracting the three secondary diagonal products.

$$\det(\mathbf{C}) = C_{11}C_{22}C_{33} + C_{12}C_{23}C_{31} + C_{13}C_{21}C_{32}$$
$$- C_{13}C_{22}C_{31} - C_{11}C_{23}C_{32} - C_{12}C_{21}C_{33}$$
$$= 5\cdot 9\cdot 2 + 5\cdot 4\cdot 2 + 2\cdot 5\cdot 4 - 2\cdot 9\cdot 2 - 5\cdot 4\cdot 4 - 5\cdot 5\cdot 2 = 4$$

3. Gauss algorithm

$$\mathbf{C} = \begin{pmatrix} 5 & 5 & 2 \\ 5 & 9 & 4 \\ 2 & 4 & 2 \end{pmatrix} \qquad \begin{aligned} &1.\ \text{column}_{\text{new}} \\ &= 1.\ \text{column}_{\text{old}} \\ &- 3.\ \text{column}_{\text{old}} \end{aligned}$$

$$\mathbf{C}_1 = \begin{pmatrix} 3 & 5 & 2 \\ 1 & 9 & 4 \\ 0 & 4 & 2 \end{pmatrix} \qquad \begin{aligned} &2.\ \text{column}_{\text{new}} \\ &= 2.\ \text{column}_{\text{old}} \\ &+ 3.\ \text{column}_{\text{old}} \cdot (-2) \end{aligned}$$

$$\mathbf{C}_2 = \begin{pmatrix} 3 & 1 & 2 \\ 1 & 1 & 4 \\ 0 & 0 & 2 \end{pmatrix} \qquad \begin{aligned} &2.\ \text{row}_{\text{new}} \\ &= 2.\ \text{row}_{\text{old}} \\ &+ 1.\ \text{row}_{\text{old}} \cdot \left(-\tfrac{1}{3}\right) \end{aligned}$$

$$\mathbf{C}_3 = \begin{pmatrix} 3 & 1 & 2 \\ 0 & \frac{2}{3} & \frac{10}{3} \\ 0 & 0 & 2 \end{pmatrix}$$

By use of elementary row or column operations the matrix is reduced to a triangular form. The product of the main diagonal elements of this triangular matrix is the determinant of the initial matrix if no exchange of rows or columns takes place.

$$\det(\mathbf{C}) = 3\cdot \frac{2}{3}\cdot 2 = 4.$$

invariants

$$I = \operatorname{tr}(\mathbf{C}) = 16,$$

$$II = \frac{1}{2}\left[I^2 - \operatorname{tr}\left(\mathbf{C}^2\right)\right] = \frac{1}{2}(256 - 200) = 28 \tag{A.6}$$

$$\therefore \quad C^2 = \begin{pmatrix} 5 & 5 & 2 \\ 5 & 9 & 4 \\ 2 & 4 & 2 \end{pmatrix} \cdot \begin{pmatrix} 5 & 5 & 2 \\ 5 & 9 & 4 \\ 2 & 4 & 2 \end{pmatrix}$$

$$= \begin{pmatrix} 54 & ? & ? \\ ? & 122 & ? \\ ? & ? & 24 \end{pmatrix} \Longrightarrow \operatorname{tr}\left(\mathbf{C}^2\right) = 200$$

$$III = \det(\mathbf{C}) = 4.$$

unit matrix

The main diagonal elements of this matrix are ones, the others are zero

$$\mathbf{E} = \mathbf{I} = \begin{pmatrix} 1 & 0 & 0 \\ 0 & 1 & 0 \\ 0 & 0 & 1 \end{pmatrix}. \tag{A.7}$$

eigenvalue problem for **C**

$$(\mathbf{C} - \lambda\mathbf{1})\,\mathbf{r} = \mathbf{0} \tag{A.8}$$

$$\nearrow \qquad \nwarrow$$

eigenvalues   eigenvectors

Solution of the problem by

$$\det(\mathbf{C} - \lambda_a\mathbf{1})\,\mathbf{r} = \mathbf{0} \tag{A.9}$$

$$\uparrow$$

a-th eigenvalue

$$\Longrightarrow \quad \begin{vmatrix} 5-\lambda & 5 & 2 \\ 5 & 9-\lambda & 4 \\ 2 & 4 & 2-\lambda \end{vmatrix} = 0.$$

This gives rise to a *characteristic equation* which can also be written in the following way:

$$\lambda^3 - I\lambda^2 + II\lambda - III = 0, \tag{A.10}$$

i.e., in the example we have to solve the equation

$$\lambda^3 - 16\lambda^2 + 28\lambda - 4 = 0.$$

The solution of the cubic equation can be found by use of a pocket calculator or a computer package as e.g. *Maple*, *Mathematica* or *Matlab*. However, often the problem reduces to quadratic equations of the form

$$\lambda^2 + p\lambda + q = 0. \tag{A.11}$$

Solutions of the quadratic equation:

$$\lambda_{1/2} = -\frac{p}{2} \pm \sqrt{\frac{p^2}{4} - q}, \tag{A.12}$$

here the solutions of the cubic equation are the three *eigenvalues*

$$\lambda_1 = 0.157, \qquad \lambda_2 = 1.820, \qquad \lambda_3 = 14.024.$$

determination of eigenvectors

Here, exemplarily, for the eigenvalue $\lambda_1 = 0.157$:

$$
\begin{array}{rcccccl}
\boxed{4.843} & & & & & & \\
(5 - \lambda_1)\, r_1 & + & 5r_2 & + & 2r_3 & = & 0 \\
 & & \boxed{8.843} & & & & \\
5r_1 & + & (9 - \lambda_1)\, r_2 & + & 4r_3 & = & 0 \\
 & & & & \boxed{1.843} & & \\
2r_1 & + & 4r_2 & + & (2 - \lambda_1)\, r_3 & = & 0
\end{array}
$$

This is a homogeneous system of equations. It can be solved by elimination (non-homogeneous problems have to be solved by the rule of Cramer).

$$
\begin{aligned}
\text{1. eq.:} \quad & r_3 = -2.4215r_1 - 2.5r_2 \\
\text{2. eq.:} \quad & 5r_1 + 8.843r_2 - 4\,(2.4215r_1 + 2.5r_2) = 0 \\
& -4.686r_1 - 1.157r_2 = 0 \\
\implies \quad & r_2 \approx -4.050r_1, \qquad r_3 \approx 7.704r_1.
\end{aligned}
\tag{A.13}
$$

The eigenvectors shall be unit vectors therefore the following relation must hold

$$r_1^2 + r_2^2 + r_3^2 = 1. \tag{A.14}$$

After insertion of (A.13)

$$
\begin{aligned}
& r_1^2 + (-4.050r_1)^2 + (7.704r_1)^2 = 1 \\
\implies \quad & 76.754r_1^2 = 1 \\
\implies \quad & r_1 = 0.114, \quad r_2 = -0.462, \quad r_3 = 0.879.
\end{aligned}
$$

The same procedure has to be carried out for the other two eigenvalues. Finally:

$$
\begin{aligned}
\mathbf{r}^{(2)} &= (0.851, -0.411, -0.326)^T, \\
\mathbf{r}^{(3)} &= (0.512, 0.786, 0.347)^T.
\end{aligned}
\tag{A.15}
$$

Therewith

$$
(\Lambda^{ij}) =
\begin{pmatrix}
\sqrt{\lambda_1} & 0 & 0 \\
0 & \sqrt{\lambda_2} & 0 \\
0 & 0 & \sqrt{\lambda_3}
\end{pmatrix}
=
\begin{pmatrix}
0.396 & 0 & 0 \\
0 & 1.349 & 0 \\
0 & 0 & 3.745
\end{pmatrix},
\tag{A.16}
$$

$$
(\psi^{ij}) =
\begin{pmatrix}
r_1^{(1)} & r_1^{(2)} & r_1^{(3)} \\
r_2^{(1)} & r_2^{(2)} & r_2^{(3)} \\
r_3^{(1)} & r_3^{(2)} & r_3^{(3)}
\end{pmatrix}
=
\begin{pmatrix}
0.114 & 0.851 & 0.512 \\
-0.462 & -0.411 & 0.786 \\
0879 & -0.326 & 0.347
\end{pmatrix},
\tag{A.17}
$$

$$
\mathbf{U} = \boldsymbol{\psi}\boldsymbol{\Lambda}\boldsymbol{\psi}^T =
\begin{pmatrix}
1.965 & 1.015 & 0.330 \\
1.015 & 2.625 & 1.040 \\
0.330 & 1.040 & 0.900
\end{pmatrix}.
\tag{A.18}
$$

<u>inverse</u>
The matrix which after multiplication with a matrix $\mathbf{U}$ yields the unit matrix is called the inverse matrix of $\mathbf{U}$, i.e., $\mathbf{U}^{-1}$.

Procedure to build $\mathbf{U}^{-1}$ for a given, regular (i.e., invertible; $\det \mathbf{U} \neq \mathbf{0}$) matrix $\mathbf{U}$

1. transpose $\mathbf{U}$ $\Rightarrow$ $\mathbf{U}^T$

2. build all subdeterminants of $\mathbf{U}^T$ and write them – using a chess board scheme – into a new matrix $\Rightarrow$ adj $(\mathbf{U})$

3. $\mathbf{U}^{-1} = \mathrm{adj}\,(\mathbf{U})\,/\det \mathbf{U}$.

Performed for the *example*:

$\det \mathbf{U} = 2 \neq 0$ $\Rightarrow$ $\mathbf{U}$ is regular.

$$1.\ \mathbf{U}^T = \begin{pmatrix} 1.965 & 1.015 & 0.330 \\ 1.015 & 2.625 & 1.040 \\ 0.330 & 1.040 & 0.900 \end{pmatrix} = \mathbf{U}$$

$$\mathbf{U}^T = \mathbf{U} \quad \Rightarrow \quad \text{symmetric matrix}$$

For comparison: for skew symmetric or anti metric matrices holds: $\mathbf{A} = -\mathbf{A}^T$

2. Subdeterminants:

$$D_{11} = \begin{vmatrix} 2.625 & 1.040 \\ 1.040 & 0.900 \end{vmatrix} = 1.281 \qquad D_{12} = \begin{vmatrix} 1.015 & 1.040 \\ 0.330 & 0.900 \end{vmatrix} = 0.570$$

$$D_{13} = \begin{vmatrix} 1.015 & 2.625 \\ 0.330 & 1.040 \end{vmatrix} = 0.189 \qquad D_{21} = \begin{vmatrix} 1.015 & 0.330 \\ 1.040 & 0.900 \end{vmatrix} = 0.570$$

$$D_{22} = \begin{vmatrix} 1.965 & 0.330 \\ 0.330 & 0.900 \end{vmatrix} = 1.660 \qquad D_{23} = \begin{vmatrix} 1.965 & 1.015 \\ 0.330 & 1.040 \end{vmatrix} = 1.709$$

$$D_{31} = \begin{vmatrix} 1.015 & 0.330 \\ 2.625 & 1.040 \end{vmatrix} = 0.189 \qquad D_{32} = \begin{vmatrix} 1.965 & 0.330 \\ 1.015 & 1.040 \end{vmatrix} = 1.709$$

$$D_{33} = \begin{vmatrix} 1.965 & 1.015 \\ 1.015 & 2.625 \end{vmatrix} = 4.128$$

$$3.\ \mathbf{U}^{-1} = \frac{1}{\det \mathbf{U}} \begin{pmatrix} D_{11} & -D_{21} & D_{31} \\ -D_{12} & D_{22} & -D_{32} \\ D_{13} & -D_{23} & D_{33} \end{pmatrix}$$

$$= \frac{1}{2} \begin{pmatrix} 1.2810 & -0.570 & 0.189 \\ -0.570 & 1.660 & -1.709 \\ 0.189 & -1.709 & 4.128 \end{pmatrix} = \begin{pmatrix} 0.641 & -0.285 & 0.095 \\ -0.285 & 0.830 & -0.855 \\ 0.095 & -0.855 & 2.064 \end{pmatrix}$$

$$\mathbf{R} = \mathbf{F}\mathbf{U}^{-1} = \begin{pmatrix} 1 & 1 & 0 \\ 2 & 2 & 1 \\ 0 & 2 & 1 \end{pmatrix} \begin{pmatrix} 0.641 & -0.285 & 0.095 \\ -0.285 & 0.830 & -0.855 \\ 0.095 & -0.855 & 2.064 \end{pmatrix}$$

$$= \begin{pmatrix} 0.356 & 0.545 & -0.760 \\ 0.807 & 0.235 & 0.544 \\ -0.475 & 0.805 & 0.354 \end{pmatrix}$$

$$\det(\mathbf{R}) = -1 \searrow$$

orthogonal matrix

The determinant of <u>orthogonal matrices</u> is either 1 or -1. If the determinant is -1, as in the present example, then a rotational reflection takes place.

A matrix $\mathbf{Q}$ is called orthogonal if its transpose is equal to its inverse $\implies \quad \mathbf{Q} \cdot \mathbf{Q}^T = \mathbf{I}.$

testing of $\mathbf{R}$:

$$\mathbf{R} \cdot \mathbf{R}^T = \begin{pmatrix} 0.356 & 0.545 & -0.760 \\ 0.807 & 0.235 & 0.544 \\ -0.475 & 0.805 & 0.354 \end{pmatrix} \begin{pmatrix} 0.356 & 0.8075 & -0.475 \\ 0.54 & 0.235 & 0.805 \\ -0.760 & 0.544 & 0.354 \end{pmatrix}$$

$$\approx \begin{pmatrix} 1 & 0 & 0 \\ 0 & 1 & 0 \\ 0 & 0 & 1 \end{pmatrix} = \mathbf{I}.$$

The result is not exact due to roundings.

<u>On the geometric interpretation of the scalar product</u>

About which angle does the vector $\mathbf{w} = (1, 2, 3)^T$ rotate due to action of $\mathbf{F}$?

Absolute value of a vector

$$|\mathbf{w}| = \sqrt{1^2 + 2^2 + 3^2} = 3.742$$
$$\uparrow$$
length of $\mathbf{w}$

$$\mathbf{w}' = \mathbf{F}\mathbf{w} = \begin{pmatrix} 1 & 1 & 0 \\ 2 & 2 & 1 \\ 0 & 2 & 1 \end{pmatrix} \begin{pmatrix} 1 \\ 2 \\ 3 \end{pmatrix} = \begin{pmatrix} 3 \\ 9 \\ 7 \end{pmatrix}$$
$$\uparrow$$
column vector

$$|\mathbf{w}'| = \sqrt{3^2 + 9^2 + 7^2} = 11.790$$

angle

$$\cos\varphi = \frac{\mathbf{w}' \cdot \mathbf{w}}{|\mathbf{w}'|\,|\mathbf{w}|} = \frac{42}{3.742 \cdot 11.790} = 0.9520$$

$$\therefore \quad \mathbf{w}' \cdot \mathbf{w} = \begin{pmatrix} 3 \\ 9 \\ 7 \end{pmatrix} \cdot \begin{pmatrix} 1 \\ 2 \\ 3 \end{pmatrix} = 3 \cdot 1 + 9 \cdot 2 + 7 \cdot 3 = 42$$
$$\uparrow$$
scalar product (inner product) of two vectors

$$\rightsquigarrow \quad \varphi = 17.82°$$

A.2 Curvilinear coordinate systems

A.2.1 General considerations

In this appendix a description of some orthogonal and one non-orthogonal curvilinear coordinate systems in the Euclidean space is introduced. Mathematical details can be found in many books on differential geometry, e.g. S. Sternberg [365]. An even broader choice of coordinate systems than here is described in the book of H. Margenau & G. M. Murphy [241] first published in 1943. This book is also the source for the presentation of the somewhat more seldom used coordinate systems presented in Subsections A.2.4 to A.2.7.

We begin with the extension of the fundamental geometrical notions for three-dimensional Euclidean spaces which were already presented in Sections A.2-A.4 of Part I. For all coordinate systems we choose a common origin and then an arbitrary point $\mathbf{x}$ of the space which we identify with the position vector whose beginning is located in the origin of coordinates. Such an identification is possible only for Euclidean spaces as vectors in more general, for instance Riemann spaces, belong to the so-called tangent spaces and cannot be identified with any fragment of a straight line in such spaces. We show an example of such a two-dimensional space in Figure A.1.

The 2D Riemann space shown in this figure consists of points, say $\mathbf{x}$, and at each such point a tangent vector space is a plane. All tangent vectors to the space are lying in this plane and, of course, one cannot connect two points from the space, say $\mathbf{x}_1$ and $\mathbf{x}_2$, with a vector belonging to those tangent spaces. The vector $\mathbf{x}_1 - \mathbf{x}_2$ connecting these points in the three-dimensional Euclidean space in which we have embedded the two-dimensional Riemann space has nothing to do with vectors connected with this two-dimensional space. One cannot even compare vectors belonging to tangent spaces in different points of the space. In this simple example we could embed the 2D Riemann space in a three-dimensional Euclidean space but this is not always possible. However, dealing with the Euclidean rather than Riemannian spaces we can identify all tangent vector spaces for different points – they are isomorphic – and, in this way, vectors which begin in the origin can be identified with vectors which begin at any other point $\mathbf{x}$ of the space. In this appendix we deal only with such situations.

If x^k are right-handed Cartesian coordinates then the functions

$$y^\alpha = y^\alpha \left(x^1, x^2, x^3\right) \equiv y^\alpha \left(x^k\right), \quad \alpha = 1, 2, 3, \quad k = 1, 2, 3, \quad \det \frac{\partial y^\alpha}{\partial x^k} \neq 0, \qquad (A.19)$$

define new coordinates provided that these functions are continuously differentiable. In this case the condition $y^\alpha = $ const. for either $\alpha = 1, 2$ or 3 describes the *parametric surface* $\mathcal{S}^\alpha$ which – in contrast to Cartesian coordinates – does not have to be a plane. The Greek index α distinguishes curvilinear coordinates from Cartesian coordinates which we label with Latin letters. This has not been done in Appendix A of Part I because certain transformations for which it is essential and which are discussed here were not presented in Part I.

The intersection of two parametric surfaces with different indices α determines curves which are denoted as *parametric lines* $\mathcal{C}^\alpha$ (see Figure A.2). Thus, for instance, the intersection of the surfaces $\mathcal{S}^2$ and $\mathcal{S}^3$ determines the parametric line $\mathcal{C}^1$: $\mathcal{C}^1 = \left\{\mathbf{x} \mid y^2 = \right.$ const., $y^3 = $ const.$\left.\right\}$.

The *position vector* in the Euclidean space can be, of course, written in the following form

$$\mathbf{x} = x^i \mathbf{e}_i, \quad \mathbf{e}_i \cdot \mathbf{e}_j = \delta_{ij}, \qquad (A.20)$$

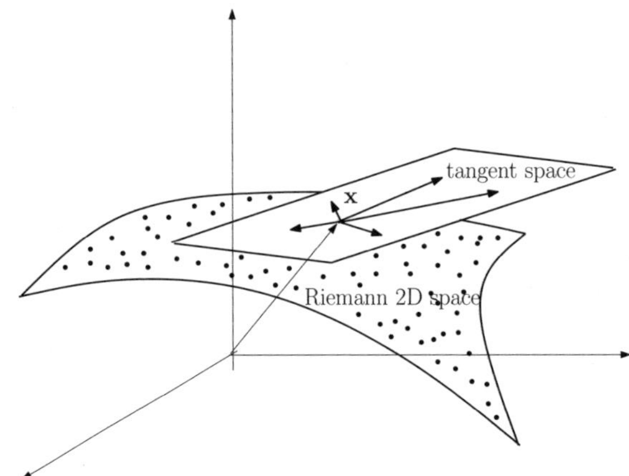

Figure A.1: *Two-dimensional Riemann space and the vector space tangent to the Riemann space at the point* **x**.

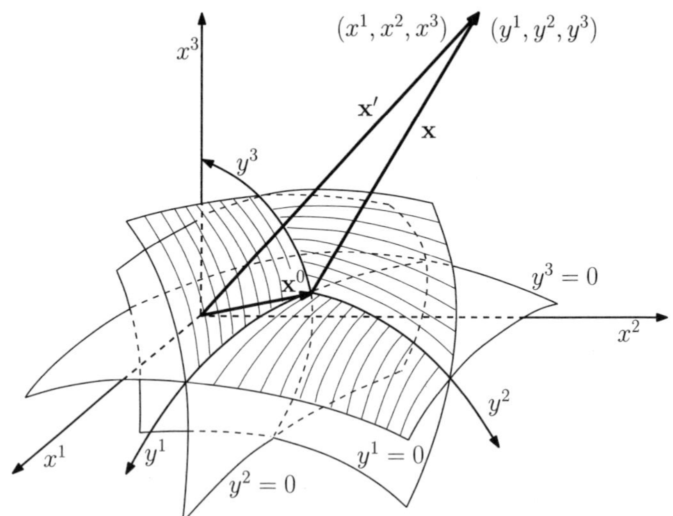

Figure A.2: *Curvilinear coordinates in the Euclidean space.*

where x^i are Cartesian coordinates of the point $\mathbf{x}$ and the vectors $\mathbf{e}_i$ are orthogonal to each other and all have unit length.

Due to condition (A.19) the position vector $\mathbf{x}$ can also be written as a function of curvilinear coordinates

$$x^k = x^k\left(y^\alpha\right) \quad \Rightarrow \quad \mathbf{x} = \mathbf{x}\left(y^\alpha\right). \tag{A.21}$$

It should be stressed that in the frequently used notation, particularly in numerical methods, the matrix $(x^1, x^2, x^2)^T$ can be identified with components of the position vector $\mathbf{x}$ in Cartesian coordinates. However, the matrix $(y^1, y^2, y^3)^T$ which is constructed from curvilinear coordinates is not a vector. We shall see that this matrix does not possess transformation properties characteristic for vectors.

We proceed to present a few basic notions of tensor analysis. Details can be found in numerous books on the subject (e.g. in the classical textbook of J. L. Synge and A. Schild [373]). Obviously, tangent spaces in which the vectors are defined must contain an infinitesimal change of the position vector $\mathbf{x}$

$$d\mathbf{x} = \frac{\partial \mathbf{x}}{\partial y^\alpha} dy^\alpha = \mathbf{g}_\alpha dy^\alpha, \quad \mathbf{g}_\alpha = \frac{\partial \mathbf{x}}{\partial y^\alpha}. \tag{A.22}$$

The vectors $\mathbf{g}_\alpha$ are called the *covariant basis vectors* of the curvilinear coordinate system. As $d\mathbf{x}$ is a vector we can say, in contrast to the above described matrix of curvilinear coordinates, that the matrix of infinitesimal increments $(dy^1, dy^2, dy^3)^T$ is a vector.

According to the definition of parametric lines the basis vectors $\mathbf{g}_\alpha$ are tangent to these lines. For instance, $\mathbf{g}_1$ is tangent to line $\mathcal{C}_1$ which is the logical consequence of the definition of partial derivatives.

An arbitrary vector $\mathbf{u}$ can be written in Cartesian coordinates in the form of a linear vector combination of its components. We expect the same in curvilinear coordinates, i.e.,

$$\mathbf{u} = u^i \mathbf{e}_i = u^\alpha \mathbf{g}_\alpha. \tag{A.23}$$

The components u^α can be easily found if we introduce *contravariant basis vectors* $\mathbf{g}^\alpha$

$$u^\alpha = \mathbf{u} \cdot \mathbf{g}^\alpha = \left(u^\beta \mathbf{g}_\beta\right) \cdot \mathbf{g}^\alpha \quad \Rightarrow \quad \mathbf{g}_\beta \cdot \mathbf{g}^\alpha = \delta^\alpha_\beta. \tag{A.24}$$

This definition means that the contravariant basis vectors are orthogonal to the parametric surfaces. For instance, the vector $\mathbf{g}^1$ is orthogonal to the parametric surface $y^1 = \text{const.}$, etc.

Often it is more convenient to calculate the vectors of the contravariant base $\mathbf{g}^\alpha$ directly by the vectors $\mathbf{g}_\alpha$. It is easy to see that the set of equations (A.24) has the following solution

$$\mathbf{g}^1 = \frac{1}{g} \mathbf{g}_2 \times \mathbf{g}_3, \quad \mathbf{g}^2 = \frac{1}{g} \mathbf{g}_3 \times \mathbf{g}_1, \quad \mathbf{g}^3 = \frac{1}{g} \mathbf{g}_1 \times \mathbf{g}_2, \tag{A.25}$$

where

$$g = \mathbf{g}_1 \cdot (\mathbf{g}_2 \times \mathbf{g}_3) = \det\left(\frac{\partial x^k}{\partial y^\alpha}\right). \tag{A.26}$$

The scalar product of an arbitrary vector with covariant basis vectors defines another set of components of this vector. We have

$$u_\alpha = \mathbf{u} \cdot \mathbf{g}_\alpha. \tag{A.27}$$

These are the so-called *covariant components of the vector* $\mathbf{u}$. Consequently,

$$\mathbf{u} = u_\alpha \mathbf{g}^\alpha = u^\alpha \mathbf{g}_\alpha. \tag{A.28}$$

Obviously, covariant and contravariant components of the vector $\mathbf{u}$ are identical in Cartesian coordinate systems. They are not in curvilinear systems. They usually do not even have the same units.

In the differential geometry it is customary to introduce the so-called linear infinitesimal element whose length is defined by the relation

$$ds^2 = d\mathbf{x} \cdot d\mathbf{x} = g_{\alpha\beta} dy^\alpha dy^\beta, \quad g_{\alpha\beta} = \mathbf{g}_\alpha \cdot \mathbf{g}_\beta. \tag{A.29}$$

The tensor $g_{\alpha\beta}$ is called the *metric tensor*. It plays an important role in the transformation of covariant components into contravariant components and vice versa. For the vector $\mathbf{u}$ we have, for example,

$$u_\alpha = \mathbf{u} \cdot \mathbf{g}_\alpha = \left(u^\beta \mathbf{g}_\beta\right) \cdot \mathbf{g}_\alpha = g_{\alpha\beta} u^\beta, \quad g_{\alpha\beta} = g_{\beta\alpha}, \tag{A.30}$$

i.e., the metric tensor is symmetrical. Its inverse, defined by the relation

$$g_{\alpha\gamma} g^{\gamma\beta} = \delta_\alpha^\beta, \tag{A.31}$$

plays a similar role in the transformation

$$u^\alpha = g^{\alpha\beta} u_\beta. \tag{A.32}$$

Of course, for Cartesian coordinates the metric tensor and its inverse are simple unit tensors (Kronecker deltas) δ_{ij} and δ^{ij}. For this reason, we do not have to distinguish in these coordinates between covariant and contravariant components – they are identical.

The transformation relations (A.19) determine uniquely the basis vectors and the metric tensors. We have

$$\mathbf{g}_\alpha = \frac{\partial \mathbf{x}}{\partial y^\alpha} = \frac{\partial \mathbf{x}}{\partial x^i} \frac{\partial x^i}{\partial y^\alpha} = \mathbf{e}_i \frac{\partial x^i}{\partial y^\alpha},$$

$$\mathbf{g}_\alpha \cdot \mathbf{g}^\beta = \left(\mathbf{e}_i \frac{\partial x^i}{\partial y^\alpha}\right) \cdot \left[\left(\mathbf{g}^\beta \cdot \mathbf{e}_k\right) \mathbf{e}^k\right] = \left(\frac{\partial x^i}{\partial y^\alpha}\right) \left(\mathbf{g}^\beta \cdot \mathbf{e}_i\right) = \delta_\alpha^\beta \quad \Rightarrow \quad \mathbf{g}^\beta \cdot \mathbf{e}_i = \frac{\partial y^\beta}{\partial x^i},$$

$$\text{i.e.,} \quad \mathbf{g}^\beta = \mathbf{e}^i \frac{\partial y^\beta}{\partial x^i}. \tag{A.33}$$

These relations indicate also the transformation rules for the components of arbitrary vectors and tensors. For the vector $\mathbf{u}$ we have

$$\mathbf{u} = u^\alpha \mathbf{g}_\alpha = u^\alpha \frac{\partial x^i}{\partial y^\alpha} \mathbf{e}_i \quad \Rightarrow \quad u^i = A^i_{.\alpha} u^\alpha, \quad A^i_{.\alpha} = \frac{\partial x^i}{\partial y^\alpha},$$

$$u^\alpha = \mathbf{u} \cdot \mathbf{g}^\alpha = u^i \left(\mathbf{e}_i \cdot \mathbf{g}^\alpha\right) \quad \Rightarrow \quad u^\alpha = A^\alpha_{.i} u^i, \quad A^\alpha_{.i} = \frac{\partial y^\alpha}{\partial x^i}, \tag{A.34}$$

$$u^i = A^{i\alpha} u_\alpha, \quad u_\alpha = A_{\alpha i} u^i,$$

$$A^{i\alpha} = A^i_{.\beta} g^{\alpha\beta}, \quad A_{\alpha i} = g_{\alpha\beta} A^\alpha_{.i}.$$

Such transformation matrices $A^i_{.\alpha}, \dots$, appear in arbitrary transformations of coordinates, Cartesian to curvilinear or curvilinear to curvilinear. As they, obviously, do not possess any symmetry properties it is convenient to distinguish indices for two coordinate systems participating in the transformation, for instance Latin indices for Cartesian coordinates and Greek indices for curvilinear coordinates.

We proceed to the transformation of differential operators. Clearly, in Cartesian coordinates we have for the vector $\mathbf{u}$

$$\frac{\partial \mathbf{u}}{\partial x^i} = \frac{\partial \left(u^j \mathbf{e}_j \right)}{\partial x^i} = \frac{\partial u^j}{\partial x^i} \mathbf{e}_j. \tag{A.35}$$

Hence, for instance, the gradient of the vector has the form

$$\operatorname{grad} \mathbf{u} = \frac{\partial u^j}{\partial x^i} \mathbf{e}_j \otimes \mathbf{e}^i. \tag{A.36}$$

It is a tensor of second order whose components are partial derivatives of the components of vector $\mathbf{u}$. This is not the case anymore in curvilinear coordinates because the basis vectors are not constant. We have

$$\frac{\partial \mathbf{u}}{\partial y^\alpha} = \frac{\partial \left(u^\beta \mathbf{g}_\beta \right)}{\partial y^\alpha} = \frac{\partial u^\beta}{\partial y^\alpha} \mathbf{g}_\beta + u^\beta \frac{\partial \mathbf{g}_\beta}{\partial y^\alpha}. \tag{A.37}$$

As for any α the derivative of the covariant basis vector is a vector again, it can be written in contravariant components

$$\frac{\partial \mathbf{g}_\beta}{\partial y^\alpha} = \frac{\partial^2 \mathbf{x}}{\partial y^\alpha \partial y^\beta} = \Gamma^\gamma_{\alpha\beta} \mathbf{g}_\gamma, \quad \Gamma^\gamma_{\alpha\beta} \equiv \left\{ \begin{array}{c} \gamma \\ \alpha\,\beta \end{array} \right\} = \frac{\partial \mathbf{g}_\beta}{\partial y^\alpha} \cdot \mathbf{g}^\gamma. \tag{A.38}$$

The quantities $\Gamma^\gamma_{\alpha\beta}$ are called *Christoffel symbols of the second kind* (for Christoffel symbols of the first kind see e.g. Part I). They are, obviously, symmetric with respect to covariant indices. This means that only 18 out of 27 components of this object are independent. By means of this object we define now the components of the gradient in curvilinear coordinates

$$u^\beta_{;\alpha} = \frac{\partial u^\beta}{\partial y^\alpha} + \Gamma^\beta_{\alpha\gamma} u^\gamma, \quad \operatorname{grad} \mathbf{u} = u^\beta_{;\alpha} \mathbf{g}_\beta \otimes \mathbf{g}^\alpha. \tag{A.39}$$

This object is called the *covariant derivative of the vector* $\mathbf{u} = u^\beta \mathbf{g}_\beta$. We may also use covariant components $\mathbf{u} = u_\beta \mathbf{g}^\beta$. Then, we have

$$\frac{\partial \mathbf{u}}{\partial y^\alpha} = \frac{\partial u_\beta}{\partial y^\alpha} \mathbf{g}^\beta + u_\beta \frac{\partial \mathbf{g}^\beta}{\partial y^\alpha} = u_{\beta;\alpha} \mathbf{g}^\beta, \quad u_{\beta;\alpha} = \frac{\partial u_\beta}{\partial y^\alpha} - \Gamma^\gamma_{\alpha\beta} u_\gamma, \tag{A.40}$$

which follows easily by the differentiation of the identity $\mathbf{g}_\alpha \cdot \mathbf{g}^\beta = \delta^\beta_\alpha$.

The above relations can be immediately extended on tensors. It can easily be shown that for a tensor of second order

$$\mathbf{T} = T^{\alpha\beta} \mathbf{g}_\alpha \otimes \mathbf{g}_\beta = T^\alpha_{.\beta} \mathbf{g}_\alpha \otimes \mathbf{g}^\beta = T_{\alpha\beta} \mathbf{g}^\alpha \otimes \mathbf{g}_\beta, \tag{A.41}$$

holds

$$T^{\alpha\beta}_{..;\gamma} = \frac{\partial T^{\alpha\beta}}{\partial y^\gamma} + \Gamma^\alpha_{\mu\gamma} T^{\mu\beta} + \Gamma^\beta_{\mu\gamma} T^{\alpha\mu},$$

$$T^\alpha_{.\beta;\gamma} = \frac{\partial T^\alpha_{.\beta}}{\partial y^\gamma} + \Gamma^\alpha_{\mu\gamma} T^\mu_{.\beta} - \Gamma^\mu_{\beta\gamma} T^\alpha_{.\mu}, \tag{A.42}$$

$$T_{\alpha\beta;\gamma} = \frac{\partial T_{\alpha\beta}}{\partial y^\gamma} - \Gamma^\mu_{\alpha\gamma} T_{\mu\beta} - \Gamma^\mu_{\beta\gamma} T_{\alpha\mu}.$$

A brief remark on Christoffel symbols is appropriate. They are closely related to the notion of the *connection*. It is obvious that the differentiation along curves in the space defines locally a translation of a vector along these curves. Consequently, by the calculation of differentials we can reach vectors in neighboring tangent vector spaces from the tangent vector space in a chosen point of the space. This is done by an equation specified by the Christoffel symbols (e.g. see A. J. McConnel [248], Chapter 12). They define the so-called parallel transport of vectors. This is possible for connections which are metric. Such is the connection identified with the above defined Christoffel symbols. Namely, we can easily prove the following relation

$$g_{\alpha\beta}\Gamma^{\beta}_{\mu\nu} = \frac{1}{2}\left(\frac{\partial g_{\alpha\nu}}{\partial y^{\mu}} + \frac{\partial g_{\alpha\mu}}{\partial y^{\nu}} - \frac{\partial g_{\mu\nu}}{\partial y^{\alpha}}\right), \tag{A.43}$$

i.e., partial differentiation of the metric tensor determines the metric connection. Some continuous models generalize this notion. In the continuous theory of dislocations, a non-metric connection is introduced which measures the density of dislocations. We explain the geometrical motivation of this generalization in Chapter 9 of this book.

The property of curvilinear coordinates that different basis vectors possess different units is often inconvenient in practice, especially if one wants to compare the orders of magnitude. This fault can be prevented by introduction of the so-called *physical components*. They differ to common curvilinear coordinates only by normalized basis vectors

$$\mathbf{e}_{\alpha} := \frac{\mathbf{g}_{\alpha}}{\sqrt{\mathbf{g}_{\alpha}\cdot\mathbf{g}_{\alpha}}}, \qquad \mathbf{e}^{\alpha} := \frac{\mathbf{g}^{\alpha}}{\sqrt{\mathbf{g}^{\alpha}\cdot\mathbf{g}^{\alpha}}}, \qquad \alpha = 1,2,3, \qquad \text{no sum!} \tag{A.44}$$

Then, for example, the vector $\mathbf{u}$ can be written as

$$\mathbf{u} = u^{(\alpha)}\mathbf{e}_{\alpha} = u_{(\alpha)}\mathbf{e}^{\alpha}, \tag{A.45}$$

and the coordinates $u^{(\alpha)}$ and $u_{(\alpha)}$ have the same units – they are comparable. Obviously, there is

$$u^{(\alpha)} = u^{\alpha}\sqrt{\mathbf{g}_{\alpha}\cdot\mathbf{g}_{\alpha}} = u^{\alpha}\sqrt{g_{\alpha\alpha}}, \qquad u_{(\alpha)} = u_{\alpha}\sqrt{\mathbf{g}^{\alpha}\cdot\mathbf{g}^{\alpha}} = u_{\alpha}\sqrt{g^{\alpha\alpha}}. \tag{A.46}$$

Similar rules can be found also for tensors. For instance, the Cauchy stress $\mathbf{T}$ has the following form

$$\begin{aligned} T^{(\alpha)(\beta)} &= T^{\alpha\beta}\sqrt{\mathbf{g}_{\alpha}\cdot\mathbf{g}_{\alpha}}\sqrt{\mathbf{g}_{\beta}\cdot\mathbf{g}_{\beta}}, \\ T^{(\alpha)}_{.(\beta)} &= T^{\alpha}_{.\beta}\sqrt{\mathbf{g}_{\alpha}\cdot\mathbf{g}_{\alpha}}\sqrt{\mathbf{g}^{\beta}\cdot\mathbf{g}^{\beta}} \qquad \text{no sum!} \\ T_{(\alpha)(\beta)} &= T_{\alpha\beta}\sqrt{\mathbf{g}^{\alpha}\cdot\mathbf{g}^{\alpha}}\sqrt{\mathbf{g}^{\beta}\cdot\mathbf{g}^{\beta}}. \end{aligned} \tag{A.47}$$

We proceed to present a few chosen curvilinear coordinate systems which find a frequent application in problems of continuum thermodynamics.

A.2.2 Cylindrical coordinates

Cylindrical coordinates y^{α} are illustrated on the left-hand side of Figure A.3 and they are defined by the relations

$$\begin{aligned} y^1 &= \sqrt{(x_1)^2 + (x_2)^2}, & x^1 &= y^1\cos y^2, \\ y^1 &= \arctan\frac{x_2}{x_1}, & \Rightarrow \quad x^2 &= y^1\sin y^2, \\ y^3 &= x_3, & x^3 &= y^3. \end{aligned} \tag{A.48}$$

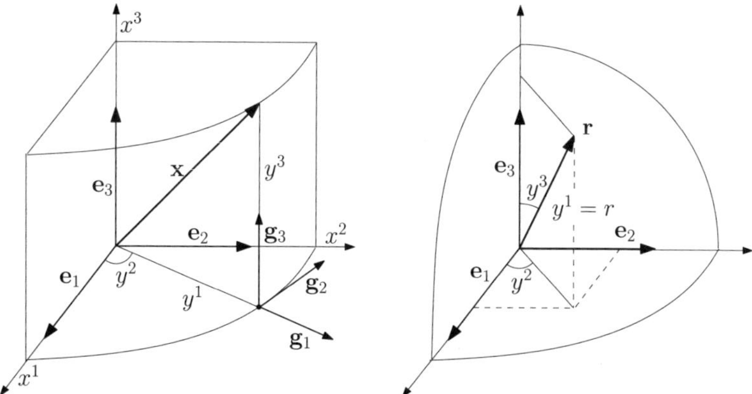

Figure A.3: *Cylindrical (left) and spherical (right) coordinates. Another notations for cylindrical coords.: (r, ϑ, z) or (R, Θ, Z), for spherical coords.: (r, ϑ, φ) or (R, Θ, Φ).*

In this case the position vector is given by

$$\mathbf{x} = x^k \mathbf{e}_k = y^1 \cos y^2 \mathbf{e}_1 + y^1 \sin y^2 \mathbf{e}_2 + y^3 \mathbf{e}_3. \tag{A.49}$$

The covariant basis vectors follow by differentiation

$$\mathbf{g}_\alpha = \begin{pmatrix} \cos y^2 \mathbf{e}_1 + \sin y^2 \mathbf{e}_2 \\ y^1 \left(-\sin y^2 \mathbf{e}_1 + \cos y^2 \mathbf{e}_2 \right) \\ \mathbf{e}_3 \end{pmatrix}. \tag{A.50}$$

It is obvious that these vectors are not only dependent on the position but also that they have different units. This is visible on components of the metric tensor $(g_{\alpha\beta}) = (\mathbf{g}_\alpha \cdot \mathbf{g}_\beta)$ which are not equal to zero if $\alpha \neq \beta$:

$$\mathbf{g}_1 \cdot \mathbf{g}_1 = 1, \qquad \mathbf{g}_2 \cdot \mathbf{g}_2 = \left(y^1\right)^2, \qquad \mathbf{g}_3 \cdot \mathbf{g}_3 = 1. \tag{A.51}$$

This means that the components of a vector $\mathbf{u}$ cannot be directly compared to each other. Using (A.25) and (A.26), the contravariant basis vectors can be determined

$$
\begin{aligned}
g &= \quad (\cos y^2 \mathbf{e}_1 + \sin y^2 \mathbf{e}_2) \cdot [(-y^1 \sin y^2 \mathbf{e}_1 + y^1 \cos y^2 \mathbf{e}_1) \times \mathbf{e}_3] \quad = y^1, \\[2ex]
\mathbf{g}^1 &= \quad \frac{1}{g}(-y^1 \sin y^2 \mathbf{e}_1 + y^1 \cos y^2 \mathbf{e}_1) \times \mathbf{e}_3 = \cos y^2 \mathbf{e}_1 + \sin y^2 \mathbf{e}_2 \quad = \mathbf{g}_1, \\[2ex]
\mathbf{g}^2 &= \quad \frac{1}{g}\mathbf{e}_3 \times (\cos y^2 \mathbf{e}_1 + \sin y^2 \mathbf{e}_2) = \frac{1}{y^1}(-\sin y^2 \mathbf{e}_1 + \cos y^2 \mathbf{e}_2) \quad = \frac{1}{(y^1)^2}\mathbf{g}_2, \\[2ex]
\mathbf{g}^3 &= \quad \frac{1}{g}(\cos y^2 \mathbf{e}_1 + \sin y^2 \mathbf{e}_2) \times (-y^1 \sin y^2 \mathbf{e}_1 + y^1 \cos y^2 \mathbf{e}_1) = \mathbf{e}_3 \quad = \mathbf{g}_3.
\end{aligned}
\tag{A.52}
$$

Again, the units are not the same. The collinearity of the covariant and contravariant basis vectors follows from the orthogonality of the parametric lines.

Differentiation of (A.50) with respect to y^β results in

$$\frac{\partial \mathbf{g}_\alpha}{\partial y^1} = \begin{pmatrix} 0 \\ -\sin y^2 \mathbf{e}_1 + \cos y^2 \mathbf{e}_2 \\ 0 \end{pmatrix}, \quad \frac{\partial \mathbf{g}_\alpha}{\partial y^2} = \begin{pmatrix} -\sin y^2 \mathbf{e}_1 + \cos y^2 \mathbf{e}_2 \\ -y^1 \left(\cos y^2 \mathbf{e}_1 + \sin y^2 \mathbf{e}_2 \right) \\ 0 \end{pmatrix}, \quad \frac{\partial \mathbf{g}_\alpha}{\partial y^3} = \begin{pmatrix} 0 \\ 0 \\ 0 \end{pmatrix}.$$

$$(A.53)$$

The scalar products of these vectors with the vectors $\mathbf{g}^\gamma$ (A.52) yield

$$\Gamma_{21}^2 = \Gamma_{12}^2 = \frac{1}{y^1}, \qquad \Gamma_{22}^1 = -y^1, \qquad \Gamma_{\beta\gamma}^\alpha = 0 \quad \text{otherwise.} \tag{A.54}$$

A.2.3   Spherical coordinates

Spherical coordinates y^α are illustrated on the right-hand side of Figure A.3 and they describe the transformation

$$\begin{aligned} y^1 &= \sqrt{(x^1)^2 + (x^2)^2 + (x^3)^2}, \\ y^2 &= \arctan \frac{x^2}{x^1}, \\ y^3 &= \arccos \frac{x_3}{\sqrt{(x^1)^2 + (x^2)^2 + (x^3)^2}}, \end{aligned} \quad \Rightarrow \quad \begin{aligned} x^1 &= y^1 \sin y^3 \cos y^2, \\ x^2 &= y^1 \sin y^3 \sin y^2, \\ x^3 &= y^1 \cos y^3. \end{aligned} \tag{A.55}$$

The corresponding covariant basis vectors are

$$\mathbf{g}_\alpha = \begin{pmatrix} \sin y^3 \cos y^2 \mathbf{e}_1 + \sin y^3 \sin y^2 \mathbf{e}_2 + \cos y^3 \mathbf{e}_3 \\ y^1 \left(-\sin y^3 \sin y^2 \mathbf{e}_1 + \sin y^3 \cos y^2 \mathbf{e}_2 \right) \\ y^1 \left(\cos y^3 \cos y^2 \mathbf{e}_1 + \cos y^3 \sin y^2 \mathbf{e}_2 - \sin y^3 \mathbf{e}_3 \right) \end{pmatrix}, \tag{A.56}$$

and the contravariant ones are connected in the following way

$$\mathbf{g}^\alpha = \begin{pmatrix} \mathbf{g}_1 \\ \dfrac{1}{(y^1)^2 \sin^2 y^3} \mathbf{g}_2 \\ \dfrac{1}{(y^1)^2} \mathbf{g}_3 \end{pmatrix}. \tag{A.57}$$

Using the following relations which are a consequence of (A.56)

$$\frac{\partial \mathbf{g}_\alpha}{\partial y^1} = \begin{pmatrix} 0 \\ \dfrac{1}{y^1} \mathbf{g}_2 \\ \dfrac{1}{y^1} \mathbf{g}_3 \end{pmatrix}, \quad \frac{\partial \mathbf{g}_\alpha}{\partial y^2} = \begin{pmatrix} \dfrac{1}{y^1} \mathbf{g}_2 \\ -y^1 \left(\mathbf{g}_1 - \cos y^3 \mathbf{e}_3 \right) \\ \cot y^3 \mathbf{g}_2 \end{pmatrix}, \quad \frac{\partial \mathbf{g}_\alpha}{\partial y^3} = \begin{pmatrix} \dfrac{1}{y^1} \mathbf{g}_3 \\ \cot y^3 \mathbf{g}_2 \\ -y^1 \mathbf{g}_1 \end{pmatrix}, \tag{A.58}$$

the Christoffel symbols are obtained:

$$\begin{aligned} \Gamma_{21}^2 = \Gamma_{12}^2 &= \frac{1}{y^1}, & \Gamma_{31}^3 = \Gamma_{13}^3 &= \frac{1}{y^1}, & \Gamma_{32}^2 = \Gamma_{23}^2 &= \cot y^3, \\ \Gamma_{22}^1 &= -y^1 \sin^2 y^3, & \Gamma_{22}^3 &= -\sin y^3 \cos y^3, & \Gamma_{33}^1 &= -y^1, & \Gamma_{\beta\gamma}^\alpha &= 0 \quad \text{otherwise.} \end{aligned}$$

$$(A.59)$$

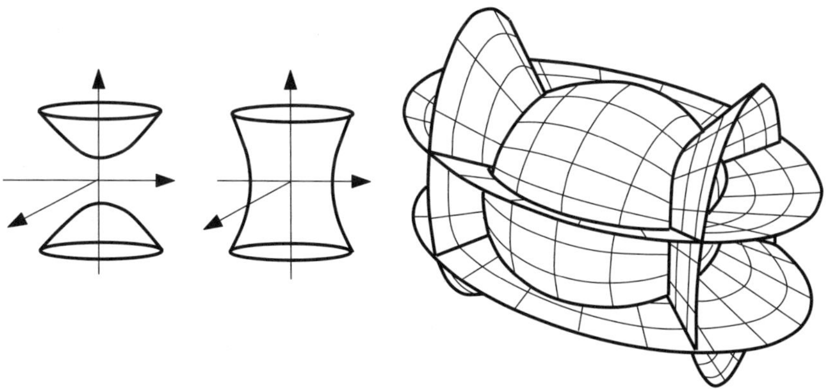

Figure A.4: *Left: Hyperboloid of two sheets, middle: hyperboloid of one sheet, right: confocal ellipsoidal coordinates – the surfaces are families of ellipsoids and hyperboloids.*

A.2.4 Ellipsoidal coordinates

A.2.4.1 Confocal ellipsoidal coordinates

In a confocal ellipsoidal coordinate system, often simply called ellipsoidal coordinate system, (illustrated on the right-hand side of Figure A.4) the coordinate surfaces are families of ellipsoids ($\lambda = $ const.), hyperboloids of one sheet ($\mu = $ const.) and hyperboloids of two sheets ($\nu = $ const.). The hyperboloids are sketched on the left-hand side of Figure A.4. They are given by the implicit equations

$$\frac{\left(x^1\right)^2}{a^2 - \lambda} + \frac{\left(x^2\right)^2}{b^2 - \lambda} + \frac{\left(x^3\right)^2}{c^2 - \lambda} = 1,$$

$$\frac{\left(x^1\right)^2}{a^2 - \mu} + \frac{\left(x^2\right)^2}{b^2 - \mu} - \frac{\left(x^3\right)^2}{\mu - c^2} = 1, \qquad \begin{aligned} \lambda &= y^1, \\ \mu &= y^2, \\ \nu &= y^3, \end{aligned} \qquad (A.60)$$

$$\frac{\left(x^1\right)^2}{a^2 - \nu} - \frac{\left(x^2\right)^2}{\nu - b^2} - \frac{\left(x^3\right)^2}{\nu - c^2} = 1,$$

where λ, μ, ν are parameters called ellipsoidal coordinates and a, b, c are constants satisfying $a^2 > \nu > b^2 > \mu > c^2 > \lambda > -\infty$. It can be shown that intersections of these three surfaces are orthogonal and that all of them have common foci. Moreover, through any fixed point there passes one and only one surface of each type.

The relation between new and old coordinates may be found by solving (A.60) directly. However, according to [241] there exists an easier way if we consider the cubic equation in a parameter q

$$\frac{\left(x^1\right)^2}{a^2 - q} + \frac{\left(x^2\right)^2}{b^2 - q} + \frac{\left(x^3\right)^2}{c^2 - q} - 1 = 0, \qquad (A.61)$$

with three real roots λ, μ, ν satisfying the above stated inequalities. Since q varies between a^2 and $-\infty$ (a consequence of the above statement that $a^2 > \nu > b^2 > \mu > c^2 > \lambda > -\infty$), Equation (A.61) describes the complete system of confocal surfaces given in (A.60).

Equation (A.61) can be transformed to

$$(x^1)^2 (b^2 - q)(c^2 - q) + (x^2)^2 (a^2 - q)(c^2 - q) + (x^3)^2 (a^2 - q)(b^2 - q)$$
$$- (a^2 - q)(b^2 - q)(c^2 - q) \equiv (q - \lambda)(q - \mu)(q - \nu) \equiv 0 \tag{A.62}$$

which must hold for every value of q. By setting $q = a^2, b^2, c^2$ in turn, we obtain in terms of Cartesian coordinates

$$(x^1)^2 = \frac{(a^2 - \lambda)(a^2 - \mu)(a^2 - \nu)}{(b^2 - a^2)(c^2 - a^2)},$$

$$(x^2)^2 = \frac{(b^2 - \lambda)(b^2 - \mu)(b^2 - \nu)}{(a^2 - b^2)(c^2 - b^2)}, \tag{A.63}$$

$$(x^3)^2 = \frac{(c^2 - \lambda)(c^2 - \mu)(c^2 - \nu)}{(a^2 - c^2)(b^2 - c^2)}.$$

In order to find the components of the metric tensor we use

$$ds^2 = dx^i dx^j \delta_{ij} = dy^\alpha dy^\beta g_{\alpha\beta}, \tag{A.64}$$

with

$$dx^i = \frac{\partial x^i}{\partial y^\alpha} dy^\alpha, \qquad dx^j = \frac{\partial x^j}{\partial y^\beta} dy^\beta, \tag{A.65}$$

so that follows

$$g_{\alpha\beta} = \frac{\partial x^i}{\partial y^\alpha} \frac{\partial x^j}{\partial y^\beta} \delta_{ij}. \tag{A.66}$$

If $\{y^\alpha\}$ is orthogonal ($\mathbf{g}_\alpha \cdot \mathbf{g}_\beta = 0$ for $\alpha \neq \beta$) then

$$(g_{\alpha\beta}) = \begin{pmatrix} g_{11} & 0 & 0 \\ 0 & g_{22} & 0 \\ 0 & 0 & g_{33} \end{pmatrix}. \tag{A.67}$$

Taking the logarithm of (A.63), differentiating partially with respect to λ and using (A.66) we obtain

$$g_{\lambda\lambda} = \frac{1}{4} \left\{ \frac{(a^2 - \mu)(a^2 - \nu)}{(a^2 - \lambda)(b^2 - a^2)(c^2 - a^2)} + \frac{(b^2 - \mu)(b^2 - \nu)}{(b^2 - \lambda)(a^2 - b^2)(c^2 - a^2)} \right.$$
$$\left. + \frac{(c^2 - \mu)(c^2 - \nu)}{(c^2 - \lambda)(a^2 - c^2)(b^2 - c^2)} \right\}. \tag{A.68}$$

Values for $g_{\mu\mu}$ and $g_{\nu\nu}$ may be obtained in a similar way. Simplification of the resulting expressions yields, reverting to y^1, y^2, y^3 instead of λ, μ, ν,

$$(g_{\alpha\beta}) = \frac{1}{4} \begin{pmatrix} \frac{(y^2-y^1)(y^3-y^1)}{(a^2-y^1)(b^2-y^1)(c^2-y^1)} & 0 & 0 \\ 0 & \frac{(y^3-y^2)(y^1-y^2)}{(a^2-y^2)(b^2-y^2)(c^2-y^2)} & 0 \\ 0 & 0 & \frac{(y^1-y^3)(y^2-y^3)}{(a^2-y^3)(b^2-y^3)(c^2-y^3)} \end{pmatrix}. \tag{A.69}$$

Due to the fact that x^1, x^2 and x^3 appear as squares in (A.63), a given point (x^1, x^2, x^3) is not uniquely determined by (y^1, y^2, y^3). In fact, eight points symmetrically located

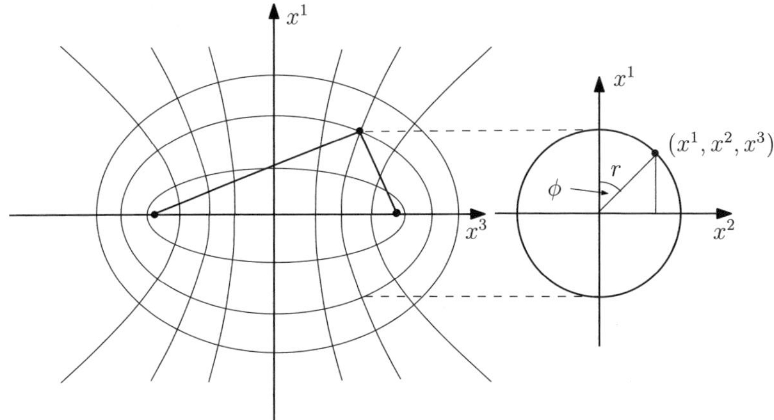

Figure A.5: *The intersection of the surfaces obtained by rotating the ellipses and hyperbolas of (A.70) is a circle of radius r.*

relative to the (x^1, x^2, x^3)-axes correspond to the set (y^1, y^2, y^3). This ambiguity may be resolved by adopting some convention concerning the signs of (y^1, y^2, y^3), or in more elegant fashion by the introduction of elliptic functions (for details see e.g. [428]).

The confocal ellipsoidal coordinate system has proved useful in problems of mechanics, potential theory, electrodynamics and hydrodynamics. E.g. in [120] the equations of stellar hydrodynamics for Eddington stellar systems with separable potentials are written in coordinates based on principal velocity surfaces. Namely, in 1915 A. S. Eddington [109] postulated the existence of a triply orthogonal family of surfaces – principal velocity surfaces – the interactions of which trace the directions of the axes of the velocity ellipsoid.

A.2.4.2 Elliptic cylindrical coordinates

Degenerate cases of the preceding one may arise if two or three of the axes in (A.61) become equal. Then, in order to have a complete coordinate system, additional surfaces are needed. To obtain them we proceed by considering the equations of an ellipse and a hyperbola

$$\frac{\left(x^3\right)^2}{a^2} + \frac{\left(x^1\right)^2}{a^2\left(1 - e_1^2\right)} = 1,$$

$$\frac{\left(x^3\right)^2}{a^2} + \frac{\left(x^1\right)^2}{a^2\left(1 - e_2^2\right)} = 1,$$

(A.70)

where a is the semi-major axis, $e_1 < 1$ is the eccentricity of the ellipse and $e_2 > 1$ the eccentricity of the hyperbola. By the simplifying definition $u := y^1$ and $v := y^2$ as well as some transformations ($a \to a\cosh u$, $e_1 \to \operatorname{sech} u$ in the ellipse, $a \to a\cos v$, $e_2 \to \sec v$

in the hyperbola and $(x^1)^2 \to (x^1)^2 + (x^2)^2 = r^2$) the following equations are obtained

$$\frac{(x^3)^2}{a^2 \cosh^2 u} + \frac{r^2}{a^2 \sinh^2 u} = 1,$$

$$\frac{(x^3)^2}{a^2 \cos^2 v} + \frac{r^2}{a^2 \sin^2 v} = 1,$$

$$\text{with} \qquad \begin{array}{l} 0 \le u \le \infty, \\ 0 \le v \le \pi. \end{array} \tag{A.71}$$

They represent families of ovary ellipsoids (also called prolate spheroids) for $u = \text{const.}$ and hyperboloids of revolution for $v = \text{const.}$ obtained by rotating the ellipses and hyperbolas of (A.70). Their intersections build a circle of radius r (see Figure A.5). Hence, if $0 \le \phi \le 2\pi$, the addition of a family of planes through the x^3-axis ($\phi = \text{const.}$) to the spheroids and hyperboloids gives three suitable coordinate surfaces.

If (A.71) is rewritten so that

$$\frac{(x^1)^2}{a^2 \cosh^2 u} + \frac{(x^2)^2}{a^2 \sinh^2 u} = 1,$$

$$\frac{(x^1)^2}{a^2 \cos^2 v} + \frac{(x^2)^2}{a^2 \sin^2 v} = 1,$$

$$\text{with} \qquad \begin{array}{l} 0 \le u \le \infty, \\ 0 \le v \le \pi, \end{array} \tag{A.72}$$

then the loci of these equations are cylindrical surfaces parallel to the x^3-axis and perpendicular to the $x^1 x^2$-plane. Their intersections with this plane are ellipses and hyperbolas. The coordinate surfaces are elliptic cylinders for $u = \text{const.}$, hyperbolic cylinders for $v = \text{const.}$ and planes parallel to the $x^1 x^2$-plane ($x^3 = \text{const.}$). We may then solve (A.72) for x^1 and x^2 and simplify the result by means of relations between trigonometric functions. We obtain

$$\begin{aligned} x^1 &= a \cosh u \cos v, \\ x^2 &= a \sinh u \sin v, \\ x^3 &= y^3, \end{aligned} \tag{A.73}$$

and, reverting to the original notions of the coordinates ($u := y^1$, $v := y^2$)

$$(g_{\alpha\beta}) = a^2 \begin{pmatrix} \sinh^2 y^1 + \sin^2 y^2 & 0 & 0 \\ 0 & \sinh^2 y^1 + \sin^2 y^2 & 0 \\ 0 & 0 & 1 \end{pmatrix}. \tag{A.74}$$

The intersection of these cylinders with the $x^1 x^2$-plane may also be inferred from Figure A.5.

Solving partial differential equations, e.g. the Laplace equation or the Helmholtz equation, elliptic cylindrical coordinates are particularly favorable since this type of coordinates allows a separation of variables. An example for a classic application is the electric field surrounding a flat conducting plate of width $2a$.

A.2.5   Paraboloidal coordinates

A.2.5.1   Confocal paraboloidal coordinates

Now a system of coordinate surfaces consisting of families of elliptic paraboloids extending in the direction of the negative x^3-axis ($\lambda = \text{const.}$), hyperbolic paraboloids ($\mu = \text{const.}$)

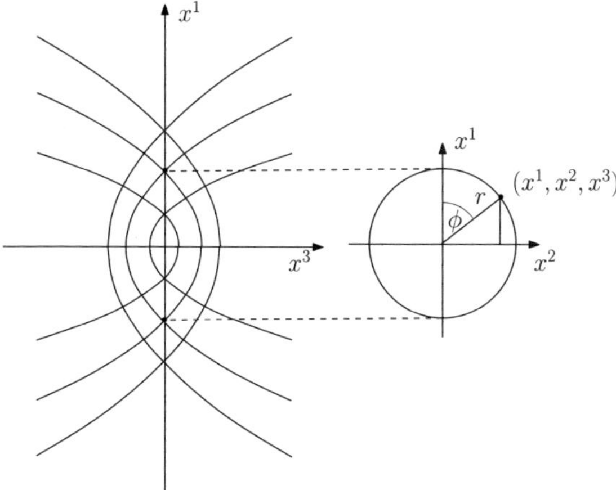

Figure A.6: *Paraboloidal coordinates.*

and elliptic paraboloids extending along the positive x^3-axis ($\nu = \text{const.}$) are considered. The equations for the surfaces are

$$\frac{(x^1)^2}{a^2 - \lambda} + \frac{(x^2)^2}{b^2 - \lambda} + 2x^3 + \lambda = 0,$$

$$\frac{(x^1)^2}{a^2 - \mu} - \frac{(x^2)^2}{\mu - b^2} + 2x^3 + \mu = 0, \qquad \begin{array}{l} \lambda = y^1, \\ \mu = y^2, \\ \nu = y^3, \end{array} \qquad (A.75)$$

$$\frac{(x^1)^2}{\nu - a^2} + \frac{(x^2)^2}{\nu - b^2} - 2x^3 - \nu = 0,$$

where $-\infty < \lambda < b^2 < \mu < a^2 < \nu < \infty$. As for the confocal ellipsoidal system we may write the cubic equation in a parameter q

$$\frac{(x^1)^2}{a^2 - q} + \frac{(x^2)^2}{b^2 - q} + 2x^3 + q = 0, \qquad (A.76)$$

with three real roots λ, μ, ν. In the same way as above we obtain finally considering $\lambda = y^1$, $\mu = y^2$, $\nu = y^3$,

$$(g_{\alpha\beta}) = \frac{1}{4}\begin{pmatrix} \dfrac{(y^2 - y^1)(y^3 - y^1)}{(a^2 - y^1)(b^2 - y^1)} & 0 & 0 \\ 0 & \dfrac{(y^3 - y^2)(y^1 - y^2)}{(a^2 - y^2)(b^2 - y^2)} & 0 \\ 0 & 0 & \dfrac{(y^1 - y^3)(y^2 - y^3)}{(a^2 - y^3)(b^2 - y^3)} \end{pmatrix}. \qquad (A.77)$$

Here a point (x^1, x^2, x^3) corresponds to four points (y^1, y^2, y^3) symmetrically located with respect to the x^1x^3- and x^2x^3-planes.

A.2.5.2 Parabolic coordinates

If two roots of (A.76) become equal, the preceding method fails since there are only two surfaces left. In this case we consider the family of parabolas

$$
\begin{aligned}
\left(x^1\right)^2 &= 2\xi^2\left(x^3 + \frac{\xi^2}{2}\right), & \xi &= y^1, \\
& & \eta &= y^2, \\
\left(x^1\right)^2 &= -2\eta^2\left(x^3 - \frac{\eta^2}{2}\right), & \phi &= y^3.
\end{aligned}
\tag{A.78}
$$

The vertices of all parabolas lie on the x^3-axis at distances $-\frac{\xi^2}{2}$ and $\frac{\eta^2}{2}$, respectively, and all of them have a common focus at the origin of the Cartesian coordinate system. If now the parabolas are rotated about the x^3-axis, the resulting intersections are circles and the paraboloids of revolution are still given by (A.78) if we replace $\left(x^1\right)^2$ by $r^2 = \left(x^1\right)^2 + \left(x^2\right)^2$, $\left(x^1\right)^2 = r\cos\phi$ and $\left(x^2\right)^2 = r\sin\phi$ (see Figure A.6). Thus, we obtain

$$
\begin{aligned}
x^1 &= \xi\eta\cos\phi, \\
x^2 &= \xi\eta\sin\phi, \\
x^3 &= \frac{\eta^2 - \xi^2}{2},
\end{aligned}
\tag{A.79}
$$

and from (A.65), coming back to the original notation

$$
(g_{\alpha\beta}) = \begin{pmatrix} \left(y^1\right)^2 + \left(y^2\right)^2 & 0 & 0 \\ 0 & \left(y^1\right)^2 + \left(y^2\right)^2 & 0 \\ 0 & 0 & \left(y^1\right)^2\left(y^2\right)^2 \end{pmatrix}.
\tag{A.80}
$$

The coordinate surfaces are paraboloids of revolution extending in the direction of the positive x^3-axis ($y^1 = $ const.), paraboloids of revolution extending toward the negative x^3-axis ($y^2 = $ const.) and planes through the x^3-axis. Intersections of these surfaces with the x^1x^3- and x^2x^3-planes are shown in Figure A.6.

Parabolic coordinates have been used e.g. in the treatment of the Stark effect. It is the splitting of atomic spectral lines as a result of an externally applied electric field and it can be explained by quantum mechanical approaches. In [379] a method for solving the Schrödinger equation in parabolic coordinates is presented. The separation in parabolic coordinates is important for the investigation of the Stark effect.

A.2.5.3 Parabolic cylindrical coordinates

A system similar to elliptical cylindrical coordinates is obtained by adding planes to the parabolic cylinders represented by (A.78). If we replace x^3 by x^2 in those equations, we have

$$
\begin{aligned}
x^1 &= \xi\eta = y^1y^2, \\
x^2 &= \frac{\eta^2 - \xi^2}{2} = \frac{\left(y^2\right)^2 - \left(y^1\right)^2}{2}, \\
x^3 &= y^3,
\end{aligned}
\tag{A.81}
$$

and

$$
(g_{\alpha\beta}) = \begin{pmatrix} \left(y^1\right)^2 + \left(y^2\right)^2 & 0 & 0 \\ 0 & \left(y^1\right)^2 + \left(y^2\right)^2 & 0 \\ 0 & 0 & 1 \end{pmatrix}.
\tag{A.82}
$$

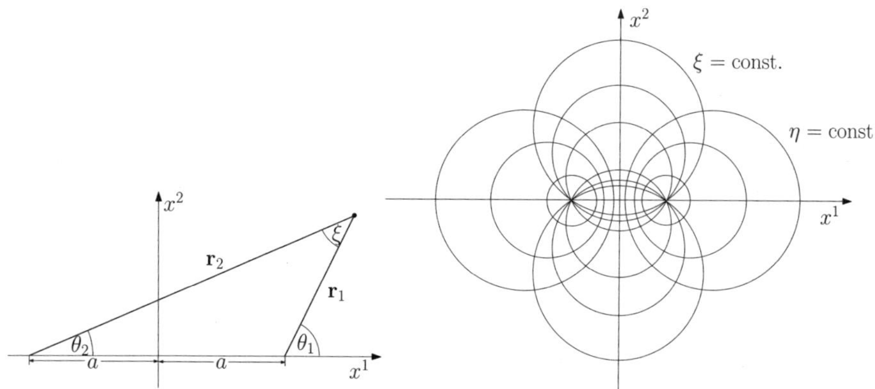

Figure A.7: *Bipolar coordinates.*

In this case, the coordinate surfaces are parabolic cylinders ($y^1 = $ const.), parabolic cylinders ($y^2 = $ const.) and planes ($y^3 = $ const.). The intersection of these surfaces with the x^1x^2-plane is also illustrated in Figure A.6.

As is the parabolic coordinate system, also the parabolic cylindrical coordinate system is important especially due to the fact that it is a coordinate system with translational symmetry along an axis known to have separable analytical solutions. For example it is used to find the dynamical variables of the electromagnetic field in [322] by solving the parabolic cylindrical Helmholtz equation.

A.2.6 Bipolar coordinates

A point (x^1, x^2) is located as shown on the left-hand side of Figure A.7 by means of two vectors $\mathbf{r}_1$ and $\mathbf{r}_2$ and two angles θ_1 and θ_2. For different positions of the point in the x^1x^2-plane, the vectors emanate from the two fixed points symmetrically located on the x^1-axis a distance a from the x^2-axis apart.

If we use exponentials instead of trigonometric functions and if we define $\rho^+ = x^1 + ix^2$ and $\rho^- = x^1 - ix^2$, then

$$x^1 = \frac{\rho^- + \rho^+}{2}, \qquad x^2 = \frac{i}{2}\left(\rho^- + -\rho^+\right). \tag{A.83}$$

The coordinates of the point (x^1, x^2) are

$$\rho^+ - a = r_1 e^{i\theta_1}, \qquad \rho^+ + a = r_2 e^{i\theta_2}, \tag{A.84}$$

and from the geometry of Figure A.7 (left) follows that

$$r_1^2 = \left(x^1 - a\right)^2 + \left(x^2\right)^2, \qquad \theta_1 = \frac{\tan^{-1} x^2}{x^1 - a},$$

$$\tag{A.85}$$

$$r_2^2 = \left(x^1 + a\right)^2 + \left(x^2\right)^2, \qquad \theta_2 = \frac{\tan^{-1} x^2}{x^1 + a}.$$

We define new quantities

$$\xi = y^1 = \theta_1 - \theta_2, \qquad \eta = y^2 = \ln\frac{r_2}{r_1}, \tag{A.86}$$

and divide the second equation of (A.84) by the first

$$\frac{\rho^+ + a}{\rho^+ - a} = e^{-i\chi}, \qquad \frac{\rho^+}{a} = \frac{e^{-i\chi} + 1}{e^{-i\chi} - 1}, \qquad \text{with} \qquad \chi = \xi + i\eta. \tag{A.87}$$

In order to find x^1 and x^2 as functions of ξ and η we substitute (A.87) in (A.83). Using the first two groups of expressions of footnote[1] results in

$$x^1 = \frac{a\sinh\eta}{\cosh\eta - \cos\xi}, \qquad x^2 = \frac{a\sin\xi}{\cosh\eta - \cos\xi}. \tag{A.88}$$

To find the form of the coordinate surfaces, we start from (A.86) and use the third group of relations of footnote 1 to obtain

$$\xi = \frac{1}{2}i\ln\frac{\left(ix^1 - ia + x^2\right)\left(ix^1 + ia - x^2\right)}{\left(ix^1 - ia - x^2\right)\left(ix^1 + ia + x^2\right)}, \tag{A.89}$$

which may also be expressed by

$$\left(x^1\right)^2 + \left(x^2\right)^2 - a^2 + 2iax^2\frac{1 + e^{2i\xi}}{1 - e^{2i\xi}} = 0. \tag{A.90}$$

From the last term of the first group of expressions of footnote 1 we observe that the last term of (A.90) equals $-2ax^2/\tan\xi = -2ax^2\cot\xi$. Hence,

$$\left(x^1\right)^2 + \left(x^2\right)^2 - a^2 - 2ax^2\cot\xi = 0, \tag{A.91}$$

or

$$\left(x^1\right)^2 + \left(x^2 - a\cot\xi\right)^2 = a^2\left(1 + \cot^2\xi\right) = a^2\csc^2\xi. \tag{A.92}$$

In the same way we find

$$e^{2\eta} = \frac{r_2^2}{r_1^2} = \frac{\left(x^1 + a\right)^2 + \left(x^2\right)^2}{\left(x^1 - a\right)^2 + \left(x^2\right)^2} \qquad \text{and} \qquad \left(x^1 - a\coth\eta\right)^2 + \left(x^2\right)^2 = a^2\text{csch}^2\eta. \tag{A.93}$$

We thus see that for $\xi = \text{const.}$, $0 \le \xi \le 2\pi$, we have a family of circles with centers on the x^2-axis at the point, $x^1 = 0$, $x^2 = a\cot\xi$, the radii of the circles being $a\csc\xi$. Each member of this family will pass through two fixed points as shown on the right-hand

---

[1]Some relations which are used in this subsection:

$$\sin x = \frac{i}{2}\left(e^{-ix} - e^{ix}\right), \qquad \cos x = \frac{1}{2}\left(e^{-ix} + e^{ix}\right), \qquad \tan x = \frac{\sin x}{\cos x} = \frac{i\left(1 - e^{2ix}\right)}{1 + e^{2ix}}.$$

By replacing x by ix, the following hyperbolic functions are obtained

$$\sin ix = \frac{i}{2}\left(e^x - e^{-x}\right) = i\sinh x, \qquad \cos ix = \frac{1}{2}\left(e^x + e^{-x}\right) = \cosh x,$$

$$\tan ix = \frac{i\left(e^{2x} - 1\right)}{e^{2x} + 1} = i\tanh x.$$

If $x = \tan u$, the inverse circular function $\tan^{-1}x = u$ follows from the above relations

$$e^{2iu} = \frac{i - x}{i + x}, \qquad 2iu = \ln\frac{i - x}{i + x}, \qquad u = \tan^{-1}x = \frac{i}{2}\ln\frac{i + x}{i - x}.$$

side of Figure A.7 and will intersect the circles $\eta = $ const. orthogonally. The members of the second family have radii of the length $a \csc \eta$ and are all situated on the x^1-axis at the points $x^1 = a \coth \eta$, $x^2 = 0$. The left intersection point with the x^1-axis is obtained when $\eta = \infty$, the right one if $\eta = -\infty$. When $\eta = 0$, the circles degenerate into points on the x^2-axis. The position of a point in the $x^1 x^2$-plane is thus fixed when we know in which quadrant it lies and furthermore the constant values of η, ξ of the circles which pass through it.

In order to use these circles as a coordinate system in space, imagine them to be moved along the x^3-axis. Then (A.92) and (A.93)$_2$ represent two families of right circular cylinders with axes parallel to the x^3-axis. Suitable coordinate surfaces are then: cylinders with centers on the x^2-axis ($\xi = $ const.), cylinders with centers on the x^1-axis ($\eta = $ const.) and planes perpendicular to the x^3-axis ($x^3 = $ const.).

From (A.88) and (A.65) we obtain, again reverting to the original notation of the coordinates

$$(g_{\alpha\beta}) = \begin{pmatrix} \dfrac{a^2}{(\cosh y^2 - \cos y^1)^2} & 0 & 0 \\ 0 & \dfrac{a^2}{(\cosh y^2 - \cos y^1)^2} & 0 \\ 0 & 0 & 1 \end{pmatrix}. \tag{A.94}$$

Such coordinates are useful in problems of hydrodynamics and electricity (see e.g. [152], [220]). The former of these books reports on an important application of bipolar coordinates in hydrodynamics: Consider a rigid particle of arbitrary shape translating through an unbounded fluid which is at rest at infinity. Appropriate boundary conditions can be given. If further particles are present, additional conditions have to be satisfied. The only exact solution of a multi-particle problem of this class is that of M. Stimson and G. B. Jeffery [366] for the slow motion of two spheres parallel to their line of centers (an axisymmetric flow). The system of bipolar coordinates employed by them is unique in that it permits simultaneously to satisfy boundary conditions on two external spheres. For larger collections of particles or for pairs of nonspherical bodies it is not generally possible to find coordinate systems which permit simultaneous satisfaction of all boundary conditions.

A.2.7 Toroidal coordinates

If (A.91) and (A.93)$_2$ are rewritten with $(x^3)^2$ substituted for $(x^2)^2$ and $r^2 = (x^1)^2 + (x^2)^2$ for $(x^1)^2$, then the resulting equations

$$\begin{aligned} 2ax^3 \cot \xi &= r^2 + (x^3)^2 - a^2, & y^1 &= \xi, \\ 4a^2 r^2 \coth^2 \eta &= \left(r^2 + (x^3)^2 + a^2 \right)^2, & y^2 &= \eta, \end{aligned} \tag{A.95}$$

represent the families of spheres and tores (or anchor rings) obtained by rotating the circles of the previous system about the x^3-axis (see Figure A.8). If we take as the third surface planes through the x^3-axis, $y^3 = \psi = $ const., then

$$\frac{x^2}{x^1} = \tan \psi. \tag{A.96}$$

The orthogonal coordinate surfaces are thus: spheres with centers on the axis of revolution at distances $\pm a \cot \xi$ from the origin and radii $a \csc \xi$ ($\xi = $ const.), anchor rings or tores

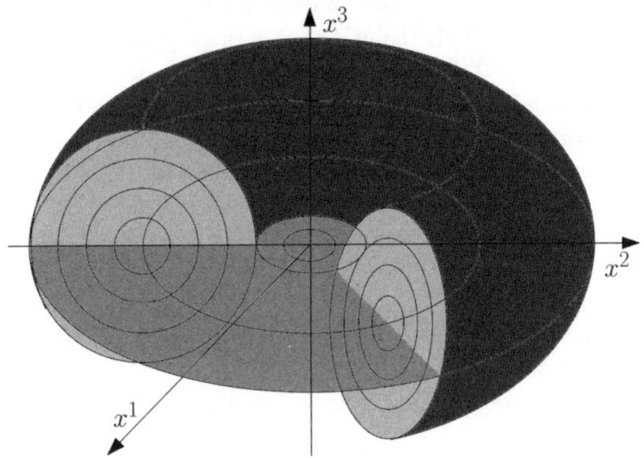

Figure A.8: _Toroidal coordinates._

whose axial circles have radii $a \coth \eta$ and whose cross sections are circles of radii $a \csc \eta$ (η = const.) and planes through the x^3-axis (ψ = const.). The spheres and anchor rings have a common circle, $r = a$, $x^3 = 0$. With methods similar to those used in the last subsection we obtain

$$x^1 = r \cos \psi, \qquad x^2 = r \sin \psi,$$

$$r = \frac{a \sinh \eta}{\cosh \eta - \cos \xi}, \qquad x^3 = \frac{a \sin \xi}{\cosh \eta - \cos \xi}, \tag{A.97}$$

and, in original coordinates

$$(g_{\alpha\beta}) = \begin{pmatrix} \dfrac{a^2}{(\cosh y^2 - \cos y^1)^2} & 0 & 0 \\ 0 & \dfrac{a^2}{(\cosh y^2 - \cos y^1)^2} & 0 \\ 0 & 0 & \dfrac{a^2 \sinh^2 y^2}{(\cosh y^2 - \cos y^1)^2} \end{pmatrix}. \tag{A.98}$$

Such systems find application in problems of electricity and potential theory (see e.g. [155], [122]). In [155] the electric potential inside and outside of a toroid is searched for. From the potential the electric field and surface charges can be found. A toroidal conductor with uniform resistivity is considered. It carries a steady current in the azimuthal direction, flowing along the circular loop. The electric potential has been calculated using toroidal coordinates. [122] is a theoretical approach which introduces a method formulated in toroidal coordinates to solve non-axisymmetric problems exactly and in closed form.

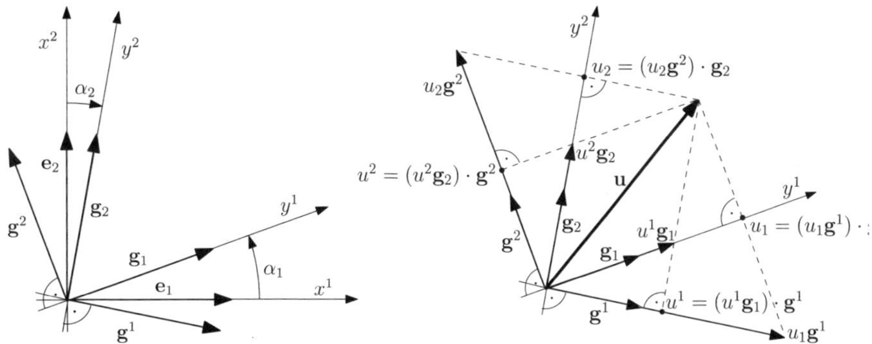

Figure A.9: *Oblique-angled Cartesian coordinates (left); covariant and contravariant co-ordinates of the vector* **u** *in the coordinate system shown on the left-hand side (right).*

A.2.8 Non-orthogonal coordinates

Finally, we examine the case of oblique-angled Cartesian coordinates y^α which are illustrated on the left-hand side of Figure A.9 and defined by the following relations

$$y^1 = \frac{1}{\cos(\alpha_1 + \alpha_2)}\left(x^1\cos\alpha_2 - x^2\sin\alpha_2\right),$$

$$y^2 = \frac{-1}{\cos(\alpha_1 + \alpha_2)}\left(x^1\cos\alpha_1 - x^2\sin\alpha_1\right), \quad \Rightarrow \quad \begin{array}{l} x^1 = y^1\cos\alpha_1 + y^2\sin\alpha_2, \\ x^2 = y^1\sin\alpha_1 + y^2\cos\alpha_2, \\ x^3 = y^3, \qquad \alpha_1 + \alpha_1 \neq \dfrac{\pi}{2}. \end{array} \qquad (\text{A.99})$$

$$y^3 = x^3,$$

It follows that

$$\left(\frac{\partial x^k}{\partial y^\beta}\right) = \begin{pmatrix} \cos\alpha_1 & \sin\alpha_2 & 0 \\ \sin\alpha_1 & \cos\alpha_2 & 0 \\ 0 & 0 & 1 \end{pmatrix}, \qquad \left(\frac{\partial y^\beta}{\partial x^k}\right) = \begin{pmatrix} \dfrac{\cos\alpha_2}{g} & -\dfrac{\sin\alpha_2}{g} & 0 \\ -\dfrac{\sin\alpha_1}{g} & \dfrac{\cos\alpha_1}{g} & 0 \\ 0 & 0 & 1 \end{pmatrix}, \qquad (\text{A.100})$$

with

$$g = \det\left(\frac{\partial x^k}{\partial y^\beta}\right) = \cos(\alpha_1 + \alpha_2). \qquad (\text{A.101})$$

Thus, covariant and contravariant basis vectors have the following form

$$\mathbf{g}_\beta = \begin{pmatrix} \mathbf{e}_1\cos\alpha_1 + \mathbf{e}_2\sin\alpha_1 \\ \mathbf{e}_1\sin\alpha_2 + \mathbf{e}_2\cos\alpha_2 \\ \mathbf{e}_3 \end{pmatrix}, \qquad \mathbf{g}^\beta = \frac{1}{g}\begin{pmatrix} \mathbf{e}_1\cos\alpha_2 - \mathbf{e}_2\sin\alpha_2 \\ -\mathbf{e}_1\sin\alpha_1 + \mathbf{e}_2\cos\alpha_1 \\ g\mathbf{e}_3 \end{pmatrix}. \qquad (\text{A.102})$$

On the right-hand side of Figure A.9 the graphical construction of the coordinates of a vector **u** is shown. It should be taken into account that the values u_1, u_2, u_3 and u^1, u^2, u^3 are determined by the length of the basis vectors which mostly are not unit vectors. Roughly speaking, the contravariant components result from parallel projection of the vector arrow on the coordinate axes and the contravariant ones from orthogonal projection.

Table A.1: *Summary of the two mostly used types of curvilinear orthogonal coordinates: cylindrical and spherical coordinates.*

Cylindrical coordinates	**Spherical coordinates**
Position vector	
$\mathbf{r} = y^1 \cos y^2 \mathbf{e}_1$ $\qquad + y^1 \sin y^2 \mathbf{e}_2 + y^3 \mathbf{e}_3,$	$\mathbf{r} = y^1 \cos y^2 \sin y^3 \mathbf{e}_1$ $\qquad + y^1 \sin y^2 \sin y^3 \mathbf{e}_2 + y^1 \cos y^3 \mathbf{e}_3,$
Covariant basis vectors	
$\mathbf{g}_1 = \cos y^2 \mathbf{e}_1 + \sin y^2 \mathbf{e}_2,$ $\mathbf{g}_2 = y^1 \left(-\sin y^2 \mathbf{e}_1 + \cos y^2 \mathbf{e}_2 \right),$ $\mathbf{g}_3 = \mathbf{e}_3,$	$\mathbf{g}_1 = \cos y^2 \sin y^3 \mathbf{e}_1 + \sin y^2 \sin y^3 \mathbf{e}_2$ $\qquad + \cos y^3 \mathbf{e}_3,$ $\mathbf{g}_2 = -y^1 \sin y^2 \sin y^3 \mathbf{e}_1 + y^1 \cos y^2 \sin y^3 \mathbf{e}_2,$ $\mathbf{g}_3 = y^1 \cos y^2 \cos y^3 \mathbf{e}_1 + y^1 \sin y^2 \cos y^3 \mathbf{e}_2$ $\qquad - y^1 \sin y^3 \mathbf{e}_3,$
Contravariant basis vectors	
$\mathbf{g}^1 = \cos y^2 \mathbf{e}_1 + \sin y^2 \mathbf{e}_2,$ $\mathbf{g}^2 = \frac{1}{y^1} \left(-\sin y^2 \mathbf{e}_1 + \cos y^2 \mathbf{e}_2 \right),$ $\mathbf{g}^3 = \mathbf{e}_3,$	$\mathbf{g}^1 = \cos y^2 \sin y^3 \mathbf{e}_1 + \sin y^2 \sin y^3 \mathbf{e}_2 +$ $\qquad + \cos y^3 \mathbf{e}_3,$ $\mathbf{g}^2 = \frac{1}{y^1 \sin^2 y^3} \left(-\sin y^2 \sin y^3 \mathbf{e}_1 + \cos y^2 \sin y^3 \mathbf{e}_2 \right),$ $\mathbf{g}^3 = \frac{1}{y^1} \left(\cos y^2 \cos y^3 \mathbf{e}_1 + \sin y^2 \cos y^3 \mathbf{e}_2 - \right.$ $\qquad \left. - \sin y^3 \mathbf{e}_3 \right),$
Metric tensors	
$\mathbf{g} = \mathbf{g}_1 \otimes \mathbf{g}_1 + \left(y^1 \right)^2 \mathbf{g}_2 \otimes \mathbf{g}_2$ $\qquad + \mathbf{g}_3 \otimes \mathbf{g}_3,$ $\mathbf{g} = \mathbf{g}^1 \otimes \mathbf{g}^1 + \left(\frac{1}{y^1} \right)^2 \mathbf{g}^2 \otimes \mathbf{g}^2$ $\qquad + \mathbf{g}^3 \otimes \mathbf{g}^3,$	$\mathbf{g} = \mathbf{g}_1 \otimes \mathbf{g}_1 + \left(y^1 \sin y^3 \right)^2 \mathbf{g}_2 \otimes \mathbf{g}_2$ $\qquad + \left(y^1 \right)^2 \mathbf{g}_3 \otimes \mathbf{g}_3,$ $\mathbf{g} = \mathbf{g}^1 \otimes \mathbf{g}^1 + \left(\frac{1}{y^1 \sin y^3} \right)^2 \mathbf{g}^2 \otimes \mathbf{g}^2$ $\qquad + \left(\frac{1}{y^1} \right)^2 \mathbf{g}^3 \otimes \mathbf{g}^3,$
Christoffel symbols	
$\Gamma^2_{12} = \frac{1}{y^1}, \quad \Gamma^1_{22} = -y^1,$	$\Gamma^1_{22} = -y^1 \sin^2 y^3, \quad \Gamma^1_{33} = -y^1, \quad \Gamma^2_{12} = \Gamma^3_{13} = \frac{1}{y^1},$ $\Gamma^3_{22} = -\cos y^3 \sin y^3, \quad \Gamma^2_{32} = \cot y^3.$

Appendix B

Integral transforms

Integral transforms are a particular kind of mathematical operator and have the following form

$$\overline{f}(s) = \int_{t_1}^{t_2} K(t, s)\, f(t)\, dt. \tag{B.1}$$

Input of the transform is a function f, output is another function $\overline{f}$. There are numerous different integral transforms which are mostly named after their inventor (Fourier transforms, Laplace transforms, Hankel transforms, Hilbert transforms etc.). Each is specified by a kernel function K of two variables. The first two types got most prominent therefore they are recalled in the following sections.

B.1 Fourier transforms

Any periodic function $f(t)$ with period T (e.g. the rectangular function plotted in Figure B.1) can be approximated by an (infinite) sum of trigonometric functions – the so-called *Fourier series*

$$f(t) = \frac{a_0}{2} + \sum_{k=1}^{\infty} \left[a_k \cos\left(2\pi k \frac{t}{T}\right) + b_k \sin\left(2\pi k \frac{t}{T}\right)\right]. \tag{B.2}$$

The coefficients a_k and b_k have to be determined for a given function $f(t)$ by

$$a_k = \frac{2}{T} \int_{t=0}^{T} f(t) \cos\left(2\pi k \frac{t}{T}\right) dx, \quad b_k = \frac{2}{T} \int_{t=0}^{T} f(t) \sin\left(2\pi k \frac{t}{T}\right) dt. \tag{B.3}$$

The more elements are taken into account the more precise $f(t)$ can be approximated. It becomes obvious from Figure B.1 that the convergence to the rectangular function gets closer the more steps are accounted for. Curves for order 1, 10 and 100 are compared. During the calculation of the coefficients it will be noticed that not all coefficients are nonzero: for 2π-periodic even functions ($f(-t) = f(t)$) $b_k = 0$ for all k, and for 2π-periodic odd functions ($f(-t) = -f(t)$) $a_k = 0$ for all k. In the corresponding Fourier series only cosine or only sine parts, respectively, appear.

In complex form the Fourier series (B.2) and the coefficients (B.3) can be written as follows

$$f(t) = \sum_{k=-\infty}^{\infty} c_k \exp\left(ikt\right), \quad c_k = \frac{1}{2\pi} \int_{-\pi}^{\pi} f(t) \exp\left(-ikt\right) dt, \tag{B.4}$$

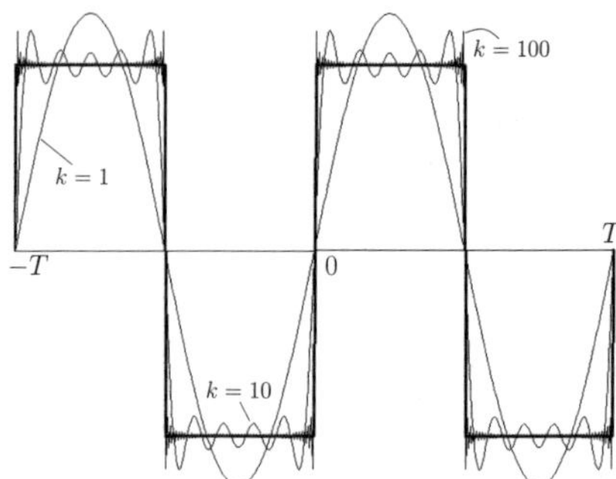

Figure B.1: *Rectangular function approximated by a Fourier series containing 1, 10 and 100 steps.*

where $f(t)$ is assumed to be a periodic function with period $T = 2\pi$ and Euler's formula $\exp(ikt) = \cos(kt) + i\sin(kt)$ has been considered. The Fourier coefficients are associated

$$a_k = c_k + c_{-k} \ (k = 0, 1, 2, ...), \qquad b_k = i(c_k - c_{-k}) \ (k = 1, 2, ...). \tag{B.5}$$

Fourier integrals and Fourier transforms extend the concept of Fourier series to non-periodic functions [375]. Then, instead of infinite but discrete sums a continuous integral is needed. The coefficients also become continuous functions of the wave number k. They have to be determined by the Fourier Integral Theorem (for the original see [129], for a discussion (in German): [266]): Let $f(t)$ be an integrable, piecewise continuous function with a finite number of jumps $\left(\int_{-\infty}^{\infty} f(t)\,dt < \infty\right)$, defined on the interval $(-\infty, \infty)$. Then

$$f(t) = \frac{1}{\sqrt{2\pi}} \int\limits_{k=-\infty}^{\infty} \bar{f}(k)\exp(-ikt)\,dk,$$

$$\bar{f}(k) = \frac{1}{\sqrt{2\pi}} \int\limits_{t=-\infty}^{\infty} f(t)\exp(ikt)\,dt. \tag{B.6}$$

It should be noted that even for real functions $f(t)$ the Fourier transform $\bar{f}(k)$ is usually complex.

B.2   Laplace transforms

We recall here the basic definitions of the Laplace transform needed in the theory of viscoelasticity. It is well known that many simple functions do not possess a Fourier transform as the defining integrals fail to converge for infinite limits. Therefore it is convenient to consider not the transform of the function $f(t)$ but rather of the function $f(t)\exp(-rt)$. Then, the transform

$$\bar{f}(s) = \int\limits_{-\infty}^{\infty} f(t)\exp(-zt)\,dt, \quad z = r + i\omega, \tag{B.7}$$

is called two-sided Laplace transform of $f(t)$. Obviously the inverse follows by multiplication of (B.7) by $\exp(rt)$ and integration

$$f(t) = \frac{1}{2\pi} \int\limits_{-\infty}^{\infty} \bar{f}(r + i\omega) \exp\left[(r + i\omega)t\right] d\omega. \tag{B.8}$$

If we are only interested in values of $f(t)$ for t positive, it is sufficient to transform the function $f(t) H(t)$ rather than $f(t)$. In this manner we avoid problems with the divergence of the above integral for the lower limit of $\exp(-rt)$. Then, the Laplace transform is the cut off of the two-sided Laplace transform

$$\bar{f}(z) = \int\limits_{0}^{\infty} f(t) \exp(-zt)\, dt. \tag{B.9}$$

Its inverse has the form

$$f(t) H(t) = \frac{1}{2\pi} \int\limits_{-\infty}^{\infty} \bar{f}(r + i\omega) \exp\left[(r + i\omega)t\right] d\omega = \frac{1}{2\pi i} \int\limits_{r-i\infty}^{r+i\infty} \bar{f}(z) \exp(zt)\, dz. \tag{B.10}$$

The first integral clarifies the meaning of the second integral. If the integral in (B.9) converges for some damping factor $r = \mathrm{Re}\, z$, it also converges for all larger damping factors. Consequently, the integral converges in some half-plane $\mathrm{Re}\, z > r_0$ in the complex z-plane. The line $r = r_0$ is called the abscissa of convergence. All values of r which lie to the right of r_0 are big enough for convergence. This is a very important property because the function $\bar{f}(z)$ has no singularity in the half-plane of convergence. However, it may have it on the abscissa of convergence. As a consequence, the integration of $\bar{f}(z)$ around any closed contour yields zero. This means that $\bar{f}(z)$ is an analytic function of z in the half-plane of convergence. This property is the basis for the calculation of inverse integrals (B.10).

Frequently used in application to viscoelasticity is the Laplace transform of the convolution integral. It is as follows [55]

$$\int\limits_{0}^{\infty} \left[\underbrace{\int\limits_{0}^{t} f(s)\, g(t - s)\, ds}_{\text{convolution integral}} \right] \exp(-zt)\, dt = \bar{f}(z)\, \bar{g}(z). \tag{B.11}$$

Example: We consider the Maxwell model, a rheological model for viscoelastic materials, described by the evolution equation

$$\dot{\sigma} + \frac{1}{\tau}\sigma = G_0 \dot{\kappa}, \quad \tau = \frac{\eta}{G_0}. \tag{B.12}$$

The Laplace transform of this equation yields

$$-\sigma_0 + z\bar{\sigma} + \frac{1}{\tau}\bar{\sigma} = G_0 \left(-\kappa_0 + z\bar{\kappa}\right). \tag{B.13}$$

At the instant of the time $t = 0$ we have the elastic reaction $\sigma_0 = G_0 \kappa_0$. This is the deformation of the spring in the Maxwell model before the dashpot had time to start moving. This yields the solution of the problem in the transformation z-plane

$$\bar{\sigma} = G_0 \frac{z}{z + 1/\tau} \bar{\kappa}. \tag{B.14}$$

Hence, the abscissa is crossing the point $r_0 = -1/\tau$, $\omega = 0$. The half-plane to the right of the vertical line $r = r_0$ contains no singular points of $\bar{\sigma}$. The inverse yields the solution

$$\sigma = \sigma_0 e^{-t/\tau} - G_0 \int_0^t \frac{d\kappa}{ds} (t - s) e^{-s/\tau} ds, \quad \sigma_0 = \sigma(t = 0). \tag{B.15}$$

Pierre-Simon Laplace	*Jean Baptiste Joseph Fourier*	*George Green*
1749 – 1827	1768 – 1830	1793 – 1841

Mistakenly in Part I the portrait of an American scientist named George Bernard Green (1822-1901) was quoted. We appreciate very much the remark by Libb Thims on the webpage http://www.eoht.info/page/George+Green for pointing out this mistake and quoting the right picture of George Green (1793-1841) whose portrait should appear and which we reproduce above.

Appendix C

Green functions

C.1   Purpose

Green functions, also known as fundamental solutions, serve the purpose of construction of analytical solutions of linear differential equations. They also form the basis for at least two important procedures of approximation. The first one yields the so-called boundary element methods (for an introduction for problems of standard elasticity see e.g. [28] or [196]). The second one evaluates average macroscopic properties of materials with microstructure. We present two examples of this procedure in this book (see Sections 9.2 and 9.4). It replaces, for instance, the method of volume averaging over the representative elementary volume (REV) – see, for example, page 262 of Part I.

The Green function allows one to solve the linear equation

$$\mathbf{L}\mathbf{u} + \mathbf{f} = 0, \tag{C.1}$$

where $\mathbf{L}$ is a linear operator, $\mathbf{u}$ an unknown vector function and $\mathbf{f}$ a given function. For the purpose of this appendix we assume the operator $\mathbf{L}$ to be of second order and the domain to be infinite. For finite domains one can obtain solutions by a simple transformation which we present further.

The formal solution of Equation (C.1) in the static case can, for instance, be written in the form of the following convolution integral (e.g. [104])

$$\mathbf{u} = \mathbf{G} * \mathbf{f} = \int \mathbf{G}\left(\mathbf{x} - \mathbf{x}_1\right) \mathbf{f}\left(\mathbf{x}_1\right) d\mathbf{x}_1, \tag{C.2}$$

where $\mathbf{G}$ is the Green tensor for the static problem.

In the following two sections we present the construction of Green functions for static and dynamical problems of linear elasticity. We follow here the presentation of T. D. Shermergor [352].

C.2   Statics of isotropic elastic materials

As a first example we present the construction of the Green function for linear elasticity in the static case. The operator $\mathbf{L}$ has in this case the following form

$$\mathbf{L} = L_{ik}\mathbf{e}_i \otimes \mathbf{e}_k, \quad L_{ik} = \frac{\partial}{\partial x_j} c_{ijkl} \frac{\partial}{\partial x_l}, \tag{C.3}$$

where c_{ijkl} is the tensor of elasticity. For heterogeneous materials it may be dependent on the point $\mathbf{x}$. We consider only homogeneous materials for which it consists of material constants. For isotropic materials it has the following explicit form

$$c_{ijkl} = \lambda \delta_{ik}\delta_{jl} + 2\mu\delta_{ij}\delta_{kl}, \tag{C.4}$$

where λ and μ are Lamé constants.

Substitution of (C.2) in (C.1) yields

$$\mathbf{L}\left(\mathbf{x}\right)\int \mathbf{G}\left(\mathbf{x}-\mathbf{x}_1\right)\mathbf{f}\left(\mathbf{x}_1\right)d\mathbf{x}_1 = -\mathbf{f}\left(\mathbf{x}\right). \tag{C.5}$$

Consequently, the Green function must satisfy the equation

$$\mathbf{L}\left(\mathbf{x}\right)\mathbf{G}\left(\mathbf{x}-\mathbf{x}_1\right) = -\mathbf{1}\delta\left(\mathbf{x}-\mathbf{x}_1\right), \quad \mathbf{1} = \delta_{ij}\mathbf{e}_i \otimes \mathbf{e}_j, \tag{C.6}$$

where $\delta\left(\mathbf{x}-\mathbf{x}_1\right)$ is the Dirac function.

In linear elasticity the vector $\mathbf{u}$ denotes the displacement. Then the components of the Green function $G_{ij}\left(\mathbf{x}-\mathbf{x}_1\right)$ define the components of the displacement $u_i\left(\mathbf{x}\right)$ at the point $\mathbf{x}$ of the infinite medium caused by the unit force acting at the point $\mathbf{x}_1$ in the direction $\mathbf{e}_j$. If the medium is finite, the Green function specifies the displacement $\mathbf{u}$ by the relation (see [104])

$$u_m\left(\mathbf{x}\right) = \int_V G_{im}\left(\mathbf{x}-\mathbf{x}_1\right)f_i\left(\mathbf{x}_1\right)dV_1$$

$$+ \int_{\partial V}\left[u_i\left(\mathbf{x}_1\right)c_{ijkl}\frac{\partial G_{km}}{\partial x_l}\left(\mathbf{x}-\mathbf{x}_1\right) + G_{im}\left(\mathbf{x}-\mathbf{x}_1\right)\sigma_{ij}\left(\mathbf{x}_1\right)\right]n_j dA_1, \tag{C.7}$$

where ∂V is the surface of the domain V and n_i are the components of the normal vector of this surface.

The form of the Green tensor G_{ij} follows from Equation (C.6). It is usually found by means of the Fourier integral transform (see: Appendix B). We have

$$\bar{\mathbf{G}}\left(\mathbf{k}\right) = \int \mathbf{G}\left(\mathbf{x}\right)e^{-i\mathbf{k}\cdot\mathbf{x}}dV_{\mathbf{x}}, \quad \mathbf{G}\left(\mathbf{x}\right) = \frac{1}{8\pi^3}\int \bar{\mathbf{G}}\left(\mathbf{k}\right)e^{i\mathbf{k}\cdot\mathbf{x}}dV_{\mathbf{k}}. \tag{C.8}$$

Application of the Fourier transform to Equation (C.6) for $\mathbf{x}_1 = 0$ (this assumption is immaterial as for the infinite domain we can always shift the origin of the coordinates to the point $\mathbf{x}_1$), i.e., to the equation

$$c_{ijkl}\frac{\partial^2 G_{kn}}{\partial x_j \partial x_l}\left(\mathbf{x}\right) = -\delta_{in}\delta\left(\mathbf{x}\right), \tag{C.9}$$

yields

$$\bar{\zeta}_{ik}\bar{G}_{kn}\left(\mathbf{k}\right) = \delta_{in}, \quad \bar{\zeta}_{ik} = c_{ijkl}k_j k_l. \tag{C.10}$$

This is a set of algebraic relations which can be solved by inverting the matrix ζ_{ik}. In the case of isotropic materials given by (C.4) it is immediate. It can also be done for materials with hexagonal symmetry (see: E. M. Lifshitz and L. N. Rosenzweig [228], E. Kröner [208],

L. Leicek [222]). In general some approximate methods such as the method of perturbation must be applied (e.g. T. Mura [267] or L. J. Gray, D. Ghosh and T. Kaplan [140][1]).

For isotropic materials we have

$$\bar{\zeta}_{ik} = \mu k^2 \delta_{ik} + (\lambda + \mu)\, k_i k_k. \tag{C.11}$$

Consequently,

$$\bar{G}_{ik} = \bar{\zeta}_{ik}^{-1} = \frac{1}{\mu}\left(\frac{1}{k^2}\delta_{ik} - \kappa\frac{k_i k_k}{k^2}\right), \quad \kappa = \frac{\lambda + \mu}{\lambda + 2\mu}. \tag{C.12}$$

It remains to invert the Fourier transform. Before we do so, let us quote an identity which follows from considerations of electrostatics. In such a case, the Maxwell equations (see Chapter 11) describing the electromagnetic field, reduce to the following two equations

$$\operatorname{div}\mathbf{E} = 4\pi\rho, \quad \operatorname{rot}\mathbf{E} = 0, \tag{C.13}$$

where $\mathbf{E}$ is the electric field and ρ is the electric charge density. The second relation implies the existence of the potential φ such that

$$\mathbf{E} = -\operatorname{grad}\varphi. \tag{C.14}$$

The choice of signs is a matter of tradition. Consequently, the equation determining the potential φ has the form of the Poisson equation

$$\Delta^2\varphi = -4\pi\rho, \quad \Delta^2 = \frac{\partial^2}{\partial x_k \partial x_k}, \tag{C.15}$$

where the last relation for the Laplace operator holds for Cartesian coordinates. An easy argument based on the balance equation of charge (e.g. [217]) yields for the charge density $\rho = e\delta(\mathbf{x})$ the following solution of (C.15)

$$\mathbf{E} = \frac{e\mathbf{x}}{r^3} \quad \Rightarrow \quad \varphi = \frac{e}{r}, \quad r^2 = \mathbf{x}\cdot\mathbf{x}, \tag{C.16}$$

where e is the electric charge. Obviously, it is the Coulomb law. Now the substitution of φ into (C.13) yields a representation of the Dirac δ-function by a function regular beyond the point $\mathbf{x} = \mathbf{0}$. We have

$$\delta(\mathbf{x}) = -\frac{1}{4\pi}\Delta^2\left(\frac{1}{r}\right). \tag{C.17}$$

We use this identity in the derivation of the Green function.

In addition to the above identity we have

$$\Delta^2 r = \frac{\partial^2 r}{\partial x_k \partial x_k} = \frac{\partial}{\partial x_k}\left(\frac{x_k}{r}\right) = \frac{2}{r}. \tag{C.18}$$

---

[1]Quotation of the abstract of the paper [140]:
A perturbation expansion technique for approximating the three dimensional anisotropic elastic Green's function is presented. The method employs the usual series for the matrix $(\mathrm{I}{-}\mathrm{A})^{-1}$ to obtain an expansion in which the zeroth order term is an isotropic fundamental solution. The higher order contributions are expressed as contour integrals of matrix products, and can be directly evaluated with a symbolic manipulation program. A convergence condition is established for cubic crystals, and it is shown that convergence is enhanced by employing Voigt averaged isotropic constants to define the expansion point. Example calculations demonstrate that, for moderately anisotropic materials, employing the first few terms in the series provides an accurate solution and a fast computational algorithm. However, for strongly anisotropic solids, this approach will most likely not be competitive with the Wilson-Cruse interpolation algorithm.

Hence, bearing (C.17) in mind,

$$-8\pi\delta\left(\mathbf{x}\right) = \frac{\partial^4 r}{\partial x_k \partial x_k \partial x_l \partial x_l}. \tag{C.19}$$

We apply to this relation the Fourier transform. It follows

$$\int \frac{\partial^4 r}{\partial x_k \partial x_k \partial x_l \partial x_l} e^{-i\mathbf{k}\cdot\mathbf{x}} dV_{\mathbf{x}} \tag{C.20}$$

$$= \int \frac{\partial}{\partial x_k} \left(\frac{\partial^3 r}{\partial x_k \partial x_l \partial x_l} e^{-i\mathbf{k}\cdot\mathbf{x}} \right) dV_{\mathbf{x}} + i k_k \int \frac{\partial^3 r}{\partial x_k \partial x_l \partial x_l} e^{-i\mathbf{k}\cdot\mathbf{x}} dV_{\mathbf{x}} = \ldots$$

$$\ldots = k^4 \int r e^{i\mathbf{k}\cdot\mathbf{x}} dV_{\mathbf{x}},$$

where we have used the Gauss Divergence Theorem and accounted for the fact that surface integrals must vanish for the infinite domain. Application of the inverse transform yields

$$r = -\frac{1}{\pi^2} \int \frac{e^{i\mathbf{k}\cdot\mathbf{x}}}{k^4} dV_{\mathbf{k}}. \tag{C.21}$$

Differentiation of this relation leads to the following identities

$$\frac{\partial^2 r}{\partial x_k \partial x_l} = \frac{1}{\pi^2} \int \frac{k_k k_l}{k^4} e^{i\mathbf{k}\cdot\mathbf{x}} dV_{\mathbf{k}}, \quad \Delta^2 r = \frac{1}{\pi^2} \int \frac{e^{i\mathbf{k}\cdot\mathbf{x}}}{k^2} dV_{\mathbf{k}}. \tag{C.22}$$

Now, we are in the position to find the Fourier inverse of Relation (C.12). Bearing (C.8) in mind, we obtain immediately

$$G_{ik}\left(\mathbf{x}\right) = \frac{1}{8\pi\mu} \left(\delta_{ik}\Delta^2 r - \kappa \frac{\partial^2 r}{\partial x_i \partial x_k} \right), \quad \kappa = \frac{\lambda+\mu}{\lambda+2\mu}. \tag{C.23}$$

This is the Green function for static equations of the linear isotropic elasticity.

C.3   Dynamical Green function for isotropic elastic materials

In the dynamical case we have to include the inertial force in the momentum balance equation. Hence, for linear elasticity Relation (C.3) for the operator $\mathbf{L}$ must be replaced by the following one

$$\mathbf{L} = L_{ik}\mathbf{e}_i \otimes \mathbf{e}_k, \quad L_{ik} = -\delta_{ik}\rho\frac{\partial^2}{\partial t^2} + \frac{\partial}{\partial x_j} c_{ijkl} \frac{\partial}{\partial x_l}, \tag{C.24}$$

where ρ is the constant mass density.

For isotropic materials defined by (C.4), we have

$$L_{ik} = -\delta_{ik}\rho\frac{\partial^2}{\partial t^2} + (\lambda+\mu)\frac{\partial^2}{\partial x_i \partial x_k} + \mu\frac{\partial^2}{\partial x_l \partial x_l}\delta_{ik}. \tag{C.25}$$

The Green tensor for the above operator will be sought again by Fourier transformation. In the dynamical case we also have to perform the transformation with respect to time.

To this aim we could use the Laplace transform or, as we do below, we have to cut the Fourier transform to the range of nonnegative time. This reflects the principle of causality (determinism) of classical mechanics. Consequently, the Green function has to satisfy the following equations

$$L_{il}G_{lj}\left(\mathbf{x},t\right) = \delta_{ij}\delta\left(\mathbf{x}\right)\delta\left(t\right) \quad \text{for} \quad t \geq 0, \tag{C.26}$$
$$G_{ij} = 0 \qquad \text{for} \quad t < 0.$$

Obviously, the Green function G_{ij} determines the displacement $u_i\left(\mathbf{x},t\right)$ in the instant of time t and at the point $\mathbf{x}$ caused by the unit force acting in the $\mathbf{e}_j$-direction in the instant of time $t = 0$ at the point $\mathbf{x} = \mathbf{0}$. Causality physically means that the displacement cannot be caused by incoming waves which should not exist yet before the force was applied at the point $\mathbf{x} = \mathbf{0}$.

As before, the displacement $u_i\left(\mathbf{x},t\right)$ for an arbitrary given external force $f_j\left(\mathbf{x},t\right)$ is then specified by the convolution integral

$$u_i = G_{il} * f_j. \tag{C.27}$$

Obviously, the time integration in (C.26) yields immediately that the static Green function $G_{ij}\left(\mathbf{x}\right)$, calculated in the previous section, should satisfy the relation

$$G_{ij}\left(\mathbf{x}\right) = \int_{-\infty}^{\infty} G_{ij}\left(\mathbf{x},t\right)dt. \tag{C.28}$$

As already mentioned, we find the Green function for the infinite medium by the double Fourier transform

$$\bar{\mathbf{G}}\left(\mathbf{k},\omega\right) = \int\int \mathbf{G}\left(\mathbf{x},t\right)e^{i(\mathbf{k}\cdot\mathbf{x}-\omega\mathbf{t})}dV_{\mathbf{x}}dt, \tag{C.29}$$
$$\mathbf{G}\left(\mathbf{x},t\right) = \frac{1}{16\pi^4}\int\int \bar{\mathbf{G}}\left(\mathbf{k},\omega\right)e^{-i(\mathbf{k}\cdot\mathbf{x}-\omega\mathbf{t})}dV_{\mathbf{k}}d\omega.$$

As already indicated in many places of this book, we call $\mathbf{k}$ the wave vector and the scalar ω the frequency. Then, after the Fourier transform the operator $L_{ij}\left(\mathbf{x},t\right)$ for isotropic materials has the following form

$$\bar{L}_{ij}\left(\mathbf{k},\omega\right) = \rho\omega^2\delta_{ij} - \left(\lambda + \mu\right)k_ik_j - \mu k^2\delta_{ij}, \tag{C.30}$$

and Equation (C.26) becomes purely algebraic

$$\bar{L}_{ik}\bar{G}_{kj} = -\delta_{ij}. \tag{C.31}$$

We have to invert the matrix (C.30). Hence, after easy calculations, the Fourier transform of the dynamical Green function is as follows

$$\bar{G}_{ij} = \frac{1}{\mu k^2 - \rho\omega^2}\left[\delta_{ij} - \frac{\left(\lambda + \mu\right)k_ik_j}{\left(\lambda + 2\mu\right)k^2 - \rho\omega^2}\right]. \tag{C.32}$$

Obviously, the static transform of the Green function (C.12) follows from (C.32) by the substitution $\omega = 0$.

It is convenient to write the above relation by means of the speeds of propagation of longitudinal and transversal waves in a linear elastic material

$$c_L^2 = \frac{\lambda + 2\mu}{\rho}, \quad c_T^2 = \frac{\mu}{\rho}. \tag{C.33}$$

We obtain

$$\rho \bar{G}_{ij} = \frac{1}{c_L^2 k^2 - \omega^2} \left[\delta_{ij} - \frac{\left(c_L^2 - c_T^2\right) k_i k_j}{c_L^2 k^2 - \omega^2} \right]. \tag{C.34}$$

The inverse of the above relation is given by the double integration prescribed by $(C.29)_2$. We perform first the integration with respect to the wave vector $\mathbf{k}$. We have

$$G_{ij}\left(\mathbf{x}, \omega\right) = \frac{1}{8\pi^3} \int \bar{G}_{ij}\left(\mathbf{k}, \omega\right) e^{-i\mathbf{k}\cdot\mathbf{x}} dV_{\mathbf{k}}. \tag{C.35}$$

Hence,

$$G_{ij}\left(\mathbf{x}, \omega\right) = I^0 \delta_{ij} - \frac{\partial I}{\partial x_i \partial x_j}, \tag{C.36}$$

where

$$
\begin{aligned}
I^0 &= \frac{1}{8\pi^3} \int \frac{e^{-i\mathbf{k}\cdot\mathbf{x}}}{c_T^2 k^2 - \omega^2} dV_{\mathbf{k}} = \frac{1}{4\pi r c_T^2} \exp\left(-\frac{i\omega r}{c_T}\right), \\
I &= -\frac{c_L^2 - c_T^2}{8\pi^3} \int \frac{e^{-i\mathbf{k}\cdot\mathbf{x}}}{\left(c_L^2 k^2 - \omega^2\right)\left(c_T^2 k^2 - \omega^2\right)} dV_{\mathbf{k}} \\
&= \frac{1}{4\pi r \omega^2} \left[\exp\left(-\frac{i\omega r}{c_L}\right) - \exp\left(-\frac{i\omega r}{c_T}\right) \right].
\end{aligned}
\tag{C.37}
$$

The calculations of the integrals I^0 and I are made using the method of residuals for complex functions. In order to use this method, we assume formally that both speeds of propagation are complex. This is, indeed, the case for viscoelastic materials. We demonstrate the calculations on the example of the integral I^0. Then

$$c_T^2\left(\omega\right) = \operatorname{Re} c_T^2\left(\omega\right) + i \operatorname{Im} c_T^2\left(\omega\right) = \left[1 + i\beta\left(\omega\right)\right] \operatorname{Re} c_T^2\left(\omega\right). \tag{C.38}$$

This extension yields the following form of the integral I^0

$$
\begin{aligned}
I^0 &= \frac{1}{8\pi^3} \int \frac{e^{-i\mathbf{k}\cdot\mathbf{x}}}{c_T^2\left(\omega\right) k^2 - \omega^2} dV_{\mathbf{k}} \\
&= \frac{1}{8\pi^3} \int_0^\infty \int_0^\pi \frac{e^{ikr\cos\theta} 2\pi \sin\theta k^2}{c_T^2\left(\omega\right) k^2 - \omega^2} dk d\theta \\
&= \frac{1}{4\pi^2 ir} \int_0^\infty \frac{k\left(e^{ikr} - e^{-ikr}\right)}{c_T^2\left(\omega\right) k^2 - \omega^2} dk = \frac{1}{4\pi^2 ir} \int_{-\infty}^\infty \frac{k e^{ikr}}{c_T^2\left(\omega\right) k^2 - \omega^2} dk.
\end{aligned}
\tag{C.39}
$$

This integral can be evaluated by the method of residuals. Obviously, it possesses two poles

$$k = \pm \frac{\omega}{\left\{\left[1 + i\beta\left(\omega\right)\right] \operatorname{Re} c_T^2\left(\omega\right)\right\}^{0.5}} = \pm \frac{\omega \exp\left[-\frac{i}{2}\arctan\beta\left(\omega\right)\right]}{\left\{\sqrt{1 + \beta^2\left(\omega\right)} \operatorname{Re} c_T^2\left(\omega\right)\right\}^{0.5}}. \tag{C.40}$$

The pole with the minus sign lies in the second quadrant of the complex plane, while the other pole lies in the fourth quadrant. Hence, we choose as the path of integration the real axis and the semicircle of the infinite radius in the upper part of the complex plane. We obtain

$$I^0 = \frac{1}{4\pi r c_T^2} \exp\left[-\frac{ir\omega}{c_T\left(\omega\right)}\right]. \tag{C.41}$$

The transition $c_T\left(\omega\right) \to c_T$ gives the desired result. In a similar manner one can calculate the integral I. We have to use obvious identities when differentiating in (C.36)

$$\frac{\partial r}{\partial x_i} = n_i, \quad \frac{\partial^2 r}{\partial x_i \partial x_j} = \frac{1}{r}\left(\delta_{ij} - n_i n_j\right), \quad n_i n_i = 1. \tag{C.42}$$

It follows

$$G_{ij}\left(\mathbf{x}, \omega\right) = \frac{1}{r}\left[h\left(\omega r\right)\delta_{ij} + g\left(\omega r\right)n_i n_j\right], \tag{C.43}$$

where

$$h\left(\omega r\right) = \frac{1}{4\pi r^2 \rho \omega^2}\left\{\left[\left(1 + \frac{ir\omega}{c_L}\right)e^{-i\omega r/c_L} - \left(1 + \frac{ir\omega}{c_T}\right)e^{-i\omega r/c_T}\right] + \frac{r^2\omega^2}{c_T^2}e^{-i\omega r/c_T}\right\},$$

$$g\left(\omega r\right) = -\frac{1}{4\pi r^2 \rho \omega^2}\left\{\left[3\left(1 + \frac{ir\omega}{c_L}\right) - \frac{r^2\omega^2}{c_L^2}\right]e^{-i\omega r/c_L} \tag{C.44}\right.$$
$$\left.-\left[3\left(1 + \frac{ir\omega}{c_T}\right) - \frac{r^2\omega^2}{c_T^2}\right]e^{-i\omega r/c_T}\right\}.$$

This is, obviously, the Green function for monochromatic waves of given frequency ω. The time dependence of the solution is then given by the factor $\exp\left(i\omega t\right)$.

It remains to perform the second inverse transformation. We use here the following relations

$$\frac{1}{2\pi}\int_{-\infty}^{\infty} e^{i\omega t}d\omega = \delta\left(t\right),$$

$$\frac{1}{2\pi}\int_{-\infty}^{\infty}\frac{1}{i\omega}e^{i\omega t}d\omega = H\left(t\right) = \begin{cases} 1 \text{ for } t > 0, \\ 0 \text{ for } t < 0, \end{cases}$$

$$-\frac{1}{2\pi}\int_{-\infty}^{\infty}\frac{1}{\omega^2}e^{i\omega t}d\omega = \Psi\left(t\right) = \begin{cases} t \text{ for } t > 0, \\ 0 \text{ for } t < 0, \end{cases} \tag{C.45}$$

$$\frac{\partial\Psi}{\partial t} = H\left(t\right), \quad \frac{\partial H}{\partial t} = \delta\left(t\right).$$

Bearing these relations in mind as well as

$$\frac{\partial^2}{\partial x_i \partial x_j}\left(\frac{1}{r}\right) = \frac{3n_i n_j - \delta_{ij}}{r^3}, \tag{C.46}$$

$$-\frac{1}{2\pi}\int_{-\infty}^{\infty}\frac{1}{\omega^2}\left(1 + \frac{ir\omega}{c_L}\right)e^{i\omega(t-r/c_L)}d\omega = \Psi\left(t - \frac{r}{c_L}\right) + \frac{r}{c_L}H\left(t - \frac{r}{c_L}\right),$$

and similarly for c_T, we finally obtain

$$4\pi\rho G_{ij}\left(\mathbf{x}, t\right) = \delta\left(t - \frac{r}{c_T}\right)\left(\frac{\delta_{ij}}{c_T^2} - \frac{x_i x_j}{c_T^2 r^3}\right) + \delta\left(t - \frac{r}{c_L}\right)\frac{x_i x_j}{c_L^2 r^3}$$
$$+ \frac{\partial^2}{\partial x_i \partial x_j}\left(\frac{1}{r}\right)t\left(H\left(t - \frac{r}{c_L}\right) - H\left(t - \frac{r}{c_T}\right)\right). \tag{C.47}$$

The first contribution describes the transversal part of the impulse which arrives with the speed c_T, the second contribution is the longitudinal part of the impulse which arrives with the speed c_L and the third contribution is the evolution of the impulse between the arrival of the longitudinal and transversal parts.

Easy integration in Relation (C.28) yields the static Green function given by Relation (C.23).

A summary of several fundamental solutions or Green functions for full space problems, half-space problems, strata and plates subjected to differently applied point and line loads has been presented by E. Kausel in [197]. The results are shown in the three major coordinate systems (Cartesian, cylindrical and spherical – see Appendix A.2) in two and three dimensions and the MATLAB codes for tested formulas are given.

C.4 Some fundamental solutions for poroelastic and thermoelastic materials

Since the beginning of the development of numerical techniques, especially of the boundary element method, for the solution of problems in poroelastic media various approaches have been published which aim to find fundamental solutions for such materials. A detailed overview on history and procedures for dynamic poroelasticity is given in [336] and references therein.

Practically all approaches are based on Biot's theory albeit in different formulations. In Chapter 13 several problems connected to porous media have been discussed and the formulation has been compared to Biot's equations. For the description of these problems a formulation including partial mass and momentum balances for each component has been used. For the numerical treatment, often another description is suitable, namely the equivalent formulation including the effective stress. We reproduce here the approach of finding fundamental solutions for poroelastic materials which uses such a formulation by A. Cheng, T. Badmus and D. Beskos [73]. It is especially vivid since not only fundamental solutions (in the frequency domain) for poroelasticity are derived but they are also compared to earlier found fundamental solutions for thermoelasticity found by W. Nowacki [278].

The dynamic equations of poroelasticity in the time domain in a formulation with effective stresses are (in the original notation of [73][2])

- constitutive relations

$$\sigma_{ij} = 2\mu e_{ij} + \frac{2\mu\nu}{1 - 2\nu}\delta_{ij}e - \alpha_B \delta_{ij} p, \tag{C.48}$$

$$\zeta = \alpha_B e + \frac{n^2}{R}p, \tag{C.49}$$

---

[2]In order to avoid confusions with the Green functions G of the last sections instead of the shear modulus G used in [73] we use the notion μ and therefore instead of μ for the fluid dynamic viscosity (see also the remark on the denotation of the viscosity on page 48) the notion μ_v and for the permeability coefficient κ we use κ_v because in Chapter 13 κ denotes the compressibility parameter of the fluid. Furthermore we continue to denote the porosity by n instead of ϕ in [73] and $\rho_f \equiv \rho^F$, $\rho_s \equiv \rho^S$. Biot's effective stress coefficient is denoted by α_B because α in Chapter 13 denotes the surface permeability parameter.

- equilibrium equations

$$\sigma_{ij,j} = \rho \ddot{u}_i + n\rho_f \ddot{v}_i - F_i, \tag{C.50}$$

- generalized Darcy law

$$q_i = -\kappa \left(p_{,i} + \rho_f \ddot{u}_i + \frac{\rho_a + n\rho_f}{n} \ddot{v}_i - f_i \right), \tag{C.51}$$

- continuity equation

$$\dot{\zeta} + q_{i,i} = \gamma. \tag{C.52}$$

In these equations which differ slightly from the original formulation of Biot [49] σ_{ij} is the total stress and p is the pore pressure. As kinematic parameters the solid displacement u_i and the specific flux q_i have been chosen. Additionally, the relative fluid to solid displacement v_i, defined by $q_i = nv_i$, where n is the porosity, is used. The solid strain tensor and dilatation are given by $e_{ij} = \frac{1}{2}\left(u_{i,j} + u_{j,i}\right)$ and $e = e_{ii}$, respectively. For the fluid Biot introduced a similar quantity, namely the change of fluid volume contents per unit reference volume $\zeta = -nv_{i,i} + \Gamma$. Therein, the possibility of a fluid source in the volume is taken into account: $\Gamma = \int_0^t \gamma dt$ is the cumulative injected fluid volume from a fluid source of strength γ. Detailed information on this quantity and also on the derivation of Green functions and their applications can be found in the book by H. Wang [416]. The material parameters appearing in the above equations are the shear modulus μ, the Poisson ratio ν, the permeability coefficient $\kappa_v = k/\mu_v$ where k is the intrinsic permeability and μ_v the fluid dynamic viscosity.[3] Solid and fluid mass densities are denoted by ρ_s and ρ_f, and their combination $\rho = (1-n)\rho_s + n\rho_f$ builds the bulk mass density. One of the parameters connected with the added mass effect and which is not included in the models used in Chapter 13 is the apparent mass density ρ_a. According to [73] it corresponds to the work done by the solid phase to the fluid phase due to the relative motion between them and can for geotechnical applications be written as $\rho_a = 0.66n\rho_f$. The factor depends on the geometry of the pores and on the frequency of excitation. Furthermore Biot's effective stress coefficient $\alpha_B = \frac{3(\nu_u - \nu)}{B(1-2\nu)(1+\nu_u)}$ and the poroelastic constitutive coefficient $R = \frac{2n^2\mu B^2(1-2\nu)(1+\nu_u)^2}{9(\nu_u-\nu)(1-2\nu_u)}$ are used. Details on these coefficients can be found in [50]. Both coefficients include the undrained Poisson ratio ν_u and Skempton's pore pressure coefficient B. For a discussion on the parameters included in the above model see [311]. Finally, the bulk body force is represented by $F_i = \rho g_i$, where g_i is the gravity component, and $f_i = \rho_f g_i$ is the fluid body force.

The comparable dynamic equations of thermoelasticity in the time domain are quoted from [73] who took them from Nowacki [278]:

$$\sigma_{ij} = 2\mu e_{ij} + \frac{2\mu\nu}{1-2\nu}\delta_{ij}e - \gamma_t \delta_{ij}\theta, \tag{C.53}$$

$$s = \gamma_t e + \frac{c_\epsilon}{T_0}\theta, \tag{C.54}$$

$$\sigma_{ij,j} = \rho\ddot{u}_i - X_i, \tag{C.55}$$

---

[3]Note that in the literature many different definitions and denotations of permeabilities and viscosities are used. E.g. in Section 13.5 a different notation is used than in [73]. For a comparison of notations see e.g. Wilmanski [447].

$$q_i = -\lambda_0 \theta_{,i}, \tag{C.56}$$

$$T_0 \dot{s} + q_{i,i} = W. \tag{C.57}$$

Here, θ is the temperature increment, T_0 the average temperature, s the entropy, q_i the heat flux, c_ϵ the specific heat at constant strain, $\gamma_t = 3K\alpha_t$ with the bulk modulus of elasticity $K = \frac{2\mu(1+\nu)}{3(1-2\nu)}$, α_t is the coefficient of linear thermal expansion, λ_0 the heat conduction coefficient, X_i the solid body force and W the heat source. Equations (C.48) to (C.52) contain parameters of the fluid body, inertial forces and the apparent mass in the poroelastic system which are not present in the thermoelastic equations (C.53) to (C.57). Thus, in this form the latter system can be deduced from the further but not vice versa.

In order to compare both sets the equations of both sets are combined ($i = 1, 2, 3$). Equations (C.48) to (C.52) can be rewritten to

$$\mu u_{i,jj} + \frac{\mu}{1 - 2\nu_u} u_{j,ji} + \frac{\alpha_B R}{n} v_{j,ji} = -F_i + \frac{\alpha_B R}{n^2} \Gamma_{,i} + \rho \ddot{u}_i + n\rho_f \ddot{v}_i, \tag{C.58}$$

$$\frac{\alpha_B R}{n} u_{j,ji} + R v_{j,ji} - \frac{n^2}{\kappa} \dot{v}_i = -n f_i + \frac{R}{n} \Gamma_{,i} + (\rho_a + n\rho_f) \ddot{v}_i + n\rho_f \ddot{u}_i, \tag{C.59}$$

and Equations (C.53) to (C.57) to

$$\mu u_{i,jj} + \frac{\mu}{1 - 2\nu} u_{j,ji} - \gamma_t \theta_{,i} = -X_i + \rho \ddot{u}_i, \tag{C.60}$$

$$\theta_{,ii} - \frac{c_\epsilon}{\lambda_0} \dot{\theta} - \frac{\gamma_t T_0}{\lambda_0} \dot{u}_{i,i} = -\frac{W}{\lambda_0}. \tag{C.61}$$

These sets of equations are transformed to the frequency domain. As before, ω denotes the frequency and quantities with bar are the Fourier transforms of the original quantity. In order to have a mathematically correct formulation, the description in the frequency domain is preferable to the description in the time domain since the coefficients κ and ρ_a are generally frequency dependent. Additionally the following equations are reduced to four variables $\bar{u}_i$ and $\bar{p}$ instead of the six variables u_i and v_i used in (C.58) and (C.59)

$$\mu \bar{u}_{i,jj} + \frac{\mu}{1 - 2\nu} \bar{u}_{j,ji} + \omega^2 (\rho - \beta_B \rho_f) \bar{u}_i - (\alpha_B - \beta_B) \bar{p}_{,i} = -\bar{F}_i + \beta_B \bar{f}_i, \tag{C.62}$$

$$\bar{p}_{,ii} + \frac{\omega^2 \rho_f n^2}{\beta_B R} \bar{p} + \frac{\omega^2 \rho_f (\alpha_B - \beta_B)}{\beta_B} \bar{u}_{i,i} = \bar{f}_{i,i} + \frac{i\omega \rho_f}{\beta_B} \bar{\gamma}, \tag{C.63}$$

where

$$\beta_B = \frac{\omega n^2 \rho_f \kappa}{i n^2 + \omega \kappa (\rho_a + n\rho_f)}, \tag{C.64}$$

and the following relations have been used

$$\bar{q}_i = -i\omega n \bar{v}_i, \qquad \bar{\zeta} = -n \bar{v}_{i,i} + \frac{i}{\omega} \bar{\gamma}. \tag{C.65}$$

The set (C.62), (C.63) can be compared with the thermoelastic equations in the frequency domain containing four variables $\bar{u}_i$ and $\bar{\theta}$

$$\mu \bar{u}_{i,jj} + \frac{\mu}{1 - 2\nu} \bar{u}_{j,ji} + \omega^2 \rho \bar{u}_i - \gamma_t \bar{\theta}_{,i} = -\bar{X}_i, \tag{C.66}$$

$$\bar{\theta}_{,ii} + \frac{i\omega}{\kappa_t}\bar{\theta} + i\omega\eta\bar{u}_{i,i} = -\frac{\bar{Q}}{\kappa_t}, \tag{C.67}$$

where

$$\kappa_t = \frac{\lambda_0}{c_\epsilon}, \qquad \eta = \frac{\gamma_t T_0}{\lambda_0}, \qquad Q = \frac{W}{c_\epsilon}. \tag{C.68}$$

Corresponding parameters of the sets (C.62), (C.63) for poroelasticity and (C.66), (C.67) for thermoelasticity are summarized in Table C.4 (reproduced from [73]).

Table C.1: *Equivalence of poroelasticity and thermoelasticity in the frequency domain (Table taken from [73]).*

Poroelasticity	Thermoelasticity
$\bar{u}_i$	$\bar{u}_i$
$\bar{p}$	$\bar{\theta}$
μ	μ
ν	ν
$\alpha_B - \beta_B$	γ_t
$(iR\beta_B)/(\omega n^2 \rho_f)$	κ_t
$-i\omega\rho_f(\alpha_B - \beta_B)/\beta_B$	η
$\rho - \beta_B\rho_f$	ρ
$\bar{F}_i \quad \beta_B\bar{f}_i$	$\bar{X}_i$
$R\bar{\gamma}/n^2 - (iR\beta_B\bar{f}_{i,i})/(\omega n^2 \rho_f)$	$\bar{Q}$
$\bar{\sigma}_{ij} + \beta_B\delta_{ij}\bar{p}$	$\bar{\sigma}_{ij}$
$\bar{t}_i + \beta_B n_i\bar{p}$	$\bar{t}_i$
$\bar{\zeta} - \beta_B\bar{e}$	$\bar{s}$
$\bar{q}_i - i\omega\beta_B\bar{u}_i - (i\beta_B\bar{f}_i)/(\omega\rho_f)$	$\bar{q}_i/T_0$

Now, the equivalence between thermoelasticity and poroelasticity is used. Nowacki [278] derived the integral equations for thermoelasticity:

$$\int \left[\bar{t}_i^{(1)}\bar{u}_i^{(2)} - \bar{t}_i^{(2)}\bar{u}_i^{(1)}\right] dS + \frac{i\gamma_t}{\omega\eta}\int \left[\bar{\theta}^{(1)}\frac{\partial\bar{\theta}^{(2)}}{\partial n} - \bar{\theta}^{(2)}\frac{\partial\bar{\theta}^{(1)}}{\partial n}\right] dS$$
$$+ \int \left[\bar{X}_i^{(1)}\bar{u}_i^{(2)} - \bar{X}_i^{(2)}\bar{u}_i^{(1)}\right] dV + \frac{i\gamma_t}{\omega\kappa_t\eta}\int \left[\bar{\theta}^{(1)}\bar{Q}^{(2)} - \bar{\theta}^{(2)}\bar{Q}^{(1)}\right] dV = 0, \tag{C.69}$$

where superscripts $^{(1)}$ and $^{(2)}$ refer to two independent stress-strain states. $\bar{t}_i = \bar{\sigma}_{ij}n_j$ is the boundary traction vector with the outward normal vector of the boundary n_i; the boundary integration is denoted by dS while the volume integration by dV. With the help of Table C.4 Equation (C.69) can be easily transformed in an equivalent for poroelasticity:

$$\int \left[\bar{t}_i^{(1)} \bar{u}_i^{(2)} - \bar{t}_i^{(2)} \bar{u}_i^{(1)} \right] dS + \frac{i}{\omega} \int \left[\bar{p}^{(1)} \bar{q}^{(2)} - \bar{p}^{(2)} \bar{q}^{(1)} \right] dS + \int \left[\bar{F}_i^{(1)} \bar{u}_i^{(2)} - \bar{F}_i^{(2)} \bar{u}_i^{(1)} \right] dV$$

$$+ \frac{i}{\omega} \int \left[\bar{f}_i^{(1)} \bar{q}_i^{(2)} - \bar{f}_i^{(2)} \bar{q}_i^{(1)} \right] dV + \frac{i}{\omega} \int \left[\bar{p}^{(1)} \bar{\gamma}^{(2)} - \bar{p}^{(2)} \bar{\gamma}^{(1)} \right] dV = 0. \tag{C.70}$$

In [73] this equation is derived without the help of earlier results and then compared to the above shown equation for thermoelasticity. The following substitutions are then made:

1. point force in the j-direction: $\bar{F}_i^{(2)} = \delta_{ij}\delta\left(\mathbf{x}', \mathbf{x}\right)$, $\bar{f}_i^{(2)} = 0$ and $\bar{\gamma}^{(2)} = 0$,

2. fluid source: $\bar{F}_i^{(2)} = 0$, $\bar{f}_i^{(2)} = 0$ and $\bar{\gamma}^{(2)} = \delta\left(\mathbf{x}', \mathbf{x}\right)$.

They result in the following Somigliana-type integral equations

$$b\left(\mathbf{x}\right) \bar{u}_j\left(\mathbf{x}; \omega\right) = \int \left[\bar{t}_i\left(\mathbf{x}'; \omega\right) \bar{u}_{ij}^f\left(\mathbf{x}', \mathbf{x}; \omega\right) - \bar{t}_{ij}^f\left(\mathbf{x}', \mathbf{x}; \omega\right) \bar{u}_i\left(\mathbf{x}'; \omega\right) \right] dS\left(\mathbf{x}'\right)$$

$$- \frac{i}{\omega} \int \left[\bar{p}\left(\mathbf{x}'; \omega\right) \bar{q}_j^f\left(\mathbf{x}', \mathbf{x}; \omega\right) - \bar{p}_j^f\left(\mathbf{x}', \mathbf{x}; \omega\right) \bar{q}\left(\mathbf{x}'; \omega\right) \right] dS\left(\mathbf{x}'\right), \tag{C.71}$$

$$b\left(\mathbf{x}\right) \bar{p}\left(\mathbf{x}; \omega\right) = i\omega \int \left[\bar{t}_i\left(\mathbf{x}'; \omega\right) \bar{u}_i^s\left(\mathbf{x}', \mathbf{x}; \omega\right) - \bar{t}_{ij}^s\left(\mathbf{x}', \mathbf{x}; \omega\right) \bar{u}_i\left(\mathbf{x}'; \omega\right) \right] dS\left(\mathbf{x}'\right)$$

$$+ \int \left[\bar{p}\left(\mathbf{x}'; \omega\right) \bar{q}^s\left(\mathbf{x}', \mathbf{x}; \omega\right) - \bar{p}^s\left(\mathbf{x}', \mathbf{x}; \omega\right) \bar{q}\left(\mathbf{x}'; \omega\right) \right] dS\left(\mathbf{x}'\right), \tag{C.72}$$

where $\mathbf{x}$ and $\mathbf{x}'$ denote a "base point" where the singularity is located and a "field point" upon which the integration is performed, respectively. The quantities marked with superscript f are singular solutions of traction, pressure, displacement and flux due to a point force in the j-direction, $\bar{t}_{ij}^f$, $\bar{p}_j^f$, $\bar{u}_{ij}^f$, $\bar{q}_j^f$, while $\bar{t}_i^s$, $\bar{p}^s$, $\bar{u}_i^s$, $\bar{q}^s$ are those due to a point source. The constant factor b equals one for $\mathbf{x} \in V$ and $\mathbf{x} \notin S$, zero for $\mathbf{x} \notin V$ and one half for $\mathbf{x} \in S$, where V is the solution domain and S the solution boundary.

Due to the correspondence between poroelastic and thermoelastic equations, the expressions for $\bar{u}_{ij}^f$, $\bar{p}_j^f$, $\bar{u}_i^s$ and $\bar{p}^s$ can be taken over from the thermoelastic literature, e.g. [278]. After conversion to the poroelastic parameters the fundamental solutions are the following (see [73])

Two-Dimensional Solutions

Point Force Solutions

$$\bar{u}_{ij}^f = \frac{i}{4\omega^2 \left(\rho - \beta_B \rho_f\right)} \left\{ \partial_i \partial_j \left[A_1 H_0^{(1)}\left(k_1 r\right) - A_2 H_0^{(1)}\left(k_2 r\right) - H_0^{(1)}\left(\tau r\right) \right] - \delta_{ij}\tau^2 H_0^{(1)}\left(\tau r\right) \right\}, \tag{C.73}$$

$$\bar{p}_j^f = \frac{i\epsilon q_t}{4mc_1^2\omega^2 \left(\rho - \beta_B \rho_f\right)\left(k_1^2 - k_2^2\right)} \partial_j \left[H_0^{(1)}\left(k_1 r\right) - H_0^{(1)}\left(k_2 r\right) \right]. \tag{C.74}$$

Point Source Solutions

$$\bar{u}_i^s = \frac{imR}{4\kappa_t n^2 \left(k_1^2 - k_2^2\right)} \partial_i \left[H_0^{(1)}\left(k_1 r\right) - H_0^{(1)}\left(k_2 r\right) \right], \tag{C.75}$$

$$\bar{p}^s = \frac{iR}{4\kappa_t n^2 \left(k_1^2 - k_2^2\right)} \left[\left(k_1^2 - \sigma^2\right) H_0^{(1)}\left(k_1 r\right) - \left(k_2^2 - \sigma^2\right) H_0^{(1)}\left(k_2 r\right)\right]. \tag{C.76}$$

Three-Dimensional Solutions

Point Force Solutions

$$\bar{u}_{ij}^f = \frac{1}{4\pi\omega^2 \left(\rho - \beta_B \rho_f\right)} \left\{\partial_i \partial_j \left[A_1 \frac{e^{ik_1 r}}{r} - A_2 \frac{e^{ik_2 r}}{r} - \frac{e^{i\tau r}}{r}\right] - \delta_{ij}\tau^2 \frac{e^{i\tau r}}{r}\right\}, \tag{C.77}$$

$$\bar{p}_j^f = \frac{\epsilon q_t}{4\pi m c_1^2 \omega^2 \left(\rho - \beta_B \rho_f\right)\left(k_1^2 - k_2^2\right)} \partial_j \left[\frac{e^{ik_1 r}}{r} - \frac{e^{ik_2 r}}{r}\right], \tag{C.78}$$

Point Source Solutions

$$\bar{u}_i^s = \frac{mR}{4\pi\kappa_t n^2 \left(k_1^2 - k_2^2\right)} \partial_i \left[\frac{e^{ik_1 r}}{r} - \frac{e^{ik_2 r}}{r}\right], \tag{C.79}$$

$$\bar{p}^s = \frac{R}{4\pi\kappa_t n^2 \left(k_1^2 - k_2^2\right)} \left[\left(k_1^2 - \sigma^2\right)\frac{e^{ik_1 r}}{r} - \left(k_2^2 - \sigma^2\right)\frac{e^{ik_2 r}}{r}\right]. \tag{C.80}$$

Here, $r = |\mathbf{x} - \mathbf{x}'|$ is the radial distance between the singularity point and the field point, ∂_i denotes the differentiation with respect to x_i' and $H_0^{(1)}$ is the Hankel function of first kind and order zero. Moreover, the following notations have been used

$$k_{1/2}^2 = \frac{1}{2}\left\{\sigma^2 + q_t\left(1 + \epsilon\right) \pm \sqrt{\left[\sigma^2 + q_t\left(1 + \epsilon\right)\right]^2 - 4q_t\sigma^2}\right\},$$

$$\sigma = \frac{\omega}{c_1}, \quad \tau = \frac{\omega}{c_2}, \quad q_t = \frac{\omega^2 n^2 \rho_f}{\beta_B R}, \quad \kappa_t = \frac{i\beta_B R}{\omega n^2 \rho_f},$$

$$\epsilon = \frac{R\left(\alpha_B - \beta_B\right)^2\left(1 - 2\nu\right)}{2n^2\mu\left(1 - \nu\right)}, \quad m = \frac{\left(\alpha_B - \beta_B\right)\left(1 - 2\nu\right)}{2\mu\left(1 - \nu\right)}, \tag{C.81}$$

$$c_1^2 = \frac{2\mu\left(1 - \nu\right)}{\left(\rho - \beta_B \rho_f\right)\left(1 - 2\nu\right)}, \quad c_2^2 = \frac{\mu}{\rho - \beta_B \rho_f},$$

$$A_1 = \frac{\sigma^2\left(k_1^2 - q_t\right)}{k_1^2\left(k_1^2 - k_2^2\right)}, \quad A_2 = \frac{\sigma^2\left(k_2^2 - q_t\right)}{k_2^2\left(k_1^2 - k_2^2\right)}.$$

In [73] it is pointed out that the poroelastic source solutions differ from the thermoelastic ones by a multiplication factor R/n^2 which arises from the difference in the definition of the poroelastic source γ and the thermoelastic one Q. Moreover, the interesting correspondence between the pressure solution of a point force and the displacement solution due to a point source, $\bar{p}_j^f = -i\omega\bar{u}_i^s$, is mentioned. Once fundamental solutions of displacement and pressure are available, the traction and flux expressions needed in (C.71) and (C.72) can be derived (see [73]).

Meanwhile there exists an extensive literature on fundamental solutions for linear two-component poroelastic media. Further approaches can be found e.g. in [74], [79], [59], [276], [53] or [335]. For an extension to partially saturated media we refer e.g. to [133] or [227].

Appendix D

Bessel functions and Bessel equation

Here, a short overview on Bessel functions and on the Bessel equation, as found in [66], is given. For detailed information see e.g. [419].

The Bessel function $J_\nu(z)$, named after Friedrich Wilhem Bessel (1784-1846), is defined by the equation

$$J_\nu(z) = \sum_{r=0}^{\infty} \frac{(-1)^r \left(\frac{1}{2}z\right)^{\nu+2r}}{r!\,\Gamma\left(\nu + r + 1\right)}, \tag{D.1}$$

where ν is real and z may be complex, its argument being given its principal value. Γ is the gamma function, an extension of the factorial function. The function $J_\nu(z)$ satisfies Bessel's equation of order ν

$$\frac{d^2y}{dz^2} + \frac{1}{z}\frac{dy}{dz} + \left(1 - \frac{\nu^2}{z^2}\right) y = 0. \tag{D.2}$$

If ν is not an integer, $J_\nu(z)$ and $J_{-\nu}(z)$ are independent solutions of (D.2), but if ν is an integer, n,

$$J_n(z) = (-1)^n J_{-n}(z). \tag{D.3}$$

In order to have a second solution of (D.2) which is available for all values of ν, the function

$$Y_\nu(z) = \frac{J_\nu(z)\cos\nu\pi - J_{-n}(z)}{\sin\nu\pi}, \tag{D.4}$$

is defined, the function of integral order $Y_n(z)$ being defined as $\lim_{\nu \to n} Y_\nu(z)$. With this definition

$$\frac{1}{2}\pi Y_0(z) = \left\{\ln\left(\frac{1}{2}z\right) + \gamma\right\} J_0(z) + \left(\frac{1}{2}z\right)^2 - \left(1 + \frac{1}{2}\right)\frac{\left(\frac{1}{2}z\right)^4}{(2!)^2} + \left(1 + \frac{1}{2} + \frac{1}{3}\right)\frac{\left(\frac{1}{2}z\right)^6}{(3!)^2} - ..., \tag{D.5}$$

where $\gamma = 0.5772...$ is Euler's constant.

Also, when n is any positive integer,

$$\pi Y_n(z) = 2\left\{\ln\left(\frac{1}{2}z\right) + \gamma\right\} J_n(z) - \sum_{r=0}^{\infty}(-1)^r \frac{\left(\frac{1}{2}z\right)^{n+2r}}{r!\,(n+r)!}\left[\sum_{m=1}^{n+r} m^{-1} + \sum_{m=1}^{r} m^{-1}\right] \tag{D.6}$$

$$- \sum_{r=0}^{n-1}\left(\frac{1}{2}z\right)^{-n+2r}\frac{(n-r-1)!}{r!},$$

where for $r = 0$ we replace $\left(\displaystyle\sum_{m=1}^{n+r} m^{-1} + \sum_{m=1}^{r} m^{-1}\right)$ by $\displaystyle\sum_{m=1}^{n} m^{-1}$.

Appendix E

Basic physical units

A. *Units of pressure and stresses*

(p and partial pressures in mixtures, Piola-Kirchhoff $\mathbf{P}$ and Cauchy $\mathbf{T}$;

pressure in SI: 1 Pascal $= 1\ \mathrm{kg/s^2 m}$)

$\mathrm{Pa} = \frac{\mathrm{N}}{\mathrm{m^2}}$	$\mathrm{at} = \frac{\mathrm{kp}}{\mathrm{cm^2}}$	atm	bar	torr	$\mathrm{mmWs} = \frac{\mathrm{kp}}{\mathrm{m^2}}$
1	$1.02 \cdot 10^{-5}$	$9.87 \cdot 10^{-6}$	10^{-5}	$75 \cdot 10^{-4}$	0.102
$9.81 \cdot 10^4$	1	0.968	0.981	736	10^4
$1.013 \cdot 10^5$	1.033	1	1.013	760	$1.033 \cdot 10^4$
10^5	1.02	0.987	1	750	$1.02 \cdot 10^4$
133	$1.36 \cdot 10^{-3}$	$1.32 \cdot 10^{-3}$	$1.33 \cdot 10^{-3}$	1	13.6
9.81	10^{-4}	$9.68 \cdot 10^{-5}$	$9.81 \cdot 10^{-5}$	$7.36 \cdot 10^{-2}$	1

B. *Units of the force*

(body forces $\rho\mathbf{b}$ have the SI unit: $[\mathrm{Newton/m^3}]$; force in SI: 1 Newton $= 1\ \mathrm{kg \cdot m/s^2}$)

N	kp	Mp	p	dyna
1	0.102	$1.02 \cdot 10^{-4}$	102	10^5
9.81	1	10^{-3}	10^3	$9.81 \cdot 10^5$
$9.81 \cdot 10^3$	10^3	1	10^6	$9.81 \cdot 10^8$
$9.81 \cdot 10^{-3}$	10^{-3}	10^{-6}	1	981
10^{-5}	$1.02 \cdot 10^{-6}$	$1.02 \cdot 10^{-9}$	$1.02 \cdot 10^{-3}$	1

C. *Units of energy and work*

(energy density $\rho\varepsilon$ has the SI unit: $[\text{Joule/m}^3]$; energy in SI: 1 Joule $= 1$ kg·m^2/s^2)

J	kpm	kWh	kcal	erg	eV
1	0.102	$2.78 \cdot 10^{-7}$	$2.39 \cdot 10^{-4}$	10^7	$6.24 \cdot 10^{18}$
9.81	1	$2.72 \cdot 10^{-6}$	$2.34 \cdot 10^{-3}$	$9.81 \cdot 10^7$	$6.12 \cdot 10^{19}$
$3.6 \cdot 10^6$	$3.67 \cdot 10^5$	1	860	$3.6 \cdot 10^{13}$	$2.25 \cdot 10^{25}$
$4.19 \cdot 10^3$	427	$1.16 \cdot 10^{-3}$	1	$4.19 \cdot 10^{10}$	$2.61 \cdot 10^{22}$
10^{-7}	$1.02 \cdot 10^{-8}$	$2.78 \cdot 10^{-14}$	$2.39 \cdot 10^{-11}$	1	$6.24 \cdot 10^{11}$
$1.6 \cdot 10^{-19}$	$1.63 \cdot 10^{-20}$	$4.45 \cdot 10^{-26}$	$3.83 \cdot 10^{-23}$	$1.6 \cdot 10^{-12}$	1

D. *Units of power*

(working of body forces $\rho\mathbf{b} \cdot \mathbf{v}$, energy radiation ρr, working of stresses $\mathbf{P} \cdot \frac{\partial \mathbf{F}}{\partial t}$
and $\mathbf{T} \cdot \mathbf{L}$ have SI unit: $[\text{Watt/m}^3]$; power in SI: 1 Watt $= 1$ kg·m^2/s^3)

W	kW	$\frac{\text{kpm}}{\text{s}}$	PS	$\frac{\text{cal}}{\text{s}}$	$\frac{\text{kcal}}{\text{h}}$
1	10^{-3}	0.102	$1.36 \cdot 10^{-3}$	0.239	0.86
10^3	1	102	1.36	239	860
9.81	$9.81 \cdot 10^{-3}$	1	$1.33 \cdot 10^{-2}$	2.34	8.43
736	0.736	75	1	176	632
4.19	$4.19 \cdot 10^{-3}$	0.427	$5.69 \cdot 10^{-3}$	1	3.6
1.16	$1.16 \cdot 10^{-3}$	0.119	$1.58 \cdot 10^{-3}$	0.278	1

Remark: Heat fluxes $\mathbf{Q}$ and $\mathbf{q}$ have SI unit: $[\text{Watt/m}^2]$;

entropy has SI unit: $[\text{Joule/m}^3 \cdot \text{K}]$;

entropy fluxes $\mathbf{H}$ and $\mathbf{h}$ have SI unit: $[\text{Watt/m}^2 \cdot \text{K}]$;

entropy radiation ρs has the SI unit: $[\text{Watt/m}^3 \cdot \text{K}]$.

Bibliography

[1] ACHENBACH, J. D. *Wave Propagation in Elastic Solids.* North-Holland, Amsterdam, 1973.

[2] ADKINS, J. E., AND RIVLIN, R. S. Large elastic deformations of isotropic materials, IX. The deformation of thin shells. *Phil. Trans. Roy. Soc. Lond., A 244* (1952), 505–531.

[3] ADKINS, J. E., AND RIVLIN, R. S. Large elastic deformations of isotropic materials, X. Reinforcement by inextensible cords. *Phil. Trans. Roy. Soc. Lond., A 248* (1955), 201–223.

[4] AGUILERA, R. *Naturally fractured Reservoirs.* Pennwell Corp, 1980.

[5] AKI, K., AND RICHARDS, P. G. *Quantitative Seismology.* University Science Books, Sansalito, 2002.

[6] ALBERS, B. Coupling of adsorption and diffusion in porous and granular materials. A 1-D example of the boundary value problem. *Arch. Appl. Mech. 70* (2000), 519–531.

[7] ALBERS, B. *Makroskopische Beschreibung von Adsorptions-Diffusions-Vorgängen in porösen Körpern.* PhD thesis, TU Berlin, Logos-Verlag, 2000. In German.

[8] ALBERS, B. Linear stability of a 1d flow in porous media under transversal disturbance with adsorption. *Arch. Mech. 54,* 5-6 (2002), 361–375.

[9] ALBERS, B. Relaxation analysis and linear stability vs. adsorption in porous materials. *Continuum Mech. Thermodyn. 15,* 1 (2003), 73–95.

[10] ALBERS, B. Modelling of surface waves in poroelastic saturated materials by means of a two-component continuum. In *Surface Waves in Geomechanics: Direct and Inverse Modelling for Soils and Rocks,* C. Lai and K. Wilmanski, Eds., CISM Courses and Lectures. Springer, Wien, 2005, pp. 277–323.

[11] ALBERS, B. On the influence of saturation and frequency on monochromatic plane waves in unsaturated soils. In *Coupled site and soil-structure interaction effects with application to seismic risk mitigation,* T. Schanz and R. Iankov, Eds. NATO Science Series, Springer Netherlands, 2009, pp. 65–76.

[12] ALBERS, B., Ed. *Continuous Media with Microstructure.* Springer, 2010.

[13] ALBERS, B. *Modeling and Numerical Analysis of Wave Propagation in Saturated and Partially Saturated Porous Media*, vol. 48 of *Veröffentlichungen des Grundbauinstitutes der Technischen Universität Berlin*. Shaker Verlag, Aachen, 2010. Habilitation thesis.

[14] ALBERS, B. On a micro-macro transition for a poroelastic three-component model. *ZAMM 90*, 12 (2010), 929–943.

[15] ALBERS, B. Linear elastic wave propagation in unsaturated sands, silts, loams and clays. *Transport in Porous Media 86* (2011), 537–557.

[16] ALBERS, B. Modeling the hysteretic behavior of the capillary pressure in partially saturated porous media: a review. *Acta Mechanica 225*, 8 (2014), 2163–2189.

[17] ALBERS, B., AND WILMANSKI, K. On modeling acoustic waves in saturated poroelastic media. *J. Engrn. Mech. 131*, 9 (2005), 975–985.

[18] ALBERS, B., AND WILMANSKI, K. Influence of coupling through porosity changes on the propagation of acoustic waves in linear poroelastic materials. *Arch. Mech. 58*, 4-5 (2006), 313–325.

[19] ALBERS, B., AND WILMANSKI, K. Continuous modeling of soil morphology - thermomechanical behavior of embankment dams. *Frontiers of Architecture and Civil Engineering in China 5*, 1 (2011), 11–23.

[20] ALBERS, B., AND WILMANSKI, K. Acoustics of two-component porous materials with anisotropic tortuosity. *Continuum Mechanics and Thermodynamics 24*, 4-6 (2012), 403–416.

[21] ALBERS, B., AND WILMANSKI, K. Propagation of sound waves in poroelastic media with anisotropic permeability. In *ASCE Conference Proceedings of BIOT-5* (2013), C. Hellmich, B. Pichler, and D. Adam, Eds., ASCE, pp. 83–91.

[22] ALSABRY, A., AND WILMANSKI, K. Iterative description of freezing and thawing processes in porous materials. *Journal of Applied Mathematics and Mechanics (ZAMM) 91*, 9 (2011), 753–760.

[23] ANTIC, A., AND HILL, J. M. The double-diffusivity heat transfer model for grain stores incorporating microwave heating. *Applied Mathematical Modelling 27*, 8 (2003), 629 – 647.

[24] ARBOGAST, T., DOUGLAS, JR, J., AND HORNUNG, U. Derivation of the double porosity model of single phase flow via homogenization theory. *SIAM J. Math. Anal. 21*, 4 (1990), 823–836.

[25] ARNOLD, J. *Mobile NMR for rock porosity and permeability*. PhD thesis, RWTH, Aachen, 2007.

[26] ASH, E. A., AND PAIGE, E. G. S. *Rayleigh-Wave theory and application*. Springer, Berlin, 1985.

[27] BALL, J. M., AND JAMES, R. D. Fine phase mixtures as minimizers of energy. *Arch. Rat. Mech. Anal. 100*, 1 (1987), 13–52.

[28] BANERJEE, P. K., AND BUTTERFIELD, R. *Boundary Element Methods in Engineering Science*. McGraw-Hill Book Co. (UK), 1981.

[29] BARENBLATT, G. I., AND JOSEPH, D. D. *Collected Papers of R. S. Rivlin, vol. I & II*. Springer, N.Y., 1997.

[30] BARENBLATT, G. I., ZHELTOV, I. P., AND KOCHINA, I. N. Basic concepts in the theory of seepage of homogeneous liquids in fissured rocks [strata]. *J. Applied Mathematics and Mechanics 24*, 5 (1960), 1286 – 1303.

[31] BATEMAN, G. *MHD Instabilities*. M.I.T. Press, Cambridge, 1978.

[32] BATRA, R. C. *Elements of Continuum Mechanics*. Amer. Inst. of Aeronautics and Anstronautics, Reston, 2006.

[33] BEAR, J. *Dynamics of Fluids in Porous Media*. Dover Publications, New York, 1988.

[34] BEAR, J. *Modeling Flow and Contaminant Transport in Fractured Rocks*. Academic Press, Inc., San Diego, 1993.

[35] BEAR, J., AND BACHMAT, Y. *Introduction to Modeling of Transport Phenomena in Porous Media*. Kluwer Academic Publishers, Dordrecht, 1991.

[36] BEATTY, M. F. Seven lectures on finite elasticity. In *Topics in Finite Elasticity*, M. Hayes and G. Saccomandi, Eds., CISM Courses and Lectures. Springer, Wien, New York, 2001.

[37] BEAVERS, G. S., AND JOSEPH, D. D. Boundary conditions at a naturally permeable wall. *J. Fluid Mech. 30*, 1 (1967), 197–207.

[38] BEN-MENAHEM, A., AND SINGH, S. J. *Seismic Waves and Sources*. Dover Publications, 2000.

[39] BÉNARD, H. Les tourbillons cellulaires dans une nappe liquide. *Revue générale des Sciences pures et appliquées 11* (1900), 1261–1271.

[40] BÉNARD, H. Les tourbillons cellulaires dans une nappe liquide. *Revue générale des Sciences pures et appliquées 11* (1900), 1309–1328.

[41] BÉNARD, H. Les tourbillons cellulaires dans une nappe liquide transportant de la chaleur par convection en régime permanent. *Annales de Chimie et de Physique 23* (1901), 62–144.

[42] BERTOTTI, G., AND MAYERGOYZ, I. D. *The Science of Hysteresis vol. 1: Mathematical modeling and applications*. Elsevier/Academic Press, 2006.

[43] BERTOTTI, G., AND MAYERGOYZ, I. D. *The Science of Hysteresis vol. 2: Physical modeling, micromagnetics, and magnetization dynamics*. Elsevier/Academic Press, 2006.

[44] BERTOTTI, G., AND MAYERGOYZ, I. D. *The Science of Hysteresis vol. 3: Hysteresis in materials*. Elsevier/Academic Press, 2006.

[45] BETSCH, D. F., AND BAER, E. Structure and mechanical properties of rat tail tendon. *Biorheology 17*, 1-2 (1980), 83–94.

[46] BHATTACHARYA, K. Comparison of the geometrically nonlinear and linear theories of martensitic transformation. *Cont. Mech. Thermodyn. 5*, 3 (1993), 205–242.

[47] BILBY, B. A., BULLOUGH, R., AND SMITH, E. Continuous distribution of dislocations - a new application of the methods of non-Riemannian geometry. *Proc. R. Soc. Lond. A 231*, 1185 (1955), 263–273.

[48] BINGHAM, E. C. An investigation of the laws of plastic flow. *J. Rheology* (1931), 10–107.

[49] BIOT, M. A. Theory of propagation of elastic waves in a fluid saturated porous solid, I. low frequency range, II. Higher frequency range. *J. Acoust. Soc. Am. 28*, 2 (1956), 168–178, 179–191.

[50] BIOT, M. A., AND WILLIS, D. G. The elastic coefficients of the theory of consolidation. *J. Appl. Mech. 24* (1957), 594–601.

[51] BITTENCOURT, J. A. *Fundamentals of Plasma Physics.* Springer, New York, 2004.

[52] BOBYLEW, D. Einige Betrachtungen über die Gleichungen der Hydrodynamik. *Math. Ann. 6*, 1 (1873), 72–84.

[53] BONNET, G. Basic singular solutions for a poroelastic medium in the dynamic range. *J. Acoust. Soc. Amer. 82*, 5 (1987), 1758–1762.

[54] BOUSSINESQ, J. Théorie Analytique de la Chaleur, Gautier-Villars, Paris, 1903; also: D.D. Gray and A. Giorgini; the Validity of the Boussinesq Approximation for Liquid and Gases. *Int. J. Heat Mass Transfer 19* (1976), 545–551.

[55] BRACEWELL, R. N. *The Fourier Transform and its Applications.* McGraw-Hill, 1986. 2nd Edition.

[56] BREKHOVSKIKH, L. M., AND GODIN, O. A. *Acoustics of Layered Media I: Plane and Quasi-Plane Waves.* Springer, Berlin, 1998.

[57] BREKHOVSKIKH, L. M., AND GONCHAROV, V. *Mechanics of Continua and Wave Dynamics (Second Edition).* Springer, Berlin, 1994.

[58] BRENNEN, C. E. *Cavitation and Bubble Dynamics.* Oxford University Press, 1995.

[59] BURRIDGE, R., AND VARGAS, C. A. The fundamental solution in dynamic poroelasticity. *Geophys. J. Royal Astron. Soc. 58*, 1 (1979), 61–90.

[60] BUSEMANN, A. *Gasdynamik.* Akad. Verlag, Leipzig, 2000. Bd. IV, in German.

[61] CAGNIARD, L. *Reflection and Refraction of Progressive Waves.* McGraw-Hill Book Co., N.Y., 1962.

[62] CALLEN, H. B. *Thermodynamics and an Introduction to Thermostatics (2nd ed.).* John Wiley and Sons, 1985.

[63] CAP, F., Ed. *Waves and Instabilities in Plasmas*. Springer, Wien, New York, 1994. CISM Courses and Lectures No. 349.

[64] CAREW, T. E., VAISHNAV, R. N., AND PATEL, D. J. Compressibility of the arterial wall. *Circulation Research 23*, 1 (1968), 61–68.

[65] CARMAN, P. C. *Flow of Gases through Porous Media*. Butterworth, London, 1956.

[66] CARSLAW, H. S., AND JAEGER, J. C. *Conduction of Heat in Solids*. Oxford Science Publications. Clarendon Press, 1959.

[67] CASAS-VÁZQUEZ, J., AND JOU, D. Temperature in non-equilibrium states: a review of open problems and current proposals. *Reports on Progress in Physics 66*, 11 (2003), 1937–2023.

[68] CATTANEO, C. Sulla conduzione del calore. *Atti Sem. Mat. Fis. Modena 3* (1948), 83–101.

[69] CAVIGLIA, G., AND MORRO, A. Acoustic and elastic scattering by continuously stratified media. *Acta Mechanica 206* (2009), 173–191.

[70] CHADWICK, P. *Continuum Mechanics, Concise Theory and Problems*. Dover Publ., Mineola, N. Y., 1999.

[71] CHANDRASEKHAR, S. *Hydrodynamic and Hydromagnetic Stability*. Clarendon Press, Oxford, 1961.

[72] CHAPMAN, S., AND COWLING, T. G. *The Mathematical Theory of Non-Uniform Gases (Third Edition)*. Cambridge University Press, Cambridge, 1970.

[73] CHENG, A. H.-D., BADMUS, T., AND BESKOS, D. Integral equation for dynamic poroelasticity in frequency domain with bem solution. *J. Engrg Mech. 117*, 5 (1991), 1136–1157.

[74] CHENG, A. H.-D., AND DETOURNAY, E. A direct boundary element method for plane strain poroelasticity. *Int. J. Num. Anal. Methods Geomech. 12*, 5 (1988), 551–572.

[75] CHRISTENSEN, R. M. *Theory of Viscoelasticity*. Academic Press, N.Y, 1971.

[76] CHRISTENSEN, R. M., AND SCHREINER, R. N. Response to pressurization of a viscoelastic cylinder with an eroding internal boundary. *AIAA J. 3* (1965), 1451.

[77] CHUONG, C. J., AND FUNG, Y. C. Residual stress in arteries. In *Frontiers in Biomechanics*, G. W. Schmid-Schönbein, S. L.-Y. Woo, and B. W. Zweifach, Eds. Springer New York, 1986, pp. 117–129.

[78] CIMMELLI, V. A., SELLITTO, A., AND JOU, D. Nonequilibrium temperatures, heat waves, and nonlinear heat transport equations. *Phys. Rev. B 81* (Feb 2010), 054301.

[79] CLEARY, M. P. Fundamental solutions for a fluid-saturated porous solid. *Int. J. Solids and Structures 13*, 9 (1977), 785–806.

[80] COLEMAN, B. D., DUFFIN, R. J., AND MIZEL, V. J. Instability, uniqueness and nonexistence theorems for the equation $u_t = u_{xx} - u_{xtx}$ on a strip. *Arch. Rat. Mech. Anal. 19* (1965), 100–116.

[81] COLEMAN, B. D., AND GURTIN, M. E. Waves in materials with memory. II. On the growth and decay of one-dimensional acceleration waves. *Arch. Rat. Mech. Anal. 19* (1965), 239–265.

[82] COLEMAN, B. D., AND GURTIN, M. E. Waves in materials with memory. III. Thermodynamic influences on the growth and decay of acceleration waves. *Arch. Rat. Mech. Anal. 19* (1965), 266–298.

[83] COLEMAN, B. D., AND GURTIN, M. E. Waves in materials with memory. IV. Thermodynamics and the velocity of general acceleration waves. *Arch. Rat. Mech. Anal. 19* (1965), 317–338.

[84] COLEMAN, B. D., AND GURTIN, M. E. Thermodynamics and wave propagation in non-linear materials with memory. In *Proceedings of the IUTAM Symposium Vienna* (Springer Wien, 1966), H. Parkus and L. L. Sedov, Eds., pp. 54–76.

[85] COLEMAN, B. D., AND GURTIN, M. E. Waves in materials with memory. V. On the amplitude of acceleration waves and mild discontinuities. *Arch. Rat. Mech. Anal. 22* (1966), 333–354.

[86] COLEMAN, B. D., GURTIN, M. E., AND HERRERA, I. R. Waves in materials with memory. I. The velocity of one-dimensional shock and acceleration waves. *Arch. Rat. Mech. Anal. 19* (1965), 1–19.

[87] COLEMAN, B. D., MARKOVITZ, H., AND NOLL, W. *Viscometric Flows on Non-Newtonian Fluids.* Springer, Berlin, 1966.

[88] COLEMAN, B. D., AND MIZEL, V. J. Breakdown of laminar shearing flows for second-order fluids in channels of critical width. *ZAMM 46* (1966), 445–448.

[89] COLEMAN, B. D., AND NOLL, W. An approximation theorem for functionals, with applications in continuum mechanics. *Arch. Rat. Mech. Anal. 6* (1960), 355–370.

[90] COTTRELL, A. H. *Dislocations and Plastic Flow in Crystals.* Oxford University Press, Oxford, 1953.

[91] COULSON, C. A., AND JEFFREY, A. *Waves; A Mathematical Approach to the Common Types of Wave Motion.* Longman, London, 1977.

[92] COURNAT, R., AND HILBERT, D. Partial differential equations. In *Methods of mathematical physics, Vol.II*, R. Courante, Ed. Interscience, New York, London, 1962.

[93] COUSSY, O. *Mechanics of Porous Continua.* John Wiley, Chichester, 1995.

[94] COUSSY, O. Poromechanics of freezing materials. *J. Mech. Phys. Solids 53*, 8 (2005), 1689–1718.

[95] COUSSY, O., AND FEN-CHONG, T. Crystallization, pore relaxation and micro-cryosuction in cohesive porous materials. *C. R. Mecanique 333* (2005), 507–512.

[96] COWIN, S. C. The relationship between the elasticity tensor and the fabric tensor. *Mechanics of Materials 4*, 2 (July 1985), 137–147.

[97] COWIN, S. C. Bone poroelasticity. *J. Biomech. 32* (1999), 218–238.

[98] COWIN, S. C., AND CARDOSO, L. Fabric dependence of wave propagation in anisotropic porous media. *Biomech Model Mechanobiol. 10* (2011), 39–65.

[99] COWIN, S. C., AND DOTY, S. B. *Tissue Mechanics.* Springer, 2007.

[100] CROCCO, L. Sulla trasmissione del calore da una lamina piana a un fluido scorrente ad alta velocité. *L'Aerotecnica. 12* (1932), 181–197.

[101] CRPP. Anual report 2010 of centre de recherches en physique des plasmas. Tech. rep., EPFL, 2010.

[102] CURRIE, I. G. *Fundamental Mechanics of Fluids.* McGraw-Hill, New York, 2005.

[103] DE PASCALIS, R. *The Semi-Inverse Method in solid mechanics: Theoretical underpinnings and novel applications.* PhD thesis, Université Pierre et Marie Curie and Universitá del Salento, 2010.

[104] DE WIT, R. Continuum theory of stationary dislocations. *Solid State Physics 10* (1960), 249.

[105] DOETSCH, G. *Handbuch der Laplace-Transformation: Theorie der Laplace-Transformation.* Lehrbücher und Monographien aus dem Gebiete der exakten Wissenschaften: Mathematische Reihe. Birkhäuser, 1950.

[106] DUNFORD, N., AND SCHWARTZ, J. T. *Linear Operators. Part I: General Teory.* Interscience Publ., New York, 1958.

[107] DUNN, E., AND FOSDICK, R. L. Thermodynamics, stability and boundedness of fluids of complexity 2 and fluids of second grade. *Arch. Rat. Mech. Anal. 56* (1974), 191–252.

[108] DURRAN, D. R., AND ARAKAWA, A. Generalizing the Boussinesq approximation to stratified compressible flow. *Comptes Rendus Mecanique 335*, 9-10 (2007), 655–664.

[109] EDDINGTON, A. S. The dynamics of a stellar system. third paper: oblate and other distributions. *Mon. Not. Royal Astron. Soc. 75* (1915), 37–60.

[110] EINSTEIN, A. Theorie der Opaleszenz von homogenen Flüssigkeiten und Flüssigkeitsgemischen in der Nähe des kritischen zustandes. *Ann. Phys. 338*, 16 (1910), 1275–1298.

[111] EL TANI, M. Hydrostatic paradox of saturated media. *Géotechnique 57* (2007), 773–777.

[112] ELMORE, W., AND HEALD, M., Eds. *Physics of Waves.* Dover, N.Y., 1985.

[113] EPSTEIN, N. On tortuosity and the tortuosity factor in flow and diffusion through porous media. *Chem. Engn. Sci. 44* (1989), 777–779.

[114] ERICKSEN, J. L. Deformations possible in every isotropic, incompressible, perfectly elastic body. *ZAMP 5* (1954), 466–489.

[115] ERICKSEN, J. L. Deformations possible in every compressible, isotropic, perfectly elastic material. *J. Math. Phys. 34* (1955), 126–128.

[116] ERICKSEN, J. L., AND RIVLIN, R. S. Large elastic deformations of homogeneous anisotropic materials. *J. Rat. Mech. Anal. 3* (1954), 281–301.

[117] ERINGEN, A., AND MAUGIN, G., Eds. *Electrodynamics of Continua I. Foundations and Solid Media.* Springer, N.Y., 1989.

[118] ERINGEN, A., AND MAUGIN, G., Eds. *Electrodynamics of Continua II. Fluids and Complex Media.* Springer, N.Y., 1989.

[119] ESHELBY, J. D. *Continuum Theory of Lattice Defects, in: Solid State Physics, vol. 3, 79.* Academic Press, N.Y., 1956.

[120] EVANS, N. W., AND LYNDEN-BELL, D. Solutions of the equations of stellar hydrodynamics for eddington systems. *Mon. Not. Royal Astron. Soc. 236* (1989), 801–816.

[121] FABBRI, A., FEN-CHONG, T., AND COUSSY, O. *Freezing/thawing curves and characterization of porous network by a capacitive based apparatus.* A. A. Balkema, 2005, pp. 567–572.

[122] FABRIKANT, V. I. Mixed boundary value problem of potential theory in toroidal coordinates. *Z. Angew. Math. Phys. 42* (1991), 680–707.

[123] FALKENHAGEN, H. *Theorie der Elektrolyte.* S. Hirzel Verlag, Leipzig, 1971.

[124] FERRANDON, J. Le lois de l'ecoulement de filtration. *Genie Civil 24-28* (1948), 777–779.

[125] FINLAY, H. M., McCULLOUGH, L., AND CANHAM, P. B. Three-dimensional collagen organization of human brain arteries at different transmural pressures. *J. Vascular Res. 32* (1995), 301–312.

[126] FLANDERS, H. *Differential Forms with Applications to the Physical Sciences.* Academic Press, New York-London, 1963.

[127] FORSYTH, A. E. On the motion of a viscous incompressible fluid. *Mess. Math. 9* (1880), 134–139.

[128] FOTI, S., LAI, C. G., RIX, G. J., AND STROBBIA, C. *Surface Wave Methods for Near-Surface Site Characterization.* Taylor & Francis, 2014.

[129] FOURIER, J. B. J., AND FREEMAN, A. *The analytical theory of heat, by Joseph Fourier. Translated, with notes, by Alexander Freeman.* The University Press; Cambridge [Eng.], 1878.

[130] FRIEDRICHS, K. O., AND LAX, P. D. Systems of conservation equations with a convex extension. *Proceedings of the National Academy of Sciences 68*, 8 (1971), 1686–1688.

[131] FUNG, Y. C. *Biomechanics: Mechanical Properties of Living Tissues.* Biomechanics. Springer, 1993.

[132] GANDHI, V., AND THOMPSON, B. S. *Smart Materials and Structures.* Springer, 1992.

[133] GATMIRI, B., AND JABBARI, E. Time-domain Green's functions for unsaturated soils. part ii: Three-dimensional solution. *International Journal of Solids and Structures 42*, 23 (2005), 5991–6002.

[134] GILMAN, A., AND BEAR, J. The influence of free convection on soil salinization in arid regions. *Transport in Porous Media 23* (1994), 275–301.

[135] GINN, R. F., AND METZNER, A. B. Measurement of stresses developed in steady laminar shearing flows of viscoelastic media. *Trans. Soc. Rheol. 13*, 4 (1969), 429–453.

[136] GINZBARG, A., AND STRICK, E. Stoneley-wave velocities for a solid-solid interface. *Bull. Seismol. Soc. Am. 48* (1958), 51–63.

[137] GOEDBLOED, J. P., KEPPENS, R., AND POEDTS, S. *Advanced Magnetohydrodynamics with Applications to Laboratory and Astrophysical Plasmas.* Cambridge University Prerss, Cambridge, 2010.

[138] GOEDBLOED, J. P., AND POEDTS, S. *Principles of Magnetohydrodynamics with Applications to Laboratory and Astrophysical Plasmas.* Cambridge University Prerss, Cambridge, 2004.

[139] GRAD, H. Principles of the kinetic theory of gases. In *Thermodynamik der Gase / Thermodynamics of Gases*, S. Flügge, Ed., vol. 3 / 12 of *Handbuch der Physik / Encyclopedia of Physics*. Springer Berlin Heidelberg, 1958, pp. 205–294.

[140] GRAY, L. J., GHOSH, D., AND KAPLAN, T. Evaluation of the anisotropic Green's function in three dimensional elasticity. *Computational Mechanics 17*, 4 (1996), 255–261.

[141] GRAY, W. G. Elements of a systematic procedure for the derivation of macroscale conservation equations for multiphase flow in porous media. In *Kinetic and Continuum Theories of Granular and Porous Media*, K. Hutter and K. Wilmanski, Eds., no. 400 in CISM Courses and Lectures. Springer, Wien - New York, 1999, pp. 67–130.

[142] GREEN, A. E., AND RIVLIN, R. S. The mechanics of non-linear materials with memory, part I. *Arch. Rat. Mech. Anal. 1* (1957), 387–404.

[143] GREEN, A. E., AND ZERNA, W. *Theoretical Elasticity (Second Edition).* Dover, N.Y., 1968.

[144] GREGG, S. J., AND SING, K. S. W. *Adsorption, surface area and porosity.* Academic Press, London, 1982.

[145] GRIMBERG, G., PAULS, W., AND FRISCH, U. Genesis of d'Alembert paradox and analytical elaboration of the drag problem. *Physica D 237* (2008), 1878–1886.

[146] GROT, R. A. Relativistic continuum physics: Electromagnetic interactions. In *Continuum Physics. Volume III: Mixtures and EM Field Theories*, A. C. Eringen, Ed., Continuum Physics. Academic Press, New York, 1976, pp. 130–219.

[147] GURSON, A. L. Continuum theory of ductile rupture by void nucleation and growth: Part I – yield criteria and flow rules for porous ductile media. *Journal of Engineering Materials and Technology 99*, 1 (1977), 2–15.

[148] GURTIN, M. E. On the thermodynamics of materials with memory. *Arch. Rat. Mech. Anal. 28* (1968), 40–59.

[149] GURTIN, M. E., AND STERNBERG, E. On the linear theory of viscoelasticity. *Arch. Rat. Mech. Anal. 11* (1962), 291–356.

[150] GUZ, A. N. *Fundamentals of the Three-dimensional Theory of Stability of Deformable Bodies.* Springer, Berlin, 1999.

[151] HAO, S., AND BROCKS, W. The Gurson-Tveergard-Needleman-model for rate and temperature-dependent materials with isotropic and kinematic hardening. *Computational Mechanics 20* (1997), 34–40.

[152] HAPPEL, J., AND BRENNER, H. *Low Reynolds Number Hydrodynamics: With Special Applications to Particulate Media.* Mechanics of Fluids and Transport Processes. M. Nijhoff, 1983.

[153] HARRIS, S. *An Introduction to the Theory of the Boltzmann Equation.* Dover books on physics. Dover Publications, 2004.

[154] HARTMANN, J. Hg-dynamics I – theory of the laminar flow of an electrically conductive liquid in a homogeneous magnetic field. *Kgl. Danske Videnskabernes Selskab, Mathematisk-Fysiske Meddelelser 15* (1937).

[155] HERNANDES, J. A., AND ASSIS, A. K. T. Surface charges and external electric field in a toroid carrying a steady current. *Brazilian Journal of Physics 34* (2004), 1738–1744.

[156] HIRSCHFELDER, J. O., CURTISS, C. F., AND BIRD, R. B., Eds. *Molecular Theory of Gases and Liquids.* John Wiley, New York, 1954.

[157] HOEVE, C. A. J., AND FLORY, P. J. The elastic properties of elastin. *Biopolymers 13*, 4 (1974), 677–686.

[158] HOLZAPFEL, G. A. *Nonlinear Solid Mechanics: A Continuum Approach for Engineering.* John Wiley & Sons, 2000.

[159] HOLZAPFEL, G. A. Section 10.11 – Biomechanics of Soft Tissue. In *Handbook of Materials Behavior Models*, J. Lemaitre, Ed. Academic Press, Burlington, 2001, pp. 1057–1071.

[160] HOLZAPFEL, G. A., AND GASSER, T. C. A viscoelastic model for fiber-reinforced composites at finite strains: Continuum basis, computational aspects and applications. *Computer Methods in Applied Mechanics and Engineering 190*, 34 (2001), 4379–4403.

[161] HOLZAPFEL, G. A., GASSER, T. C., AND OGDEN, R. W. A new constitutive framework for arterial wall mechanics and a comparative study of material models. *Journal of elasticity and the physical science of solids 61*, 1-3 (2000), 1–48.

[162] HOLZAPFEL, G. A., GASSER, T. C., AND STADLER, M. A structural model for the viscoelastic behavior of arterial walls: Continuum formulation and finite element analysis. *European Journal of Mechanics - A/Solids 21*, 3 (2002), 441–463.

[163] HOLZAPFEL, G. A., AND OGDEN, R. W. *Biomechanics of Soft Tissue in Cardiovascular Systems.* CISM International Centre for Mechanical Sciences. Springer, 2003.

[164] HOLZAPFEL, G. A., AND OGDEN, R. W. *Biomechanical Modelling at the Molecular, Cellular and Tissue Levels.* CISM International Centre for Mechanical Sciences. Springer, 2009.

[165] HONEYCOMBE, R. W. K. *The Plastic Deformation of Metals.* E. Arnold Ltd., 1968.

[166] HORI, M., AND MORIHIR, H. Micromechanical analysis on deterioration due to freezing and thawing in porous brittle materials. *Int. J. Eng. Sci. 36*, 4 (1998), 511–522.

[167] HULL, D. *Introduction to Dislocations.* Pergamon, New York, 1975.

[168] HUMPHREY, J. D. Mechanics of the arterial wall: review and directions. *Crit. Rev. Biomed. Eng. 23*, 1-2 (1995), 1–162.

[169] HUNT, J. N. *Incompressible Fluid Dynamics.* John Wiley, New York, 1964.

[170] HUTTER, K. On thermodynamics and thermostatics of viscous thermoelastic solids in the electromagnetic fields. a lagrangian formulation. *Arch. Rat. Mech. Anal. 58* (1975), 339–368.

[171] HUTTER, K. A thermodynamic theory of fluids and solids in the electromagnetic fields. *Arch. Rat. Mech. Anal. 64* (1977), 269–298.

[172] HUTTER, K. *Fluid- und Thermodynamik: Eine Einführung.* Springer, Berlin, 2003.

[173] HUTTER, K., AND JÖHNK, K. *Continuum Methods of Physical Modeling. Continuum Mechanics, Dimensional Analysis, Turbulence.* Springer, Berlin, 2004.

[174] HUTTER, K., VAN DE VEN, A. A. F., AND URSESCU, A. *Electromagnetic Field Matter Interactions in Thermoelastic Solids and Viscous Fluids.* Springer, Berlin, 2006.

[175] HUYGHE, J. M., AND JANSSEN, J. Quadriphasic mechanics of swelling incompressible porous media. *International Journal of Engineering Science 35*, 8 (1997), 793–802.

[176] HUYGHE, J. M., AND BOVENDEERD, P. H. M. Swelling media: concepts and applications. In *Chemo-Mechanical Couplings in Porous Media Geomechanics and Biomechanics*, B. Loret and J. M. Huyghe, Eds., vol. 462 of *International Centre for Mechanical Sciences*. Springer Vienna, 2004, pp. 57–124.

[177] HUYGHE, J. M., AND CAMPEN, D. H. V. Finite deformation theory of hierarchically arranged porous solids: I. Balance of mass and momentum. *International Journal of Engineering Science 33*, 13 (1995), 1861–1871.

[178] HUYGHE, J. M., AND CAMPEN, D. H. V. Finite deformation theory of hierarchically arranged porous solids: II. Constitutive behaviour. *International Journal of Engineering Science 33*, 13 (1995), 1873–1886.

[179] HUYGHE, J. M., OOMENS, C. W., AND VAN CAMPEN, K. H. Low Reynolds number steady state flow through a branching network of rigid vessels: II. A finite element mixture model. *Biorheology 26* (1989), 73–84.

[180] HUYGHE, J. M., OOMENS, C. W., VAN CAMPEN, K. H., AND HEETHAAR, R. M. Low Reynolds number steady state flow through a branching network of rigid vessels: I. A mixture theory. *Biorheology 26* (1989), 55–71.

[181] HUYGHE, J. M., SCHRÖDER, Y., AND BAAIJENS, F. P. T. Bioengineering: the future of poromechanics? In *17th ASCE Engineering Mechanics Conference* (United States, University of Delaware, Newark, 2004), pp. 1–9.

[182] INGARDEN, R. S., AND JAMIOLKOWSKI, A. *Classical Electrodynamics*. Elsevier, Amsterdam, 1985.

[183] JACKSON, J. D. *Classical Electrodynamics*. Wiley, New York, 1965.

[184] JOHNSON, W. E., AND BRESTON, J. N. Directional permeability measurements on oil sandstone from various states. *Producers Monthly 14* (1951), 10–19.

[185] JOHNSON, W. E., AND HUGHES, R. V. Directional permeability measurements and their significance. *Producers Monthly 13* (1948), 17–25.

[186] JOSEPH, D. D. *Stability of Fluid Motion, I and II*. Springer, Berlin, 1976. Springer Tracts in Natural Philosophy, vols. 27 and 28.

[187] JOSEPH, D. D. Instability of the rest state of fluids of arbitrary grade greater than one. *Arch. Rat. Mech. Anal. 75* (1981), 251–256.

[188] JOU, D. Fluctuation theory and extended irreversible thermodynamics. *Physica A 155*, 2 (1989), 221–231.

[189] JOU, D., AND CASAS-VÁZQUEZ, J. Fluctuations of dissipative fluxes and the onsager-machlup function. *Journal of Non-equilibrium Thermodynamics 5*, 2 (1980), 91–102.

[190] JOU, D., CASAS-VÁZQUEZ, J., AND CRIADO-SANCHO, M. *Thermodynamics of Fluids Under Flow*. Springer, 2011. Second edition.

[191] JOU, D., CASAS-VÁZQUEZ, J., AND LEBON, G. *Extended Irreversible Thermodynamics.* Springer, Berlin. Second edition (1996) and fourth edition (2010).

[192] JOU, D., RUBI, J. M., AND CASAS-VÁZQUEZ, J. Hydrodynamical fluctuations in extended irreversible thermodynamics. *Physica A 101*, 2-3 (1980), 588–598.

[193] JOU, D., SCIACCA, M., AND MONGIOVI, M. S. Vortex dynamics in rotating counterflow and plane couette and poiseuille turbulence in superfluid helium. *Physical Review B 78*, 2 (2008), 024524.

[194] KADIC, A., AND EDELEN, D. G. B. *A Gauge Theory of Dislocations and Disclinations.* Springer, Berlin, 1983.

[195] KADOMTSEV, B. B. Hydromagnetic stability of a plasma. *Reviews of Plasma Physics 2* (1963), 153–199.

[196] KANE, J. H. *Boundary Element Analysis in Engineering Continuum Mechanics.* Prentice Hall, 1994.

[197] KAUSEL, E. *Fundamental Solutions in Elastodynamics: A Compendium.* Cambridge University Press, 2006.

[198] KAZEMI, H. Pressure transient analysis of naturally fractured reservoirs with uniform fracture distribution. *Trans. Soc. Pet. Eng. 246* (1969), 463–472.

[199] KHALED, M. Y., BESKOS, D. E., AND AIFANTIS, E. C. On the theory of consolidation with double porosity - III a finite element formulation. *International Journal for Numerical and Analytical Methods in Geomechanics 8*, 2 (1984), 101–123.

[200] KLIMONTOVICH, Y. L. *Kinetic Theory of Electromagnetic Processes.* Springer, Heidelberg, 1982.

[201] KNOPS, R. J., AND WILKES, E. W. *Theory of Elastic Stability, vol. VIa/3.* Springer, Berlin, 1973. Flügge, S. (ed.).

[202] KOCKS, U. F., ARGON, A., AND ASHBY, M. *Thermodynamics and Kinetics of Slip*, vol. 19. Pergamon Press, Oxford, 1975. Chalmers, B. and Christian, J.W. and Massalski, T.W. (eds.).

[203] KOHN, R. V. The relaxation of a double-well energy. *Continuum Mechanics and Thermodynamics 3*, 3 (1991), 193–236.

[204] KOLSKY, H. *Stress Waves in Solids.* Dover, 1963.

[205] KOSEVICH, A. M. *Crystal Lattice: Photons, Solitons, Dislocations.* John Wiley, New York, 1999.

[206] KOZENY, J. Über kapillare Leitung des Wassers in Böden (Aufstieg, Versickerung und Anwendung auf Bewässerung). *Sber. Akad. Wiss., Wien, Abt. IIa 136* (1927), 271–306.

[207] KRALL, N., AND TRIVELPIECE, A. *Principles of Plasma physics.* McGraw-Hill, New York, 1986.

[208] KROENER, E. Das Fundamentalintegral der anisotropen elastischen Differentialgleichungen. *Z. Phys. 151* (1958), 504.

[209] KUCZMA, M. S. *Application of Variational Inequalities in the Mechanics of Plastic Flow and Martensitic Phase Transformations.* Politechnika Poznanska, Poznan, 1999. Habilitation thesis.

[210] KUCZMA, M. S. Composite beams with embedded shape memory alloy. In *Continuous Media with Microstructure*, B. Albers, Ed. Springer, 2010.

[211] KUCZMA, M. S., AND MIELKE, A. Influence of hardening and inhomogeneity on internal loops in pseudoelasticity. *ZAMM, Z. Angew. Math. Mech. 80*, 5 (2000), 291–306.

[212] LAGOUDAS, D. C. *Shape Memory Alloys: Modeling and Engineering Applications.* Springer ebook collection / Chemistry and Materials Science 2005-2008. Springer, 2008.

[213] LAI, C. G. Surface waves in dissipative media: Forward and inverse modelling. In *Surface Waves in Geomechanics, Direct and Inverse Modelling for Soils and Rocks*, C. G. Lai and K. Wilmanski, Eds., CISM Courses and Lectures. Springer, Wien, New York, 2005.

[214] LAI, C. G., AND WILMANSKI, K., Eds. *Surface Waves in Geomechanics, Direct and Inverse Modelling for Soils and Rocks.* Springer, Wien, New York, 2005. CISM Courses and Lectures No. 481.

[215] LAL, G. K. *Introduction to Machining Science.* New Age International (P) Limited, 1996.

[216] LAMB, H. *Hydrodynamics (6th ed.).* Cambridge University Press, Cambridge, 1994.

[217] LANDAU, L. D., AND LIFSCHITZ, E. M. *Course of Theoretical Physic, vol. 2: The Classical Theory of Fields.* Butterworth-Heinemann, 1980. 4th ed.

[218] LANDAU, L. D., AND LIFSHITZ, E. M. *Statistical Physics.* Pergamon, Oxford, 1980.

[219] LANDAU, L. D., AND LIFSHITZ, E. M. *Course of Theoretical Physics, Fluid Mechanics*, vol. 6. Butterworth-Heinemann, 1987.

[220] LEBEDEV, N. *Worked Problems in Applied Mathematics.* Dover Publications, 2010.

[221] LEBON, G., AND JOU, D. *Understanding Non-equilibrium Thermodynamics: Foundations, Applications, Frontiers.* SpringerLink: Springer e-Books. Springer, 2008.

[222] LEICEK, L. The Green function of the theory of elasticity in an anisotropic hexagonal medium. *Czechosl. J. Phys. B19*, 6 (1969), 799.

[223] LEMAITRE, J. *Handbook of Materials Behavior Models: Nonlinear Models and Properties.* Academic Press, 2001.

[224] LEMAITRE, J., AND CHABOCHE, J. *Mechanics of Solid Materials.* Cambridge Univ. Press, Cambridge, 1990.

[225] LEMAITRE, J., AND DESMORAT, R. *Engineering Damage Mechanics. Ductile, Creep, Fatigue and Brittle Failures.* Springer, Berlin, 2005.

[226] LEVITAS, V., STEIN, E., AND LENGNICK, M. On a unified approach to the description of phase transitions and strain localization. *Arch. Appl. Mech. 66*, 4 (1996), 242–254.

[227] LI, P. *Boundary Element Method for Wave Propagation in Partially Saturated Poroelastic Continua.* PhD thesis, TU Graz, Monographic Series TU Graz, 2012.

[228] LIFSHITZ, E. M., AND ROSENZWEIG, L. N. O postrojenii tensora grina dla osnownogo urawnienia teorii uprugosti w sluczaje nieograniczenoj uprugo-anisotropnoj sredy. *JETF 17*, 9 (1947), 783. In Russian.

[229] LIN, C. C., AND SEGEL, L. A. *Mathematics Applied to Deterministic Problems in the Natural Sciences.* Macmillan Publishing Co., New York, 1974.

[230] LIU, G., AND LIU, J. Q. Guided circumferential waves in a circular annulus. *J. Appl. Mech. 65* (1998), 424–430.

[231] LIU, I.-S. *Continuum Mechanics.* Springer, Berlin, 2002.

[232] LIU, I.-S. On well-posedness of classical boundary conditions in extended thermodynamics. In *Trends in Applications of Mathematics to Mechanics, STAMM 2004* (Aachen, 2005), K. H. Y. Wang, Ed., Shaker, pp. 225–233.

[233] LIU, I.-S. A method of differential iteration for boundary value problems in extended thermodynamics. *Nonlinear Analysis: Real World Application 8*, 4 (2007), 1113–1131.

[234] LIU, I.-S. *Constitutive Theories: Basic Principles.* Eolss Publishers, Oxford, 2009, pp. 1–38. Continuum Mechanics, J. Merido and G. Saccomandi (eds.); http://www.eolss.net.

[235] LIU, I.-S., AND MÜLLER, I. Extended thermodynamics of classical and degenerate gases. *Arch. Rat. Mech. Anal. 83*, 4 (1983), 285–332.

[236] LIU, I.-S., AND RINCON, M. A. A boundary value problem in extended thermodynamics. one-dimensional steady flows with heat conduction. *Continuum Mech. Thermodyn. 16* (2004), 109–124.

[237] LOVE, A. E. H. *Some Problems in Geodynamics, Cambridge University Press, 1911.* Dover, N.Y., 1967.

[238] LUDWIG, W. *Festkörperphysik.* Akademische Verlagsgesellschaft, Wiesbaden, 1978.

[239] LUZZI, R., VASCONCELLOS, Á., AND RAMOS, J. *Predictive Statistical Mechanics: A Nonequilibrium Ensemble Formalism.* Fundamental Theories of Physics. Springer, 2002.

[240] MACHLUP, S., AND ONSAGER, L. Fluctuations and irreversible process. II. systems with kinetic energy. *Phys. Rev. 91* (1953), 1512–1515.

[241] MARGENAU, H., AND MURPHY, G. M. *The Mathematics of Physics and Chemistry; 2nd ed.* Van Nostrand, New York, 1956.

[242] MARSDEN, J. E., AND HUGHES, T. J. R. *Mathematical Foundations of Elasticity.* Dover, N.Y., 1983.

[243] MAUGIN, G. A. Relativistic continuum physics: Micromagnetism. In *Continuum Physics. Volume III: Mixtures and EM Field Theories*, A. C. Eringen, Ed., Continuum Physics. Academic Press, New York, 1976, pp. 222–312.

[244] MAUGIN, G. A. Theory of nonlinear waves and solitons. In *Surface Waves in Geomechanics, Direct and Inverse Modelling for Soils and Rocks*, C. Lai and K. Wilmanski, Eds., CISM Courses and Lectures. Springer, Wien, New York, 2005.

[245] MAUGIN, G. A. *Continuum Mechanics Through the Twentieth Century: A Concise Historical Perspective.* Solid Mechanics and Its Applications. Springer, 2013.

[246] MAUGIN, G. A., POUGET, J., DROUOT, R., AND COLLET, B. *Nonlinear Electromechanical Couplings.* John Wiley, New York, 1992.

[247] MAXWELL, J. C. On the dynamical theory of gases. *Philos. Trans. Roy. Soc. Lond. 157* (1867), 49–88.

[248] MCCONNELL, A. J. *Application of Tensor Analysis.* Dover, N.Y., 1957.

[249] MIHASHI, Z., ZHOU, Z. Y., AND TADA, S. Micromechanics model to predict macroscopic behavior of concrete under frost action. In *RILEM Proceedings PRO 024: Frost Resistance of Concrete* (2005), H.-J. K. M. J. Setzer, R. Auberg, Ed., pp. 235–241.

[250] MILTON, G. *The Theory of Composites*, vol. 6 of *Cambridge Monographs on Applied and Computational Mathematics*. 2002.

[251] MINDLIN, R. D. On the equations of motion piezoelectric crystals. *Problems of Continuum Mechanics, SIAM, Philadelphia* (1961).

[252] MINDLIN, R. D., AND TIERSTEN, H. F. Effects of couple-stresses in linear elasticity. *Arch. Rat. Mech. Anal. 11*, 1 (1962), 415–448.

[253] MINKOWSKI, H. *Die Grundgleichungen für die elektromagnetischen Vorgänge in bewegten Körpern.* Nachrichten von der Gesellschaft der Wissenschaften zu Göttingen, Göttingen, 1908.

[254] MOENCH, A. F. Double-porosity models for a fissured groundwater reservoir with fracture skin. *Water Resources Research 20*, 7 (1984), 831–846.

[255] MOLENAAR, M. M., HUYGHE, J. M., AND BAAIJENS, F. P. T. Constitutive modeling of charged porous media. In *IUTAM Symposium on Theoretical and Numerical Methods in Continuum Mechanics of Porous Materials*, W. Ehlers, Ed., vol. 87 of *Solid Mechanics and Its Applications*. Springer Netherlands, 2002, pp. 409–414.

[256] MØLLER, C. *The Theory of Relativity.* Clarendon, Oxford, 1972.

[257] Mow, V. C., Holmes, M. H., and Lai, W. M. Fluid transport and mechanical properties of articular cartilage: a review. *J. Biomech. 17*, 5 (1984), 377–394.

[258] Mow, V. C., Kuei, S. C., Lai, W. M., and Armstrong, C. G. Biphasic creep and stress relaxation of articular cartilage in compression: theory and experiments. *J. Biomech. Engrg 102*, 1 (1980), 73–84.

[259] Mow, V. C., and Ratcliffe, A. Structure and function of articular cartilage and meniscus. *Basic Orthopaedic Biomechanics 2* (1997), 113–177.

[260] Müller, I. Zum Paradoxon der Wärmeleitungstheorie. *Z. Phys. 198* (1967), 329–344.

[261] Müller, I. *Thermodynamics.* Pitman, N.Y., 1985.

[262] Müller, I., and Ruggeri, T. *Extended Thermodynamics*, vol. 37 of *Springer Tracts in Natural Philosohy.* Springer, N.Y.-Berlin-Heidelberg, 1993.

[263] Müller, I., and Ruggeri, T. *Rational Extended Thermodynamics (Second Edition).* Springer, N.Y., 1998.

[264] Müller, I., and Wilmanski, K. A model for phase transition in pseudoelastic bodies. *Il Nuovo Cimento 57B*, 2 (1980), 283–318.

[265] Müller, I., and Xu, H. On the pseudo-elastic hysteresis. *Acta Metallurgica et Materialia 39*, 3 (1991), 263–271.

[266] Müller, W., and Ferber, F. *Technische Mechanik für Ingenieure.* Fachbuchverl. Leipzig im Carl-Hanser-Verlag, 2005. In German.

[267] Mura, T. *Micromechanics of Defects in Solids.* Kluwer Academic Publishers, Dordrecht, 1987.

[268] Murphy, W. F. Effects of partial water saturation on attenuation in Massilon sandstone and Vycor porous glass. *J. Acoust. Soc. Am. 71* (1982), 1458–1468.

[269] Needleman, A., and Tvergaard, V. An analysis of ductile rupture on notched bars. *J. Mech. Phys. Solids 4*, 32 (1984), 461–490.

[270] Nelson, R. A. *Geologic analysis of naturally fractured reservoirs*, 2nd ed. Gulf Professional Publ., Boston, 2001.

[271] Nesvijski. On a possibility of Rayleigh transformed sub-surface waves propagation. *NDT.net* (September 2000).

[272] Nettleton, R. E. Thermodynamics of viscoelasticity in liquids. *Physics of Fluids (1958-1988) 2*, 3 (1959), 256–263.

[273] Nicholas, T. Tensile testing of materials at high rates of strain. *Exp. Mech. 21* (1981), 177–188.

[274] Noll, W., Toupin, R. A., and Wang, C.-C., Eds. *Continuum Theory of Inhomogeneities in Simple Bodies.* Springer, N.Y., 1968.

[275] NORRIS, A. N. Back reflection of ultrasonic waves from liquid-solid interface. *J. Acoust. Soc. Am. 73*, 2 (1983), 427–434.

[276] NORRIS, A. N. Radiation from a point source and scattering theory in a fluid-saturated porous solid. *The Journal of the Acoustical Society of America 77*, 6 (1985), 2012–2023.

[277] NORRIS, A. N., AND SINHA, B. K. The speed of a wave along a fluid/solid interface in the presence of anisotropy and prestress. *J. Acoust. Soc. Am. 98*, 2 (1995), 1147–1154.

[278] NOWACKI, W., Ed. *Dynamic Problems of Thermoelasticity.* Noordhoff, Leyden, 1975.

[279] NOWACKI, W. *Electromagnetic Effects in Deformable Solids.* PWN, Warsaw, 1983. in Polish.

[280] NYE, J. F. Some geometrical relations in dislocated crystals. *Acta Metallurgica 1* (1953), 153–162.

[281] OGDEN, R. W. *Non-Linear Elastic Deformations.* Dover, Mineola, N.Y., 1984.

[282] OLEJNIK, O. A., AND SAMOCHIN, V. N. *Mathematical Models in Boundary Layer Theory.* Vhapman & Hill, Boca Raton, 1999.

[283] OLSEN, C. O., ATTARIAN, D. E., JONES, R., HILL, R. C., SINK, J., LEE, K., AND WECHSLER, A. S. The coronary pressure-flow determinants left ventricular compliance in dogs. *Circulation Research 49* (1981), 856–865.

[284] ONSAGER, L. Reciprocal relations in irreversible processes. I. *Phys. Rev. 37* (1931), 405–426.

[285] ONSAGER, L. Reciprocal relations in irreversible processes. II. *Phys. Rev. 38* (1931), 2265–2279.

[286] OROWAN, E. Zur Kristallplastizität. *Z. Phys. 89* (1934), 605–659.

[287] OROWAN, E. *Notch, brittleness and the strength of metals.* Institution of Engineers and Shipbuilders in Scotland, Glasgow, 1945.

[288] OTSUKA, K., AND WAYMAN, C. *Shape Memory Materials.* Cambridge University Press, 1999.

[289] PARKER, D. F., AND MAUGIN, G. A., Eds. *Recent Developments in Surface Acoustic Waves.* Springer, Berlin, 1988.

[290] PENFIELD, P., AND HAUS, H. A. *Electrodynamics of Moving Media.* M.I.T. Press, Cambridge, 1967.

[291] PENNER, E. Pressures developed during the unidirectional pressing of water-saturated porous materials. In *Proceedings of the International Conference on Low Temperature Science, 1966, 1, part 2* (1967), H. Oura, Ed., pp. 1401–1412.

[292] PENTTALA, V. Freezing-induced strains and pressures in wet porous materials and especially in concrete mortars. *Adv. Cement Bas. Mat. 7* (1998), 8–9.

[293] PHINNEY, R. A. Propagation of leaking interface waves. *Bull. Seismol. Soc. Am. 51*, 4 (1961), 527–555.

[294] PIPKIN, A. *Lectures on Viscoelasticity Theory.* Springer, N.Y., 1972.

[295] PIRSON, S. J. Performance of fractured oil reservoirs. *AAPG Bulletin 37* (1953), 232–244.

[296] POLANYI, M. Über eine Art Gitterstörung, die einen Kristall plastisch machen könnte. *Z. Phys. 89* (1934), 660–664.

[297] PRANDTL, L. Über Flüssigkeitsbewegung bei sehr kleiner Reibung. In *Verhandlungen des III. Internationalen Mathematiker-Kongresses, Heidelberg, 1904.* Teubner, 1905, pp. 484–491.

[298] PRANDTL, L., AND TIETJENS, O. *Applied Hydro- and Aeromechanics.* Dover Publications, New York, 1957. Reproduktion von: United Engineering Trustees Inc., 1934.

[299] PUROHIT, P. K., AND BHATTACHARYA, K. On beams made of a phase-transforming material. *Int. J. of Solids and Structures 39*, 13 (2002), 3907–3929.

[300] RACHEV, A., AND HAYASHI, K. Theoretical study of the effects of vascular smooth muscle contraction on strain and stress distributions in arteries. *Annals of Biomedical Engineering 27*, 4 (1999), 459–468.

[301] RADKEVICH, E. V. *Mathematical Questions of Nonequilibrium Processes.* Tamara Roschkovska, Novosibirsk, 2007. In Russian.

[302] RANIECKI, B., REJZNER, J., AND LEXCELLENT, C. Anatomization of hysteresis loops in pure bending of ideal pseudoelastic {SMA} beams. *International Journal of Mechanical Sciences 43*, 5 (2001), 1339–1368.

[303] RAYLEIGH, J. W. On waves propagated along the plane surface of an elastic solid. *Proceedings of the London Mathematical Society 17* (1885), 4–11.

[304] RAYLEIGH, J. W. On convective currents in a horizontal layer of fluid when the higher temperature is on the under side. *Phil. Mag. 32* (1916), 529–546.

[305] READ, W. T. Stress analysis for compressible viscoelastic materials. *J. Appl. Phys. 21* (1950), 671–674.

[306] REECE, P. L. *Smart Materials and Structures: New Research.* Nova Science Publishers, 2007.

[307] REIF, F. *Statistical and Thermal Physics.* McGraw-Hill, Aukland, 1985.

[308] REINER, M. *Twelve Lectures on Theoretical Rheology.* North-Holland, Amsterdam, 1949.

[309] REINER, M. *Rhéologie Théorique.* Dunod, Paris, 1955.

[310] RICE, J. R. Mechanics of earthquake rupture. In *Physics of the Earth's Interior*, E. B. A. M. Dziewonski, Ed. North-Holland, 1980.

[311] RICE, J. R., AND P., C. M. Some basic stress-diffusion solutions for fluid saturated elastic porous media with compressible constituents. *Rev. Geophys. Space Phys. 14* (1976), 227–241.

[312] RICKEN, T., AND BLUHM, J. Modeling fluid saturated porous media under frost. *GAMM-Mitt. 33*, 1 (2010), 40–56.

[313] RIVLIN, R. S. Large elastic deformations of isotropic materials, I. some uniqueness theorems for pure homogenous deformation. *Philos. Trans. Roy. Soc. A 240* (1948), 459–490.

[314] RIVLIN, R. S. Large elastic deformations of isotropic materials, II. some uniqueness theorems for pure homogenous deformation. *Philos. Trans. Roy. Soc. A 240* (1948), 491–508.

[315] RIVLIN, R. S. Large elastic deformations of isotropic materials, III. some uniqueness theorems for pure homogenous deformation. *Philos. Trans. Roy. Soc. A 240* (1948), 509–525.

[316] RIVLIN, R. S. Large elastic deformations of isotropic materials, IV. further developments of the general theory. *Philos. Trans. Roy. Soc. A 241* (1948), 379–397.

[317] RIVLIN, R. S. Large elastic deformations of isotropic materials, V. the problem of flexture. *Philos. Trans. Roy. Soc. A 195* (1949), 463–473.

[318] RIVLIN, R. S. Large elastic deformations of isotropic materials, VI. further results in the theory of torsion, shear and flexture. *Philos. Trans. Roy. Soc. A 242* (1949), 173–195.

[319] RIVLIN, R. S., AND SAUNDERS, D. W. Large elastic deformations of isotropic materials, VII. experiments on the deformation of rubber. *Philos. Trans. Roy. Soc. A 243* (1951), 251–258.

[320] RIVLIN, R. S., AND THOMAS, A. G. Large elastic deformations of isotropic materials, VIII. strain distribution around a hole in a sheet. *Philos. Trans. Roy. Soc. A 243* (1951), 289–298.

[321] RIX, G. J. Surface testing for near-surface site characterization. In *Surface Waves in Geomechanics, Direct and Inverse Modelling for Soils and Rocks*, C. Lai and K. Wilmanski, Eds., CISM Courses and Lectures. Springer, Wien, New York, 2005.

[322] RODRÍGUEZ-LARA, B. M., AND JÁUREGUI, R. Dynamical constants of structured photons with parabolic-cylindrical symmetry. *Phys. Rev. A 79* (May 2009), 055806.

[323] ROHRLICH, A. F. *Classical Charge Particles. Foundations of Their Theory.* Addison-Wesley Publ., Reading, Mass., 1965.

[324] ROMANO, A., AND MARASCO, A. *Continuum Mechanics. Advanced Topics and Research Trends.* Birkhäuser, Boston, 2010.

[325] ROUBIČEK, T. Models of microstructure evolution in shape memory alloys. In *Nonlinear Homogenization and its Applications to Composites, Polycrystals and Smart Materials*, P. P. Castañeda, J. J. Telega, and B. Gambin, Eds., vol. 170 of *NATO Science Series II: Mathematics, Physics and Chemistry*. Springer Netherlands, 2005, pp. 269–304.

[326] RUBINSTEIN, L. I. On the problem of the process of propagation of heat in heterogeneous media. *Izv. Akad. Nauk SSSR Ser. Geogr. 1* (1948). In Russian.

[327] RUGGERI, T. Symmetric-hyperbolic system of conservative equations for a viscous heat conducting fluid. *Acta Mechanica 47*, 3-4 (1983), 167–183.

[328] RUGGERI, T. Multi-temperature mixture of fluids. *Theor. Appl. Mech. 36* (2009), 207–238.

[329] RUGGERI, T., AND LOU, J. Heat conduction in multi-temperature mixtures of fluids: the role of the average temperature. *Phys. Lett. A 373* (2009), 3052–3055.

[330] RUGGERI, T., AND SIMIC, S. *Mixture of Gases with Multi-Temperature: Maxwellian Iteration*. World Scientific, New Jersey, 2007, pp. 186–194. Ruggeri, T. and Sammartino, M. (eds.).

[331] RUGGERI, T., AND SIMIC, S. Average temperature and maxwellian iteration in multitemperature mixtures of fluids. *Phys. Rev. E 80* (2009), 0263171 –263178.

[332] RUGGERI, T., AND STRUMIA, A. Convex covariant entropy density, symmetric conservative form, and shock waves in relativistic magnetohydrodynamics. *J. Math. Phys. 22*, 8 (1981), 1824–1827.

[333] SACCOMANDI, G. Universal solutions and relations in finite elasticity. In *Topics in Finite Elasticity*, M. Hayes and G. Saccomandi, Eds., CISM Courses and Lectures. Springer, Wien, New York, 2001.

[334] SATO, H., AND FEHLER, M. C. *Seismic Wave Propagation and Scattering in the Heterogeneous Earth*. Springer, 1998.

[335] SCHANZ, M. *Wave Propagation in Viscoelastic and Poroelastic Continua, A Boundary Element Approach*. Springer, 2001.

[336] SCHANZ, M. Poroelastodynamics: linear models, analytical solutions, and numerical methods. *Appl. Mech. Rev. 62*, 030803 (2009).

[337] SCHEIDEGGER, A. E. Directional permeability of porous media to homogeneous fluids. *Geofis. Pur. Appl., Milano 28* (1954), 75–90.

[338] SCHLICHTING, H. *Boundary Layer Theory*. McGraw-Hill, New York, 1979. See also the 8th revised and enlarged edition of this book by H. Schlichting and K. Gersten (2001).

[339] SCHLICHTING, H., AND GERSTEN, K. *Grenzschicht-Theorie*. Springer, Berlin, Heidelberg, 1997.

[340] SCHMID, E. Über die Schubverfestigung von Einkristallen bei plastischer Deformation. *Z. Phys. 40* (1924), 54–74.

[341] SCHMID, E., AND BOAS, W. *Kristallplastizität mit besonderer Berücksichtigung der Metalle.* Springer, Berlin, 1935.

[342] SCHMITZ, S., SCHWARZ, W., AND LÖFFLER, M. Die Bergung einer verstürzten Tunnelbohrmaschine unter Anwendung der Bodenvereisung. In *Vorträge zum 6. Hans Lorenz Symposium.* Shaker, TU Berlin, 2010, pp. 83–98. In German.

[343] SCHOLTE, J. The range of existence of rayleigh and stoneley waves. *Mon. Not. R. Astr. Soc., Geophys. Suppl. 5* (1947), 120–126.

[344] SCHRÖDER, C. T., AND SCOTT, W. R. A finite-difference model to study the elastic-wave interactions with buried land mines. *IEEE Trans Geosci. Remote Sensing 38*, 4 (2000), 1505–1512.

[345] SCHRÖDER, C. T., AND SCOTT, W. R. On the complex conjugate roots of the Rayleigh equation: The leaky surface wave. *J. Acoust. Soc. Am. 110*, 6 (2001), 2867–2877.

[346] SEDOV, L. I. *Similarity and Dimensional Methods in Mechanics.* Academic Press, 1959.

[347] SEGAL, L. A. *Mathematics Applied to Continuum Mechanics.* Macmillan Publishing Co., New York, 1977.

[348] SELLITTO, A., ALVAREZ, F. X., AND JOU, D. Temperature dependence of boundary conditions in phonon hydrodynamics of smooth and rough nanowires. *J. Appl. Phys. 107*, 11 (2010).

[349] SEMBLAT, J. F., AND PECKER, A., Eds. *Waves and Vibrations in Soils: Earthquakes, Traffic, Shocks, Construction works.* IUSS Press, Pavia, 2009.

[350] SERRIN, J. B. *Mathematical Principles of Classical Fluid Mechanics*, vol. 8. Springer, Berlin, 1959, pp. 125–263. part 1.

[351] SETH, B. R. Generalized strain measure with applications to physical problems. In *Second order effects in elasticity, plasticity and fluid dynamics* (New York, 1964), M. Reiner and D. Abir, Eds., McMillan, pp. 162–172.

[352] SHERMERGOR, T. *Elasticity Theory of Microheterogeneous Media.* "Nauka", Moscow, 1977. In Russian.

[353] SHOWALTER, R. E., AND MOMKEN, B. Single-phase flow in composite poroelastic media. *Mathematical Methods in the Applied Sciences 25*, 2 (2002), 115–139.

[354] SIENIUTYCZ, S., AND SALAMON, P., Eds. *Extended Thermodynamic Systems.* Taylor and Francis, New York, 1992.

[355] SILHAVY, M. *The Mechanics and Thermodynamics of Continuous Media.* Springer, Berlin, 1997.

[356] SILLERS, W., FREDLUND, D., AND ZAKERZADEH, N. Mathematical attributes of some soil-water characteristic curve models. *Geotechnical and Geological Engineering 19* (2001), 243–283.

[357] SIMMONDS, J. G., AND MANN, J. E. J. *A First Look at Perturbation Theory.* Dover, N.Y., 1997.

[358] SMITH, R. C. *Smart Material Systems: Model Development.* Frontiers in Applied Mathematics. Society for Industrial and Applied Mathematics, 2005.

[359] SOKOLOWSKI, M., Ed. *Technical Mechanics. Vol. IV: Elasticity.* PWN, Warsaw, 1978. in Polish.

[360] SOMIGLIANA, C. Sulla teoria delle distorsioni elastiche. Nota I. *Rend. Acad. Lincii 23* (1914), 463–472.

[361] SONNET, A. M., AND VIRGA, E. G. *Dissipative Ordered Fluids. Theories for Liquid Crystals.* Springer, New York, 2012.

[362] SPAAN, J. A. E. Coronary diastolic pressure-flow relation and zero flow pressure explained on the basis of intramyocardial compliance. *Circulation Research 56* (1985), 293–309.

[363] SPENCER, A. Constitutive theory for strongly anisotropic solids. In *Continuum Theory of the Mechanics of Fibre-Reinforced Composites*, A. J. M. Spencer, Ed., vol. 282 of *International Centre for Mechanical Sciences*. Springer Vienna, 1984, pp. 1–32.

[364] STEKETEE, J. A. Some geophysical applications of the theory of dislocations. *Can. J. Phys. 36* (1958), 1168–1198.

[365] STERNBERG, S. *Lectures on Differential Geometry, 2nd edition.* Chelsea Pub. Co., 1982.

[366] STIMSON, M., AND JEFFERY, G. B. The motion of two spheres in a viscous fluid. *Proc. Roy. Soc. Lond. Ser. A 111*, 757 (1926), 110–116.

[367] STOLL, R. D. *Sediment Acoustics*, vol. 26 of *Lecture Notes in Earth Sciences.* Springer, New York, 1989.

[368] STONELEY, R. Elastic waves at surface of separation of two solids. *Proc. Roy. Soc. London A 106* (1924), 416.

[369] STRICK, E. III. The pseudo-Rayleigh wave. *Philos. Trans. Roy. Soc. (London), Ser. A 251* (1959), 488–522.

[370] STRICK, E., AND GINZBARG, A. Stoneley-wave velocities for a fluid-solid interface. *Bull. Seismol. Soc. Am. 46* (1956), 281–292.

[371] SWAINGER, K. H. Large strains and displacements in stress-strain problems. *Nature 160*, 399–400. Letter to the Editor.

[372] SWAINGER, K. H. *Experiment and Applied Theory*, vol. 2 of *Analysis of Deformation.* Chapman & Hall, 1954.

[373] SYNGE, J. L., AND SCHILD, A. *Tensor Calculus*. University of Toronto Press, Toronto, 1959. Also: Dover Publications, 1978.

[374] TANAKA, E., YAMADA, H., AND MURAKAMI, S. Inelastic constitutive modeling of arterial and ventricular walls. In *Computational Biomechanics*, K. Hayashi and H. Ishikawa, Eds. Springer Japan, 1996, pp. 137–163.

[375] TANEJA, H. *Advanced Engineering Mathematics:Volume II*, vol. 2. I.K. International Publishing House Pvt. Ltd., 2007.

[376] TAYLOR, G. I. Stability of a viscous liquid contained between two rotating cylinders. *Phil. Trans. Roy. Soc. (London) A 223* (1923), 289–343.

[377] TAYLOR, G. I. The mechanism of plastic deformation of crystals. Part I. Theoretical. *Phil. Trans. Roy. Soc. (London) A 145* (1934), 362–387.

[378] TEMAM, R. *Navier-Stokes Equations*. North-Holland, Amsterdam, 1979.

[379] THALLER, B. *Advanced Visual Quantum Mechanics*. Springer, New York, 2005.

[380] TITCHMARSCHI, E. C. *Eigenfunction Expansions Associated with Second-Order Differential Equations, Part I and II (published 1958)*. Clarendon Press, Oxford, 1946.

[381] TOLSTOY, I., Ed. *Acoustics, elasticity and thermodynamics of porous media: Twenty-one papers by M. A. Biot*. Acoustical Society of America, 1992.

[382] TOUPIN, R. A. The elastic dielectrics. *J. Rat. Mech. Anal. 5* (1956), 849.

[383] TOUPIN, R. A. Elastic materials with couple-stresses. *Archive for Rational Mechanics and Analysis 11*, 1 (1962), 385–414.

[384] TOUPIN, R. A. A dynamical theory of elastic dielectrics. *Int. J. Engn. Sci. 1* (1963), 101.

[385] TREFIL, J. *A Scientist at the Seashore*. Macmillan Publ. Co., N.Y., 1984.

[386] TRELOAR, L. R. G. *The Physics of Rubber Elasticity*. Clarendon Press, Oxford, 1975.

[387] TRUESDELL, C. A. *The Kinematics of Vorticity*. Indiana University, Bloomington, 1954.

[388] TRUESDELL, C. A. Sulle basi della termomeccanica. *Accademia Nazionale dei Lincei, Rendiconti della Classe die Scienze Fisiche, Mathematiche e Naturali (8) 22*, 8 (1957), 33–38, 158–166.

[389] TRUESDELL, C. A. Mechanical basis of diffusion. *J. Chem. Phys. 37* (1962), 2336–2344.

[390] TRUESDELL, C. A. *A First Course in Rational Continuum Mechanics, part I, Fundamental Concepts*. Academic Press, New York, 1977.

[391] TRUESDELL, C. A. *A First Course in Rational Continuum Mechanics*, vol. 1. Academic Press, 1979.

[392] TRUESDELL, C. A. *Rational Thermodynamics (2nd Edition)*. Springer, 1984.

[393] TRUESDELL, C. A., AND NOLL, W. *The Non-Linear Field Theories of Mechanics*, vol. III/3. Springer, Berlin, 1965. Flügge, S. (ed.).

[394] TRUESDELL, C. A., AND RAJAGOPAL, K. R. *An Introduction to the Mechanics of Fluids*. Birkhäuser, Boston, 2000.

[395] TRUESDELL, C. A., AND TOUPIN, R. A. *The Classical Field Theories, vol. III/1*. Springer, Berlin, 1960. Flügge, S. (ed.).

[396] TUNCAY, K., AND CORAPCIOGLU, M. Y. Body waves in poroelastic media saturated by two immiscible fluids. *J. Geophys. Res. 101*, B11 (1996), 25149–25159.

[397] TVERGAARD, V., AND NEEDLEMAN, A. Analysis of the cup-cone fracture in a round tensile bar. *Acta Metallurgica 32* (1984), 157–169.

[398] UDIAS, A. *Principles of Seismology*. Cambridge University Press, 1999.

[399] VAISHNAV, R. N., AND VOSSOUGHI, J. Estimation of residual strains in aortic segments. In *Biomedical Engineering II, Recent Developments*, C. W. Hall, Ed. Pergamon Press, New York, 1983, pp. 330–333.

[400] VALLE, C., QU, J., AND JACOBS, L. Guided circumferential waves in layered cylinders. *Int. J. Engn. Sci. 37* (1999), 1369–1387.

[401] VAN DEN ABEELE, K. E.-A., CARMELIET, J., JOHNSON, P., AND ZINSZNER, B. Influence of water saturation on the nonlinear elastic mesoscopic response in earth materials and the implications to the mechanism of nonlinearity. *J. Geophys. Res 107*, B6 (2002), 2121–2131.

[402] VAN DONKELAAR, C. C., HUYGHE, J. M., VANKAN, W. J., AND DROST, M. R. Spatial interaction between tissue pressure and skeletal muscle perfusion during contraction. *J. Biomech. 34* (1961), 631–637.

[403] VAN GENUCHTEN, M. T. A closed-form equation for predicting the hydraulic conductivity of unsaturated soils. *Soil Sci. Soc. Am. J. 44* (1980), 892–898.

[404] VAN LOON, R., HUYGHE, J. M., WIJLAARS, M. W., AND BAAIJENS, F. P. T. 3D FE implementation of an incompressible quadriphasic mixture model. *International Journal for Numerical Methods in Engineering 57*, 9 (2003), 1243–1258.

[405] VANKAN, W. J., HUYGHE, J. M., JANSSEN, J. D., AND HUSON, A. Poroelasticity of saturated solids with an application to blood perfusion. *International Journal of Engineering Science 34*, 9 (1996), 1019–1031.

[406] VANKAN, W. J., HUYGHE, J. M., JANSSEN, J. D., HUSON, A., HACKING, W. J. G., AND SCHREINER, W. Finite element analysis of blood flow through biological tissue. *International Journal of Engineering Science 35*, 4 (1997), 375–385.

[407] VIKTOROV, I. A. Rayleigh-type waves on curved surfaces. *J. Acoust. Soc. Am. 4* (1958a), 131–136.

[408] VIKTOROV, I. A. Wolny tipa relejewskich na cilindrichnostiach. *Akust. Zurnal 4*, 2 (1958b), 131–136. in Russian.

[409] VIKTOROV, I. A. *Rayleigh and Lamb waves. Physical theory and applications.* Plenum Press, New York, 1967.

[410] VIKTOROV, I. A. *Zwukowyje powierchnostnyje wolny w twiordych telach (Acoustic surface waves in solids).* Nauka, Moskwa, 1981. In Russian.

[411] VOLTERRA, V. Sull'equilibrio dei corpi elastici più volte connessi. *Rend. Accad. Lincei 14* (1905), 193–202.

[412] VOLTERRA, V. Sur l'equilibre des corps elastiques. *Ann. Ecole Norm. 24* (1907), 401–517.

[413] VVEDENSKAYA, A. V. Determination of displacement fields for earthquakes by means of the dislocation theory. *Izv. Akad. Nauk SSSR, Geofiz. 3* (1956), 277–284. in Russian.

[414] WANG, C.-C. Universal solutions for incompressible laminated bodies. *Arch. Rat. Mech. Anal. 29* (1968), 161–192.

[415] WANG, C.-C., AND TRUESDELL, C. *Introduction to Rational Elasticity.* Noordhoff, Groningen, 1973.

[416] WANG, H. F. *Theory of Linear Poroelasticity with Applications to Geomechanics and Hydrogeology.* Princeton University Press, Princeton, 2000.

[417] WANG, R., PAVLIN, T., ROSEN, M. S., MAIR, R. W., CORY, D. G., AND WALSWORTH, R. L. Xenon nmr measurements of permeability and tortuosity in reservoir rocks. *Magn. Res. Im.* (2004), 1–9.

[418] WARREN, J. E., AND ROOT, P. J. The behavior of naturally fractured reservoirs. *Trans. Soc. Pet. Eng. 3* (1963), 245–255.

[419] WATSON, G. N. *A Treatise on the Theory of Bessel Functions.* Cambridge Press, 1944.

[420] WEERTMAN, J., AND WEERTMAN, J. R. *Elementary Dislocation Theory.* Macmillan, New York, 1967.

[421] WEHAUSEN, J. V., AND LAITONE, E. V. *Surface Waves, vol. IX, Fluid Dynamics III.* Springer, Berlin, 1960, pp. 446–815. Flügge, S. (ed.).

[422] WESOLOWSKI, Z. *Nonlinear Dynamics of Elastic Bodies,* vol. 227. Springer, Wien, New York, 1978.

[423] WESOLOWSKI, Z. Nonlinear elasticity theory. In *Technical Mechanics IV: Elasticity (in Polish),* M. Sokolowski, Ed. PWN, Warsaw, 1978. In Polish.

[424] WHITE, F. M. *Viscous Fluid Flow.* McGraw-Hill, Inc., New York, 1991.

[425] WHITE, J. E. *Underground sound. Application of seismic waves, Methods in Geochemistry and Geophysics*, vol. 18. Elsevier, Amsterdam, New York, 1983.

[426] WHITE, R. Surface elastic waves. *Proc IEEE 58*, 8 (1970), 1238–1276.

[427] WHITHAM, G. B. *Linear and nonlinear waves.* John Wiley & Sons, New York, 1974.

[428] WHITTAKER, E. T., AND WATSON, G. N. *A Course of Modern Analysis*, fourth ed. Cambridge University Press, 1927. Reprinted 1990.

[429] WILHELM, T. *Piping in Saturated Granular Media.* PhD thesis, University of Innsbruck, 2000.

[430] WILMANSKI, K. Propagation of the interface in the stress-induced austenite-martensite transformation. *Ingenieur-Archiv 53* (1983), 291–301.

[431] WILMANSKI, K. Thermodynamics of a heat conducting Maxwellian fluid. *Arch. Mech. 40* (1988), 217–232.

[432] WILMANSKI, K. Symmetric model of stress-strain hysteresis loops in shape memory alloys. *Int. J. Engn. Sci. 31*, 8 (1993), 1121–1138.

[433] WILMANSKI, K. Lagrangean model of two-phase porous material. *J.Non-Equilibrium Thermodyn. 20* (1995), 50–77.

[434] WILMANSKI, K. Dynamics of porous materials under large deformations and changing porosity. In *Contemporary Research in the Mechanics and Mathematics of Materials* (Barcelona, 1996), R. C. Batra and M. F. Beatty, Eds., CIMNE, pp. 343–356. Jerald L. Ericksen's Anniversary Volume.

[435] WILMANSKI, K. Porous media at finite strains – the new model with the balance equation for porosity. *Arch. Mech. 48*, 4 (1996), 591–628.

[436] WILMANSKI, K. A thermodynamic model of compressible porous materials with the balance equation of porosity. *Transport in Porous Media 32* (1998), 21–47.

[437] WILMANSKI, K. *Thermomechanics of Continua.* Springer, Berlin, N.Y., 1998.

[438] WILMANSKI, K. Waves in porous and granular materials. In *Kinetic and continuum theories of granular and porous media*, K. Hutter and K. Wilmanski, Eds., no. 400 in CISM Courses and Lectures. Springer, Wien New York, 1999, pp. 131–186.

[439] WILMANSKI, K. Sound and shock waves in porous and granular materials. In *Proceedings "WASCOM 99", 10th Conference on Waves and Stability in Continuous Media* (2000), V. Ciancio, A. Donato, F. Olivieri, and S. Rionero, Eds., World Scientific, pp. 489–503.

[440] WILMANSKI, K. Some questions on material objectivity arising in models of porous materials. In *Rational continua, classical and new*, M. Brocato, Ed. Springer, Italia Srl, Milano, 2001, pp. 149–161.

[441] WILMANSKI, K. Thermodynamics of multicomponent continua. In *Earthquake Thermodynamics and Phase Transitions in the Earth's Interior*, R. Teisseyre and E. Majewski, Eds. Academic Press, San Diego, 2001, pp. 567–655.

[442] WILMANSKI, K. Thermodynamical admissibility of Biot's model of poroelastic saturated materials. *Archives of Mechanics 54*, 5-6 (2002), 709–736.

[443] WILMANSKI, K. On microstructural tests for poroelastic materials and corresponding Gassmann-type relations. *Géotechnique 54* (2004), 593–603.

[444] WILMANSKI, K. Elastic modelling of surface waves in single and multicomponent systems. In *Surface Waves in Geomechanics, Direct and Inverse Modelling for Soils and Rocks*, C. Lai and K. Wilmanski, Eds., no. 481 in CISM Courses and Lectures. Springer, Wien-New York, 2005.

[445] WILMANSKI, K. Tortuosity and objective relative accelerations in the theory of porous materials. *Proc. R. Soc. A 461* (2005), 1533–1561.

[446] WILMANSKI, K. A few remarks on Biot's model and linear acoustics of poroelastic saturated materials. *Soil Dynamics & Earthquake Engineering 26*, 6-7 (2006), 509–536.

[447] WILMANSKI, K. Permeability, tortuosity, and attenuation of waves in porous materials. *Civil and Environmental Engineering Research (CEER), Zielona Gora 5* (2010), 9–52.

[448] WILMANSKI, K. Monochromatic waves in saturated porous materials with anisotropic permeability. *ZAMM* (2012). DOI: 10.1002/zamm.201100054.

[449] WILMANSKI, K., AND ALBERS, B. Acoustic waves in porous solid-fluid mixtures. In *Dynamic response of granular and porous materials under large and catastrophic deformations*, K. Hutter and N. Kirchner, Eds., vol. 11 of *Lecture Notes in Applied and Computational Mechanics*. Springer, Berlin, 2003, pp. 285–314.

[450] WOOD, A. B. *A Textbook of Sound*. G. Bell and Sons, London, 1957.

[451] WOODS, L. *An Introduction to the Kinetic Theory of Gases and Magnetoplasmas*. Oxford science publications. Oxford University Press, 1993.

[452] WYCKOFF, R. D., AND BOTSET, H. G. The flow of gas-liquid mixtures through unconsolidated sands. *Physics 7* (1936), 325–345.

[453] XIE, J., ZHOU, J., AND FUNG, Y. C. Bending of blood vessel wall: Stress-strain laws of the Intima-media and adventitial layers. *J. Biomech. Engrg. 117* (1995), 136–145.

[454] ZAHORSKI, S. *Mechanics of Viscoelastic Fluids*. Martinus Nijhoff, The Hague, 1982.

[455] ZENG, Y. Q., AND LIU, H. L. Acoustic detection of buried objects in 3-d fluid saturated porous media: Numerical modeling. *IEEE, Trans. Geosci. Remote Sensing 39*, 6 (2001), 1165–1173.

[456] ZENG, Y. Q., AND LIU, H. L. A staggerd-grid finite-difference method with perfectly matched layers for poroelastic wave equations. *J. Acoust. Soc. Am. 109*, 6 (2001), 2571–2580.

[457] ZHDANOV, V. M. *Transport Phenomena in Multicomponent Plasma*. Energoizdat, Moscow, 1982. In Russian.

[458] ZORAWSKI, M. *La Théorie Mathématique des Dislocations*. Dunod, Paris, 1967.

[459] ZORSKI, H. Theory of discrete defects. *Arch. Mech. Stos. 18*, 3 (1966), 301–372.

Subject Index

Series on Advances in Mathematics for Applied Sciences

Aims and Scope

This Series reports on new developments in mathematical research relating to methods, qualitative and numerical analysis, mathematical modeling in the applied and the technological sciences. Contributions rlated to constitutive theories, fluid dynamics, kinetic and transport theories, solid mechanics, system theory and mathematical methods for the applications are welcomed.

This Series includes books, lecture notes, proceedings, collections of research papers. Monograph collections on specialized topics of current interest are particularly encouraged. Both the proceedings and monograph collections will generally be edited by a Guest editor.

High quality, novelty of the content and potential for the applications to modern problems in applied science will be the guidelines for the selection of the content of this series.

Instructions for Authors

Submission of proposals should be addressed to the editors-in-charge or to any member of the editorial board. In the latter, the authors should also notify the proposal to one of the editors-in-charge. Acceptance of books and lecture notes will generally be based on the description of the general content and scope of the book or lecture notes as well as on sample of the parts judged to be more significantly by the authors.

Acceptance of proceedings will be based on relevance of the topics and of the lecturers contributing to the volume.

Acceptance of monograph collections will be based on relevance of the subject and of the authors contributing to the volume.

Authors are urged, in order to avoid re-typing, not to begin the final preparation of the text until they received the publisher's guidelines. They will receive from World Scientific the instructions for preparing camera-ready manuscript.